Latin America
Geographical Perspectives

Latin America

Geographical Perspectives

Edited by Harold Blakemore
and Clifford T. Smith

SECOND EDITION

METHUEN
London and New York

First published in 1971 by
Methuen & Co. Ltd
11 New Fetter Lane, London EC4P 4EE
Second edition 1983
Published in the USA by
Methuen & Co.
in association with Methuen, Inc.
733 Third Avenue, New York, NY 10017
© *1971 and 1983 Methuen & Co. Ltd*
Printed in Great Britain at the
University Press Cambridge

British Library Cataloguing in Publication Data
Latin America.—2nd.—(University paperbacks; 511)
1. *Latin America—Description*
I. *Blakemore, Harold*   II. *Smith,*
Clifford T.
918   *F1409.2*

ISBN 0-416-32830-X

Library of Congress Cataloging in Publication Data
Main entry under title:
Latin America—geographical perspectives.
Includes bibliographies and index.
1. *Latin America—Economic conditions—1945-*
—Addresses, essays, lectures.   2. *Latin America—Social*
conditions—1945-   *—Addresses, essays, lectures.*
I. *Blakemore, Harold.*   II. *Smith, Clifford T. (Clifford*
Thorpe), 1924-
HC125.L344   1983   *980'.038   82-20877*

ISBN 0-416-32830-X

Contents

Notes on the contributors		vi
Acknowledgements		viii
Preface		ix
1	Introduction The Editors	1
2	Mexico D.J. Fox	25
3	The Caribbean David L. Niddrie	77
4	Central America, including Panama D.J. Fox	133
5	Colombia and Venezuela David Robinson and Alan Gilbert	187
6	The Guianas D.J. Robinson and the Editors	241
7	The Central Andes Clifford T. Smith	253
8	Brazil J.H. Galloway	325
9	The River Plate countries J. Colin Crossley	383
10	Chile Harold Blakemore	457
11	Conclusion: unity and diversity in Latin America The Editors	533
	Index	543

Notes on the contributors

HAROLD BLAKEMORE is Secretary of the Institute of Latin American Studies and Reader in Latin American History at the University of London. He is a specialist on modern Chilean history and is particularly interested in contemporary developments in the continent. Co-editor of *The Journal of Latin American Studies* since its inception in 1969, he is also the author of numerous books and articles.

J. COLIN CROSSLEY is Senior Lecturer in the Department of Geography at the University of Leicester. His research and publications primarily concern rural development: land settlement in Bolivia and Argentina, agrarian reform and the cattle industry in the Plate countries. He is a long-time member of the Oxfam Latin America Committee, which he chaired from 1970 to 1979, and has acted as consultant to the Overseas Development Administration. He also served as Chairman of the Society for Latin American Studies from 1975 to 1977.

D. J. FOX is Senior Lecturer in Geography at the University of Manchester, host to the 44th International Congress of Americanists in 1982. His published work is mainly in the field of contemporary economic and social change in Latin America. He has served as Vice-Chairman of the Oxfam Latin American Committee. He is an Honorary Consul for Chile and President of the Manchester Consular Association.

J. H. GALLOWAY is Professor of Geography at the University of Toronto. He has visited Brazil several times since he first went there as a graduate student and has travelled over most of the country. His research interests are the historical geography of Brazil and the development of the cane sugar industry. He has been editor of the *Canadian Geographer* and review editor for the Americas of the *Journal of Historical Geography*.

ALAN GILBERT lectures at University College and the Institute of Latin American Studies, University of London. He is author of *Latin American Development: A Geographical Perspective* (1974), co-author of *Cities, Poverty and Development: Urbanization in the Third World* (1982) and has edited two collections of papers on urban and regional development. His recent research has focused mainly on Colombia, Mexico and Venezuela and he has been involved in a major research project concerned with the relationships between urban housing, the poor and the state.

DAVID L. NIDDRIE is Professor of Geography at the University of Florida, Gainesville. His principal interests include the study of land utilization, population distribution and agricultural settlement both in the Caribbean Basin and Southern Africa. He is the author of *Land Use and Population in Tobago* (1961), *When the Earth Shook* (1962), *South Africa - Nation or Nations?* (1968) and *Tobago* (1981). He is at present preparing a book on South Africa to be published in 1984.

DAVID ROBINSON is Professor of Geography at the University of Syracuse, NY. Initially concerned with the historical geography of Venezuela, then later with settlement in north-west Argentina, he has since carried out and guided substantial research into the demographic history of Latin America, and has published substantive work on the social fabric and spatial structure of colonial Latin America.

CLIFFORD T. SMITH is Emeritus Professor of the University of Liverpool where he was, until October 1982, Professor of Latin American Geography and Director of the Centre for Latin-American Studies. He has been particularly interested in the geography of Peru, and is currently involved in studies of regional development in Peru in the late nineteenth and early twentieth centuries, and in the growth of Lima. He has also published on various topics concerned with the historical geography and agrarian development of that country.

Acknowledgements

The authors and publisher would like to thank the following for permission to reproduce copyright material: The University of Wisconsin Press for Table 3.1; Inter-American Development Bank for Tables 4.1, 8.3 and 8.4 and Figure 4.2; Banco Central de Venezuela for Table 5.1; Louisiana State University Press for Figure 5.5; Moscow Academy of Sciences for Figures 8.4 and 8.5; D.C. Heath and Company for Table 8.5; *Hispanic American Historical Review* for Figure 8.8; Fondation Nationale des Sciences Politiques for Figures 8.9 and 8.11; Serviço Gráfico do Instituto Brasileiro de Geografia e Estatística for Figure 8.10; John Bartholomew & Son Ltd for Figure 8.15; *El Mercurio* for Table 10.2; The World Bank for Tables 10.3 and 10.4 and Figure 10.3; Chilean Forestry News for Table 10.7. Every effort has been made to contact copyright holders; where this has not been possible we apologize to those concerned.

Preface

Since the Second World War international attention has been focused most directly on a number of regions and a series of issues closely related to that event. The rise of the super-powers, the United States and the Soviet Union, was itself a product of victory in war, of military might and technological advance derived from a conflict which weakened or shattered other potential rivals. The emergence of Communist China was no less a direct consequence of the world-wide struggle, as was the dissolution of colonial control in Asia and Africa, and the appearance there of a host of new countries led by nationalist leaders. No less compelling in the last four decades, the remarkable resurgence from post-war prostration of western Europe and of Japan has shown the resilience of old civilizations imbued with modern dynamism. Alone among the world's major regions, at least until very recently, Latin America remained a passive rather than an active agent in an era of revolutionary upheaval. In Latin American experience this was nothing new: from the dawn of the colonial era to the very recent past the continent had been accustomed to a degree of remoteness from the mainstream of world events, and to an isolation shaped by historical circumstances, underlined by ignorance of Latin America in the outside world. Today that isolation no longer obtains, and Latin America's peripheral status in the world community diminishes day by day. The revolutions in technology and communications, which are so much a part of the modern world, and the increasing momentum of change in Latin America itself, combine to give the continent long-delayed recognition not only as a significant element in international affairs, but also as a distinctive region from which the rest of mankind has much to learn.

The scene of the first, large-scale, sea-borne empires of European civilization, Latin America produced, in the conflict of cultures and racial fusion which characterized its origins, societies which were *sui generis*, and forms of political culture and economic organization which have remained significant to the present day. In their revolutions for independence in the early years of the nineteenth century, the various parts of Spanish America and Portuguese Brazil anticipated the contemporary process of decolonization, and their subsequent search for viable constitutional frameworks, political expression and economic life provides much relevant material for the nations of Asia and Africa today. In the lessons of economic development, in the process of urbanization, which is so much a feature of the modern world, in the search for wider unities transcending national boundaries, and in countless other ways, the geographical and historical experience of Latin America repays attention, while the current re-shaping of the continent will concern the world as a whole and not only those who live there.

This collection of geographical essays is offered as a contribution to the understanding of Latin America. The authors, all university teachers in the Latin American field, have been selected for their detailed knowledge of, and experience in, the various parts of the continent, but no attempt has been made to impose a uniform treatment, so as to allow each contributor the personal freedom to emphasize in his individual study the aspects and issues which seem to him to deserve particular emphasis. Nor does this book represent any attempt at a comprehensive geography of Latin America if that, indeed, were possible. It is intended above all to present individual analyses of the various countries and regions of Latin America, and provide some guidance, for those who require it, through the complexities of the current Latin American scene.

London, 1971   H.B. and C.T.S.

For the second edition extensive re-writing has been undertaken throughout, in order to take account of the dramatic changes which have occurred in the continent in the 'development decade' of the 1970s. While the basic structure of the original work has been retained, in the light of its apparent popularity, the reader will find much new material, which also takes cognisance of the very substantial secondary literature on Latin America that has appeared in recent years.

London, 1983   H.B. and C.T.S.

1 · *Introduction*

The Editors

By common consent Latin America is a recognizable entity in the modern world. Unity is expressed in ways which need little elaboration: in the existence of the Andean Pact and the Latin American Free Trade Association; in the relationships which are expressed in the Organization of American States and the Inter-American Development Bank, albeit dominated by the USA; in such formal organs as the Economic Commission for Latin America, and in a number of other organizations. Participation in formal organizations undoubtedly reflects a feeling of common interest and cultural identity which in turn rests on common traits which are sometimes elusive and intangible, but which certainly exist: the Iberian heritage of language and the emergence of a corpus of peculiarly Latin American literature; the common historical experience of conquest and colonization from Spain and Portugal, and of successive independence movements in the nineteenth century. There are common elements in political and social structures which have been repeatedly stressed in the literature, and there is a shared general concern about economic development, and yet at the same time a universal preoccupation with the preservation or creation of national and cultural identity in the face of economic change.

Within this broad unity, however, there is a vast range of diversity. Common themes have called forth a growing volume of literature on Latin American affairs which has itself tended to reinforce the impression of uniformity, not least in matters such as agrarian reform, urban growth and industrialization. By its very existence this literature contributes to the reality of an intellectual and cultural unity. A feedback process operates, but it is easy to exaggerate the extent to which uniformities really do exist. There is always a very real danger of generalizing from one or two parts of Latin America to the whole, just as there is sometimes a tendency to see as 'typically Latin American' situations and problems which are common in the developing world as a whole. It is obviously impossible to summarize in a short introduction the complex variety of Latin America. One may hope to pick out themes which help to differentiate and identify the areas which have been made the subject of the essays which follow, and to attempt to draw together some of the common elements.

Environment and resources

It is, indeed, the diversity within a larger unity that is the main theme of this book, and has suggested the type of regional division that has been adopted.

Environmental contrasts are at least as great as in any other major world division (Dorst, 1967, pp. 6–10). The latitudinal extent of Latin America, from 56°S at Cape Horn to almost 33°N in Mexico, is greater than that of any other major world division, including mainland Asia. Climatic regimes are correspondingly diverse. Analogues can be found in Latin America for areas as different, for example, as the Congo basin, the Namib Desert, the Tibetan Plateau, the so-called 'Mediterranean' regions of California or south-west Australia, the Middle West of the USA or the Norwegian coast, and a few climates can be added for which counterparts cannot be easily found, such as the Patagonian Desert or the problem climate of north-east Brazil.

The major axis of the continent is the Andean chain and its northern continuations in Central America and Mexico, together with the island arcs of the Caribbean. It is a region of complex folded and faulted ranges and crystalline batholiths which are largely the product of a Cretaceous orogeny and of later phases of uplift. It is still a region liable to occasional disastrous earthquakes such as those which shook Peru in June 1970 or Nicaragua in 1972, and it is also a zone of volcanic activity. The landscapes of the Andes and of Mexico or Central America are infinitely variable, ranging from high glaciated massifs and almost perfect volcanic cones through high-level rolling plains to deeply cut river gorges, structurally guided troughs and the intricately dissected ranges of hot, humid regions or the stark chaos of arid foothills. Intermontane basins provide almost the only large areas of uniformity, constituting the heartland of Mexican settlement in the north and of Indian settlement in the *altiplano* of Bolivia and southern Peru, though smaller intermontane basins elsewhere are the basis for nuclei of populations in north-west Argentina, Peru, Ecuador, Colombia and Central America. Latitude, aspect and altitude produce a complex variety of local climates and contrasting vegetation patterns: the high-altitude tundra vegetation of the *puna* in Bolivia and Peru and its more humid equivalent further north in the *paramos* of Ecuador and Colombia; the temperate climates of intermediate altitudes and the highly varied ecologies of inter-Andean basin at low altitudes, where hot, humid forests often alternate over surprisingly short distances with hot, dry regions of xerophytic vegetation where rainfall is low and unreliable. Only at the highest altitudes and at the highest latitudes, however, are the Andes uncompromisingly hostile to man. Elsewhere the variety of ecological niches has nourished human settlement with access to a range of potential resources, but characteristically in islands of settlement separated by steep slopes, difficult terrain or unirrigable desert. It is precisely the fragmentation of the Andean environment that has favoured a varied local economy, but has inhibited large-scale specialization and cheap, efficient communication. The plateau of Mexico, the largest of the intermontane basins, is the major exception within the mountain region of Latin America as a whole.

Elsewhere in Latin America there are uniformities of physical environment on a much larger scale. Eastern South America is built around the ancient massifs of the Guiana and Brazilian Highlands, the residual massifs of the Pampas and Patagonia. Ancient rocks of the crystalline basement complex are exposed over large areas and are elsewhere overlain by sedimentary or volcanic rocks which are often the major relief-forming features, as for example in southern Brazil. But between the Andes and the eastern massifs are the great lowland basins, filled with debris largely derived from the weathering and erosion of the Andes, and thus reflecting in their general character the nature of the dominant erosional process according to latitude and climate: the thick alluvial spreads of the Orinoco, Amazon and Paraná-Paraguay systems; the loess-covered plains of the Pampas; and the glacial moraines and boulder clays of Patagonia. There are corresponding uniformities of climate and vegetation over wide areas which are spelled out in subsequent regional chapters. But perhaps the most general point to make is that intensive, large-scale settlement has effectively been restricted to the coastal regions, south-east and southern Brazil, Uruguay and the Pampas of Argentina. For many different reasons the empty heart of South America has repelled intensive settlement in the historical past, though much of it has suffered waves of exploitation for Indian slaves, for rubber or cinchona, for gold and diamonds or for extensive cattle ranching. Perhaps the most obvious and basic reason has been simply the lack of obvious opportunities for intensive settlement in the production of commodities for export or for the satisfaction of urban markets nearer the coast. The difficulties and cost of communication across the Andes or away from marginally navigable rivers in the rain forests, or the gallery forests of the Amazon basin and the difficult thorn scrub lands of the Gran Chaco have inhibited settlement, but there are the many other environmental difficulties of climatic unreliability, such as drought and flood, the fragility or infertility of soils which vary from one region to another, and these are described in more detail below.

In terms of significance for economic development, however, the diversity of climate, terrain, vegetation and soils over the continental canvas is less important than the range of variation within each state, and it is in this respect that most Latin American countries can display a remarkable complementarity of potential resources within their boundaries. Almost all the Latin American countries contain environments ranging from arid to humid or from temperate to tropical; some by reason of their latitudinal extent, such as Brazil or Chile, some by reason of altitudinal range, such as Venezuela, Colombia, Ecuador and Bolivia, and some by reason of both, such as Mexico and Argentina. There are some states to which this generalization does not apply, of course, notably the smaller states such as Paraguay, Uruguay and the Central American countries, but most do have the capacity to produce a vast range of raw materials and foodstuffs.

There is a similar, though less systematic variety in the mineral wealth and power resources of Latin America. Highly mineralized regions of the Andes, of Mexico and of the ancient massifs of Brazil and the Guianas are less valued now for gold and silver than they once were in colonial times, but deposits of ferrous and non-ferrous metal ores, more widely distributed, have generated investment and created sources of external revenue, increasingly so as lower costs of transport and new technologies have brought new resources, such as iron ores and low-grade disseminated copper ores, within the range of economic exploitation. Fifty years ago the lack of coal resources was thought to be a major handicap to Latin American prospects of industrialization, but in terms of modern energy requirements it is fairly well served: Venezuela and Mexico are the major oil producers of Latin America, but Argentina, Brazil, Ecuador, Colombia, Peru, Bolivia and Chile all produce significant quantities, and several of these have high hopes of future oil prospects. Most Latin American countries are well endowed with potential resources of hydroelectricity, and they are well developed in Argentina, Brazil and Venezuela. Nuclear power is a reality in Argentina, and nuclear power stations are under construction in Brazil and Mexico and contemplated elsewhere.

The Latin American heritage

Most of the regional essays which follow stress an historical approach to the geography of Latin America, and it may be useful to introduce a few of the most fundamental themes which have profoundly affected the spatial organization and the character of Latin American society.

Conquest and settlement by the Spanish and Portuguese from the sixteenth century is perhaps the most outstanding single process in the evolution of Latin American society. The linguistic, cultural and religious unities which were imposed have only marginally been eroded. The fabric of society rested on a hierarchical structure which is still a central fact in Latin American geography, and which to some extent reflected the contemporary social stratification of the Iberian homeland; except that the Indian peasant or the black slave lay at the base of a social pyramid in which ethnicity and status underlined social divisions, even though racial mixture blurred the outlines, in comparison, for example, with North America.

Over much of Latin America the Indian populations, who may once have numbered something of the order of 53 million, have been eliminated as a result of miscegenation, the depopulation of Indian societies following the impact of European diseases against which they had little or no immunity, or as a result of military campaigns; and they survive as identifiable cultural groups mainly in southern Mexico, parts of Central America, and the Central Andes, chiefly in areas of formerly high Indian culture where Spanish occupation or later immigrations were not so intensive, or in remote areas of the Amazon basin where they are even now under threat, or in occasional reserves elsewhere. In most areas of Latin America, until the nineteenth century, slaves accompanied the European settlement, massively so in those areas where plantation agriculture provided the basis for export production of sugar, cotton, indigo, coffee or cacao. An African ethnic and cultural dimension was added to the cultural complex, most clearly surviving in parts of Brazil and the Caribbean, but in most other areas subordinated to a mestizo culture or, as in the temperate south, absorbed within a numerically greater European immigration in the late nineteenth century.

The hierarchical social structure, based essentially on the exploitation of Indian or slave labour or on the exaction of tribute from Indian communities, was also at the root of a polarization in the occupation of the land between the peasant farmer on the one hand and the great estate on the other, most clearly so in the juxtaposition of Indian peasant communities and the classic hacienda in the densely settled areas of Indian occupation, but also apparent in the slave plantations and in the massive cattle-ranching estates, both of which represented a major expansion of the agricultural frontier into areas of new settlement during the colonial period. But it was not until the late nineteenth and early twentieth centuries that the problem of the *latifundia-minifundia* complex reached its modern proportions.

The Spanish settlement, though not the Portuguese, was from its inception essentially urban based. Most of the important cities of Spanish Latin America were founded before the end of the sixteenth century. Indeed, the foundation of towns was an essential instrument of settlement and of control over Indian labour and the surrounding countryside. It was also highly standardized, the Crown codifying the formalities and the privileges involved in urban foundation and dictating the pattern of streets, the location of the plaza and the disposition of important public buildings. This colonial tradition, replicated in later urban foundations, is one of the most striking features of the urban geography of Spanish Latin America.

The second great watershed of Latin American development spans a broader period, and is less easily defined. It involved political independence from Spain, and then later in the nineteenth century the generation of export economies and an accelerated participation in international commerce. In the early nineteenth century the colonies achieved their independence from Spain and Portugal to embark upon the long and arduous process of nation-building, and the political geography of Latin America began to take its modern shape. Territorial boundaries were largely based on the colonial administrative units which had been created by the Bourbon reforms of the late colonial period, and national identities, sometimes already rough-hewn, were forged in the struggle for independence. Paraguay and Bolivia broke away from La Plata, the unit from which Argentina was carved. Uruguay was founded as a buffer state between Spanish and Portuguese spheres of influence. The Caribbean and the northern margin between Spanish and Portuguese claims in Guiana, was already fragmented by the colonial acquisitions of other European powers: the British, the French, the Dutch and the Danes, and this was continued into the early nineteenth century. It was here, too, that Spain retained the last fragment of its empire in Cuba and Puerto Rico until the end of the century. Central America, too, shattered into a series of separate political units in spite of ill-fated attempts at federation, which were also unsuccessful in drawing together Colombia, Venezuela and Ecuador.

Boundaries proved to be fluid, indefinite and disputed, particularly in the little-known and unsettled heart of the continent and on the northern margins at the boundary with the USA, but also in more sensitive areas, involving for example, the re-drawing of boundaries between Bolivia, Chile and Peru after the War of the Pacific (1879–82). Disputed boundaries still remain: between Peru and Ecuador in the Amazon basin; between Bolivia and Chile over the former's access to the Pacific; between Venezuela and Guiana; and, most recently, over the question of the Beagle Channel between Argentina and Chile. But, in general, basic questions of national identity were resolved by the middle of the nineteenth century, except in the remaining colonial territories of the Caribbean, though local particularisms kept alive the issue of federative rather than unitary states, notably in Colombia and Argentina.

The political identities created by the independence movement reflect, to some extent, the lack of integration between colonial administrative units and their development in relative isolation from each other, separated by distance, and bound mainly by links through Spain itself. But in the second phase of radical change in the nineteenth century, characterized by the development of export economies linked with the expanding industrial and commercial world of

Britain, continental western Europe and later the USA, the emerging nation states of Latin America were again linked more closely to the external commercial world across the Atlantic than they were to each other. It is only recently that movements towards economic integration have made significant progress.

The export economies of the nineteenth century rested on the recognition of opportunities to meet the needs for raw materials and foodstuffs in the expanding western economies, on the technological changes, particularly in transport, which made such production practicable, but also, to a greater or lesser extent, on the availability of capital, some of it foreign, but much of it internally accumulated. They also rested on immigration from Europe, massively so in the case of Argentina and Brazil in the late nineteenth century, but significantly, too, in Chile and most other Latin American countries. The dominance of the export economies to the mid-twentieth century represents the most formative stage in the development of the modern social and economic geography of Latin America, and receives attention in some detail in most of the chapters which follow. Two general points may, perhaps, be made at this introductory stage.

First, the nineteenth century witnessed a major change in the space relations and spatial organization of the area as a whole. To the late colonial period the Spanish empire had been most firmly based on its focal possessions in Mexico and Peru, with access at least officially through the Caribbean. But during the course of the commercial expansion of the nineteenth century it was the southern temperate plains, the sources of hides, wool and then beef and grain which were the major recipients of capital and migrants. The South Atlantic route to Buenos Aires and Cape Horn replaced the Caribbean as the route of access, at least until the opening of the Panama Canal in 1914. In Brazil, too, there was a southward shift towards Rio and then São Paulo associated primarily with the expansion of coffee production. In short, the temperate regions of Latin America, once sparsely occupied by nomadic Indians, largely ignored by Spanish and Portuguese, became the major zone of expansion, playing a role comparable to that of Australia or the Middle West in relation to the industrialized world.

Secondly, the potential complementarity of resources referred to above had important implications for the development of the external sector. It has been suggested that in the assessment of resource endowment for economic development, complementarity is an important element in the growth of the external sector, as well as in internal growth, provided that resources can be brought into play successively as a response to changing market structures or technological change, making possible the broadening of an export base and shifts of activity from one region to another as the succession proceeds. Many such shifts can be identified in Latin America: from agricultural development and grain production for export in central Chile to the mining of nitrates and then of copper; from the exploitation of guano in Peru to coastal plantations producing sugar and cotton, or from the exploitation of wool in the sierra to the production of copper and other minerals; from minerals to tobacco and then coffee in Colombia; from cacao and coffee to oil in Venezuela; and in Brazil, over a longer period, the shift from sugar in the north-east to gold and diamonds in the eighteenth century to coffee in the nineteenth. Yet such complementarities, even when fully exploited, have rarely led to progressive regional development, sometimes because of the lack of linkages between export production and the rest of the economy, so that export-oriented enclaves of development failed to stimulate growth elsewhere, sometimes because of the export of profits overseas to the foreign-based enterprises; and sometimes because of failures by government and local enterprise to use their resources for sound development. And the dangers of reliance on a single product, or a narrow range of products, for export revenues were dramatically realized by the crisis of the 1930s.

Recent trends

The last twenty or thirty years have seen the beginnings of what is likely to prove a major stage in the development of Latin America's social and economic geography. It is marked, in one sense, by the shift from externally oriented development (*desarrollo hacia afuera*) towards internally oriented development (*desarrollo hacia adentro*). Emphasis has been placed increasingly on policies aimed at national integration in terms of the provision of physical infrastructure (power, communications, construction), the provision of education and social welfare, the creation of national consciousness and the assimilation of cultural minorities. Above all, it has involved an emphasis on industrialization aimed primarily at the substitution of domestic manufactures for imported goods, and also on the extension of processing industries involving foodstuffs and raw materials for export (e.g. the domestic refining of non-ferrous metals, canning and freezing of fruit and vegetables for export). Increasingly, industrialization has moved towards the limits set by the size of national markets, which must be seen in the context of a persisting inequality of income distribution as well as in terms of the size of populations. It is apparent that the larger countries, particularly Argentina, Mexico and Brazil, have been able to move furthest in this direction. Moreover, progress in import-substitution tends to meet increasingly high capital needs in relation to the provision of employment. Difficulties and controversies surround the issues involved in providing adequate employment, the transfer of technology, economic integration, and the penetration of foreign capital into service industries such as banking, or even retail distribution, as well as in industry, mining and agriculture.

The internalization of economies has involved a massive expansion in the role of the state, and not only in economic and physical planning, welfare and education, but also in basic investment in power production, communications or irrigation works. Strategic industries, e.g. iron and steel, oil, mining, have been established or taken over by the state or by state-controlled organizations. Control over the state apparatus assumes a new and critical importance, especially where the new bureaucracies are dependent to a greater or lesser extent on the regime in power. These, however, are trends by no means unique to Latin America; nor are the problems they create or attempt to solve.

Demographic expansion and explosive urbanization have accompanied the modernization of whole sectors of the economy, and both are closely associated with a crisis in employment and the enlargement of what has been called, perhaps with doubtful accuracy, the marginal sector of the economy: in the towns the self-employed petty traders, the small-scale artisans, the unemployed or underemployed casual labourers, the domestic servants; and in the countryside, the labourers and the peasant farmers with inadequate land. Spatial organization and the use of resources are subjected to mechanisms of state control, though not always with the success or the effects that are anticipated.

Population

Of all the major world regions, Latin America is one of those in which population is growing most rapidly. The estimated population in 1979 was some 360 million, and is increasing at the rate of 2.8 per cent per year, a figure well above the world average of 1.9 per cent per year, and far above the rates of growth experienced by western Europe during its industrial revolution in the nineteenth century or by Japan at its most rapid phase of growth. Death-rates have been falling rapidly, and expectation of life is tolerably good throughout almost all of Latin America (*Table 1.1*). Yet infant mortality is unacceptably high in a number of countries, and medical services leave much to be desired outside the major cities. Birth-rates are, however, very high and are generally over 40 per 1000. There are exceptions, of course, particularly in Argentina and Uruguay and to a lesser extent in Chile, where demographic characteristics conform more closely to those of industrialized countries, and where a low death-rate and a relatively low birth-rate result in rates of increase substantially lower than those for much of the rest of Latin America (*Table 1.1*). It has been suggested that demographic trends in many Latin American countries have not conformed to the classic pattern whereby fertility declines with rising incomes; the spread of family limitation from upper-income groups to the rest of society is limited by a relative rigidity in the class structure, to the high values attached to relationships within the extended society, to the attitude of the church, and to the image of masculine virility embodied in the term *machismo*. Yet there are hints of falling birth-rates, especially in the major cities, and it may be that very rapid rates of growth may ultimately slow down.

The rapidity of population growth poses difficult problems. The high proportion of young people means that the potential labour force is usually between 30 and 40 per cent of the total population, in contrast to a figure of 40 to 50 per cent in Europe and North America. An excessive burden is laid on the community as a whole in the provision of education, health and other social services. In areas of peasant farming the growth of rural population has often led to under-employment, to the fragmentation of farm holdings, to a search for alternative sources of income from temporary labour, and above all, to migratory movements, mainly to the cities, where populations have often swollen far beyond the capacity of manufacturing industry to employ them.

Urban expansion

The growth of cities has been one of the most outstanding features of Latin American experience in this generation. In 1930 the twenty-two largest cities of Latin America had a total population of some 10 million; by 1970–1 they contained over 55 million people. The percentage of population in towns of over 20,000 has risen from 21 per cent in 1950 to over 35 per cent in 1970. The average rate of urban population growth is such that towns may be expected to double their size in 15 or 16 years. The process of urbanization has been widespread, but the most dramatic increases have been in the largest cities, where rates of growth since 1945 have been over twice the national average in Brazil, Colombia, Peru, Venezuela and Mexico. With the exception of Brazil and Ecuador, Latin American countries all have a primate city distribution (following Mark Jefferson's index of primacy, by which the population of the largest city is larger than the sum of the populations of the second and third largest towns, together with one-sixth of the fourth largest). Montevideo, Buenos Aires, Mexico City, Santiago de Chile, Caracas and Lima are the most obvious examples of such primate cities, but the tendency is apparent in the smaller countries, though in Ecuador, Guayaquil is followed closely by Quito and in Colombia, the elevation of Bogotá to the rank of a primate city above Medellín and Cali is relatively recent. The tendency for the larger cities, usually the capital cities, to increase their leadership over other cities has been apparent, however, from about 1880 and particularly since the 1920s, and it is therefore clear that the beginnings of rapid growth almost everywhere preceded industrialization on any significant scale.

The urban explosion has attracted attention by the very scale of the problems which growth has created. The flow of migrants into the cities is universal, and their role in relation to urban growth is much debated. Yet it should not be forgotten that natural increase within the cities themselves often contributes a substantial part, and in many cities the greater part, of total growth. The provision of electricity, water supply, sewage, roads and housing has called forth heavy investment and plans which must constantly be revised in the light of new growth. Shanty towns, squatter settlements or marginal settlements, known by various terms in different Latin American countries, have grown up in or near most large cities. Again, the significance of these squatter settlements is debated, and although generalization is hazardous, it would be wrong to dismiss them as hopeless slums or semi-rural enclaves in an urban environment. Many of them were informally planned by existing urban residents, many show a surprisingly wide cross-section in social and economic terms, and many have become respectable and responsible suburban settlements as the urban fabric has been improved by the initial squatters and those who have succeeded them. The squatter settlements certainly cannot be identified with new immigrants from the countryside, squatting on the fringes of the cities.

Urban growth has usually outpaced the capacity of manufacturing industry to provide employment, and in some cases the percentage of urban population engaged in industry is even declining. Underemployment and unemployment are often distressingly high, and the cities contain a high proportion of casual labour, a marginal industrial sector engaged in small-scale workshop industry, sometimes as an adjunct to large-scale industry (in shoes and clothing industries, for example), and a large number of people in the service sector – domestic servants, self-employed street traders and the like.

The urban explosion is a dominant fact of Latin America. Opinions differ as to its significance and its potential consequences. Many authors have written of over-urbanization or over-centralization in relation to Latin American cities. Epithets such as 'parasitic cities', 'suction pumps' or 'macrocephalism' recall Cobbett's more succinct and forceful description of London as the 'Great Wen'. It is undisputed that the primate cities of Latin America generally receive more than a proportionate share of national income, that they constitute the largest concentrations of demand within their national territories, and that they are generally the preferred locations for industries which have been encouraged by policies aimed at the substitution of domestic for imported manufactures. In general, wages and working conditions are better than in other parts of the country, and levels of provision for health, education and social services are sometimes very much higher.

Processes are certainly at work by which income, revenue and skills are drained from the provinces to the capital or primate cities. Migration tends to draw from the provinces young men with levels of education and skills that are higher than the provincial average. Internal migration often involves a 'brain drain' to the capital. Absentee landowners, preferring to live in the capital, or firms with headquarters in the capital, draw from the provinces a proportion of the income earned there. Taxation and public expenditure, in very diverse ways, may lead to a diversion of income earned in the provinces to the capital city. The trends may be identified, but balanced quantitative assessments are difficult to make, and even if a drain of resources from the provinces to the primate cities can be shown to be significantly large, does this necessarily imply parasitism or over-urbanization? It may be that the concentration of investment, services and industry in the primate city represents the most economic use of resources in countries poorly endowed with capital and skills. But the most economic use of resources may not, in the conventional sense of economic growth, in fact represent the most socially beneficial use of resources.

Analysts of regional growth have drawn attention to the idea of the city as a 'growth pole' that acts in various ways to stimulate productivity within surrounding regions through the demands they make for labour, foodstuffs, raw materials and manufactured products, and through the demonstration effects they generate. These 'spread' effects represent a countervailing force to set against the drain of resources to the primate cities – the 'backwash' effects. The measurement of such 'spread' effects is difficult and it is certain that they do not coincide in spatial terms with the areas affected by the 'backwash' effects. And they are likely to differ in relative strength at different stages in economic growth. Problems of urbanization, the stimulation of 'growth poles' and the issues involved in regional development are therefore very closely linked.

The rural sector

In spite of the recent trends towards urbanization and industrialization, and the emphasis given to them in various ways by government, rural and agricultural development is still fundamental in Latin American economies and societies, though it is in these sectors that the greatest weakness seems to lie. Agricultural products still account for almost exactly a half of all exports from Latin American countries as a whole, and for most individual countries they make up over two-thirds of the value of exports, with the notable exception of those countries dependent upon exports of oil and minerals, namely Venezuela, Chile and Bolivia. In Latin America as a whole 40 per cent of the economically active population is engaged in agriculture, and in spite of the urban explosion nearly half the population lives in rural settlements of less than 2000 people. Yet there are basic problems in the agricultural sector and in rural society.

In many Latin American countries agricultural production is failing to grow as rapidly as population. In 1979 agricultural production per head was less than it had been in the period 1969–71 in no less than fourteen Latin American and Caribbean countries, including Peru, Uruguay, Venezuela, Mexico, Haiti and Honduras. Yet this is not a new problem, for in the mid-1960s it was already being claimed that in a number of countries food production had not kept pace with the growth of population in the previous decade, notably in Argentina, Colombia, Chile, Ecuador and Uruguay. Furthermore, there is some evidence that the relatively weak performance in agriculture is most apparent in the production of foodstuffs for domestic consumption, though it is not easy to distinguish very clearly between the proportion of agricultural output going into the home market from that which is exported. For example, the output of basic food grains and root crops fell in absolute terms between 1970 and 1974, though it has since recovered, and over a long period from *c.* 1950 to 1974 barely managed to hold its own with population growth. One of the major difficulties, of course, is the weakness of many of the statistics relating especially to the production of basic foodstuffs, and many countries rely primarily on estimates rather than careful and continuous data collection. But it is quite clear that there is widespread concern over the rising total of food imports and the need to import substantial quantities of foodstuffs, even when there is clearly a potential capacity to produce more internally. Between 1967 and 1973 rising imports of foodstuffs as a percentage of total imports were characteristic of Argentina, Chile, Colombia, Mexico, Nicaragua, Panama and Peru, and amounted to over 20 per cent of total imports for Chile and Peru. To some extent the need for imported foodstuffs undoubtedly reflects growing incomes and therefore a demand for a wider variety in diet and imports of luxury foods. But it is also evident in some cases that imports of essential foods reflect the failure of agricultural policies to stimulate an expansion in farming and particularly of production for the domestic market. It is also widely held that there are deeper structural reasons which are to be sought in prevailing agrarian structures, themselves a reflection of the social and economic inequalities of the countryside. It is a characteristic feature of many Latin American countries that much land is concentrated in the hands of a very small number of owners, and that a very large number of smallholders own or occupy a small percentage of the agriculturally useful land. There are few areas, though Argentina is one of them, where there is a large proportion of substantial middle-class farmers with enough land, skill and capital to undertake innovations freely. An abundant literature has described the failure of large landowners to use their broad estates efficiently, the anti-social tenures by which they are often worked, the drainage of profits from agriculture for investment in other, more profitable activities, and the drainage of revenues from the countryside to the towns. At the other end of the scale, the small farmers lack the skills and the capital, but above all the land to undertake improvements or to accept innovations which involve risk. Population growth has involved the fragmentation to an even greater degree of the *minifundia*, to the extent that peasant farmers are frequently even less willing to involve themselves in the uncertainties of commercial farming, and acquire exiguous cash incomes by selling their labour rather than the products of the farm.

In the 1960s it was widely argued that land reform was needed to secure a more equitable society by eliminating the crushing burden of anti-social tenure systems, by the redistribution of land from the fortunate few in such a way as to benefit the small peasant and the tenants on the large estates. By creating a more equitable distribution of land, and therefore of income, internal markets would be created for nationally produced manufactures, and land reform was also expected to increase agricultural productivity. In its Latin American context, agrarian reform has been a term with a very wide connotation, implying not only land reform in its strictest sense, involving a redistribution of land ownership, but also colonization, reform of tenancies, the provision of extension services, credit and technical assistance, together with the provision of roads, irrigation facilities and the like. In short, it has been a term so broad that very many different social and political groups felt able to give support to a general movement for 'integral' agrarian reform while aiming in fact at very different targets. In the event, relatively little has been achieved. Mexico's land reform occurred over a long period after the revolution of 1910, with a wave of activity particularly in the 1930s. Cuba's reform came quickly and dramatically after the revolution of 1959. Peasants seized land and expelled the former hacienda owners in the highlands of Bolivia in the years immediately following the revolution of 1952. Mild attempts at reform in the mid-1960s in Peru and Chile were radicalized at the turn of the decade, but in both countries reform has been halted, and even reversed to some extent in Chile. In these countries there has been substantial change, though many of the basic problems of raising agricultural production still remain. Elsewhere legislation has raised hopes or fears, but achievements have been relatively small, except perhaps in Venezuela, where oil revenues provide a substantial backing to agricultural improvement which is not available in most other countries.

It is sometimes argued that Latin America is relatively fortunate among the developing regions of the world in that it has large areas of under-exploited lands and low average densities of population (see *Table 1.1*), and it is true that the population of Latin America is very unevenly distributed and that large areas of the interior lands remain sparsely populated. Over half the population of Latin America lives within 300 km of the coast. The reasons for this distribution are complex, and are by no means wholly to be associated with the progress of colonization from a coastal entry by European peoples and a continuing emphasis on external trade and therefore on easy accessibility to ports. In western South America the more advanced Indian cultures, carrying a high density of population, were mainly in the Andean region within this broad zone near the coast. The main urban foci of Spanish settlement were usually within reach of the coast, though often not directly on it (e.g. Lima, Caracas, Santiago). The interior lands may be sparsely populated, but they are not virgin territory. The backlands of Brazil have been worked over for minerals and for cattle ranching, and the forests of Amazonia have been ransacked for their rubber and cinchona, and have been worked for their valuable hardwoods. But national governments have from time to time been fascinated by the promise of new lands to exploit or to settle, and above all, they have felt the need to integrate their unpopulated and vulnerable frontier regions more firmly into the national polity for strategic as well as economic reasons.

In the nineteenth century there was almost unbounded optimism about the possibilities of these unpopulated regions, thought to need only railways or steamboat navigation to open up for them a glowing future for European settlement. There were many experiments and few successes, and by the late 1920s optimism had come to an end, eroded by the tale of failures in tropical areas, realization of the fragility of tropical soils, the difficulties of access, and the problems of world trade in the depression years. Since the 1940s, however, nationalistic attitudes have rekindled interest. There has been heavy investment in the building of roads; air transport and modest airstrips have put remote settlements within reach, and colonization has been moving forward to the south of the Andes in Venezuela and to the Orinoco delta, and along the eastern slopes of the Andes from Colombia to Bolivia, into the remoter subtropical regions of Argentina, and into the Paraguayan Chaco. The foundation of Brasília, a new national capital located in the interior backlands of Brazil was, in one sense, an attempt to stimulate interior development and settlement, but has been followed up by even more dramatic attempts to construct a system of major long-distance highways in the Amazon basin, intended to act as penetration routes for colonization and large-scale exploitation of resources; though it is a policy which is not without its critics, particularly from those who fear the consequences in terms of environmental destruction.

The empty lands seem to offer the promise of relief from the overcrowding of rural population on long-cultivated lands, but this has rarely been realized. Organized colonization schemes have often been expensive, sometimes ill-run and occasionally have failed. Too often they have been the victims of abrupt policy changes by government, and even when they have been relatively successful, success is gauged by the contribution they make to agricultural output. Colonization schemes have rarely, if ever, successfully relieved a problem of overpopulation, and spontaneous settlement of new land, on the other hand, has often tended to reproduce in a new environment the poverty and social divisions of the old. Indeed, the concentration of population in areas already densely settled, particularly in the neighbourhood of large urban centres, is going on more rapidly than the spread of population into the empty areas. The distribution of population is becoming more uneven, not less so. The exuberant growth of the cities is mainly responsible, of course, but in the rural sphere the intensification of agriculture for urban markets or for export has often yielded a much greater return on capital invested and a greater return in the form of wages and profits than the colonization of new land, except where new roads or irrigation schemes have made possible new settlement within easy access of profitable and growing markets, as in parts of Colombia, coastal Ecuador and coastal Peru.

Levels of development

How does Latin America fit into the spectrum of world development? Since the 1950s Latin America has been grouped among the poor or 'underdeveloped' regions of the world, and although it has since become more usual to refer these regions as the 'developing' countries or the 'less-developed' countries (LDCs), the fundamental distinction is between the rich and the poor, the industrialized

Table 1.1 Some indicators of levels of development

	Area		Population			General	Agriculture			Industry			Communications		Health		Education	
	1	2	3	4	5	6	7	8	9	10	11	12	13	14	15	16	17	18
1 Argentina	2,777	27,064	10	1.3	70	1,627	140	3,069	1,804	225	528	28	88	4.6	68	19	93	52
2 Uruguay	177	3,244	18	0.3	65	1,357	196	2,496	1,000	217	208	24	41	4.1	69	14	91	33
3 Venezuela	899	14,134	16	3.6	64	2,127	39	1,532	2,838	314	244	23	78	7.7	65	11	77	22
4 Chile	757	11,547	15	1.7	66	1,361	42	1,386	1,377	94	182	22	26	4.1	63	5	90	25
5 Costa Rica	51	2,286	45	2.5	28	1,099	23	1,234	488	181	155	19	33	5.4	68	7	89	25
6 Panama	76	1,927	25	2.7	43	1,254	18	1,280	885	185	154	14	42	1.9	66	7	82	30
7 Mexico	1,967	69,965	36	3.5	39	1,016	21	644	1,227	214	220	22	44	3.9	65	7	84	17
8 Brazil	8,457	126,389	15	3.0	45	1,122	19	787	731	174	179	17	54	1.5	59	6	68	29
9 Peru	1,280	17,711	14	2.9	45	848	7	670	642	128	155	17	18	1.8	59	6	72	22
10 Colombia	1,139	30,215	27	2.4	48	637	11	1,099	685	139	80	14	15	1.8	61	5	74	20
11 Nicaragua	118	2,733	23	3.5	35	779	5	1,004	478	90	139	16	15	1.6	53	6	57	12
12 Paraguay	407	3,085	8	3.0	23	576	6	786	189	59	57	18	7	1.1	62	8	79	23
13 Ecuador	271	8,303	31	3.1	36	627	6	567	455	87	71	16	7	1.5	56	5	68	23
14 Dominican Rep.	48	6,053	126	2.9	37	754	3	640	683	114	99	10	15	0.5	58	3	51	12
15 El Salvador	21	4,813	229	3.1	22	615	4	650	260	78	81	10	11	2.3	58	3	58	15
16 Guatemala	109	7,100	65	3.2	16	880	4	909	257	56	90	18	13	1.3	49	2	47	11
17 Honduras	112	3,595	32	3.6	21	451	2	543	264	77	48	14	6	0.7	53	3	52	15
18 Bolivia	1,099	6,162	6	2.7	29	480	1	244	318	43	43	9	6	1.0	47	3	38	24
19 Haiti	28	6,665	238	2.4	14	201	0.3	176	28	39	13	6	3	0.2	49	1	20	5

Note: Countries are ranked in this table in order of collective ranking on all indicators.

Identity of columns in Table 1.1

1 Area, '000 km²
2 Population, mid-1980, '000 inhabitants
3 Density of population, per km²
4 Annual rate of growth of population, 1970–80
5 Percentage of population in towns of 20,000 or more, 1975
6 GDP *per capita*, 1978 (US $ of 1976)
7 Tractors in use, per '000 economically active population in agriculture, 1977
8 Value added in agriculture, per head of economically active population in agriculture, 1975 (US $ of 1973)
9 Energy consumption, per head, kg coal equivalent, 1976
10 Cement production, kg per head, 1976
11 Value added by manufacture, per head of population, 1975 (US $ of 1973)
12 Percentage of economically active population in manufacture and construction, 1970–6
13 Passenger cars in circulation, per '000 population
14 Newsprint consumption, kg per head, 1977
15 Life expectancy at birth, 1970–6
16 Physicians per 10,000 population, 1970–6
17 Adult literacy rate, percentage (level of literacy unknown), 1976
18 Teachers per '000 persons aged 7–14, 1975

Sources: Ed. J.W. Wilkie, *Statistical Abstract of Latin America*, vol. 20, UCLA Latin American Center Publications, 1980; *United Nations Statistical Yearbook, 1978*, United Nations, 1979, for columns 13 to 16.

Table 1.2 Selected indicators of development: ranking order

	General		Agriculture			Industry			Communications		Health		Education		Total
	5	6	7	8	9	10	11	12	13	14	15	16	17	18	
Argentina	1	2	2	1	2	2	1	1	1	3	2	1	1	1	1
Uruguay	3	4	1	2	5	3	4	2	6	4	1	2	2	2	2
Venezuela	4	1	4	3	1	1	2	3	2	1	5	3	8	10	3
Chile	2	3	3	4	3	11	5	5	8	4	7	15	3	6	4
Costa Rica	14	7	5	6	11	6	7	6	7	2	2	7	4	5	5
Panama	7	5	8	5	6	5	9	15	5	8	4	5	6	3	6
Mexico	9	8	6	14	4	4	3	4	4	6	6	6	5	13	7
Brazil	6	6	7	10	7	7	6	10	3	12	10	10	11	4	8
Peru	6	10	10	12	10	9	7	9	9	9	11	8	10	11	9
Colombia	5	13	9	7	8	8	14	13	10	9	9	11	9	12	10
Nicaragua	12	11	13	8	12	12	10	12	11	11	15	9	14	17	11
Paraguay	15	16	11	11	18	16	16	8	16	15	8	4	7	9	12
Ecuador	11	14	12	16	13	13	15	11	15	12	14	13	11	8	13
Dominican Rep.	10	12	16	15	9	10	11	16	12	18	13	12	16	16	14
El Salvador	16	15	14	13	16	14	13	17	14	7	12	17	13	14	15
Guatemala	18	9	15	9	17	17	12	7	13	14	17	18	17	18	16
Honduras	17	18	17	17	15	15	17	14	16	17	15	16	15	15	17
Bolivia	12	17	18	18	14	18	18	18	18	16	19	14	18	7	18
Haiti	19	19	19	19	19	19	19	19	19	19	18	19	19	19	19
Correlation coefficient	0.824	0.911	0.979	0.872	0.868	0.904	0.904	0.804	0.906	0.876	0.924	0.798	0.918	0.714	

Note: Column headings are as for *Table 1.1*. The final column gives the average ranking of each country.

and the non-industrialized, or, more accurately perhaps, between those countries in which the majority of the population are reasonably well provided with material goods, and those in which the majority of the population habitually live close to the margins of subsistence. Alternatively, Latin America may be seen as a part of the Third World, a term based originally on a three-fold classification in which the First World comprised the industrial countries of the non-communist world, the Second World that of centrally planned communist economies (including Cuba) and the Third World the remaining developing countries. This is not the place to discuss these categories in any detail (Wolf-Phillips, 1979), but it is important to recognize the difficulties of compressing the complex variety of the real world into simple categories. Levels of income, material welfare and of productive activities vary continuously from the richest to the poorest countries and division into two or three categories is bound to be arbitrary and to a certain extent misleading (Berry, 1960, p. 93). Indeed, the recent wealth of the oil-exporting countries has made it necessary to think of a different category of oil-rich countries, and the differences in material wealth and welfare between the poorest and the relatively poor has suggested the need to divide the 'developing' world into two: a low-income group and a middle-income group, divided at a notional level of average *per capita* income of US $250 (1976 prices), thus creating a new category, the Fourth World!

If this kind of division is adopted, all the Latin American countries fall into the group of middle-income developing countries with the single exception of Haiti. According to the relative crude measure of national income *per capita*, Argentina, Venezuela and Uruguay rank highly within this group, and with average incomes which place them at a comparable level with some of the countries conventionally ranked as among the industrialized economies.

In order to examine more closely the variations in development within Latin America, a league table of development has been drawn up in *Tables 1.1* and *1.2*. A few countries have been omitted, chiefly because of the lack of available and comparable statistics, but nineteen countries are included. In *Table 1.1* the first four columns give general information on area, population, the density of population and its rate of growth. The remaining fourteen columns show selected indices of development, which are intended to give some impression of relative performance in general terms (levels of urbanization and *per capita* gross domestic product (GDP), and then in agriculture, industry, communications and welfare. Welfare, in terms of health and educational levels, has been given considerable emphasis, but it must be recognized that any group of indices, such as those chosen here, are to some extent arbitrary and sometimes based on statistics of doubtful comparability. The order in which countries are placed in *Table 1.1* reflects their general ranking position for each of the criteria listed for columns 5 to 18. Argentina and Uruguay thus lead the league table, with Honduras, Bolivia and Haiti bringing up the rear. The degree of correlation between each column and the final order was calculated by Spearman's technique of rank correlation, and the correlation coefficients are shown at the foot of each column. All the results were significant to a level of probability better than 0.01, but Gross National Product (GNP), Gross Domestic Product (GDP), the number of tractors in use, energy consumption and value added in manufacture, are all very close indeed to the overall ranking.

Argentina, Uruguay, Venezuela and Chile are the leaders. They have high income levels, but have suffered slow rates of economic growth and even stagnation in recent years. Their rates of population growth are substantially lower than almost all other Latin American countries, as a result of family limitation and relatively low birth-rates rather than high mortality-rates. The average expectation of life is high. They are highly urbanized, with 60 per cent or more of their populations in towns of over 20,000; and they have, correspondingly, well-developed communications and industry. Above all, they are characterized by high rates of literacy and fairly good educational systems. They are the temperate countries of Latin America in which economic expansion during the nineteenth century was accompanied by European immigration, massive in Argentina, but significant, too, in Uruguay and Chile. In many respects they could be classed among the developed countries of the world, though average figures certainly conceal great variations in income distribution. They are also the countries in which regional disparities of wealth and commercialization, though important and discussed in detail in later chapters, are not so strikingly obvious as in most other Latin American countries.

The relatively high development of the southern temperate countries of Latin America is rooted deeply in the past. Argentina and Uruguay were undoubtedly the best developed of the Latin American countries in the early twentieth century when contemporaries could compare them with New Zealand or Australia in terms of their prosperity and prospects. Chile's economy was well sustained by an export economy based on rich nitrate resources and later on its copper deposits. In comparison, Venezuela appears as a *nouveau riche*, its modern leadership in the development stakes being based on the rapid development of oil resources, especially since 1930, and even more spectacularly following the organization of the oil-exporting countries (in which Venezuela itself played a leading part) and the rise in the price of oil after 1973. Urban and industrial development have been relatively recent, but agriculture has lagged, and the unevenness of rapid development has created fairly sharp contrasts in social and material welfare as between different classes in society and between urban and rural sectors. Levels of literacy and the proportion of youth in secondary schools, as well as a poor showing in terms of agricultural development, reflect the uneven and recent pattern of growth, and it is also clear that the rate of population growth is among the highest in Latin America. Among the major countries of Latin America, Mexico has sustained high rates of economic growth over most of the period following the Second World War, and although its future promise appears to be bound up with recent discoveries of oil, Mexican growth has been more closely associated with the successful establishment of a complex industrial base in the recent past. This would also be true of Brazil, in which the mature development of industry in the south-eastern region of the country is offset by the profound poverty of the north-east, so that in spite of the size of its industrial sector and the so-called 'economic miracle' of the recent past Brazil's overall average position also reflects a high degree of social and regional inequality.

Peru and Colombia take up an intermediate position among the major countries of Latin America. Colombia takes an average position in most respects, though it has had only a modest rate of growth in *per capita* income since 1950. Incomes are certainly higher in the central zone of Colombia, but problems of regional diversity are perhaps less pressing than in many other Latin American countries. There is a strong regionalist tradition and Bogotá has never achieved overwhelming dominance as a primate city, though it has certainly increased more rapidly than other cities in recent decades. Regional urban centres such as Medellín, Cali, Barranquilla and Bucaramanga have a prosperity and vitality of their own, and are well distributed to serve as growth poles for the regions in which they lie. Since 1945 the improvement of communications, especially by road and by air, has gone far to break down the isolation in which provincial cities and regional identities initially developed. In Peru, as in Brazil, national statistics conceal a very high degree of social and regional inequality. Coastal Peru, including the metropolitan region of Lima–Callao, has long been the major focal area of growth in agriculture and industry, but differences in income and degree of development between Peru's coastal region and the sierra are very considerable. It is in the sierra of Peru that isolation and the pressure of a predominantly Indian population on exiguous resources of fertile and cultivable land, have combined with social inequalities to produce a region of great poverty and social and economic stress. The sparsely populated eastern regions have yet to make a major contribution to national development. Ecuador and Bolivia share with Peru the problems of integrating a predominantly Indian population in the sierra, though in Ecuador, as in Peru, the expansion of settlement in the coastal region from the nineteenth century was associated with an active participation in external trade, and in more recent years Ecuador has yielded oil in significant quantities. For Bolivia, however, the difficulties imposed by isolation, the harsh sierra environment of the *altiplano* and the subordination of an overwhelmingly Indian peasantry are compounded by the lack of a coastal outlet and the rather marginal competitiveness of Bolivian minerals on the world market. Its eastern regions suffer from the inaccessibility which is also one of the factors, though not perhaps the most important, which gives Paraguay one of the poorest set of indices among the South American countries. Finally, the Central American countries with Hispaniola (Haiti and the Dominican Republic) replicate the variety of development to be found in Latin America as a whole. The reasons for this variety of performance, from the utter poverty of Haiti to the relative prosperity of Costa Rica and Panama, are outlined later in this book, but it is worth stressing that in this area, as in Latin America as a whole, disparities in development in environmentally comparable areas, even within the confines of the single island of Hispaniola, have much more to do with history and politics than with the opportunities and restraints presented by the physical environment.

The parameters of development used in this edition are not identical with those used a few years ago in the first edition, but there is sufficient comparability to reach a few conclusions about the direction of change. Rates of population growth continue to be among the highest in the world in some Latin American countries, but in a number of countries a decline has been registered compared with the mid-1960s. Colombia, Honduras and Haiti are the only countries to register increases. There has been no change in Brazil, Peru, Mexico, Ecuador, the Dominican Republic and Bolivia, but all other Latin American countries in the list have registered a decline. Urbanization of the population continues at a fairly rapid rate, and correspondingly the proportion of population in towns of over 100,000 continues to increase. Agricultural production has, in general, failed to respond to growth elsewhere in the Latin American economies, and this must in part be due to the emphasis given by national investment programmes and policies to the encouragement of industry. In 8 out of 19 Latin American countries agricultural output *per capita* has actually fallen between 1961–5 and 1975, and has substantially increased (by 20 per cent or more) in only four (Bolivia, Costa Rica, Guatemala and Venezuela). On the other hand, the larger Latin American countries, and especially Brazil, Mexico, Argentina, Venezuela, Colombia and Peru, have made rapid progress in their industrialization during the last decade, though difficulties have been experienced in maintaining the impetus of industrial expansion as the limits of feasible import-substitution policies have been reached. And only Brazil, Argentina and Mexico, among the major Latin American countries, have succeeded in finding external markets for their industrial produce on anything like a substantial scale. Indices of welfare show modest progress, whether in terms of expectation of life, medical services, food supplies, and in education; though once again the indices used in this edition make direct comparison difficult with those of 8 years ago in the first edition.

Finally, in *Figure 1.1* the trends in *per capita* GNP for the major Latin American countries since 1950 are brought together. The overall pattern is evident, with a modest increase in *per capita* GNP for most of the countries involved. Indeed, over the whole period, the average growth in *per capita* GNP for all Latin American countries from 1950 to 1974 has been of the order of 80 per cent, representing in percentage terms a slightly higher rate of growth than that for the USA at 65 per cent, and although it is abundantly obvious that the US economy starts from a very much higher base in 1950, it is also worth some comment that the rate of growth of Latin American GNP *per capita* has not in fact fallen behind that of the USA, even though the absolute gap has widened very considerably. The patterns of growth are the product of many forces which have to do with the response of economies to the fluctuations in international trade and prices for export commodities, internal policies towards investment, trade and industry, and, of course, to political policies. Mexico has been able to sustain a fairly steady rate of growth over a long period; Brazil has shown astonishing rates of growth since the mid-1960s; Uruguay has stagnated, and it was not until the mid-1960s that Bolivia recovered the level of GNP it had achieved by 1950. Yet there is a fairly high degree of stability in the overall pattern. Among the relatively prosperous countries of Latin America, Uruguay, Argentina, Chile and Venezuela have changed their positions since 1950, followed fairly closely by Mexico, which has retained its position with respect to Peru but widened the gap between them. Brazil is one of the few countries to have changed position drastically in the league table since 1950, when its *per capita* GNP was closely comparable with those of Bolivia and Ecuador.

The problem of regional development

The trends and levels of development considered above relate to national averages, but these average figures conceal internal variations within each country which are frequently at least as great as those between countries.

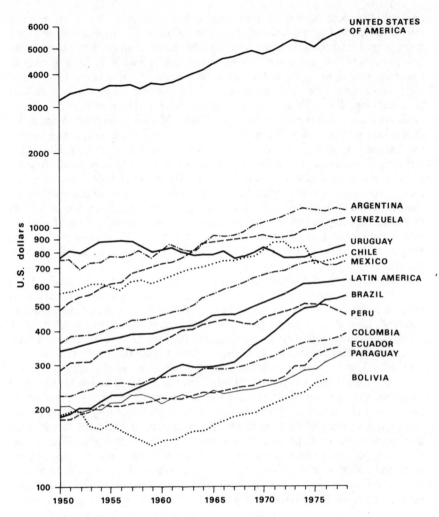

Figure 1.1   Trends in *per capita* GNP for the major Latin American countries since 1950

Inequalities of regional development are among the most striking features in the geography of most Latin American countries, but they, in turn, are closely associated with a more fundamental degree of social and economic inequality as between social groups. The juxtaposition of ostentatious wealth and unrelenting poverty is one of the most obvious features of Latin America even to the most casual traveller. Some crude measure of this inequality is yielded by statistics relating to income distribution. In Mexico, for example, in 1965 the richest 5 per cent of the population received 29 per cent of national income, the richest 20 per cent of the population received 58 per cent of national income, but the poorest 20 per cent received only 3 per cent. In Argentina, the corresponding figures were 31 per cent for the richest 5 per cent, 54 per cent for the richest 20 per cent, and 5.5 per cent for the poorest 20 per cent. The proportions for Costa Rica, El Salvador and Venezuela are broadly comparable and, with the exception of Cuba, similar figures would be encountered for other Latin American countries. In Brazil there is some evidence that the degree of inequality has, in fact, been increasing rather than decreasing in recent years. Patterns of income distribution have an obvious relationship to demand and thus to the nature of industrialization, but it is also evident that patterns of income distribution are related very directly to patterns of regional development. Almost everywhere the richest groups are urban dwellers, most frequently in the capitals and the larger cities: the industrialists, the property owners, the high-level officials and bureaucrats, the owners of large estates, the professionals and the merchants. The poorest groups are certainly to be found in the cities and are numerically the most important group, but in the rural areas outside the cities the richest groups are rarely encountered. The small-town élites are rich in comparison with their rural neighbours, but only modestly prosperous in the context of the large city. In the countryside the most common characteristic is the polarization between the large estates and a peasant population with access to tiny holdings incapable of supporting a family at an adequate level of living.

Inequalities of regional development are, indeed, considerable. In Peru the divergence from the average *per capita* income in the richest department was + 136 per cent and that of the poorest – 43 per cent. In the early 1960s the ratio between the richest and the poorest state or department was 4.1 in Peru, 9.9 in Mexico, approximately 8.8 in Brazil, and perhaps 11 in Colombia. There is some evidence that the ratios are declining slightly, i.e. the relative differentials are decreasing, but it must also be said that the absolute differences in regional incomes are tending to increase. The poorest areas are, for example in Mexico, showing some improvement in basic living standards, so that the base from which ratios are calculated is rising. But the richer regions are also becoming more prosperous and the absolute gap is increasing.

The processes which have been operating to produce regional inequalities within, and also between, Latin American countries have been, and no doubt will continue to be, the subject of heated debate and controversy. Reduced to its simplest terms, the debate focuses on the extent to which stress is placed on the forces which have been making for the modernization of 'traditional' Latin American economies: the development of export sectors, the growth of industry and mining, the development of services such as banking, finance, transport, and the role of the state as an agent of modernization in the provision of infrastructure, education, social services and as industrial entrepreneur. The location and the spread of such activities are frequently sufficiently well geographically defined as to enable the identification of growth regions or core areas: the metropolitan regions as foci of industry and modern services, including government, and regions associated with successful and profitable production for export which have sometimes also become associated with a degree of industrial and urban development. The core areas may have benefited from a circular and cumulative process of growth such as that described by Gunnar Myrdal. But the debate also focuses on the extent to which the processes of expansion in one area have, of necessity, been accompanied by an impoverishment of peripheral areas, or at the very least, by the imposition of constraints which have effectively prevented or inhibited growth in the peripheral areas. Surplus income above that which is needed for subsistence is effectively syphoned to the metropolitan regions, and so too is a significant proportion of the skilled and educated population. In this sense, the peripheral regions are bound by a chain of exploitation which extends through the urban hierarchy of the country to the national metropolitan region and beyond that to the advanced industrial countries. The mechanisms involved, at regional and international levels, are variously interpreted as the result of inequalities in the terms of trade between agricultural production and industrial goods, the exploitative policies of trans-national enterprises, the syphoning of profits from regions to metropolis or to foreign-owned enterprises, or as the inevitable consequences of industrial capitalism.

Regional planning policies aimed at reducing regional disparities are relatively recent in Latin America, and it is difficult to measure their success. They have tended to follow precedents set in advanced industrial countries by giving concessions and privileges to encourage the establishment of industry in selected non-metropolitan regions; Venezuela, Peru, Brazil, Colombia, Chile and Mexico have experimented in various ways to encourage the decentralization of industry, and each has met with varying degrees of success. Some of the problems which have arisen, however, lie in the disappointing capacity of decentralized industry to act effectively as growth poles, at least in terms of the employment created, the tendency for peripheral industry of this kind to be subordinated to the interests of the metropolitan headquarters of the mother company, and also in the reluctance of government to give a needed stimulus to agriculture, the main occupation of the peripheral regions, by favourable pricing policies. Indeed, as long as preference is given to urban interests and industrialization (and it is frequently politically impossible for governments to do otherwise), it is difficult to envisage significant advance in those rural areas which produce essentially for the domestic market, and in some areas it is even unrealistic to assume that peasant farmers will willingly abandon their preoccupation with subsistence farming.

Patterns of regional development in Latin America have evolved over a long period. In some areas it is relevant to hark back as far as the pre-Colombian period in order to understand modern problems; in most areas the colonial period left a legacy in terms of urban settlement, agrarian structures, mining activities and cultural traditions which were in turn overlain by the accelerated growth of export production in the nineteenth and early twentieth centuries before the modern trend towards the internalization of economic growth and industrialization. The essays which follow are concerned to work out some of these major themes in a regional context, and against the environmental background of specific areas.

Bibliography

BERRY, B.J.L. (1960) 'An inductive approach to the regionalization of economic development', in Ginsburg, N. (ed.), *Essays on Geography and Economic Development*, Chicago, pp. 78–107.
COLE, J.P. (1975) *Latin America: An Economic and Social Geography, 1970*, London.

COLLIER, S. (1974) *From Cortes to Castro, An Introduction to the History of Latin America, 1492–1973*, New York.

DORST, J. (1967) *South America and Central America: A Natural History*, London.

FURTADO, C. (1970) *Economic Development of Latin America*, Cambridge.

GILBERT, A. (1974) *Latin American Development, A Geographical Perspective*, Harmondsworth.

JAMES, P. (1969) *Latin America*, New York.

MORRIS, A. S. (1979) *South America*, London.

MORRIS, A. S. (1981) *Latin America, Economic Development and Regional Differentiation*, London.

ODELL, P. R. and PRESTON, D. A. (1978) *Economies and Societies in Latin America: A Geographical Interpretation*, Chichester.

PENDLE, G. (1971) *A History of Latin America*, Harmondsworth.

WILKIE, J. W. (ed.) (1977) *Statistical Abstract of Latin America, Vol. 18 (1977)*, Los Angeles.

WOLF-PHILLIPS, L. (1979) 'Why Third World?', *Third World Quarterly*, 1, 1, 105–14.

2 · Mexico

D.J. Fox

Mexico has a greater range of natural conditions and a richer history than any other country in Latin America. The deserts of the north-west, the rain forests of the south, the volcanoes of the centre, the assortment of fresh foodstuffs in Mexican markets: all are marks of the natural variety of this large country. The many monuments of the Maya, the Aztec, the Zapotec and of other organized societies are a reminder of a long history of civilization; archaeologists now claim that agriculture had been practised for at least 7000 years before the arrival of the Spaniards in 1519 (Smith in Byers, 1967, vol. 1, p. 220; Niederberger, 1979, p. 136). The long pre-Columbian past, the colonial era, the nineteenth century and the post-revolutionary present all contribute to the cultural and material equipment of modern Mexico. The Indian and the European have joined to produce the characteristic Mexican – the mestizo. For some Mexicans life is still cast in the same mould as it was for their ancestors before the Spanish Conquest; for others it is truly cosmopolitan.

With such variety Mexico has her problems. Many of the phrases that are common currency elsewhere in Latin America are also heard in Mexico. The metropolitan problem, the population problem, the Indian problem, the agrarian problem, the problem of industrialization all exist, and all are discussed. Many of the Mexican attempts to solve these problems have been relatively successful and some have been novel; as a result, Mexico has frequently been held up as an example for the rest of Latin America.

The people of Mexico

Mexicans are fond of saying that people are Mexico's most valuable resource. The more thoughtful temper their pride, recognizing the changes and stresses that arise from present population trends. Economic development must not only match the phenomenal growth in population, but also satisfy the hopes for an improved standard of living held by an ever-increasing proportion of the population. In recent years for the majority of Mexicans life has improved and continues to improve. The consensus of opinion is that Mexico is making a qualified success of meeting the demands of change – more so, in fact, than are most other Latin American countries.

Mexico is the most populous Spanish-speaking country. It surpassed Spain in mid-1954, and 25 years later there were twice as many Mexicans as Spaniards. With a population of about 68 million in 1980, Mexico ranked tenth in world lists. It is, after Brazil, the most populous country in Latin America, and Mexicans outnumber the combined total of their Central American and West Indian neighbours. Although one is reminded of pre-Conquest days, when the population of what is now Mexico may have exceeded 20 million and was larger than that of any other comparable American area of the period, more useful comparisons can be made with the other years of the twentieth century. Mexico entered the revolutionary decade (1910–20) with a population of about 15 million, a population that had been growing at a yearly rate of 1 per cent during the 30 settled years of the dictatorship of Porfirio Díaz. The toll of the revolution was so great that in 1921 the census population was actually slightly less than it had been in 1910 (see *Figure 2.1b*). With the return of settled conditions, however, not only has the population tripled, but the annual rate of increase has become greater; in the 1920s the annual growth-rate was 1.6 per cent, in the 1930s 1.8 per cent, in the 1940s 2.7 per cent, in the 1950s 3.0 per cent, in the 1960s 3.4 per cent, and in the 1970s 3.5 per cent. The population of Mexico virtually doubled between 1960 and 1980. A carefully calculated forecast made by the World Bank in 1981 postulates a population of 109 million by the end of the century.

A decline in the death-rate has been the most critical change in the three factors – death-rate, birth-rate and net migration-rate – that in combination determine the overall rate of population growth. The death-rate in 1976 of 6.5 per 1000 was one-half of the figure 20 years earlier and one-fifth of that (33.6) at the beginning of the century. Greatly improved medical treatment reaching a larger proportion of the population is, of course, one reason for this change. The death-rate from some of the principal infectious diseases has fallen dramatically: from the gastro-enteritic group it is only one-fifth of what it was in 1930, from influenza and pneumonia one-third, from tuberculosis one-quarter. Malaria and smallpox, which in 1930 occupied third and fifth places in the list of mortal diseases, have been virtually eradicated. The respiratory diseases have replaced the gastro-enteritic as the leading causes of death.

In Mexico, as elsewhere, extremely detailed mortality statistics are published, but it is unwise to extend their analysis too far. The statistics are based upon death-certificate information of variable quality, transcribed and processed in an inexact manner. In 1960, for example, 37 per cent of the death certificates in the Republic were completed without reference to a doctor, the cause of death being determined by the parish registrar on the strength of, if possible, an eye-witness description; in the Federal District the certificate must be completed by a doctor, although he may not have seen the corpse. These practices continue today, and are undoubtedly amongst the reasons why the infantile mortality from gastro-enteritis and the death-rate from unknown causes are so high. Not unnaturally, post-mortem examination often does not confirm the stated cause of death, but by that time the death certificate is sacrosanct. Although deaths are required to be recorded at the municipal offices in the place of habitual residence, a very useful provision in circumstances where the domestic environment is often the key to the cause of death, this is often not done. The death-rate in Mexico City is inflated by those from the provinces who seek, without avail, a cure in the city hospitals. Within the city the misallocation of deaths results in the seeming paradox that those residents living in boroughs with the most extensive hospital facilities suffer the highest mortality rates (Fox, 1972a).

The public health services have been transformed over the last 30 years, but they are not uniformly available. On the medical side their main success has been in creating an infrastructure of clinics, district nurses and medical aides to reduce calls upon the hard-pressed doctors and hospitals. The capital is much better served than elsewhere, and this is true of the more specialized agencies. The Institute of Social Security (IMSS) had extended coverage to one-third of the population by 1978, or three times as many people as in 1965; and whereas half the metropolitan population was covered, only 5 per cent of the rural population enjoyed the same protection. Of those covered by the civil servants' health services (ISSSTE), the second most important national insurance scheme, 45 per cent lived in Mexico City in 1978.

The Mexicans are very conscious of the role of preventive medicine in reducing the death-rate. One of the most successful campaigns (CNEP) has been fought as part of a larger World Health Organization programme against malaria; and repetitive spraying over the last twenty years has eradicated it from all parts except the southern Pacific coast region. Here the new cotton-growing and oil-producing areas of Chiapas and adjoining Guatemala have been so thoroughly dosed in chemicals that the drainage canals – the nursery for the larvae of the malarial mosquito – now flow not with water but with a dilute disinfectant, and the larvae have developed a degree of immunity against the insecticide. National campaigns have been mounted against other diseases, for example smallpox, and immunization against diphtheria, tetanus and polio is becoming more widespread.

Many of the enteric diseases are caused by vectors transmitted in unclean water and food or by inadequate sewage provision, while the respiratory diseases can be linked to poor housing conditions. Some improvements have been made in the supply of public utilities, and have doubtless contributed to a lower death-rate. The provision of a safe water supply has probably been the task most energetically tackled in recent years, and outside Mexico City and Monterrey, is a federal responsibility shared between several government agencies. As recently as 1960 only 34 per cent of the non-metropolitan population of Mexico could count on a safe drinking supply; by 1975, thanks partly to US aid, about two-thirds of the urban population living outside Mexico City and one-third of those in the villages had access to a safe water supply. But 38 per cent of Mexicans, or 23 million people, could still not claim this elementary right – more than in 1960 when the programme began. In addition to the improvement in the provision of utilities, a more imaginative and effective propaganda programme has been waged, most effectively in the towns, to raise basic standards of hygiene. However, these remain low, even if not as low as previously, and the correlation between mortality-rates and the badly educated, ill-housed, poorer segments of the population, and especially the infant population, indicates the potential that remains for a further fall in the death-rate.

The birth-rate, unlike the death-rate, has changed little since 1929, when reasonably complete registration figures first became available. In the 1930s it was about 43 per 1000; during the 1950s and 1960s it rose to about 46 per 1000, but in 1979 was down to about 36 per 1000. Such minor variations are possibly more apparent than real, bearing in mind such factors as late registration and

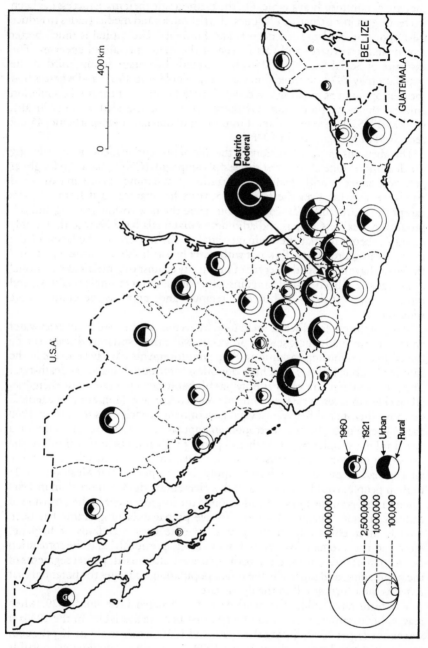

Figure 2.1(a) Mexico: urban and rural population 1921 and 1960

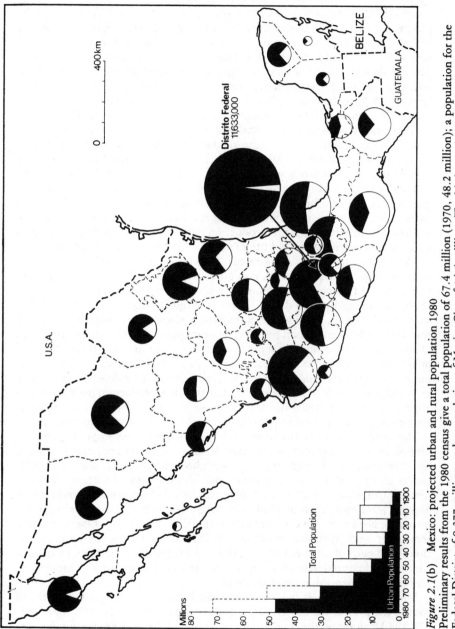

Figure 2.1(b) Mexico: projected urban and rural population 1980
Preliminary results from the 1980 census give a total population of 67.4 million (1970, 48.2 million); a population for the Federal District of 9.377 million; and a population of Mexico City of 14.4 million. The birth-rate has been lower than anticipated.

non-registration of births, especially in rural areas. There was a disparity estimated at 913,000 between the population under the age of 5 as given to census
enumerators in 1960 and as subsequently calculated from birth-certificate
returns (Benítez Zenteno and Cabrera Acevado, 1966, p. 26). Such a disparity
plays havoc with infantile-mortality statistics. Nevertheless it may be significant
that the recent fall coincides with the introduction in 1972 of a government-
sponsored 'responsible parenthood' policy. At that time the total fertility rate
(the number of children a woman would have in her child-bearing career if she
followed the current reproductive characteristics of the national population at
large) was 6.5, or three times that of the industrialized countries. The policy was
cautiously introduced by President Echeverría, himself the father of eight
children, and was given added impulse under the presidency of López Portillo,
who set the ambitious target of reducing the birth-rate to 25 per 1000 by 1982. It
is now quite acceptable for Mexican politicians to state that the demographic
explosion is the country's greatest obstacle to progress, while the Catholic
Church has raised only token resistance to the implementation of the programme. It may also be significant that it is the more affluent and urbanized
parts of the country – the capital and the northern border states – which have
the lower birth-rates and net reproduction rates. If current trends do indicate a
difference between classes and urban and rural behaviour, the implications for
the future are important, since Mexico is becoming both more affluent and more
urbanized. In 1930 only 5.5 million Mexicans, one-third of the total, lived in
towns (defined, somewhat generously, as administrative units with a population
of 2500 or more); by 1960 the figure was 16.7 million or one-half; and by 1978,
44 million or two-thirds of all Mexicans lived in towns. It is possible that the true
fraction may be even higher: some urban immigrants, like the Mixtec Indians in
Mexico City, make a point of registering for census purposes in their villages of
origin, their 'true home' (Orellana, 1973, p. 276). Comment upon the enormous geographical changes implied by these figures will be deferred, but it may
be noted that most forecasts assume a fall in the net reproduction rate induced by
a more urbanized population; if the distinction apparent today is of no long-
term consequence, the urban population may be a million more than estimated.

  The third factor in the equation yielding population is net migration abroad.
One of the interesting contrasts between Mexico and other populous countries of
the Americas is that immigration has always been insignificant, and the population is almost entirely an indigenous one. This was true in colonial days and
remains true today. In 1970 there were only 191,234 foreign-born residents of
Mexico, 8091 more than in 1950, and almost one-half of these were born just
across the borders in the USA or Guatemala (IX *Censo*, 1970). One-half of the
remainder were Spanish.[1] Mormon and Mennonite immigrant communities
flourish in parts of northern Mexico, especially Chihuahua, and descendants of
Chinese indentured labour brought into the north-west in the nineteenth
century are recognizable; negroes are rare.

  Immigration, such as it is, is more than balanced by emigration, with, of
course, the prosperous United States as the main attraction. In 1975 there were
878,198 Mexicans on the resident-aliens register of the US Immigration Service.
Over 16 million Mexicans have settled legally in the United States during the last
100 years; between 1946 and 1965 the figure was 580,000; and in 1975 62,205.

The more important reasons why Mexicans seek residence in the United States are fairly obvious and have been confirmed by statistical analysis (Frisbie, 1975). Agricultural conditions on either side of the border are basically why people move: the disparity in farm wages, the levels of crop prices and of agricultural output, and the general trends of capital investment explain half of the volume of migration. Most of the Mexicans who respond to these push-and-pull factors are farm labourers with minimal skills. Under the impetus of a shortage of farm labour in the United States during the Second World War, this flow was officially sanctioned under the *braceros* programme, which allowed farm-hands into the country on a temporary, seasonal basis; and the *braceros* scheme continued until 1964. Since then the demand for entry to the United States has continued to rise, and is far in excess of the official US immigrant quota for Mexico. Since the Mexican–US border is not an insurmountable barrier, and since sanctuaries amongst the substantial Spanish-speaking population of the south-western United States are widespread, the real figures of Mexican immigration into the United States are well above those given in the official statistics. One recent study suggests that illegal migrants outnumbered legal migrants by a factor of eight between 1946 and 1965. Over 4.6 million *braceros* entered the United States, but only 3.2 million returned to Mexico. Between 1924 and 1972 8 million Mexicans were deported from the United States (presumably many not for the first time). In recent years a more energetic surveillance of the border has been mounted, with the result that the annual number of Mexicans deported or voluntarily repatriated has risen from 500,000 in 1975 to over a million in 1977. Most of such *mojados* or 'wetbacks' intend to stay out of Mexico for less than 6 months at a time, but it is thought that a rising proportion intend to settle illegally in the United States. The American authorities believe that there are 5 million Mexicans living illegally in the United States at the present time. Such a figure is a substantial one, and the traffic it represents is a bone of political contention between the two countries. Yet this, and other considerations, suggest that the net annual migration rate of Mexicans leaving to settle in the United States by legal or illegal means in the late 1970s did not exceed a quarter of a million. In other words, such migration represents only one-tenth of the amount by which the overall population is rising each year, and makes only a minor contribution to resolving the problems consequent upon the fecundity of the average Mexican mother.

It is, therefore, the balance between a very high birth-rate and an ever-decreasing death-rate that has largely determined the characteristics of the Mexican population. As a consequence, despite an increase of 25 years in life expectancy since 1940, the population is an astonishingly young one. The average Mexican is a 17-year-old, and three-quarters of the population are under 30. Only one in twenty-seven is over the age of 65. In 1980 children below the age of 15 outnumbered old-age pensioners by 10 to 1; between them they made up half the population. By way of comparison the ratio in the United Kingdom is 2 to 1, and together they represent only 38 per cent of the total population. In practice, the average Mexican of working age has twice as many dependants to support as does his British counterpart.

Other stresses are placed upon Mexico by the uneven distribution of population. Central Mexico is the most populous part of the country, and in this respect little has changed since Aztec or colonial days. Within central Mexico the valley of Mexico, the basin of interior drainage housing the capital city, remains not only the most heavily urbanized part of the country, but also the centre of a large rural population. Today, as in 1930, 1950 and 1970, half the population of the country lives within 300 km of the capital – in an area which is only one-sixth of the national total. Over half the remaining population lives in the border states, more in the north now than in the south (reversing the 1930 situation). The lowest population densities (of under three persons per km^2) are found in the two peninsulas remote from the seat of affairs: Baja California and Yucatán. Baja California Sur has never supported anything but a sparse population, and made the transition from being a territory to a state only in 1974; but Quintana Roo (also declared a state in 1974) and Campeche at the foot of the Yucatán peninsula are littered with the ruins of substantial archaeological sites, witnesses to substantial populations in the Mayan past.

About 16 million people now live in Mexico City. It overtook Buenos Aires in the 1960s to become the most populous city in Latin America, and overtook New York city in the 1970s to be, once again, the premier city of the Americas. Such a distinction is no new thing to Mexico. The site of the first city to flower in the western hemisphere at Teotihuacán is on the northern fringe of the present metropolis; Tenochtitlán, the capital of the Aztecs and upon whose ruins the present capital is built, was the largest city any of the Conquistadores from Europe had seen; and Mexico City remained the largest city in the New World until the beginning of the last century. But it did not share in the widespread commercial development nor in the great waves of immigration that characterized temperate America, and in 1930 it only just found a place in the top ten American cities. Recent events rather than a continuation of historic circumstances account for its present outstanding position.

The rate of growth of the city has been startling, and has created numerous problems for the authorities. At the turn of the century the city had a population of about half a million. It joined the ranks of the 'millionaire' cities in the late 1920s, and by 1950 its population exceeded 3 million; between 1950 and 1970 it increased by 5 million, and the 1980 census gave the metropolitan region 14.4 million. It is growing at a rate estimated at about half a million a year (or one a minute). Natural increase accounts for most of the current growth, but immigrants from the provinces probably number well over 1000 a week. The revolution of 1910–17 not only disrupted the existing order of things, but also made the Mexican peasantry mobile. The extension of rural bus and lorry services even into parts of the countryside extremely difficult of access, the omnipresence of radio, and the generally more prosperous economy have made the countryman increasingly aware of the supposed opportunities that await him outside his immediate neighbourhood. Few migrants to the towns are now without friends or relations already there to cushion the shock of a move. Mexico City is not the only magnet, but on the national scale it is the dominant one. *Figure 2.2a* makes it clear that the closer migrants live to the capital, the more likely they are to move to it rather than elsewhere; it acts as the strongest magnet over three-quarters of the country, drawing migrants from their native states. *Figure 2.2b* shows that the north-western states, particularly Baja California, and the upper Gulf states in the east and north-east are locally more attractive than the capital.

The national movement to Mexico City is an exaggerated version of what is happening on a lesser scale elsewhere in the country. The provincial towns and cities are powerful draws. Naturally enough, the larger the town the more attractive it is for the migrant. But the relationship is not an exact one. For example, Monterrey, the third city of Mexico, has grown more rapidly since 1950 than Guadalajara, the second city; neither, however, has anything like the same attraction for the immigrant as the capital, over six times as populous as its nearest rival. The old mining states of the centre-north have lost large numbers of people, and the zone between 150 and 300 km from Mexico City has suffered substantially. In recent years the new oil discoveries in the southern lowlands of Tabasco and Tehuantepec have reversed the outward flow of foot-loose labourers and enhanced the attractiveness of the Gulf coast (*Figure 2.3*).

A powerful suggestion of the overall changes in Mexico is given by contrasting urban and rural growth-rates over the last 50 years (*Figure 2.1*). The urban population rose at an ever-increasing rate until the end of the 1960s, when it reached a figure of 5 per cent annually; since then it has fallen to about 4.5 per cent. In contrast, the rural population has increased much more modestly and at a remarkably steady rate of 1.5 per cent per annum. Whatever the precise limitations of using census definitions, the figures show that most of the adjustment to the extraordinary growth of population over the post-revolutionary period has been made in the towns. While the countryside has been able to absorb a limited amount of change, the towns have proved even more elastic. These figures emphasize not only the increasingly more urban way of life of the average Mexican, but also the declining importance of rural Mexico in the economic and political life of the country. Such a change in emphasis is common to most Latin American countries and is due, to some extent, to explicit government actions. Today, it is claimed, the interests of the countryside are being deliberately sacrificed to meet the costs of industrialization, the benefits of which largely accrue to the towns; economic advance is being won at the expense of the proletariat. This situation seems more ironic in Mexico than it would elsewhere in Latin America, since Mexico has been governed continuously since the revolution by a party owing its strength and its origins to the efforts of the rural masses.

The quality of the land

Steep slopes, drought, periodic flooding and low temperatures are some of the natural features that limit the ability of the countryside to absorb more of the rising population; nature has not ideally endowed Mexico for modern agriculture.

Cortés's reputed comparison of Mexico to a crumpled handkerchief exaggerates the prevalence of steep slopes in the country, though it may well strike a responsive chord in the mind of the observant air passenger. More recently, Mexican geographers and agronomists, by rough-and-ready calculations, hold that 26 per cent of the country lies in areas where the dominant relief is in slopes of 25 per cent steepness or more (González Santos, 1957, p. 24). Areas of steep slopes, *par excellence*, include the fretted margins of the central mesa or plateau of northern Mexico – the rugged eastern and western Sierra Madres (West, 1964b). Many of the canyons that score the Sierra Madre Occidental, such as the

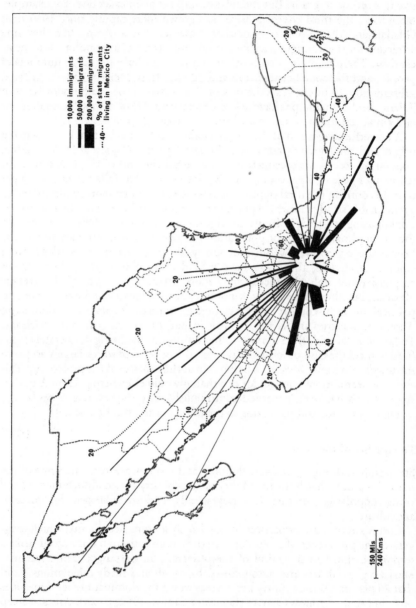

Figure 2.2 Population migration in Mexico 1950–60
(a) Immigration to Mexico City

150 Mls
240 Kms

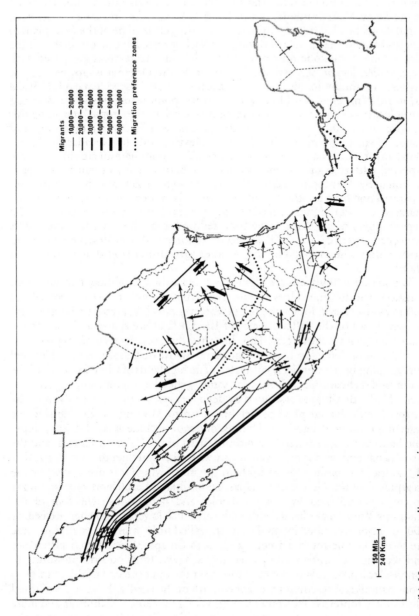

150 Mls
240 Kms

(b) Non-metropolitan migration

Barranca del Cobre, are 1500–2000 m deep and, although flattish outliers of the plateau may be preserved on their interfluves, the mass of the Sierra combines very steep slopes and a degree of inaccessibility unmatched elsewhere in the country; it has formed a refuge area for Indian groups like the Yaqui. The relief of the Sierra Madre Oriental is a little less chaotic; it is, in essence, a series of parallel, steep-sided ridges, which rise from the gentle slope of the Gulf plain to heights of 4000 m and are cut through by such spectacular canyons as that of the Moctezuma. Limestone outcrops occur widely and large areas are pitted with steep-walled karstic depressions. The Sierra Madre Oriental is crowned by the highest mountain in Mexico, the beautiful snow-capped cone of Orizaba (5750 m). This New World Fujiyama is at the point where the Sierra intersects the volcanic axis that runs across the country south of the Tropic, and along this axis lie other such splendid volcanic sights as Ixtaccíhuatl (5286 m) and Popocatépetl (5452 m) in the Valley of Mexico, and the Nevadas de Toluca (4392 m) and Colima (4265 m) to the west. They dominate an area in which the repeated pattern is one of enclosed tectonic basins, many of which were once occupied by Pleistocene lakes. Vestiges of some lakes remain to this day, separated one from another by forested uplands of medium-to-sleep slope. The Balsas River valley in the south is in reality a series of such basins (for example, the Tepalcatepec) whose slopes have been reduced and whose harsher lineaments have been softened by an infill of volcanic effluent and river alluvium; steep and shallow slopes lie in juxtaposition – easier agricultural land than in the Sierra Madre.

The southern highlands are highly dissected although of moderate elevation. Characteristically, the ridges rise to about 2000 m with the river beds 600 to 800 m below. Tracks follow the ridges wherever possible, avoiding the steep and broken slopes of the valleys. Areas of flattish land are sparse and that of the Valley of Oaxaca is the largest. This, like the other, smaller tectonic depressions, such as that of Tehuacán between the Sierra Madre de Oaxaca and the Mesa del Sur, has long been intensively cultivated. The highlands of Chiapas, south of the isthmus of Tehuantepec, present the same, generally unpromising aspect. The Sierra Madre de Chiapas is scoured by deep *barrancas* and rises steeply from the narrow Pacific coastal plain to heights of up to 3100 m; on the lee side the descent is more gentle and small basins, or *llanos*, include areas of shallow slope. The basin of Chiapas is a southern counterpart of the Valley of Oaxaca, and the Grijalva has endowed it more generously with alluvium than the Atoyac has that of Oaxaca. The basin ends abruptly in the north where it is overlooked by the escarpment of the Meseta de Chiapas, an undulating tableland with limestone outcrops pockmarked by small sinkholes and depressions. The foot of the adjoining Yucatán peninsula, although very much lower and nowhere exceeding 200 m, is honeycombed by similar, steep-walled sinks; most are small, but some are several kilometres in diameter and, as in Chiapas, have offered sustenance and seclusion to Indian groups seeking a haven from outside pressures. In contrast, the peninsula of Baja California, at the opposite end of the country, is mountainous and resembles the ranges north of the border. It is a series of tilted blocks with impressive escarpments overlooking the Gulf of California and more gentle westerly slopes. Only small pockets of flat alluvial land are found in the east, whereas these are more abundant in the west and include extensive marine terraces fronting the Pacific.

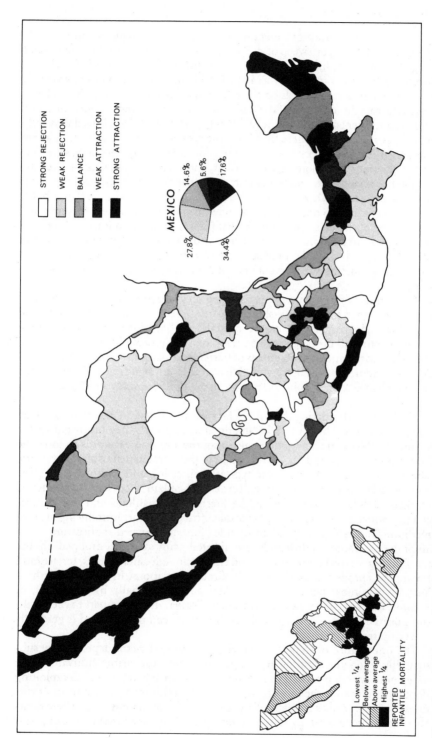

Figure 2.3 Areas attractive and unattractive to migrants in the 1970s

When we turn to areas of generally gentle slopes, three regions stand out. The largest is the great open sweep of the Mesa Central. In the north this is basin-and-range country familiar throughout the world as the setting of the classical Western film. Close to the Sierra Madre Occidental, uplift and recent volcanism have placed greater emphasis on the ranges and the basins tend to be enclosed, but in the east the alluvium of the basins or *bolsones* threatens to bury the ranges entirely and over large areas slopes are barely perceptible. The alluvial fans at the mouths of the canyons where they leave the ranges have proved the most attractive agricultural sites. Further south, the general level rises, the ranges are reinforced by older and more recent volcanic outbursts, the network of river dissection becomes closer, and flat land rarer. In the zone running from the Llanos of San Juan, west of Orizaba, to the basin of Tepic in the west, the flatter basins occupy between only one quarter and one third of the total area. Nevertheless, this is some of the most important agricultural land in the country and includes the Bajío of Guanajuato, a series of linked basins on the Lerma north-west of Mexico City, and the Valley of Mexico itself, basins in which volcanic fall-out and lacustrine sediments supplement the alluvium washed in from the adjoining slopes.

The isthmus of Tehuantepec is a broken area, despite its low elevation, and moderate rather than gentle slopes are most characteristic. In this respect it resembles most parts of the narrow coastal belt that fringes the Pacific. In the north the outwash of the peripatetic rivers of the Sierra Madre has partly buried the irregularities and it is these areas of flatter land along the courses of the Lower Sonora, Yaqui, Mayo, Fuerte and Culiacán that are significant. Elsewhere the coastal plain is either non-existent or, as in the Soconusco of Chiapas, very narrow.

The flat lands bordering the Gulf of Mexico are much more extensive. They are 300 km broad south of the Río Grande and, although they pinch out in the vicinity of Vera Cruz, they widen again southwards and eastwards to form the lowlands of Tabasco and the northern part of the Yucatán peninsula. Very gentle slopes of less than one degree are typical of these lowlands, with rivers slightly entrenched below the general level, breaking the plain into a series of segments. Outliers of hill country, such as the Sierra de Tamaulipas and the Huastec country north of Vera Cruz, and volcanic intrusions occur, introducing areas of medium and steep slopes unsuitable for mechanized agriculture. The amphibious Tabascan lowlands are littered with detritus dropped in the meanderings of the Lower Grijalva and Usumacinta; fossil sand bars, terraces and flood plains produce a local relief full of minor variations. In contrast, northern Yucatán is almost perfectly flat, broken only occasionally by sink-holes, or *cenotes*, and not cut by a single river. Like Florida, its counterpart to the north, the Yucatán peninsula is essentially of limestone composition and is gradually emerging from the sea.

The importance of slope as a limiting agricultural factor depends very substantially upon the farming techniques used and the plants cultivated. The indigenous system practised widely in Mexico is much better suited to exploiting sloping land than is the modern system. In the traditional system seeds or cuttings of plants or trees are planted by hand in holes or mounds and, where desirable, slopes are reduced by various terracing methods; as a consequence of planting a variety of species the soil may retain vegetation throughout the whole year. In this manner slopes are cultivated which are too steep and inaccessible for mechanical working or even for draught animals. By careful selection of plant combinations and the use of only a hoe the soil surface is little disturbed or exposed and the dangers of soil erosion are reduced. The introduction of plough-ing and domestic grazing animals in the sixteenth century and the movement towards monoculture brought this relatively conservative system under attack. In spite of a prodigious fall in population in early colonial days soil erosion was aggravated; a rising rural population thereafter has meant that today large areas of Mexico suffer from degraded soils. This is especially true of the southern part of the Mesa Central and in the highlands of Oaxaca and Chiapas. In the basin of Toluca and the Valle de Mezquital even shallow slopes are ravaged and the situation in Tlaxcala is notorious; in no other state is such a high proportion (over 40 per cent) of land cultivated, in spite of the prevalence of steep slopes, and in no other state is agricultural productivity so low (Yates, 1962, p. 54). In Oaxaca, bald surfaces, corrugated slopes and gashed arroyos are almost as extensive as land that is vegetated, and the landscape is a patchwork of green and red or buff. Although there is a strong correlation between steepness of slope and degree of soil erosion, the flat lands of northern Yucatán retain only a very thin and fragmentary soil cover, reminding us that another strong correlation exists between erosion and dense Indian populations and suggesting the way in which the traditional system has broken down.

In the drier north, where the vegetation cover is less complete and where rains fall more violently, soil erosion is more natural. Even so, it has been aggravated by the grazing and browsing of goats and cattle, and the Conservation Fund has found that moderate-to-severe erosion (that is, at least 10 per cent of the land surface severely eroded) characterizes half of non-tropical Mexico. H.H. Bennett estimated in 1945 that 50 per cent of the arable land of Mexico has been 'ruined for cultivation, nearly ruined, or severely affected' by soil erosion (Stevens, 1964, p. 313). No other Latin American country has suffered so much despoliation.

Another obstacle to successful commercial agriculture may be the moisture regime. Much of Mexico is too dry for field cultivation, and some of it is too waterlogged. The Ministry of Water Resources has estimated that over 73 per cent of the country is arid or semi-arid with a rainfall of below 1000 mm per year, and under 2 per cent is classed as very humid; the rest is largely classified as sub-humid. Other estimates have been made using somewhat different criteria, but all emphasize the climatic deficiencies of Mexico for agriculture.

Although less extensive, the very wet areas and their agricultural problems have attracted considerable attention recently. The heaviest rainfall in the coun-try is on the edges of the Tabascan lowlands, where the Sierra de San Cristóbal and the Tuxtla highlands are drenched by mean annual rainfalls in excess of 4000 mm. The lowlands have 2000 to 3000 mm of rain a year; unlike the rest of Mexico they experience no dry season, only a less wet one in February and March. The area is largely under forest, even in places that supported the Olmec and Maya cultures of the past. Drainage control is the principal agricultural problem; vast areas are inundated after the heavy summer rains and the swollen rivers (perhaps 5 km wide when in flood) disrupt communications. The control of flooding has been one of the objectives of the official Commission of the River

Grijalva, one of several river basins to have been created special development areas. As in other designated basins dams have been built to modify the natural regime of the river and one, the Malpaso, is not only the largest in Mexico, but one of the most ambitious civil engineering constructions in Latin America. The adjoining Papaloapan basin suffers similar problems, although total rainfall is less. The coincidence of the normal wettest season with the occasional September or October hurricane, however, means that the whole of the lower basin, in places 55 km wide, was periodically flooded (Poleman, 1964). A particularly severe flood in 1944 led to the creation of the Comisión del Papaloapan in 1947. Since then the Alemán dam has been built across the Tonto, the main tributary of the Papaloapan, and it is claimed that, as a result, the whole of the western margin of the Papaloapan as far north as Tlacotalpan enjoys protection. These attempts to reclaim what has been called 'the last unconquered frontier in Mexico' have been very costly: more capital has been invested in the Grijalva Commission than in any other comparable agency, and only the Papaloapan Commission runs it close. Judged by the targets that the commission planners originally set themselves under, for example, the Limón plan, later re-drawn as the Chontalpa plan, the land reclaimed for agriculture has been won at too high a price. Admittedly, 40,000 people have been settled under the plan, an infrastructure of roads, public services and villages built, and a commercial economy based on cattle and sugar cane established. But it had to await the unforeseen discovery of vast oil reserves to make the newly won accessibility of these wet lands pay an adequate return on investment.

Elsewhere the rainfall map of the country (Vivó Escoto, 1964, pp. 200, 204) shows several patterns. In general, rainfall declines northwards from an average of about 2000 mm on the southern border to one of below 400 mm on the northern; it is less, latitude for latitude, in the west than it is to the east. The map also reveals the pronounced effect of the coastal ranges casting heavy rain shadows inland. The net result is an average rainfall of about 1000 mm, and problems of drought are much more familiar to Mexican farmers than the problems of too much water met within the Grijalva and Papaloapan basins. The driest part of Mexico is the north-west – in the interior basins of Chihuahua and Coahuila and, more particularly, in Sonora and Baja California; the Altar and Vizcaino deserts are bone dry, and are as hot as anywhere in the Americas. Even in other areas the coincidence of the rainy season with the longer summer days means that losses through evaporation of such soil moisture as is absorbed are high. Thus, about 75 per cent of the rain that falls in the Valley of Mexico, on the semi- and sub-humid boundary, is lost by evaporation and not retained in the soil for possible agricultural use. Nor is the rain that falls in dry Mexico reliable: in most parts of the north the usual year-to-year variation is over 30 per cent of the long-term average rainfall, and in the drier parts of Baja California it exceeds 60 per cent. On the arid margins crops may be snatched in wet years, but it is the frequency of dry years rather than average rainfall that is significant.

The moisture balance between rainfall and evaporation is partly a question of temperature, and temperature is as much a question of altitude as of latitude in Mexico. Most of the tropical lowlands (the *tierra caliente*) have mean annual temperatures of over 23 °C and are virtually frost-free. A killing frost in southern Tamaulipas is an extremely rare event, and the citrus crop has been lost only once in the last 20 years. Yucatán has never registered a temperature below freezing point. The *tierra fría* coincides approximately with the land over 2000 m and, if defined by the 15 °C isotherm, includes the volcanic axis of the Mesa Central, the southern part of the Sierra Madre Oriental, extensive parts of the Sierra Madre Occidental and the higher ranges of Chiapas and Oaxaca. The intermediate zone (the *tierra templada*) lying roughly between 1000 and 2000 m is the most densely settled of the altitude zones, with daily temperatures ranging through about 10°C, and rarely suffering frosts or very high evaporation rates.

The areas best favoured climatically, not too dry, too wet or too cold, amount to perhaps no more than a quarter of the country. They include almost all the Yucatán peninsula (except, in fact, the only part to carry a heavy agricultural population, that around Mérida), most of the Isthmus of Tehuantepec and the Gulf plains south of the Tropic, the southern part of the Mesa Central including a narrowing zone following the eastern margin of the Sierra Madre Occidental, and the narrow Pacific coastal plain south of Tepic.

Agriculture

The paucity of gentle slopes and a largely unsuitable climate have severely limited agricultural opportunities in Mexico. These limitations are extremely important in a country where one-third of the labour force still works on the land, and perhaps help to explain why agriculture produced only one-eleventh of the national wealth in 1981, and why the average Mexican employed outside agriculture contributes six times as much to the gross national product as does the average farmer. Until recently the pattern of commercial agriculture accurately reflected these conditions, being essentially confined to the well-drained foot-hills of the Sierra Madre Oriental in Vera Cruz, to the Soconusco strip in Chiapas and, in particular, to the higher and cooler basins of the southern Mesa Central (West and Augelli, 1975, pp. 277–92). Recently, however, the stranglehold imposed by drought upon agriculture in the north has been broken, and large capital investments and improved technological methods have permitted the development of the most extensive irrigation programme in Latin America.

The course of development of irrigation is suggested by some figures. In 1926, before the entry of government effort into this field, there were probably a little under 1 million ha under irrigation and, although the ownership of land may have changed, the irrigated area in 1950 was only a little over this figure. The major increase has come as the results of development schemes financed by the government have begun to show fruit. By 1960 the area under irrigation had doubled to 2.2 million ha, and in the late 1970s the Ministry of Agriculture returns show that 3.5 million ha were under irrigation. The precise area given by official and academic sources varies, and this is partly because not all the areas within the officially designated irrigation districts may be irrigated, partly because it is difficult to reconcile figures for privately owned irrigated farms with total figures published by the Ministry of Water Resources, and largely because the area is dependent upon the amount of rain that has accumulated in the reservoirs and aquifers during previous seasons. Annual variations in the volume of water stored in the feeder reservoirs are not uncommonly of the order of 30 per cent. Such uncertainty from year to year has stimulated the Ministry to carry out

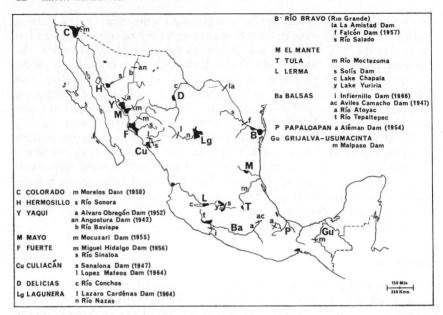

Figure 2.4   Major areas of irrigation in Mexico, 1970

cloud-seeding programmes in, for example, Sonora, but with only inconclusive results to date (Kraemer, Latorre and Cervantes, 1977). In 1950 only one-eighth of Mexico's agricultural land was irrigated; today the fraction is nearer one-quarter. During the 1950s and 1960s 10 per cent of the federal budget was spent on irrigation works, and it is only with the more recent requirements of the oil industry that this proportion has fallen. The rate of expansion of irrigated land has fluctuated, partly because once the government had committed itself to large-scale reclamation schemes it naturally tackled the easier and cheaper areas first. The importance of the irrigated lands does not rest solely upon size and location (*Figure 2.4*). They are also highly productive and more likely to produce for the commercial and overseas markets. Yields are three times those from non-irrigated farms and from accounting for about one-quarter of the value of Mexico's annual harvests in 1950, the irrigated farms are now responsible for close to half.

The largest irrigation districts are on the mainland side of the Gulf of California, on the alluvial fans at the foot of the Sierra Madre and on the delta of the depleted Colorado; here, the three driest states of the country produced, by 1967–8, half the value of all sales from irrigated areas under the control of the Ministry of Water Resources. The single largest district is the Yaqui district in Sonora where about 300,000 ha are under irrigation. Like the nearby Fuerte and Mayo areas irrigation works predate the revolution, and the first large canal dates from the 1890s. Agriculture was delayed by the Yaqui Indian uprising but by 1937, when the government decided upon expropriation, some 66,000 ha were being farmed. Since then the building of the Angostura dam and the opening of the more important Alvaro Obregón dam 10 years later, in 1952, have quadrupled the acreage.

Almost as many hectares are irrigated in the Fuerte valley, three times as many as in 1938 when the government assumed control. The Miguel Hidalgo dam (1956) has been the key factor in the Fuerte River Basin Commission's comprehensive development programme. Here, further south, water is more abundant and reliable: with only half the catchment area of the Yaqui the volume of water available is twice as large. A wider range of plants can be grown, and even such water-demanding crops as tomatoes flourish. The Mayo district, between the Fuerte and the Yaqui, is only one-third as large; it is mainly dependent upon waters stored by the Mocuzari dam (1955). Unfortunately, engineering mistakes have made this one of the problem areas. The irrigated area in the Culiacán district (120,000 ha) on the other side of the Fuerte has been cut by one-third in the last 10 years, yet it enjoys much the same advantages as the Fuerte, and the value per ha of its crops in recent years has been the highest of any of the major irrigated districts.

The history of irrigation in the Colorado delta began in pre-Spanish days, but the antecedents of the present system include works of mainly American companies begun in the nineteenth century with the help of Chinese coolie labour. In 1937, when the land was expropriated, about one-quarter of the present cultivated area was under crops. Since then the introduction of an elaborate system of control dams and pumping stations, the completion of the Morelos dam (1950) and the tapping of deep well waters to supplement surface flow have increased the total area cultivated to about 180,000 ha and have made the delta area one of the most productive of all the irrigation districts in the country.

The Río Grande basin is a second area of localized irrigation. The Conchos and other upstream tributaries have been dammed to hold back the drainage of the eastern slopes of the Sierra Madre Occidental; downstream the Río Grande is itself dammed, as are some of the shorter rivers, like the Saltado and the San Juan, which drain the Sierra Madre Oriental. The net result is the cultivation of some 350,000 ha of otherwise barren land, three-quarters of which lies in the 200 km reach of the river valley downstream from the Falcón dam (1957) as far as Matamoros. A newer dam, La Amistad, has been built across the middle Río Grande above Ciudad Acuña, and has created a greater storage capacity than any other dam in northern Mexico. At the present time the lower Río Grande produces about as much as the Fuerte or the Yaqui districts.

The Lagunera region centred on the town of Torréon in the Mesa Central was the first large-scale irrigation district to be organized, but since the completion of the Cárdenas dam across the Nazas in 1946 there has been only a modest extension of the area cultivated (about 120,000 ha) in spite of recourse to underground water supplies. The region is one of interior drainage and has suffered over-optimistic estimates of the water available to sustain agriculture; at the time of expropriation the area believed capable of irrigation, and distributed, was twice that now productive.

Away from the very dry lands of the north there are large areas of sub-humid land which have been made more productive by irrigation. The colonial granary of Mexico, the Bajío of Guanajuato and Jalisco, is irrigated by the waters of the Middle and Lower Lerma, the Ministry of Water Resources alone controlling about 150,000 ha. A storage dam exerting some control over Lake Yuriria was built as early as 1648 and the shore of Lake Chapala has a long history of irrigation; nevertheless, 100,000 ha of irrigated land have been added since 1926. The Solís dam (1949), serving the area above Irapauto, stores more water than the rest of the works on the Lerma put together. The Tula area, with 50,000 ha under irrigation, has a similar history; since 1900 it has received and, after the initial shock, welcomed the sewage of Mexico City as a rich supplement to the local reservoirs of water. The 20-year-old Aviles Camacho dam supplies water to irrigate the Valsequillo area south of Puebla in the Atoyac valley. The Atoyac is a headwater of the Balsas and so falls within the jurisdiction of the Balsas Commission, another of the multi-purpose river basin development authorities. Large areas of the Balsas basin are promised water in the future, but equally important is the generation of hydroelectric power and the creation of an infrastructure of economic and social services. Other irrigation districts of local importance are found in the drier northern sections of the Gulf of Mexico plain, for example, around Ciudad Mante.

Although the irrigated lands are a visible and very important sign that the problem of drought is not insurmountable, they nevertheless suffer from a number of difficulties and deficiencies. The amount of water available has an economic as well as a physical limit. The economic limit has fluctuated from time to time with changing techniques and priorities but an estimate of 8.2 million ha given by the Ministry of Water Resources is probably on the high side. The cost per ha irrigated in real terms rises each year (for example, the cost of water stored by the Falcón dam was 0.82 pesos/m^3, whereas the estimate for La Amistad dam was 0.95 pesos/m^3) and the pace of change must be expected to fall.

Forecasts have proved wrong in the past partly because unforeseen difficulties have been experienced in certain of the irrigation districts. More land has been distributed than the water resources have been able to serve: for this reason 12 per cent of the Colorado delta area has had to be withdrawn from cultivation, and similar over-extension is reported from the other districts fronting the Gulf of California. In many districts an increasing reliance upon underground sources to make up surface-water deficits proving only a short-term palliative, for more water is being drawn off than is being replenished, and reserves are falling. In the Costa de Hermosillo in Sonora, for example, which relies entirely upon underground supplies, the water-table is falling and the future of this district is in jeopardy. Another linked problem has been the accumulation of salts in the soil. This is a condition well known in other irrigated parts of the world, and now affects about 15 per cent of the total area under Ministry direction in Mexico. Detailed field studies in the late 1950s showed that over one-third of the Colorado delta area was seriously affected by salt accumulation, and less than one-fifth was entirely free of adverse salinity effects; in the Yaqui zone 15,000 ha on the Gulf side near the limits of the irrigation channels have become too saline to cultivate, and another 25,000 ha are affected. Only the Costa de Hermosillo was thought to be unaffected by salinity in the whole of the north-west, until it was recognized that sea water was spreading inland as excessive pumping had lowered the natural water-table (Jardines Moreno, 1976). In the Río Grande–San Juan area 80,000 ha are affected, and production in the Lagunera region has been interrupted. The cure is a threefold one: existing salinized soils are flushed with a liberal application of fresh water to remove existing salts, a satisfactory drainage system is imposed to prevent salinization recurring, and often the area irrigated must be reduced to ensure adequate natural drainage of the soil and to reduce excessive evaporation. These measures are expensive (for example, 100,000 ha in the Río Grande districts require levelling to improve their drainage) and locally unpopular and disruptive.

Although Mexico can be justifiably proud of the way in which she has mastered some of the problems of exploiting land naturally too dry for cultivation, the major part of the arable farmland is cultivated without the aid of irrigation, and is obliged to suffer the trials of climate and slope that make agriculture difficult in most parts of the Republic.

The distribution of farmed land in Mexico is an accurate reflection of the distribution of population: a statistical comparison by states of their proportion of the national population (excluding the Federal District) with their proportion of cultivated land gives a 90 per cent correspondence. The areas of most intensive cultivation are, therefore, the central states of Tlaxcala, Puebla and Guanajuato, in which more than one-third of the land is cropland; in contrast, the average proportion both for Mexico north of the Tropic and for peninsular Mexico east of the isthmus of Tehuantepec is well below one-tenth. In practice, only about half the cropland is cultivated in any given year, and only about one-third yields crops; yet, in spite of these low figure, there are obvious signs that the soil is being overstrained in many places.

The distinction between the traditional type of subsistence farming and modern commercial agriculture is worth stressing. The traditional type is of over 5000 years standing. It relies very heavily on the cultivation of maize, beans and the squashes, which, when flavoured by the ubiquitous pepper, still form the staple diet of rural Mexico. In a classic combination the three are planted in the same hole; the maize stalk grows first, later serving as the bean pole, while the large-leafed squashes blanket the ground below, the whole forming a useful conservation system. The extent to which the traditional system prevails is suggested by the extraordinary importance of these indigenous crops in Mexico. One-half of all the cropped land is still devoted to maize, and a further one-eighth is under beans. Other less well-known plants with a long tradition of cultivation remain important today. One is the agave, the spiky 'century-plant' frequently used as a symbol of the country outside Mexico. It is a useful plant partly because it will prosper under conditions too dry for others, and it has been successfully turned to commercial uses. The maguey agave yields *pulque*, a nourishing and widely consumed drink of moderate alcoholic content; grey-green maguey plantations are a common sight in the northern and eastern reaches of the Basin of Mexico. Unlike beer, pulque must be drunk fresh and does not travel well; if experiments in canning pulque currently under way prove successful, there will be a large new market for the drink, and it will add another industrialized product to the impressive list already derived from the lowly agaves. A spirit, *mescal*, is distilled from another type of agave especially grown in Jalisco (and notably in the vicinity of Tequila). In Yucatán (Fox, 1961) the henequen agave yields a sisal-like fibre, and this plant forms the backbone of the economy of the state; probably one-half of the population is dependent upon henequen, and it is not surprising that competition from Tamaulipas (Fox, 1965a), from the other hard fibres and perhaps from an artificial substitute are of current concern. Another measure of the extent of traditional elements is the widespread practice of slash-and-burn shifting agriculture. About one-fifth of the cropped area is cultivated in this manner. Such cultivation is not restricted to flat lands, and the hand tilling that accompanies it has allowed many slopes too steep for the plough to be heavily used. Not uncommonly these steep slopes, often broken by terraces, *trincheras* and other soil-checking devices used since before the Spanish Conquest, overlook flatter alluvial land on which commercial crops of, say, sugar cane or wheat or maize are being grown on the basis of mechanized, fertilized monoculture.

This situation is well seen in Vera Cruz, the most productive agricultural state of the nation with its great variety of agricultural conditions and its proximity to the Central Mexican market. Here agricultural methods range from the most modern to the most traditional. One of the newer commercial crops to gain a foothold in the state is rice, now grown in the swampy lowlands of the Papaloapan, and likely to become a more important crop in the other lowland regions of tropical Mexico. Sugar cane is the dominant commercial crop elsewhere in the Papaloapan basin, as it is on the Gulf plain bordering the railway lines running up to Córdoba and Jalapa, along the foot of the Sierra and, beyond the Panuco, in the irrigated districts of southern Tamaulipas. The state of Vera Cruz produces more sugar cane than any other, and so suffers from some of the well-known economic and social problems common to cane-growing areas elsewhere in Mexico (notably in Morelos and coastal Sinaloa) and Middle America. Commercial banana plantations are also spreading in the *tierra caliente* of Vera Cruz, but it will be a long time before the state recovers from the havoc wrought by Sigatoka disease in the 1930s, which led (as elsewhere in Latin America) to a shift of the main banana-producing areas of the country to the west and south-west coasts. In the *tierra templada* of the lower and middle slopes of the Sierra Madre coffee and citrus fruits replace the tropical plants as the major commercial ones. It was on the slopes between Córdoba and Jalapa that these were first established as commercial crops, and production continues to rise; today coffee is also grown elsewhere, particularly in Oaxaca and Chiapas (Ellis, 1971), and Nuevo León is a more important citrus-growing area. The Vera Cruz–Mexico City axis has been the heart of many things in Mexico, and it is only relatively recently that other parts of Mexico endowed with comparable or alternative advantages for commercial agriculture have become flourishing.

In spite of the importance of commercial crops maize remains the leading crop of Vera Cruz, as of the country, even though relatively less significant than in the past. It is obviously difficult to obtain statistical data for a subsistence crop (and official figures differ considerably), but two points are worth recording. First, Ministry of Agriculture figures show that national maize production changed little during the earlier part of the century, but more than tripled between 1945 and 1965, culminating in a few years when Mexico began to export production surplus to national requirements. Since the 1960s the area under maize has entered a downward trend (while total land cultivated has settled down at a figure which is about 15 million ha), and recourse has had to be made to imports once again to give the Mexicans their daily tortilla. Secondly, the yields of maize are very low; only one-quarter of those normal in the United States and substantially lower than in the Argentine. Thus, although yields of this basic staple have doubled since the mid-1930s, there is still ample scope for improvement. In fact, yields of other crops have improved at a more rapid rate than maize. Indeed, it is the traditional farmers' usually correct perception of the risk, not of a larger crop during climatically good years, but of a poorer harvest during bad years that has prevented more widespread adoption of agricultural innovations.

One of the most important mechanisms for creating crop improvements has been the introduction of improved seeds. This is obviously easier to do amongst farmers orientated to a cash crop than amongst the more conservative subsistence farmers; nevertheless, it is claimed that 20 per cent of the maize crop is now from hybrids. Another factor in improvement has been the more intensive use of fertilizers; consumption has been increasing at an annual rate in excess of 15 per cent since the middle 1950s, aided by a booming domestic fertilizer industry as yet incapable of satisfying home demand (*The Puebla Project*, 1970).

An important factor which limits agricultural production is the land-tenure system, and on this issue Mexican agronomists find themselves on the horns of a dilemma. On the one hand there is clear evidence that land cultivated in larger units is more productive than land cultivated in smallholdings: for example, yields of wheat are one-third higher on holdings over 5 ha in size as compared with those of 5 ha or below (Durán, 1966, p. 85). On the other hand, for social and political reasons, the proportion of agricultural land in smallholdings has risen substantially. In 1930 93 per cent of the agricultural land was in holdings of over 5 ha in size (González Navarro, 1965, p. 217), whereas today the corresponding figure is only 67 per cent, and the average size is dropping. The Mexican government, in the 60 years following the revolution, has, in rather spasmodic fashion, followed a programme of land reform which, though socially desirable, appears to run counter to the economic demands of modern commercial agriculture. Over 77 million ha (over a third of the national territory) has been expropriated and redistributed since 1924. Large private estates (haciendas) still remain, especially in the dry country of the north, but elsewhere they have usually been carved up, the owner (*hacendado*) sometimes being allowed to retain a small part of his estate and to become a so-called *pequeño propietario*. The expropriated land has been redistributed in a variety of ways. Ownership of most has been vested in the nation, and the use of it entrusted either to individuals as smallholdings or, once rarely but now as part of a new agricultural policy, to a group and farmed co-operatively. Such lands, called *ejidos*, account for half the cultivated area today. In recent years the government has actually sold the title of lands whose development it has been anxious to promote (for example, in the irrigation districts of the north and in the pioneer lands of southern Yucatán and the Tabascan lowlands), and moderate-sized *colonias* of 20–50 ha in size have added a further element to the land tenure pattern.

Ejidal lands tend to be less productive than those in private ownership; this is demonstrable on a crop-for-crop basis (Flores, 1967), and is also expressed in the more conservative attitude of *ejidatarios* to the production of industrial and commercial crops. To some extent this distinction is due to the poorer average quality of *ejidal* lands. In Yucatán (Fox, 1961, p. 222) for example, where yields of henequen are lower on the *ejidal* land than on private lands, many of the *hacendados* were permitted to choose the 150 ha they were allowed to retain after expropriation; they naturally chose the better lands of their former estates.

In the Fuerte and Mayo valleys and on the Colorado Delta (Henderson, 1965, p. 305) the *ejidos* are on the lower valley lands that had been farmed for decades and have been worst hit by salinization. In general, there is a higher proportion of *ejidal* land in the centre and south of the country than in the more productive north and north-west. Further, the average *ejidal* holding is much smaller than the private holding, the educational standard of the *ejidatario* is poor, and he is in a relatively disadvantageous position with respect to access to capital, credit and machinery. The government food system policy (SAM), introduced in 1980, provided credits for *ejidatarios* to purchase agricultural inputs, subsidies and support prices, and had immediate impact. The 1981 initiative to link *ejidos* to privately-owned farms was a further move to blur the economic contrasts between the public and private agricultural sectors.

Official information is not always reliable when land-tenure matters are in question. Although the reform programme has served its prime objective – control of the countryside has been taken from the hands of the *hacendados* and spread over a much wider spectrum of rural society – anomalies remain or have arisen. Large estates still persist, even in the vicinity of the capital. More widespread, however, has been the fraudulent conversion of *ejidal* land. It is said that many of the *ejidatarios* in the Basin of Mexico lease out their land, or make it over on a share-cropping basis, whilst they take factory jobs; and that between one-third and one-half of the *ejidatarios* in the Yaqui valley lease their land to others. Another illegal situation which is apparently tolerated by the government, relates to the size of holdings, both private and *ejidal*: in the Colorado delta some *ejidal* holdings are twice and others in the Yaqui valley six times the legal maximum; in the Hermosillo region private wheat farms are twice the maximum of 100 ha allowed under the Agrarian Code. In places one person may effectively farm several *ejidal* plots or private farms by controlling the extra holdings through members of his family or other front men: private farms units of this type of 500 and 1000 ha exist in Sonora and Sinaloa. The economic advantages which accrue from these large holdings and receive a degree of tacit official recognition today may be more difficult to maintain in the future as social and political pressures against them become stronger and more insistent.

In 1970 there were about 2.8 million *ejidatarios* working lands distributed to them or their parents; in addition there were about 3 million landless *campesinos* eligible for *ejidal* dotations in 1975, and their number rises year by year while Mexico is running out of land available to create new *ejidos*. The programme of land redistribution is expected to be completed in the 1980s. Between 1976 and 1982, 15 million ha, mostly in the dry north, were distributed, though to only 305,000 families. Individual grants have to be of at least 50 ha to be economically viable. Thus, despite the social achievements of the land-reform programme, it has not removed the landless peasant from the Mexican rural scene and, in fact, one can expect day-labourers, the most lowly members of rural society, to become more and more numerous. A stratified rural society is emerging in which the day-labourer remains in status a solid element below the *ejidatario*; the *ejidatario* is in turn inferior to the colonist; above him, near the top of the ladder is the *particular* or private landowner. The growing awareness of the fate awaiting many rural Mexicans is out of keeping with expectations inculcated at schools and by political leaders; it has already led to some unrest, especially in those areas where alternatives, such as migration to the factories of the cities or to the pioneer lands, are more difficult. It is hard, for example, for a Mayan peasant in Yucatán, used to the peculiar, specialized and limited requirements of work in the henequen plantations, to acquire the radically different skills and outlook required of a pioneer farmer in the south of the peninsula or of an artisan in distant Mexico City. Mérida has been no stranger to mass demonstrations of impoverished *campesinos* in recent years.

The contribution of the pastoral industries to the national economy is relatively slight, in spite of the fact that over one-third of the country was classified as grazing land in the 1970 agricultural census. Three different types of animal husbandry are practised. In most parts of the country goats browse land which would otherwise be unproductive, except perhaps of firewood, and much has been written on the acceleration of soil erosion caused by their omnivorous and voracious appetites. Pigs, sheep, mules and the occasional multi-purpose cow are common domestic animals in the ménage of many peasant establishments, and supply essentially family or local requirements. These animals, like most domestic animals in Mexico, are, of course, post-Columbian introductions into the traditional agricultural scene.

Elsewhere, and particularly in the northern states of Sonora and Chihuahua, great cattle ranches still exist, a reminder of the days early in this century when Mexico rivalled Argentina as an exporter of meat. The revolution, the uncertain position of the *hacendados* in the 1920s and 1930s, the loss of the profitable US market, the decimation of the herds by a disastrous epidemic of foot-and-mouth disease between 1947 and 1952, and the expropriation of much of the ranchland, especially since 1958, have all effected notable changes in the industry. For example, a change in government policy in the last 10 years has led to the promotion of large co-operative cattle *ejidos*, and now almost half the cattle are *ejidal*. Many of the *ejidos* carved from the large estates are themselves large – up to 30,000 or 40,000 ha in semi-desert conditions. The government is also promoting more intensive management by improving the quality of the pasturage, by encouraging the growth and purchase of supplemental feedstuffs where appropriate, and by improving the quality of the stock; it does this by direct action on the *ejidos* and by offering incentives and inducements (including confirmation of ownership of their estates or guaranteeing limited freedom from expropriation) to private ranchers. Some *ejidos* are fortunate in having inherited quality animals: for example, the seven *ejidos* created from the former 150,000 ha ranch of the Cananea Copper Company in Sonora are each stocked with 3500 Herefords, and breeding stock is maintained. Elsewhere great efforts are being made to switch to more productive beasts, including crossed zebu, and only one in three is now of the traditional *criollo* breed. The proportion of *criollo* is higher on *ejidos* than on private ranches.

Thirdly, there are several areas of intensive cattle-raising. Sale of milk to the burgeoning towns has created dairy zones in the vicinity of the cities and around rural pasteurizing plants. The eastern shore of Lake Texcoco opposite Mexico City is one such area, the Mezquital further north is another. The Huasteca area of northern Vera Cruz and southern Tamaulipas is the most important beef-fattening area. Intensive cattle-raising, for meat or milk, is increasingly becoming an adjunct to irrigation farming; this is particularly true of the Bajío and the Lerma valley where more and more alfalfa (lucerne) and other fodder crops are being grown. But there is great room for improvement, despite Mexico's long-standing cattle tradition: the average dairy cow of the *ejidos* yields only 1000 li per year, half that of the private dairy farmer and a quarter of the milk from the best herds in the country.

The forests and the fishing grounds

The forests and fishing grounds of Mexico (*México*, 1966, pp. 85–96; West and Augelli, 1975, pp. 340–2; Tamayo, 1962, pp. 107–8) are as yet of very little economic importance, but both may be potentially valuable resources. The old forests that covered much of Mexico have long been attacked by the agriculturalist, the fuel gatherer and the man searching for constructional timber; in this century the wood-pulp miller has joined the exploiters. Few conservation measures have been applied, although the pine forests of the highlands of the centre and the western Sierra Madre and the tropical forests of the south-eastern lowlands are now being worked on a more satisfactory basis. But the problems – judicial, economic, ecological, social – that await any more intensive exploitation of the forests are legion, and it was these, rather than the opportunities, that provided the burden of the 1978 government report on forestry policy.

Commercial fisheries are limited, partly, like tropical forestry, by the great mixture of species, and partly by the small number of preserving centres, by the distance to the domestic markets of the interior and by the absence of a fish-eating tradition. The only commercial fishing of significance is on the shrimp grounds of the Gulf of California and the Gulf of Campeche, encouraged by the demand of the US market and the improved freezing facilities which have developed during the last twenty years, and on the Pacific tuna and sardine grounds. During the 1970s the industry – much of it still on a non-industrial basis – received a fillip when a large number of para-state processing plants amalgamated and won federal investment funds as part of a government policy to promote the fishing industry. The size of the catch doubled in the 1970s. Shrimps remain one of the most valuable products, and sales in the United States market contribute substantially to Mexico's export trade. It is interesting to note that the law restricts the fishing of shellfish to co-operative societies (the maritime equivalent of *ejidos*) and these supply half the fishermen and 80 per cent of the value of the national catch. There are a number of government boards to promote commercial fishing, but although *per capita* consumption of fish is rising, it is still only a third of that in the United Kingdom.

The mines and the oilfields

The great days of the Mexican mining industry are doubtless over, but from them much of the geographical structure which affects the pattern of life in Mexico today has been inherited (Brading, 1971; West, 1949; Bernstein, 1964). The modern economy is not likely to return to the situation which persisted until this century, in which over one-half of the earnings from exports were from the sale of metallic ores. Not that the former staples of gold, and particularly silver, have disappeared – Mexico is still the leading world producer of silver, and new mines are being actively developed and old ones revived – but rather that other minerals have superseded the precious metals, and other sectors of the economy have expanded more rapidly than the mining sector. By 1900 precious minerals had been overtaken in value by the production of non-ferrous minerals. Copper was the most valuable until 1930, lead until 1950, and zinc today; Mexico now occupies tenth, sixth and eighth place respectively amongst world producers of these minerals. But production of silver and the non-ferrous minerals has changed little in recent years; on the other hand production of iron ore and coal, although insignificant by international standards, has risen. But all developments in the traditional mining field are eclipsed by the remarkable rejuvenation of the country's petroleum industry. Crude oil is the new liquid gold of the national economy, and it promises to have an impact as substantial and dramatic as any of the other much-prized minerals which gild the pages of Mexico's history.

Changing patterns of mineral production have benefited some areas, but damaged others. A crude division of Mexico into three mineral zones may help in making the picture clear. A western zone running from north-west Mexico (Baja California, Sonora and Chihuahua) to Chiapas is the major source of copper and gold; an eastern zone comprising the Gulf lowlands produces all the oil, natural gas and sulphur of the country; and a central, overlapping zone produces silver, lead and zinc. The north has been replaced by the Gulf coast and the southern lowlands as the focus of mineral activity, and shrunken or ghost towns and mining camps have posed social and economic problems to the authorities in the dry interior. The population of the famous silver town of Guanajuato (Atunez Echagaray, 1964), one of the wealthiest cities of the Americas in the nineteenth century and briefly capital of the country, is now less than the 1880 figure; it has turned to its fine colonial architecture, its nineteenth-century civic buildings and its picturesque setting for a second life as a tourist attraction. Taxco in the south, where copper was first mined in Mexico, has followed a similar path. Yet mining at both Guanajuato and Taxco is undergoing a revival. But other old mining centres have been less fortunate: towns like Zacatecas and Pachuca survive largely because of their inherited status as administrative centres; others have had nothing to replace failing mines (Randall, 1972). The problem is most serious in the dry, north-western states, where local alternatives to mining often do not exist.

There are, however, some highly flourishing mines in the north-west. The great open-cast workings of Cananea and the mines of Nacozari in Sonora produce 40 to 45 per cent of all Mexico's copper; the new La Caridad mine near Cananea opened in 1979 and promises to be the largest copper mine in Mexico. More than 60 per cent of her zinc and lead and one-third of her silver come from the mines of Chihuahua where the ores are commonly mixed, and production of one metal depends partly upon the state of the markets for the other metals. Durango, Coahuila, Zacatecas, Baja California Sur and San Luis Potosí are the other leading metallic-ore-producing states, all vulnerable to price fluctuations largely outside national control. Until recently many of the mining companies were owned by foreign interests, perhaps increasing the vulnerability of the mining areas to external circumstances, but two changes have reduced the element of risk in recent years. The first has been the policy of progressive 'Mexicanization' of the mining industry, which has been pursued since 1961 through legislation providing incentives to companies in which Mexican citizens or a trust hold a controlling share of the stock. Only Mexicanized companies can obtain new mining concessions, and Mexico believes her internal economy is strong enough to make good any resulting loss of external investment capital. The process is almost complete. Allied to this change has been the growing involvement of the state through the Mining Development Commission (CFM). By 1979 the state held 40 per cent of the equity in all mining activities in the country, and the proportion was expected to continue to increase. The second change has been the increasing proportion of metallic ores absorbed by domestic industry. One-fifth of the zinc, one-third of the lead, and five-sixths of the copper were consumed within Mexico in 1966 as compared with less than one-tenth, less than one-fifth and less than one-half in 1960, and this trend has continued. A considerable proportion of the metallic ores, however, is still smelted or refined outside the country, especially in the USA.

In contrast to her wealth in precious and non-ferrous minerals Mexico is not an important producer of the basic minerals, iron ore and coal. Nevertheless, production (5 million tonnes of both coal and iron ore in 1981) is rising. The coal comes from the Sabinas basin, 100 km north of Monterrey, is the best coking coal in Latin America, and is mainly consumed by the iron and steel industry of Monterrey. So is the iron ore mined outside Durango City at Cerro el Mercado. Over half Mexico's iron-ore reserves are in such deposits as those at Las Truchas and Peña Colorado in the Colima–Michoacan–Guerrero region, and this explains the location of the new Lazaro Cárdenas iron-and-steel complex on the Pacific coast at the mouth of the Río Balsas.

Sulphur is another relative newcomer to the mining picture (Seawall, 1961). Since 1954 Mexico has risen to be second only to the USA as a world producer. Sulphur domes discovered in the isthmus of Tehuantepec during the course of exploration for oil in 1915 were rediscovered in the 1950s, and have yielded a handsome return; their ownership became a political issue and they have now been Mexicanized, partly to ensure the conservative working of a wasting resource.

The long experience that Mexicans have of the all-too-transitory gains that may be won from mineral resources, is standing them in good stead as they develop the biggest bonanza of them all: their newly discovered and enormous oilfields of the south. The technical challenges offered by exploitation of the fields are small compared with the political challenges posed by the need to make wise use of the national wealth they will soon be creating. The latter set of challenges are new; the others have been with the country in one form or another throughout the twentieth century.

The stage for the emergence of Mexico's oil industry was set in the nineteenth century (Fox, 1977). In 1884 legislation was passed transferring mineral rights, including those to hydrocarbons, from the state to the individual owners of the surface-land rights. This was a switch from the old Spanish custom to those practised north of the border, and the move was part of the relatively successful policy of President Porfirio Díaz to attract entrepreneurs and capital into Mexico, modelling herself on the United States. Very soon after the great oil discoveries were made at the beginning of the century in Texas, covetous eyes were turned towards neighbouring Mexico. By 1907 drilling in the Gulf coastal plain had hit oil in the hinterlands of Tampico (the Panuco-Ebano field) and of Tuxpan (the Golden Lane field), and an oil rush began which lasted 15 years.

Spearheaded by forceful individuals, it took form as holdings were consolidated into large British and American oil companies. The oil was directed to the Gulf coast ports, where some of it was refined before being exported. The outbreak of the First World War stimulated overseas demand, and Mexico found she could sell all the oil she could produce. The oilfields were little touched by the revolutionary turmoil that swept other parts of the country during this period, remote as they were from the powers of central government and independent of the local market. Such was the spectacular growth of the industry, that for a brief period in the early 1920s Mexico was second only to the United States in the international oil stakes, and was responsible for one-quarter of the world's output. By the late 1920s a decline almost as spectacular as the earlier rise had set in, and by 1930 oil production was only one-fifth of that of a decade earlier. Refineries closed, concessions were abandoned and three-quarters of the labour force lost their jobs. The slump was partly because new discoveries were not forthcoming, partly because of uncertainty about the attitudes of the new revolutionary government (whose new constitution had returned all mineral rights to the state), and partly because oil discoveries in Venezuela made that country a more attractive outlet for overseas investors. Credit for halting the fall in production during the 1930s must go to the Royal Dutch Shell interests, who discovered the rich Poza Rica field, an extension of the Golden Lane field, a discovery which made them responsible for two-thirds of the country's output in 1937. It was in 1938, however, that the most significant event of the 1930s occurred. This was the abrupt expropriation of the oil industry by the government. Vastly popular in Mexico, this action was seen to crown the revolution and to symbolize the coming-of-age of Mexico's economic independence. But it took Petroleros Mexicanos (Pemex, the nationalized oil corporation) almost a generation to recover from the flight abroad of capital and technical skills, and the loss of half the market consequent upon the international boycott organized to deter other oil producers from following what history has since shown to be a much emulated precedent.

International rehabilitation following the Second World War, the discovery of natural-gas fields near Reynosa close to the Texas border, and a steady rise in production sufficient to satisfy much of the growing internal demand characterized the industry until the 1970s. A significant shift in the 1960s was a growing dependence upon the newer fields discovered by airborne seismic exploration methods in the swampy forest lands of central and eastern Tabasco; by 1970 these fields were responsible for half the country's oil and natural-gas production (*Figure 2.5*). But production was beginning to lag behind national demand, and Mexico found herself in the ignominious and expensive position of importing oil when the world oil crisis broke in the early 1970s.

The rejuvenation of the industry dates from 1972 when the discovery of the new Reforma fields in Chiapas and Tabasco heralded a complete re-assessment of the oil potential of south-eastern Mexico. Since then a whole family of new fields

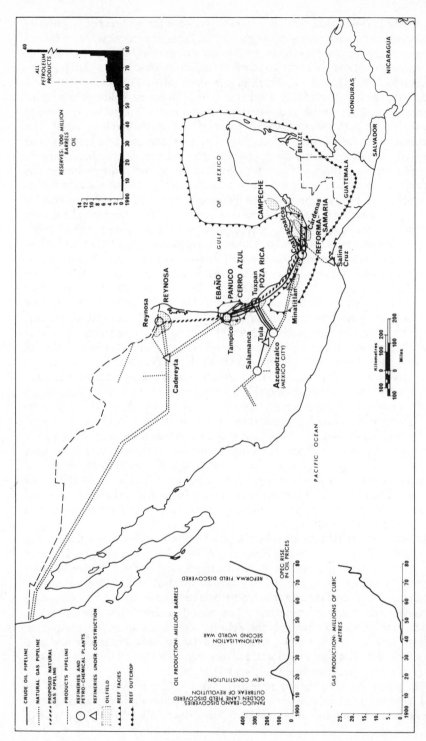

Figure 2.5 Oil and natural gas

has been discovered, extending below the Gulf of Mexico in one direction and possibly into Guatemala in the other. The fields tap three reservoirs: the Bermundez, north-west of Villahermosa, the Cactus in northern Chiapas, south of Villahermosa, and the Agave, east of Cactus. The oil is trapped in the Cretaceous sediments that underly the Tertiary subsoil. All previous production came from shallower wells, but rising oil values made deeper drilling economic, with vastly rewarding results. Almost every new well drilled strikes oil. Proven reserves have doubled and re-doubled. They stood at 5567 million barrels (equivalent) at the beginning of 1971; by the beginning of 1983 they were over 72,000 million barrels. A further 90,000 million barrels of probable reserves were claimed, and potential reserves were placed at about 250,000 million barrels by both the President of Mexico and international oil experts. Such figures placed Mexico in the same class as Iran. Rhetoric and the sober calculations of engineers have been supported by a dramatic rise in output. In 1974 daily production averaged about 0.5 million barrels, while at the beginning of 1979 it was 1.5 million; the daily production target for the end of 1980 was set at 2.25 million barrels (a progress more rapid than that other oil phenomenon of the late 1970s, the North Sea), and double that by 1985. World recession lowered this optimistic plan, but 2.8 million barrels per day were produced in 1982.

Not all the new reserves are in the fresh fields. Secondary recovery methods, pumping water under pressure into flagging wells, has improved recovery rates in some of the older fields. But 90 per cent of the new production comes from the south, where the migrant drilling crews and pipeline gangs have trampled roughshod over the lands and into the lives of people and communities, many of them Indian and previously little touched by the values of modern Mexico. A more explicit impact is being registered in the towns: Villahermosa and Coatzacoalcos are boom cities with rapidly growing populations and all the attendant problems of sudden prosperity.

Concomitant with the acceleration in the flow of oil has been the need to build additional refining capacity. Existing refineries have been expanded, and new ones built and planned. Refineries at Reynosa, Tampico, Tuxpan and Minatitlán, built either on the oilfields or at the Gulf ports, have been supplemented by new ones at Salina Cruz, Tula and Cadereyta (Monterrey). The last two, like the older refineries at Azcapotzalco (Mexico City) and Salamanca, are orientated to the domestic market and receive supplies of crude oil by pipeline. The pipeline system has been substantially extended since the new oil discoveries, and now forms a national network linking 90 per cent of the production and refining capacity of the country. The natural-gas distribution system has also been expanded and consolidated. These new distribution systems serve both the Mexican and the United States markets, and allow the balance of flow between the domestic and external markets to be adjusted as circumstances make desirable. In the late 1970s and early 1980s both the need to finance externally the huge capital requirements of the petroleum development programme and the shortage of domestic refining capacity meant that Mexico, loath to be cast once again in the role of a mere supplier of raw materials, was swallowing her pride and becoming a significant exporter of crude oil and natural gas. In the longer term the intention is to expand greatly the processing of her petroleum along the lines which are already making an impact on the manufacturing sector of the Mexican economy.

The system of distributing electricity has also been extended and rationalized in recent years, particularly since the major foreign-owned utility companies were Mexicanized and the Federal Electricity Commission created in the early 1960s. Today all public electricity supplies are generated by state enterprises, and decisions on new investment centralized in Mexico City. The Commission transforms 5 per cent of the country's oil and natural gas into electricity, and the supply from these thermal power stations is now twice as important as that from the better publicized hydroelectric stations. The latter are frequently built as part of the many integral valley development plans, but their efficiency and economics have not always been very clear. As the real unit cost of these installations and of the water-control schemes has risen, so the rate of increase in hydroelectric capacity has slackened (to about 5 per cent per annum in the 1970s). The last year in which water generated more electricity than petroleum was 1971, and hydro plants have represented only between one-quarter and one-third of new capacity installed since then. By the end of the 1970s Mexico was utilizing about one-quarter of the water-power deemed potentially economic to harness, and the new-found abundance of oil and natural gas meant that this fraction was not expected to be exceeded during the rest of the century. In the longer term Mexico plans to invest in nuclear generators, and construction work has already begun.

Manufacturing

A strong manufacturing industry has grown up in Mexico in recent years, and it is now clear that future growth is not going to be inhibited by any lack of indigenous energy. Manufacturing production first became more important than agriculture in 1951, and has remained one of the most dynamic elements of the economy. Between 1961 and 1980 industrial production increased fourfold, while agricultural output rose by only 60 per cent. Manufacturing accounts for two-thirds of the contribution of industry to Mexico's GNP. In 1980 this meant 30 per cent, as compared with agriculture's 9 per cent; twenty-five years before both were making an equal contribution (16 per cent) to a national product only one-third the size. This change in emphasis has had a strong impact on the economic geography of the country. Whereas the expansion of commercial agriculture to the arid northlands and the tropical wetlands has spread rural wealth more widely over the country, and increasing emphasis on the Gulf lowlands and the isthmus of Tehuantepec has allowed the benefit accruing from oil exploitation to be enjoyed more generally, the emerging geographical pattern in the secondary and tertiary sectors of the economy is one of concentration in the cities and especially in the capital. An ambivalent attitude by the Federal government has added to the political and social problems created by this redistribution of geographical activity.

Certain industries reliant on bulky raw materials – iron and steel and other metal smelting and refining, chemicals, oil refining, certain of the agricultural processing industries – are, however, well represented outside the capital. Monterrey is the iron-and-steel centre of the country, and has been so since the first works opened in 1903. Together with the nationalized industry at Monclava, this area accounts for one-half of Mexico's steel, and production is rising. It consumes nearby Sabinas basin coal, newly discovered natural gas and iron ore from

Durango, and it enjoys a booming local market. Elsewhere scrap forms the basic raw material for most plants including those on the border and in Mexico City. The newest mill is on the Pacific coast in Las Truchas at the mouth of the River Balsas. It is the first stage of a projected iron-and-steel complex which will draw upon 100 million tonnes of local iron ore, electric power from the nearby Infiernillo and La Villita dams, and a deep-water port. It was planned before the world oil crisis and the turndown in demand for steel, and with the encouragement of the World Bank. Eight possible sites, four on the coast and four inland (including Toluca) were studied before considerations of raw-material costs and other transport factors led to the selection of Las Truchas. Raw-material costs at the most favoured site on the Meseta (at Acambaro, Gto) were calculated to be 35 per cent higher than at Las Truchas. The first stage of the complex was opened in 1976, but the changed climate in the world steel economy has meant the subsequent stages are being delayed: the current plan is to produce at an annual rate of about 10 million tonnes by the mid-1990s. The changed climate has also meant that Mexico's first attempt to extend the market for Mexican steel outside Mexico has been similarly postponed. Her total national production − 7.1 million tonnes in 1980 − and capacity − 9 million tonnes − remain relatively low, and the small scale of most operations makes costs relatively high. The development of a direct reduction process using natural gas to produce sponge iron has been pioneered by Mexico in an attempt to bring costs down. The state now owns 60 per cent of the equity in the Mexican steel industry, and is backing fresh research in this direction.

The chemical industry is also well represented outside the metropolitan area. It covers a wide range of products and claims to contribute about 6 per cent to the GNP. Both government and private capital have been attracted to it, and investment has been at a high annual rate and rising. The sulphuric acid plants are off-shoots of Mexico's metal-refining industries, and are found generally in the north of the country. A new segment of the industry is the petrochemical one, now responsible for over 1 per cent of Mexico's GNP. Its modern development dates from 1958 when Pemex was given a monopoly of all basic petrochemical development. The first child of the union was a benzene plant adjoining the Azcapotzalco refinery in Mexico City. Since then the family has multiplied' hugely: in 1977 Pemex had 63 plants in operation, 10 under construction and 36 projected. Between 1976 and 1982 national installed capacity tripled. Earlier plants were all sited in the shadows of refineries − at Ciudad Madero, Minatitlán, Salamanca, La Venta and Reynosa. New plants are rising at Cosoleacaque, Salina Cruz and Salamanca specializing in ammonia-based chemicals. The most impressive developments are the petrochemical complexes being built at Cactus, the new oil centre in Chiapas, where sulphur-based chemicals will be the main product, and at La Cangrejera in the state of Vera Cruz. The latter site will eventually house a large number of different plants making a whole range of organic chemicals and equivalent in output to the entire national production of petrochemicals in 1976. Such plants supply the raw materials for a whole new breed of establishments manufacturing secondary petro-chemicals. At this stage manufacturing is shared by Pemex and, in essence, the international chemical companies whose names are household words around the world. Their market is the growing demand for fertilizers, detergents, acrylic resins, polyester fibres, emulsifying agents, sulphur, ammonia and many of the other modern necessities of life. It is the Mexican government's intention to make the country self-sufficient in such products by the early 1980s, and then to enter the export market. The important artificial-fertilizer industry exemplifies this policy. During the 1970s the domestic demand for phosphatic and nitrogenous artificial fertilizers rose by 9 per cent per year as agriculturalists became more commercially minded; production rose less rapidly, imports bridged the gap and the balance of payments suffered. By 1982 capacity had almost reversed the relationship between supply and demand thus saving valuable dollars. The impact of this change is reflected in the location of the industry. The early plants were built close to the sources of supply of their raw materials – near the oil refineries, the sulphuric acid plants, the phosphate rock quarries – while the more recent preference has been close to the market, particularly in central Mexico. The newest plants, however, are choosing coastal locations in view of the new role that petrochemicals are expected to play in the national economy. In contrast, that sector of the chemical industry based upon cellulose remains firmly centred in the Guadalajara–Mexico City belt, feeding the paper and textile mills of the populous core of the country.

Despite the importance of new industrial developments on the oilfields and elsewhere, the manufacturing industry of the country remains heavily concentrated in the capital city. Figures from recent industrial censuses show that over half the industrial activity of the country and over half the employment in manufacturing is in Mexico City and its suburbs. In 1975 the figures were 52 per cent and 56 per cent; in the mid-1950s the comparable figures were close to 30 per cent. Different sources give different figures, but all show both the high degree of concentration and the growing concentration of employment and production in the capital. Nowhere else in Mexico is manufacturing such an important element in the local economy, employing as it does one in three of all employees in the capital; Monterrey offers the closest parallel, but elsewhere proportions of one in six, or lower, are normal. Goods manufactured in Mexico City contribute more to the GNP than do all the farms throughout the entire country. The precise industrial structure of the capital is open to different interpretations (Yates, 1962; Bassols Batalla, 1966) depending upon the source used, but in the Federal District metal-producing and metal-using industries account for about one-quarter and the chemical, food and textile industries about one-half of the total. In the outlying districts, beyond the Federal District border, heavy industry, such as paper mills, tyre factories, electrical machinery, basic chemicals and cement works, is important. Among national industries almost exclusively concentrated in or near Mexico City are confectionery, tobacco, pharmaceutical chemicals, soaps and detergents, electrical, tyre and motor-car industries.

The attractions of the metropolitan area to manufacturers in general spring not from any abundance of notable raw materials; rather they lie at other stages in the manufacturing process.

In the first place there is no shortage of unskilled labour. In population Mexico City is six times bigger than its nearest neighbour, Guadalajara, and 43 per cent of the 20 million Mexicans living in towns with over 100,000 inhabitants in 1970 lived in Mexico City. There is, by Mexican standards, a relatively large supply of skilled labour, for the educational standards of the city are notably higher than those of the other cities of the country, and specially so in the highest levels. For example, the proportion of the labour force with professional qualifications is four times higher in the capital than in Monterrey, which is itself an enlightened industrial city. The actual figures in 1964 – 216,000 and 7500 respectively – perhaps best emphasize the enormous difference between Mexico City and the rest of the country (*La población*, 1964, vol. 1, pp. 56–7; vol. 2, pp. 156–7).

Mexico City is the outstanding consumer market of the country. Figures from the 1960s are worth recording because these were the ones available to the new manufacturers on entering the metropolitan market. In the 1960s the average income in the Federal District was 40 per cent above the national average and 25 per cent above that in the other cities over 50,000 in size; *per capita* income outside these cities was only one-half of that in the Federal District. In a sample study carried out by the Secretaría de Industria y Comercio in 1962 of the sixteen largest cities of the Republic (*Las 16 ciudades*, 1962, pp. 29, 33, 97) the Federal District, with 58 per cent of the families, accounted for about two-thirds of family expenditure on food, footwear and clothing, and four-fifths of money spent on such consumer durables as furniture, vehicles and electrical goods. Most of the electricity used in the country is consumed in Mexico City. Two-thirds of the income remaining after meeting the cost of food, housing and clothing was in the Federal District. More than three-quarters of all the urban families with incomes in excess of 3000 pesos a month lived in the Federal District (Monterrey was closest with 6 per cent and Mexicali, on the US border, third with 2 per cent). The spending power of Mexico City has been recognized by government and industrialist alike.

Mexico City is also very central to the national market, and has been even closer; the population centre of the country, which in 1930 was only 170 km from the capital, remains only 250 km away (near Salamanca) and the centre of purchasing power only a little further north. This slight demographic shift away from Mexico City has been more than negated by the greatly improved accessibility of the metropolis from all parts of the country. Recent improvements have confirmed the capital as the rail, road and air centre of the country, giving it low distribution costs and an unequalled service. The most important improvements in the last two decades have been in the radical development of the system of paved highways: mileage has quadrupled, and since 1958 has exceeded that of the railways (Fox, 1972b). The new road system has reinforced Mexico City's nodal position, one already established by its focal position in the Guadalajara–Vera Cruz railway network. A detailed comparison of theoretical minimum distribution costs (using rail, road or a combination of both, whichever was cheapest in 1963) showed that the area tributary to Mexico City was larger than the area tributary to any other transport centre in the country (Togno, 1963). This advantage flows mainly from the superior facilities in and out of the capital. It is augmented by the economies of bulk traffic permitted by the large volume of goods generated by the city. Both underline the attractions of Mexico City as a manufacturing centre for the national as well as the local market.

It is less easy to measure other factors which make the area around Mexico City attractive to the industrialist. Certainly private capital is more readily available in Mexico City than outside, and more people are prepared to invest in local than more distant ventures. For example, only about 40 per cent of the new private capital investment in the country over the last 10 years has been made in areas outside Mexico City. The cosmopolitan qualities of life in the capital and the mild climate must play a part in attracting and holding Mexican and foreign businessmen and their enterprises. The powers of state governments are relatively weak, and Mexico City is the seat of important political decisions.

Political decisions have affected the concentration of industry in the capital in several ways. On the one hand it is argued that the government has encouraged industry to locate in the capital by providing such facilities as public utilities, loans, etc., not available elsewhere, and by subsidizing certain foodstuffs, notably maize, in the capital, has kept the cost of living down, the effective wage-rates low and the labour market easy. On the other hand, the government has paid lip service in recent years to the social disadvantages of over-centralization, and has made some attempt to dissuade industry from settling itself in the Federal District by devices such as new regulatons, the strictest enforcement of existing regulations, higher land taxes reinforcing higher land values, and the like. Industrialists have also been offered fiscal incentives to locate outside the Federal District, and individual states have set aside industrial parks in attempts to attract their share of the new employment in manufacturing (Lavell, 1972). In 1977 a national commission on urban development (PNDU) was created to promote, amongst other things, a reduction in the dominance of the capital in the industrial life of the country. The most notable beneficiaries under these policies have been areas which lay on the fringes of Mexico City, beyond the boundary of the Federal District, and which now form part of the suburban sprawl of the 'great wen'. Such areas are of greater industrial importance than, say, Guadalajara and Monterrey combined. The location of the new automobile industry exemplifies this situation. When it was decided to set up the car-assembly plants in the 1960s, none was permitted to take a site in Mexico City. Instead, Ford built at Cuantitlán; General Motors, American and Chrysler-Dodge at Toluca; Volkswagen at Puebla; Datsun at Tejalpa; Renault at Ciudad Sahagún – all places within 80 km of the capital, and together accounting for over four-fifths of the cars built in the country.

The car industry is also illustrative of another aspect of Mexican government policy in as far as it affects manufacturing; a decree of 1962 required assembly plants to conform to a national plan which demanded, in effect, the substitution of imported parts by domestically produced parts; a minimum national content of 60 per cent of the total cost was the short-term aim. In return, imports of vehicles were severely restricted, import regulations and duties on parts for companies reaching the desired proportion were eased, and the whole industry benefited from the general exemptions granted under the act for development of new and essential industries.

The operation of an import-substitution policy has, of course, become the classical way of encouraging domestic industrial growth in Latin America, but several factors have recently persuaded the Mexican government to adopt a less rigorous policy. In the first place the internal market has grown, and the resulting economies of scale open to domestic manufacturers have reduced the need for protection. Secondly, there has been a growing realization that the higher costs of nationally made products (19 per cent in 1975) and the loss of import revenue may be too great to be worth sustaining. Further, there are, of course, fewer appropriate industries (labour-intensive serving a large market) left to help in this manner. The government has been adopting a more selective approach to mechanisms designed to shelter domestic industries and, in general, reduce the level of protection. The emphasis is being placed upon those industries which have been trying to go beyond the domestic market and extend into the export sphere. The encouragement of the petrochemical industries is a clear example of this policy. A good measure of the distance that Mexico has already come along the road of industrial self-sufficiency, despite a heavy reliance upon imported technology and sophisticated equipment, is that two-thirds of the capital goods being used in the modern petroleum-related developments are made in Mexico. A further measure of Mexico's achievements in manufacturing over the last twenty years is the share that manufactured goods won in the export trade of the country: from 12 per cent in 1959 to 62 per cent in 1982, if petroleum is excluded.

Economic regions and growth points

The net result of the developments described above is a very uneven pattern of economic production, with striking contrasts between one part of the country and another, and between cities, towns and the countryside. In 1960 over 43 per cent of the GNP came from the Mexico City area alone (Denis, 1965), and the proportion today is probably higher. If one divides the rest of the country into three tiers of approximately equal area, the northern tier of the six border states accounted for approximately one-half of the remaining production, the middle tier for a further one-third and the southern tier, south and south-east of the Guadalajara–Mexico City–Vera Cruz axis, for only one-sixth. At a more detailed scale, three-quarters of the national wealth is created in the fifteen small regions, occupying no more than 8 per cent of the national territory.

Mexico City has the most varied economic structure and the highest absolute rate of growth. As well as being the major industrial centre of the country, it also accounts for a disproportionate share of employment in the commercial and service trades. More men, women and children are employed in this sector of the metropolitan economy than in manufacturing. Almost one-third of all the commercial establishments in the country are in the capital. One-third of all the civil servants work in the capital, and these include almost all in the higher echelons. The dominance of the capital in some service spheres is much stronger: higher education, the arts, medical services, and finance are clear examples of very heavy levels of concentration. Many of the service industries are unrepresented away from the larger cities and the tourist resorts (Gormsen, 1977). The importance of the capital is being bolstered by the incorporation of adjoining areas into the region. The thrusting industries of Toluca in the west, the more traditional industries of Puebla in the east and the spontaneous and planned industries of the new satellite towns (for example, those of the new railway engineering centre of Ciudad Sahagún in Hidalgo) supplement the industrial income of the area, and irrigation helps an intensive agricultural effort. The Monterrey–Saltillo–Monclava triangle is second to the metropolitan region, but in spite of its heavy industry and financial importance is only one-eighth as significant. Monterrey is famous in Mexico as a brewing centre, for clothing and textiles, shoe, glass, paper and electrical industries as well as for iron and steel. In the Guadalajara region much of the economic development is based on the processing of local agricultural products, or manufacturing goods for the regional market; though a region of significance, the pace of change is slower than in, for example, Monterrey. The new oil discoveries are creating specialized growth poles around the new processing complexes in Vera Cruz, Chiapas and Tabasco. Although the direct requirements of the petroleum and petrochemical industries do not create many jobs once construction has finished, their indirect impact is significant: for every worker directly employed by Pemex in 1982 there were fourteen others indirectly employed.

A national resource of another kind which serves as a focus of economic activities is the border with the United States. First and foremost are those activities which cater to the many and varied needs of the visitors from the United States. (Price, 1973). Towns such as Tijuana, Mexicali, Ciudad Juárez and Nuevo Laredo have grown tenfold in the last 30 years. Mexicans, too, have been drawn to sample the peculiarities of life on the border: one of the more tangible being the existence, since 1933, of a free-trade zone into which United States goods can be imported free of duty. The border is a permeable one, and both goods and people flow across it at an ever-increasing rate. Mexico has registered a net benefit from these transactions: in 1977 they contributed $500 million to the Mexican exchequer. Another range of activities that have been fostered by the border situation is the manufacturing of goods for the United States market. In 1965 the Mexican government established a frontier industrialization programme which was designed to take advantage of sections of the US Commerce Code allowing goods made from American components, but assembled overseas, to enter the United States on payment of a duty related not to the value of the article, but only to the value added in the manufacturing process. Mexico has encouraged the establishment of in-bond factories (*maquiladoras*), often twinned with parent factories just across the border, and run by managers who commute from the United States each day. By the late 1970s, 450 such plants had been established, 400 of them in the border towns, employing 75,000 people and generating an annual income of about $300 million. One-sixth of the plants are in Ciudad Juárez (opposite El Paso, Texas), and they generate one-third of the income. One-third of the plants house electrical and electronic manufacturers, and they yield two-thirds of the income from the frontier programme; half the remaining income comes from the clothing and footwear factories. The force which maintains the *maquiladoras* is the difference in wage-rates payable on either side of the border: they differ by a factor of four. But the growth experienced in the first 10 years of the programme has not been maintained, as other off-shore assembly sites, particularly in Asia, have offered even lower wage-rates to tempt United States manufacturers to look further afield than Mexico. The vulnerability of these assembly industries to changes in American customs regulations, perhaps in response to American trade union pressure, and to other external forces, make them an insubstantial basis upon which to found a lasting economy. And, although they do offer employment in a region which has always suffered very badly from unemployment, it is unfortunate that 80 per cent of it is of females. The pattern of daughters being the main financial support of the family is contrary to the long-established order of things in Mexico, and has brought many social problems in its wake. Further, the multiplier effect of the $200 million (1977) paid in salaries and wages is small: economists estimate that between 60 and 75 per cent of this money is spent in the United States. The authorities recognize that in the longer term the export of out-of-season fruits and vegetables, the processing of cotton and other industrial crops grown on the irrigated farms, and the exploitation of the natural-gas fields of the lower Río Grande, will prove a more substantial foundation for the local economy.

Some social questions

The patterns of economic production are out of harmony with the distribution of population, and this naturally has severe repercussions on the well-being of the people. A comparison of data drawn from the 1970 census for the northern state of Baja California and the southern state of Oaxaca, is suggestive of the nature of the relationship and of the range of regional disparities within the country. In Baja California only 22 per cent of the economically active population was employed in agriculture as compared with 18 per cent in manufacturing and 24 per cent in the service trades; in Oaxaca agriculture employed 72 per cent of the working population and industry and services only 9 and 6 per cent respectively. Whereas seven out of eight Baja Californians had a monthly income of over 500 pesos, only one in three reached this figure in Oaxaca. Productivity, as measured in *per capita* contribution to the GNP, was ten times higher in the northern state than in the southern one. Social statistics point in the same direction as economic ones. The population of Baja California is better educated: 11 per cent of the working population had professional or superior qualifications compared with only 3 per cent in Oaxaca; 79 per cent had been to school in the north as compared with 59 per cent in Oaxaca; and only 12 per cent of the population of Baja California was illiterate compared with 42 per cent in the southern state. The average Baja Californian is very much better housed: most (77 per cent) had houses of two or more rooms; in Oaxaca most (59 per cent) lived in one-roomed dwellings. Two-thirds of the houses in Baja California had a supply of piped drinking water, and four-fifths had electricity: in Oaxaca such houses were in a minority of between one-third and one-quarter. Of those living in Baja California 41 per cent had been born outside the state (as compared with only 3 per cent in Oaxaca), and this helps explain why the population of the one had grown fourfold in 20 years and the other by only a half. Another part of the explanation lies in infantile mortality-rates, which are twice as high in Oaxaca as in Baja California, and in a life expectancy of between 5 and 10 years longer in the north than in the south.

The social and economic contrasts are strong between the brash, affluent, frontier state of Baja California with its footloose, opportunistic population receptive to political and cultural innovations on the one hand and the long-settled, time-worn, conservative state of Oaxaca, whose rural population still lives close to the land and clings to its Indian traditions. Such contrasts can be duplicated elsewhere in Mexico.

Maps showing the distribution of economic and social measures of well-being tend to be very similar, and the degree of correlation between the patterns they show is high. Factor analysis has been used to synthesize the many separate

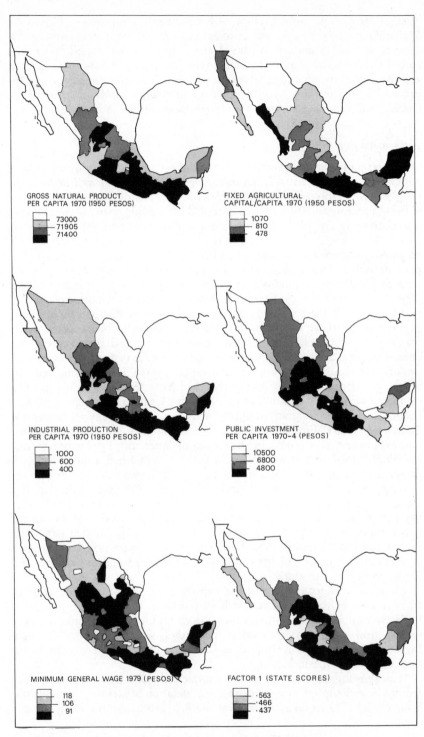

Figure 2.6 Some economic and social characteristics of Mexico

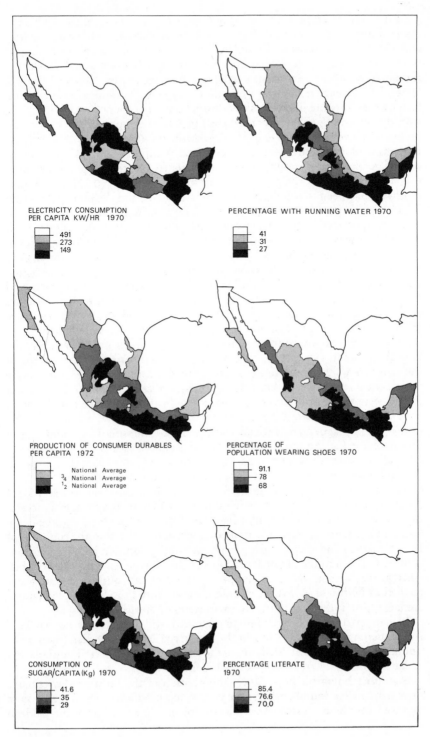

ELECTRICITY CONSUMPTION
PER CAPITA KW/HR 1970

491
273
149

PERCENTAGE WITH RUNNING WATER 1970

41
31
27

PRODUCTION OF CONSUMER DURABLES
PER CAPITA 1972

National Average
$\frac{3}{4}$ National Average
$\frac{1}{2}$ National Average

PERCENTAGE OF
POPULATION WEARING SHOES 1970

91.1
78
68

CONSUMPTION OF
SUGAR/CAPITA (Kg) 1970

41.6
35
29

PERCENTAGE LITERATE
1970

85.4
76.6
70.0

Figure 2.6   continued

measures: one attempt is illustrated in *Figure 2.6*. The factor illustrated is the result of analysing sixteen social and economic characteristics of each of thirty-two Mexican states drawing from data from the 1970 census and including all those mentioned in the comparison between Baja California and Oaxaca; it accounted for 47 per cent of the overall variability exhibited (Cole and Mather, 1974). The characteristics that were most completely subsumed into the new factor were a mixture of economic and social ones: in order of importance (and with the weighting given in parentheses) they were the proportion of the economically active population engaged in agriculture (−0.98) in services (0.97), the proportion of the population with professional qualifications (0.94), the proportion of houses with electricity (0.92) and water (0.90), the proportion earning over 500 pesos (0.82), the proportion employed in industry (0.77), the proportions who had been to school and were literate (0.74), and the proportion living in places with a population of 500 or more (0.69). The resulting map showing the relative importance of the states is shown in *Figure 2.6a*. The Federal District and the border states rank very high; the southern regions and old mining states of the centre are at the bottom of the list. The second factor extracted in the analysis gives greater emphasis to education, and the third factor more to the structure and dynamics of the population. All factors measure change or development and *Figure 2.6* distinguishes between those states which are consistently in the top and bottom quartiles in lists of all three factors, and those that are more moderate in behaviour. A second attempt at synthesis used a similar approach, but a wider range of measures to establish a socio-economic index for each of the federal entities in 1940, 1950, 1960 and 1970. A comparison of the ranking of the states shows that the average move up or down the ladder was only three places: the state of Mexico (up fifteen places), Tabasco, Durango, Sinaloa and Puebla were the most dynamic, Quintana Roo (with a negligible population in 1940), Yucatán (down nine places) and Nayarit slipped furthest in the development league. The Federal District, Baja California Norte and Nuevo León remained at the top, Guerrero, Chiapas and Oaxaca at the bottom. The comparison over the years tends to confirm the view that the regional disequilibrium between the states is growing (Barkin, 1975, p. 278).

Such analyses are a clear indication of the way in which economic and social progress go hand in hand. As such it is interesting to note that *per capita* production in Mexico is about the average figure for Latin America, and that the range found in the states of Mexico is approximately equal to that in the countries of Latin America. The Mexican living in the well-to-do northern border states contributes as much to the national economy as the average Argentine does to his, whereas the Mexican living in the poorer states of the south finds his closest parallel in Bolivia or Paraguay. In both these countries the Indian populations are large, just as they are in the southern states of Mexico.

In many parts of Mexico the Indian presence is still obvious, in most parts the Indian heritage is a reality, and in the country at large the Indian myth a not unimportant ideological weapon in the cause − often directed towards the Indian himself − of creating a national identity.

Notwithstanding his importance, the position of the Indian today remains unenviable. He is found amongst the poorer, less productive, less literate, more deprived people in Mexico. An Indian problem is discussed as in the Andean countries, but remains largely unresolved, partly because the Indian himself is not easy to identify in a predominantly mestizo country, and partly because many of the problems he suffers from are common to other members of the lower stratum of society.

The strictest available measure of the number and distribution of Indians in Mexico is that of language. In 1970 about 870,000 Mexicans over the age of 4 spoke only an Indian language, while a further 2.3 million spoke an Indian language and Spanish: together they made up 8 per cent of the Mexican population. On the criterion of language the northern half of the Yucatán peninsula (Maya), the interior highlands of Chiapas (Tzeltal, Tzotzil), the southern Tehuantepec lowlands and the Mesa del Sur of Oaxaca (Zapotec and Mazatec), the Sierra Madre del Sur west of Oaxaca City (Mixtec), the refuge areas of the Tuxtlas in Vera Cruz and the southern Sierra Madre Oriental (Nahuatl), the basins around the Valley of Mexico (Nahuatl and Otomi) and the highlands north of Uruapan (Tarasco) are predominantly Indian; only 5 per cent of the Indians, so defined, live outside these areas. The Indians are far from homogeneous: for example, forty-five or forty-six different languages are spoken, some, it is claimed, as different from each other as Chinese is from English. This linguistic variety is symptomatic of the great wealth of cultural diversity grouped under the portmanteau label of Indian.

But a linguistic definition is too exclusive to measure satisfactorily the Indian presence in Mexico. Alfonso Caso prefers a social definition, and has written that an Indian is one who feels he belongs to and is part of an Indian community, that is, one in which somatic non-European elements are predominant (Rubín de la Borbolla, 1964, p. 123). Census enumerators note certain key features of dress and diet, and these may be used to suggest how widespread is the Indian influence in Mexico. In 1970 9.35 million Mexicans over the age of one (20.0 per cent of the population) wore either sandals (*sandalias, huaraches*) or went barefoot; 22.0 million Mexicans (45.8 per cent) did not normally eat wheaten bread; 25.6 million (52.9 per cent) did not normally eat meat, fish or eggs or drink milk – that is, their diet was essentially the traditional Indian one. These figures give credence to the visual impression a traveller has in central and southern Mexico and to the estimate of Brand (1966) that there are some 10 million Indians in Mexico today – more, in fact, than in any other Latin American country. These figures measure the size of the less dynamic, less mobile, portion of Mexican society; they are the people who contribute least to the energy which drives the westernized, partly alien, social and economic system dominating Mexican life today. So long as the system runs relatively successfully (as it has for the last generation) the significance of the Indian in Mexico will decline; should the system falter this decline will slacken, and perhaps even reverse.

Most Indians are poor. It is in this context that some useful work has been done by, for example, the Instituto Nacional Indígenista. Income and production *per capita* in the predominantly Indian-speaking *municipios* are between a third and a half of the national average; federal expenditure per head is only one-fifth of the average. In fact, the tenor of much recent sociological work has been to view the limited case of the Indians as part of a wider culture of poverty found not only in the rural backwoods of southern Mexico, but in the slums of the larger cities and in developing countries generally. The Indian problem may thus be seen as part of a larger problem common to many Mexicans: the problem of adjustment to a rapidly changing environment, a problem often magnified for individuals by migration. Many of these problems are seen at their most striking in Mexico City. The shock of translation to the metropolitan environment can usually be softened by resort to a relative or friends from the same village or even to Indian enclaves speaking the same language (Orellana, 1973). Such relief may be of short duration for housing is scarce, jobs for the poorly skilled are at a premium, old standards have changed, and former ties may have been weakened and replaced by new loyalties.

Trustworthy unemployment statistics are not available; even if they were, statistics on occasional, part-time and casual employment would be needed to paint a full picture. Genuine unemployment is probably relatively insignificant. Most people of working age in Mexico City do something and earn something, and both the construction industry and the service trades have proved very elastic in their capacity to absorb casual labour. Together with the sweat-shops of the inner city and home industries (perhaps producing goods which compete successfully with the products of their rural cousins), they have taken up much of the slack that might otherwise have developed.

Eduardo Flores estimates, however, that in 1967 between 30 and 40 per cent of the statistically employed population of Mexico was in effect under-employed (5 Seminario, 1967, p. 363). A more recent calculation by the International Labour Office placed the real unemployment rate in Mexico in the early 1970s at 29 per cent, higher in the country (at 44 per cent) than in the towns (12 per cent), with under-employment amongst non-agricultural workers in employment running at 38 per cent. And these figures were based upon the industrial census, through whose coarse mesh many casual workers slip, and so underestimate the real gravity of the situation. In 1979 the National Chamber of Industries estimated that only 45 per cent of the economically active population had full-time jobs, leaving 9.5 million people of working age unemployed or under-employed. These data suggest that about 1.5 million live in the jungle world of marginal unemployment in the capital city, and agree with other informed estimates of the situation. In the 1970s about 2 million people moved into the city from the provinces, and it is reasonable to suppose that they are amongst those who bear the brunt of the increasing problem of a surplus labour supply.

Marginal and irregular employment brings in its wake, for Mexico as for elsewhere, a number of social problems. One is the housing problem. This is particularly obvious in Mexico City where 7 million extra people have found roofs over their heads since 1970, and great numbers of sub-standard houses have risen to meet this need. Most parts of the city suffer deficiencies of some nature or degree. Nevertheless, three main types of housing problem deserve specific mention. The crumbling core of the old city is the scene of the worst case of overcrowding. A sample survey (*Investigación*, 1965) made in 1962–3 of the central 10 km² around the Zócalo, roughly the area of the city in 1850, showed that only 10 per cent of the accommodation was of a satisfactory standard. Most families lived, ate and slept in the same room, a room commonly giving access to a courtyard or inner well shared by dozens of other families. Although five out of six had easy access to running water, fewer than half had any kind of sewage outlet. Oscar Lewis (1964) gives a graphic account of life in one such *vecindad* in his *Children of Sanchez*. The dwellings were once substantial buildings, but now, thanks partly to long periods of frozen rents, they are degenerating beyond redemption into slums.

Squalor is more apparent in the zone of poor-quality housing tracts which mar the periphery of the city, although the population density is only one-eighth of that in the city centre. A distinction must be drawn between the regular estates, subdivisions or *colonias* of detached single-storey, concrete and adobe, one and two-roomed houses found on land reclaimed from Lake Texcoco to the east of the city, and the littered landscape of semi-finished shacks scattered, for example, on the dry hill slopes beyond La Villa in the north-east. In the former area most public services are absent, and this deficiency is particularly marked beyond the Federal District boundary in the state of Mexico. These *colonias* are best seen after the summer rains; then the ground becomes an opaque jelly of mud and accumulated debris, of uncertain depth but distinctive smell. Improvements are occurring in the supply of public facilities, however, and in these areas of owner-occupiers and private initiative, improvement is apparent in places. Conditions are worse in the squatter areas where the country origin of many of the inhabitants – called *paracaidistas* or parachutists – is very obvious. Most have no kitchens and cook in the open air, animals and children are equally numerous, house floors are of earth; and building materials were mixed, being composed largely of whatever lay closest at hand. There is a complete absence of municipal services.

Lastly, the uttermost in squalor is found in the clusters of shacks or lean-tos that pockmark the fabric of the city proper. They cling to the walls of factories, they appear on land awaiting building contracts, they are parasites on the rubbish dumps, and they mantle the sides of ravines.

It is difficult to envisage a solution to the housing problem (Ward, 1976). Three-quarters of the existing houses are sub-standard, and about 70,000 new families have to be housed each year. The British answer would be public-authority housing. In fact, Mexican authorities are building about 40,000 family units a year, many of them in the capital. But, apart from being too few to make much of an impact, these units are not available to most people. The spectacular Nonoalco–Tlatelolco Scheme, built on a formerly notorious slum site north of Buenavista Station, is for middle-class families able to afford prices beyond the means of 90 per cent of the population. The public housing estates built by the state social security bodies, ISSSTE and IMSS, are in the same category and, even though they are attractive additions to the townscape and house productive segments of the populations, they make little contribution to meeting the mass housing needs of the city. Even the utilitarian two-roomed houses on the large San Juan de Aragón estates in the north-east are available only to those in regular employment capable of carrying a mortgage. Ten years residence in the Federal District is a requisite for participation in such schemes. In practice, financial and other demands place housing like this outside the range of perhaps one-half of the population needing housing. Studies, such as that made by the Secretaría de Hacienda y Crédito Público in 1964 (*Programa*, 1964), although designed to offer a solution to the problem, instead rather highlight the plight of the urban poor; the striking new public buildings that grace many parts of the city lose something of their appeal when viewed soberly in the context of the greater need. A practical solution, if it exists, will not be a glamorous one: do-it-yourself kits of prefabricated parts, allowing first a one-roomed dwelling to be acquired, and then rooms added as circumstances permit, assembled on a concrete house lot served by basic sanitary facilities, may just fall within the strict financial constraints. This would require efficient organization and would produce a suburban landscape of regimented ugliness and uniformity. Such an unexciting solution naturally runs counter to the image of Mexico that the Mexicans in power like to project. Similar, though in some respects less severe, housing problems face almost all the smaller cities of the country.

Parallel to the housing problem in Mexico City is the wider one of the provision of public services. The question of an adequate water supply in this semi-arid environment to meet the enormously increased domestic and industrial demand has been a taxing one (Fox, 1965b; Moore, 1968). Until a century ago the water of the springs at Chapultepec were led by open aqueduct to the erstwhile island site. In the first part of this century wells tapping the aquifers below the city proliferated to supplement this supply. This brought two consequences. First, extraction exceeded the natural restoration rate, the water-table fell and supply declined. Secondly, the desiccation of the extremely hydrous clays of the upper subsoil resulted in their contraction and in severe subsidence at the surface: the central axis of the city has sunk by up to 7 m since 1930. The city has had to go outside the Basin of Mexico to slake its thirst, and since 1951 the headwaters of the Lerma have been diverted into the metropolitan water system by tunnels through the western ranges. But water remains a perennial problem and more and more distant sources, some 400 km away, are being costed as potential suppliers; pumping water from the wet Gulf plains to the 2500 m altitude of the capital city, using off-peak hydroelectric power generated at the Sierra Madre escarpment, has been a recent suggestion; and more intensive use and re-use of existing supplies is being practised in a limited manner. Distribution within the ever-expanding city is another problem. About one-third of the population of the city is without a domestic supply, and private and public water tankers supply many of the suburbs. The drainage system of the city is also inadequate, conforming to a plan that is basically geared to the city of 1900. With a twelvefold expansion to date and with subsidence diminishing capacity, the existing sewage system is reduced both in extent and capability, and has resulted in unpleasant and insanitary conditions in large areas of the city, especially in the amphibious zone on the eastern edge. New engineering works are in progress to supplant the Gran Canal, the open sewer that takes the city's effluent northwards through the divide to the Mezquital, by two deep main sewers constructed below the level affected by subsidence; but the system will remain limited.

Poor housing, insanitary conditions, marginal incomes and a largely ill-educated populace make large parts of the city unhealthy places to live. A large number of epidemiological studies and detailed mapping of patterns of morbidity and mortality of individual cases confirm this view. Typhoid (Gonzalez Gutiérrez *et al.*, 1962) is over three times more prevalent in children living in shacks than it is those living in the *vecindades*, and six times higher than in those living in what are officially regarded as adequate housing conditions. Infantile and premature mortality-rate and deaths from certain diseases, when checked against place of habitual residence, also mirror patterns of poor housing (Fox, 1972a).

But the plight of the urban poor (Bergsman, 1980) is, in some respects, less severe than that of the rural *campesino*. Social services are more developed in the capital than in the provinces, educational facilities more abundant and, perhaps especially important, the opportunities to improve one's position, both financial and social, are there for the strong in muscle and spirit. Such opportunities may become fewer as the ranks of the poor in the towns continue to be swollen by immigrants from the countryside; a feeling of frustration, added to a growing awareness of the political power held by the urban poor, could lead to the kind of economic and political debacle from which Mexico has been singularly free since the days of the revolution.

Conclusion

This essay has dwelt on contemporary Mexico. Any conception of a people living purely in the past and dozing today beneath the hot afternoon sun is a caricature; rather, change and activity are the order of the day. Not that Mexicans are unaware of their past. The extraordinary remains of prehistoric civilizations, the relics in the landscape of colonial days, and the anachronisms of modern society, make this an impossibility; to be an archaeologist is to belong to one of the most admired professions. Nor is this awareness of the past of purely academic interest, for substantial government expenditure on restoring and preserving ancient monuments is reaping handsome dividends in receipts from the lucrative tourist trade. Furthermore, the long history of the country has been appealed to by politicians anxious to mould a national identity; the vivid murals that adorn so many public buildings carry unequivocally the message that the mainsprings of Mexican culture lie in the pre-Columbian past. Raised high amongst the post-Conquest folk heroes are those with Indian connections – Father Hidalgo, who led an Indian army in an abortive insurrection in 1810, Benito Juárez, the Zapotec Indian who was president of the country before Porfirio Díaz, and, more recently, a motley band of revolutionary generals amongst whom Emiliano Zapata and Pancho Villa are perhaps the most widely known. Cynics may claim that such emphasis is merely a sop to comfort the underprivileged, and to deflect them from the knowledge that the revolution has neither reduced the wide disparities in the distribution of wealth nor made a reality of the idea of equality of opportunity. The majority believes, however, that Mexico is in a stronger position today to master existing and future problems than almost any other of her Latin American colleagues, and that more than lip-service should be rendered to past events.

One strength of Mexico over the last 30 years has been a stable and generally responsible government. Anti-clerical movements since independence have removed any question of significant conflict between Church and State, and skilful management of opposition groups has made one-party rule popular and acceptable. Since the exceptional presidential term of Lázaro Cárdenas, which revived the radical aims of the revolution and made the revolution itself irreversible, there has been a growing national and, more recently, international confidence in Mexico. But the valuation of oil reserves has suffered a downward revision in the more sober days of the early 1980s, and Mexico's huge external debts, underwritten by oil, have forced her to reassess optimistic development plans of the late 1970s and, indeed, the future development of the oil industry itself. But Mexico remains in a good position to make the most of this most generous gift. History has taught her that in the oil industry entrepreneurship and calculated risk-taking can pay handsome dividends; that boom can give way to slump almost overnight; and that resources are to be husbanded if they are to last. An experience longer than that of almost any other country in managing her own oil affairs has shown her how difficult it is to be self-reliant in an age of high technology and multi-nationals. She has tasted the difference between supplying a captive domestic market and meeting the tougher requirements of international trade. She has experience, too, of accommodating the technical and fiscal requirements of a nationalized industry to the wider political and social aims of good government.

Confidence at home and abroad is reflected in Mexico's increasing international stature. She is one of the leading members of the Latin American Free Trade Association (although this membership may be more important from the political standpoint than from the economic), and has encouraged the creators of the Central American Common Market. She maintains friendly relationships with the Organization of Petroleum Exporting Countries (OPEC) and has a tacit agreement not to undercut export prices fixed by its members. But she has resisted invitations to join OPEC, thus preserving her privileged trading position with the United States. Today Mexico's proximity to the United States brings her undoubted benefits, refuting, at least in part, Porfirio Díaz's famous aphorism 'Poor Mexico, so far from God, so near to the United States.' Her new bonanza in oil is bringing reassessments on both sides of the border, and a new pragmatism is abroad. She has taken an independent line in the Organization of American States, and is less beholden to the United States than most other important Latin American countries. She has increasing influence at the United Nations.

The image Mexico presents to the casual visitor is undeniably attractive; below the surface and away from the beaten tourist track the image is more enigmatic. The extent to which indigenous ideas and solutions are being found to meet the changing situation is important for a country whose history has given it good reason to be xenophobic; it is especially difficult for a foreigner to diagnose solutions for a people whose warmth and strangeness, violence and restraint, optimism and cynicism make them easy to like, but difficult to understand.

Notes

[1] Including many refugees from the Civil War, warmly welcomed by the Mexican government, which continued to give official recognition to the Republican government-in-exile until General Franco died.

Bibliography

ATÚNEZ ECHAGARAY, F. (1964) *Monografía histórica y minera del Districto de Guanajuato*, Consejo de Recursos Naturales no Renovables, Pub. 17-E, Mexico.

BARKIN, D. (1975) 'Regional development and interregional equity: a Mexican case study', in CORNELIUS, W.A. and TRUEBLOOD, F.M. (eds) *Urbanization and Inequality. Latin American Urban Research* 5, 277–99.

BARKIN, D. and KING,T. (1970) *Regional Economic Development: the River Basin Approach in Mexico*, Cambridge.

BASSOLS BATALLA, A. (1966) *La ciudad de México y su region económica*, Mexico, Prim. Conf. Reg. Latinoamericana, Union Geográfica International, 4, 113–36.

BATAILLON, C. (1968) *Régions géographiques au Mexique*, Paris, Trav. et Mem. de l'Inst. des Hautes Etudes de l'Amér. Lat.

BATAILLON, C. (1971) *Ville et campagnes dans la région de México*, Paris.

BENÍTEZ ZENTENO, R. and CABRERA ACEVADO, G. (1966) *Proyecciones de la población de México 1960–1980*, Mexico, Dept de Investig. Industriales, Banco de Mexico.

BERGSMAN, J. (1980) *Income Distribution and Poverty in Mexico*, Washington, DC, World Bank Staff Working Paper 395.

BERNSTEIN, M.D. (1964) *The Mexican Mining Industry, 1890–1950*, New York.

BOLSA (Bank of London and South America) *Review*, (cited in text as BOLSA *Review*, with relevant dates) published monthly by Lloyds Bank International Limited, London.

BONINI, W., HEDBERG, H. and KALLIOKOSKI, J. (1964) *The Role of National Governments in Exploration for Mineral Resources*, Ocean City, NJ.

BRADING, D. (1971) *Miners and Merchants in Bourbon Mexico, 1763–1810*, Cambridge.

BRAND, D.R. (1966) *Mexico: Land of Sunshine and Shadow*; Princeton, NJ.

BYERS, D.S. (ed.) (1967) *The Prehistory of the Tehuacan Valley*, Austin, Texas.

IX *Censo general de población 1970: resúmen general (1973)*, Mexico, Dir. Gen. de Estad., Sec. de Industria y Comercio.

CENTRO DE INVESTIGACIONES AGRARIAS (1974) *Estructura agraria y desarrollo agrícola en México*, Mexico.

COLE, J.P. and MATHER, P.M. (1974) 'Mexico 1970: estudio geográfico usando análisis de factores', *Bol. del Instituto de Geografía del UNAM*, 5, 177–86.

Comercio Exterior, published monthly by the Banco Nacional de Comercio Exterior SA, Mexico.

CÓRDOVA, R. (1964) *Urbanismo y desarrollo de la comunidad*, Mexico, Sección urbanismo, Sec. de la Presidencia de la República.

DENIS, P.Y. (1965) 'Une dimension nouvelle au Mexique: l'espace économique', *Rev. Géogr. de Montreal*, 19, 1–2, 3–42.

DURÁN, M.A. (1966) 'Perspectivas de la producción y del commercio del trigo y del maiz', *Comercio Exterior*, 16, 83–8.

ELLIS, P.B. (1971) 'Changes in agriculture and settlement in coastal Chiapas, Southern Mexico', *Occasional Papers*, Institute of Latin-American Studies, University of Glasgow, 2, 1–25.

FLORES, E. (1967) 'Cómo funciona el sector agropecuario de México', *Comercio Exterior*, 17, 701–5.

FOX, D.J. (1961) 'Henequen in Yucatan: a Mexican fibre crop', *Transactions, Inst. Brit. Geogr.*, 29, 215–29.

FOX, D.J. (1965a) 'Henequen in Tamaulipas', *J. Trop. Geogr.*, 21, 1–11.

FOX, D.J. (1965b) 'Man-water relationships in metropolitan Mexico', *Geogr. Rev.*, 55, 523–45.

FOX, D.J. (1972a) 'Patterns of morbidity and mortality in Mexico City', *Geogr. Rev.*, 62, 151–85.

FOX, D.J. (1972b) 'Mexico: the transport system', BOLSA *Review*, 6, 62, 60–9.

FOX, D.J. (1977) 'The Mexican Oil Industry', BOLSA *Review*, 11, 10, 520–32.

FRISBIE, P. (1975) 'Illegal migration from Mexico to the United States: a longitudinal analysis', *International Migration Review*, 9, 1, 3–14.

GONZÁLEZ GUTIÉRREZ, T.G., BENAVIDES, L., KUMATE, J. and RANGEL, R. (1962) 'Encuesta immunológica en la población infantil', *Bol. Med. del Hosp. Inf. de Méx.*, 19, 102–16.

GONZÁLEZ NAVARRO, M. (1965) 'Mexico: the lop-sided revolution', in VELIZ.C. (ed.)

Obstacles to Change in Latin America, New York, (1965), pp. 206–29.
GONZÁLEZ SANTOS, A. (1957) *La agricultura*, Mexico.
GORMSEN, E. (1977) *El turismo como factor de desarrollo regional en México*, Geogr. Inst., Johannes Gutenberg Univ. Mainz.
GUZMÁN, P. (1964) 'Exploration by Petróleos Mexicanos', in BONINI, HEDBERG and KALLIOKOSKI, op. cit.
HENDERSON, D. (1965) 'Arid lands under agrarian reform in northwest Mexico', *Econ. Geogr.*, 41, 300–12.
Inventario de la información basica para la programación del desarrollo agricola en la América Latina: Mexico (1964) Washington, DC, Pan American Union.
Investigación de Vivienda (1965) Mexico, Instituto Mexicano de Seguro Social, vol. 1.
JARDINES MORENO, J.L. (1976) 'Los distritos de riego por bombeo del centro y norte de Sonora', *Recursos Hidraulicos*, 5, 1, 8–25.
KRAEMER MALINOWSKY, D., LATORRE DIEZ, C. and CERVANTES SANCHEZ, O. (1977) 'Proyecto Sonora', *Recursos Hidraulicos* 6, 2, 104–26.
La población económicamente activa de México en junio de 1964 (1964) Mexico, Dir. Gen. de Muestreo, Sec. de Industria y Comercio, vols 1 and 2.
Las 16 ciudades principales de la república Mexicana: ingresos y egresos familiares 1960 (1962) Mexico, Dir. Gen. de Muestreo, Sec. de Industria y Comercio.
LAVELL, A.M. (1972) 'Regional industrialization in Mexico: some policy considerations', *Regional Studies*, 6, 343–62.
LEWIS, O. (1964) *The Children of Sanchez*, London.
MANGELSDORF, P.C., MACNEISH, R.S. and WILLEY, G.R. (1964) 'Origins of agriculture in Middle America', in WEST, R.C. (ed.) *Natural Environment and Early Cultures*, Austin, Texas, pp. 427–45.
México 1966: Hechos, cifras, tendencias (1966) Mexico, Banco Nacional de Comercio Exterior (*México, 1970*).
MOORE, W.B. (1968) *Industry and Water for the Valley of Mexico*, Mexico, Depto. de Investig. Industriales, Banco de Mexico.
NIEDERBERGER, C. (1979) 'Early Sedentary Economy in the Basin of Mexico', *Science*, 203, 4376, 131–42.
ORELLANA, C.L. (1973) 'Mixtec migrants in Mexico City: a case study of urbanization', *Human Organization*, 32, 3, 273–83.
POLEMAN, T.T. (1964) *The Papaloapan Project*, Stanford.
PRICE, J.A. (1973) *Tijuana: Urbanization in a Border setting*, Notre Dame, Indiana. *Programa financiero de vivienda* (1964) Mexico, Sec. de Hacienda y Crédito Público.
RANDALL, R.W. (1972) *Real del Monte: a British Mining Venture in Mexico*, Austin, Texas.
Recursos Hidraulicos Published quarterly by the Secretaria de Agricultura y Recursos Hidraulicos, Mexico.
REYNOLDS, C.W. (1970) *The Mexican Economy: Twentieth-Century Structure and Growth*, New Haven, Conn.
RUBÍN DE LA BORBOLLA, D.F. (1964) 'The Mexican Indian today', in WILGUS, A.C. (ed.) *The Caribbean: Mexico Today*, Gainesville, pp. 121–31.
SEAWALL, F. (1961) 'Recent developments in Mexican sulphur production', *J. Trop. Geogr.*, 15, 39–45.
STEVENS, R.C. (1964) 'The soils of Middle America and their relation to Indian peoples and cultures', in WEST, R.C. (ed.) *Natural Environment and Early Cultures*, Austin, Texas, pp. 26–315.
STEVENS, R.P. (1966) 'Algunos aspectos de la migración interna y la urbanización en México, 1950–1960', *Comercio Exterior*, 16, 570–5.
TAMAYO, J.L. (1962) *Geografía general de México*, 4 vols, Mexico.
The Puebla Project 1967–9 (1970) Mexico, Cent. Internat. de Mejoramiento de Maiz y Trigo.

TOGNO, F.M. (1963) 'Planeación ferroviaria – métodos simplificados para completar la red de México', *Communicaciones y Transportes*, 5, 27, 16–63.

UN FOOD AND AGRICULTURE ORGANIZATION (1957) *The Selective Expansion of Agricultural Production in Latin America*, (E/CN 12.378/Rev. 8), New York.

UTRIA, R. (1966) 'The housing problem in Latin America in relation to structural development factors', *Econ. Bull. for Lat. Amer.*, 11, 2, 81–110.

VÉLIZ, C. (ed.) (1965) *Obstacles to Change in Latin America*, New York.

VIVÓ ESCOTO, J.A. (1964) 'Weather and climate of Mexico and Central America', in WEST, R.C. (ed.) *Natural Environment and Early Cultures*, Austin, Texas, pp. 187–215.

WARD, P.M. (1976) 'The squatter settlement – a slum or housing solution: evidence from Mexico City', *Land Economics*, 52, 3, 330–45.

WERNER, G. and SCHÖNHALS, E. (1977) 'La destrucción de los suelos en la region de Puebla-Tlaxcala', *Communicaciones Proyecto Puebla-Tlaxcala*, 14, 9–14.

WEST, R.C. (1949) 'The mining community in northern New Spain: the Parral mining district', *Ibero-Americana*, 30.

WEST, R.C. (ed.) (1964a) *Natural Environment and Early Cultures*, Austin, Texas, Handbook of Middle American Indians (ed. R. Wauchope), vol. 1.

WEST, R.C. (1964b) 'Surface configuration and associated geology of Middle America', in WEST (1964a) op. cit., pp. 33–83.

WEST, R.C. and AUGELLI, J.P. (1976) *Middle America, its Lands and Peoples*, Englewood Cliffs, NJ.

WHETTEN, N. (1948) *Rural Mexico*, Chicago.

WORLD BANK (1979) *Mexico: Manufacturing Sector*, Washington, DC.

YATES, P.L. (1962) *El desarrollo regional de Mexico*, 2nd ed. México, Depto. de Investig. Industriales, Banco de México.

3 · The Caribbean

David L. Niddrie

The physical setting

The Caribbean Sea, bounded on the west by the Central American isthmus and on the south by the South American mainland, is less easily defined in the north and east, except by a wide arc of islands, some forty in number. Within this sea there are several basins separated by ridges and swells whose peaks do not, however, break the surface anywhere, except on the Aves Swell and south of the Yucatán basin (*Figure 3.1*). Some of the world's great deeps or oceanic troughs are also found in the northernmost stretches of the basin, suggesting a geotectonic origin for the islands themselves. Such disposition of land and sea is from the outset a serious disadvantage, since there is no centrally located site for a regional centre. A study of the Caribbean islands becomes in effect a description and analysis of peripheries.

Linked tectonically with the eastern flanks of the Yucatán peninsula on the one hand and the west-east folds of Honduras and Nicaragua on the other, the larger islands (the Greater Antilles) stretch some 2400 km as far as Puerto Rico. From this point a string of much smaller islands takes a southward course, while yet another group projects the Venezuelan mountains eastward and northward to complete the arc.

A number of disparate geological episodes is therefore responsible for the creation of the islands themselves, none of which is older than the Jurassic period. Some of the early structural trends suggest a link with large-scale movements involving the tectonic plates of South and North America during the Cretaceous period. Old volcanic cores embedded within successive marine limestones, shales and sandstones intercalated with widespread volcanic detritus, indicate several palaeographic outlines differing from those of the present day until the Upper Miocene period, after which eustatic changes fringed many of the islands with coral shores and created widespread coral platforms beyond the boundaries of the Caribbean Sea. A phase of volcanic activity, bringing the youngest of the Lesser Antilles into being, marks the Holocene period. So many events crammed into a brief space of geological time created a variety of island land forms, whose polygenetic characteristics make subdivision of the islands possible (*Figure 3.2*).

The Greater Antillean complex mountains

This is the oldest core region in the Caribbean. Strongly folded mountain ranges, steeply sloped and strongly dissected, pass from west to east across southern Cuba (Sierra Maestra, Sierra de Nipe and Cuchillas de Toar), parts of Jamaica,

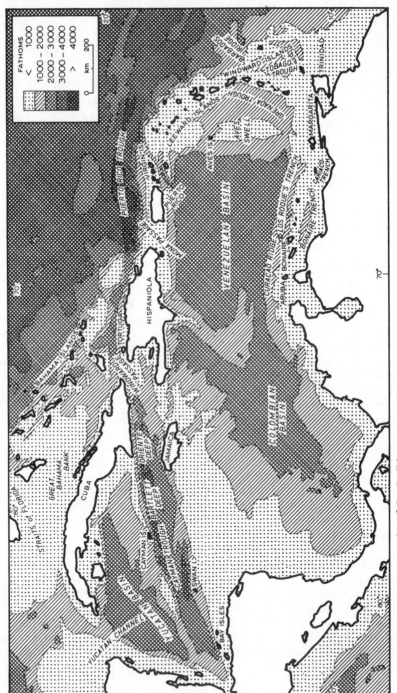

Figure 3.1 Submarine topography of the Caribbean

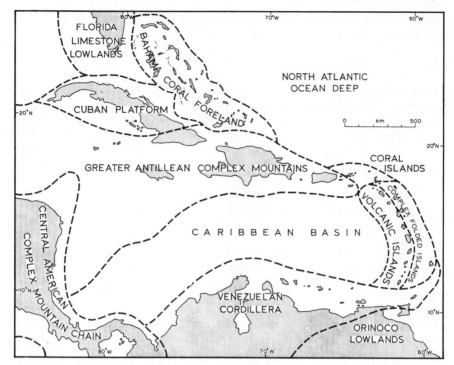

Figure 3.2 Physical regions of the Caribbean

Hispañola (Cordillera Central), Puerto Rico (Cordillera Central and Sierra de Cayey) and the Virgin Islands.

The Cuban platform
The remainder of Cuba is composed of several mountain ranges trending south-west to north-east (Sierra de Trinidad and Sancti Spiritus) in the central area and the Sierras de los Organos in the north. Mostly composed of Cretaceous limestones, they are today strongly affected by karstic erosion.

Complex folded islands
Several islets (Barbuda, Marie Galante), Antigua, the eastern half of Guadeloupe and possibly parts of Barbados consist of uplifted, moderately folded sedimentary deposits. These have been eroded sub-aerially for a long time, yielding a flattish, gently rolling countryside devoid of major heights, an important climatological factor.

Venezuelan cordillera
The Netherlands Antilles, Santa Margarita, Trinidad and Tobago and parts of Barbados are an extreme extension of the west-east folded mountains linking the South American mainland. They are composed of crystalline, metamorphic, and occasionally volcanic rocks. These narrow ranges are well dissected and moderately sloped. To the south of the Trinidad ranges vast thicknesses of

sediment from a proto-Orinoco river engulfed the Trinidad mountains and hills, creating an extensive area of ill-drained lowland.

The volcanic islands

A chain of small islands, coming into existence during the Pleistocene period, includes Grenada, the Grenadines, St Vincent, St Lucia, the northern half of Martinique, Dominica, the western half of Guadeloupe, Montserrat, St Kitts, Nevis, St Eustatius, Saba and St Martin. These consist of typical volcanic peaks, some of lava and some of ash, all presenting evidence of dormancy or activity. All have a fringing apron of gently sloping land from the base of the actual volcanic mountain to the seashore, which is often edged by black, sandy beaches.

The coral foreland, coral islands and coral fringes

Local instability and world-wide eustatic sea-level changes have combined with a favourable oceanic environment to create the extensive Bahama island group, which are today flat, inconsequential stretches of land to the north of the Caribbean Sea itself, based on a crystalline platform. Then there are the coral islands that lie scattered across the area, as well as the broad stretches of shallow sea round most islands, whose white, sandy beaches are derived from such coral deposits; and finally, the occasional Holocene coral platform lying about 6 m above present sea-level.

Evidence for recent eustatic changes may be seen on most islands, where wave-cut beaches and cliffs well above the present sea-level may be traced. The best-known examples are those on Tobago, Aruba, Curaçao and Bonaire in the south, and along the southern coastlines of the Dominican Republic and Cuba. Attempts to correlate such widely separated phenomena have not met with much success. In fact, very little is known of the cyclic history of the Caribbean, apart from the scattered studies of polycyclic landscapes in western Puerto Rico and the Dominican Republic. Geomorphologists have yet to apply their ingenuity to such studies.

Environmental influences and hazards

Although the Caribbean islands may appear to the casual observer 'full of noises, sounds and sweet airs that give delight and hurt not', there are few elements in their favour except a benign temperature regime, considerable sunshine and the trade winds. There are, built into the landscape, certain other elements the more hazardous because they are uncertain in their time, place and degree.

Earthquakes have always been a natural accompaniment to tectonic events such as shaped the islands. Along the borders of the abyssal trenches north of Puerto Rico and south of Cuba, and following the trend lines of the Greater Antillean complex mountains, there is abundant evidence of heavy shocks, mainly from shallow earthquakes, from the first European penetration of the area. Santiago de Cuba was reduced to rubble on a number of occasions (1678, 1755, 1766, 1932 and 1947), while Port Royal (1692) and Kingston (1907) in Jamaica were virtually destroyed.

Volcanic activity is confined to the Lesser Antilles, along the inner arc of volcanic islands (*Figure 3.2*). All the present craters are at least dormant, although fumaroles, hot springs, solfataras and other minor phenomena of this kind may be seen in the immediate neighbourhood. There are, furthermore, events from the recent and historical past that confirm the ever-present menace on such islands. The scars left by the eruptions of Mont Pelée in Martinique and of Soufrière on St Vincent in 1902, as well as those on Dominica in 1930, are still visible today. Volcanic ash ejected from Soufrière in April 1979 has sterilized most of St Vincent's agricultural land.

More frequent, unpredictable in behaviour, and by far the greatest natural hazard in the Caribbean, are the tropical cyclones known as hurricanes (*Figure 3.3*). The full blast of a well-developed hurricane has been experienced many times by all the Caribbean islands except Trinidad, where only one storm of any strength has been recorded in the past 46 years (1933). Destructive winds, reaching velocities of 240 km/h and changing direction as the circulatory anticlockwise system passes an island, are associated with heavy, damaging rains, which bring floods, landslides and all-pervading mud. All economic activity can be brought to a halt within a few hours. Cash crops lie broken, trees are uprooted, food gardens are flattened, houses are scattered in fragments across the countryside and within days starvation may be followed by infectious plagues, looting and complete despondency in the community. No better example can be found of the devastating consequences of hurricane damage on the island communities than David and Frederic which swept successively through the Caribbean in August-September 1979.

Until the nineteenth century the effect, though devastating, could be shrugged off. Once sugar lost its primacy in the Lesser Antilles, however, a hurricane was enough to break the back of a small island's precarious economy. Yet it is remarkable to observe how a community can recover from a violent hurricane within a short period and rehabilitate its cash crops. The latter are all too often highly vulnerable to high winds, and many changes in agriculture have been effected by substituting another cash crop for those destroyed. By this means the Windward Islands were converted to banana cultivation after Hurricane Janet in 1956.

The normal trade winds that sweep the Caribbean make life bearable for the islanders by providing a cooling breeze and fair sailing conditions. None the less, they produce excessive evaporation and aridity whenever they sweep across the plains and lowlands of Aruba, Antigua, Tobago or Grand Terre in Guadeloupe. Only the presence of high hills and mountains saves the Caribbean from aridity, since these provide the necessary orographic lift to bring high rainfall to the upper slopes, with perennial streams even in the dry season. Even the steady force of a normal trade wind necessitates the building of wind-breaks to protect cash crops such as cocoa and banana from serious physiological damage, and its ability to fan a fire, be it forest, garden or building, is too well known to require comment.

Natural vegetation patterns are also strongly influenced by such wide variations in rainfall and evaporation. On most islands possessing some relief, there would normally occur a series of climax communities ranging from lowland, tropical rain-forest, through lower montane forests and deciduous forests adapted to markedly dry seasons, to xerophytic associations. Few of these vegetation types are generally correlated with soil differences. One exception, however,

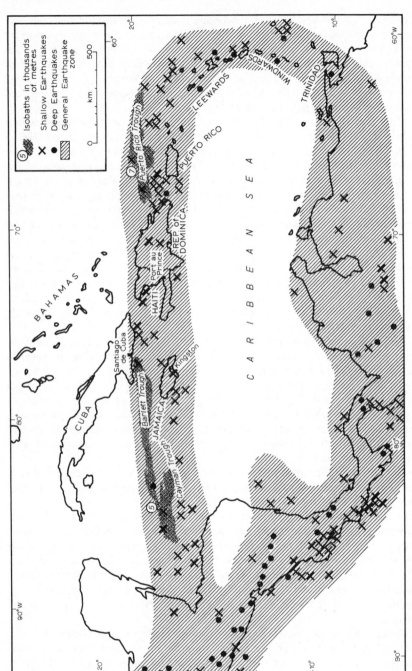

Figure 3.3 General tracks of hurricanes in the Caribbean June to October

is the limestone landscape, at present in the karstic stage of development. In such areas surface water disappears, leaving a desert-like forest scrub growing on karstic land forms. The latter are found in northern Cuba, central Jamaica, eastern Hispañola, northern Puerto Rico, Guadeloupe and the Netherlands Leeward Antilles.

For all their manifest physical disadvantages, the Caribbean islands have been inhabited continuously for many centuries. There are today on most islands the remains of once impregnable defences – forts, walls, moats and cannons – all in their heyday appendages of the great empires of Spain, Holland, France and England. Maritime rivalries during the three centuries after Columbus gave them added importance as naval bases safe from enemy fleets. As a result, various islands changed hands at the whim of the treaty-makers, not once, but several times, leaving them with a confused succession of cultural, linguistic, religious and imperial influences from Europe. To this must be added the influence of Africa, through slavery.

The types of government that have evolved among the Caribbean islands include cruel despotism, classical Hispanic American oligarchies, the subsidized paternalism of a United States protectorate, the well-known British systems of evolutionary colonial governments leading to independence, and the French and Dutch forms whereby the islands are considered to be part of the metropolitan area, but with some degree of devolution and autonomy. Only the British islands have ever attempted some form of regional or federal government, the last of which failed 4 years after its creation in 1958. Such diverse physical conditions as size and distance, and the varying political institutions that are found today, discourage closer association except in the economic sphere.

Pre-Colombian occupation of the islands

Before the arrival of the Europeans there was a considerable aboriginal population of Amerindians, whose numbers, estimated in several hundreds of thousands, were reduced to a tiny fraction of the population within a few years of the arrival of the conquistadores. Any influence that these communities might have exerted upon European practices was therefore dissipated by genocide and newly introduced diseases (*Figure 3.4*). Much archaeological evidence is now pointing to an earlier pre-Columbian occupation of the islands. Among the first tribes known to have occupied the area were the Ciboney, a group practising a Stone Age culture along the Guaicayarima peninsula in south-western Haiti and throughout the greater part of western Cuba.

The second group of aboriginal peoples, speaking a language known as Arawakan, is thought to have originated in the South American mainland, where remnants are found today. These tribes are believed to have migrated from the eastern slopes of the Andes along the Amazon and its tributaries (where they are still fairly numerous), and thereafter along the Orinoco, the coasts of Venezuela, eastern Colombia and Guiana, and finally into the Lesser and Greater Antilles. Some were known to be settling along the Florida Cays during the first stages of Spanish exploration of this area.

The Arawaks were the first people encountered by Columbus and his crew. They possessed a more advanced culture than the Ciboney, with stratified

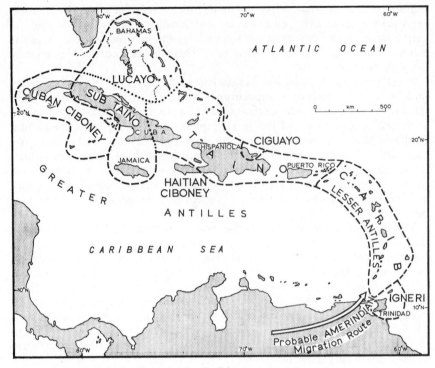

Figure 3.4 Aboriginal cultures in the Caribbean at contact

societies and an elaborate religion; they were active farmers and expert fishermen, with some knowledge of metallurgy. Few Arawak-speaking peoples survived the European invasion, although as late as 1900 some 400 Taino were
recorded in eastern Cuba.

The third and last of the Caribbean migrations was undertaken by the
Cariban-speaking Amerindians, who were moving through the eastern
Caribbean arc and into the Greater Antilles contemporaneously with European
exploration. The Caribs, some of whom are found today scattered through the
Amazon basin and the Caribbean, also varied in racial, ethnic and cultural
characteristics, sharing only the Cariban language in common. In the manner of
many advancing conquerors, they slew the men among their enemies and carried
off the women. In this way the language of the latter (Arawakan) passed to the
children, causing seventeenth-century observers to report that the island Caribs
spoke only Arawakan with an overlay of Cariban words.

The Caribs were the most mobile of the Amerindian aboriginals. Their long
canoes with sails could cover vast distances, so that Puerto Rico, Cuba and
Trinidad were never wholly out of reach of these adventurous warriors. They were
skilled in growing and preparing bitter cassava, and could weave and spin the
cotton they grew into sails and clothing. Among early explorers and adventurers
the Caribs were notorious for resisting all attempts of the white man to befriend
or enslave them, attacking whenever possible with every intention of destroying

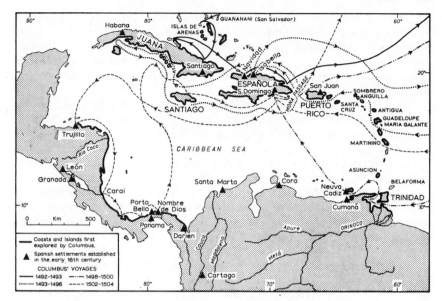

Figure 3.5 Discovery and early settlement in the Caribbean

all intruders. Many sub-tribes, ruled by 'kings' or chiefs, survived in Trinidad, Tobago and other islands of the Lesser Antilles well into the first half of the eighteenth century, while compact groups in St Lucia, St Vincent and Dominica managed to survive the most rigorous pressures for another 75 years. Recent archaeological research in the islands however, has supported the view that many of the distinguishing characteristics of these Amerindian groups recorded by early chroniclers were unreal, and largely a matter of subjective, ethnocentric description.

The voyages undertaken by Christopher Columbus have received considerable attention in the literature of exploration and discovery, and need little elaboration (*Figure 3.5*). Between 1492, when he set out on his first voyage, and 1504 when in disgrace and ill-health he finally returned to Europe, this hardy sailor had etched the coastlines of the Greater Antilles and the Lesser Antilles. Those who followed him completed the picture by delineating the Gulf of Mexico and filling in the details of the shore of Guiana and Venezuela. The conquistadores and those who followed them founded settlements, and evolved a peculiarly Spanish system of colonial administration, land tenure and land usage, in strong contrast to the more casual techniques of the English, French and Dutch who came into the area a century later.

Organization of the Spanish islands

Two objectives were reflected in Spanish forms of settlement. The first lay in the need to defend the windward approaches into the Caribbean Sea. This is demonstrated by the fortress towns that were built at various vital spots – the early landfalls of the sailing ship riding down the trade winds. The mighty fortress of San

Felipe del Morro, together with its associated fortification system enclosing San Juan (Puerto Rico), was the embodiment of the Spanish will to resist all invaders. Furthermore, it protected large forest reserves whose timbers were readily available for ship repairs and shipbuilding, a factor which delayed the full-scale development of Puerto Rico for more than 250 years. Havana was similarly fortified as a naval base.

The first Governor of Cuba, Don Diego de Velasquez, was ordered to set up cities and distribute crown lands to the first colonizers. Seven major cities were founded in the island. Each was granted an area of jurisdiction only vaguely defined at the time, and was laid out according to a plan recommended from Spain, with a centre (plaza), a market place, sites for public buildings and building lots for houses. Round the city a strip of land was reserved for common use by all citizens. Beyond this were distributed the larger rural holdings known as *mercedes*. Once established, the councils of each city (*cabildos*) took upon themselves the right to grant these lands from as early as 1536, and abandoned the privilege only when forced to do so by a *cedula* (royal decree) of 23 November 1729.

Apart from the growing of necessary food crops, such land was used primarily for stock-raising of all kinds, particularly of cattle to supply salt beef for ships and the city, and horses to supply remounts for the armies of the conquistadores in Mexico. Consequently there were few permanent settlers on the island during the next 50 years, and much of Cuba's land remained unoccupied and unalienated, as the island was primarily a staging post for those proceeding to the mainland.

The Cuban system of land tenure during the next two centuries evolved from the Spanish refusal to allow more than a minimal decentralization or delegation of authority. It had clearly been the intention of the Spanish Crown at the beginning of the great westward surge into the Americas to curb the personal ambitions of its captains-general and conquistadores who, using the *encomienda* and *repartimiento* (allotment of land together with the residing native community), had hoped to enslave the aboriginal peoples and carve out great empires for themselves. The patterns laid upon the landscape of Cuba by the merced and its successor, the polygonal holding, persisted for nearly 300 years, until new forms of land utilization drove it from the contemporary scene.

Stock-raising on very large ranches persisted throughout the sixteenth and seventeenth centuries in Cuba, providing exports of hides, tallow and salted beef. Without the stimulus of population growth and immigration the island continued to be harried by repressive trade policies on the part of the Spanish crown and by piracy along its coasts. The second half of the eighteenth century brought many reforms to colonies such as Cuba. Imperial laws, long out-dated, were rescinded, most trade restrictions were removed, and the emigration of large numbers of Spaniards encouraged. Monopolies on slave imports and tobacco sales were broken up, so that sugar cane soon spread across the island, hastening the break-up of the cattle ranches. Coffee and tobacco continued to thrive on smaller holdings in certain parts of the island. Other Europeans, refugees from South American countries embroiled in wars of liberation, from Haiti and Martinique, added to the richness of Cuban life during this peaceful and prosperous era. Unfortunately, the Spanish Crown, fearing that its authority was declining, returned to absolutism and direct rule in the nineteenth century. For Cuba it meant a permanent breach between the metropolitan administration (*gachupine*) and the native *criollo*, and continual strife until the Spanish American War brought the necessary changes. Near-bankruptcy followed upon the failure of the sugar industry to cope with the end of slavery and the innovative sugar-beet industry in Western Europe. A sugar economy continues, none the less, to dominate Cuba's physical and cultural landscape to the present day.

As might be expected, Santo Domingo and Puerto Rico came under the same Spanish law, but their historical geography is slightly different from that of Cuba. Spain ceded the western part of Hispañola (Saint Domingue, later Haiti) ·in 1697 and in 1795 its eastern part, to be physically occupied by Haitian invaders in 1801. It was not until 1861 that direct Spanish rule eventually returned, short lived, to Santo Domingo. Four years later the country freed itself from Spanish hegemony to become an independent republic.

After the gold rush had ended in 1520, Hispañola divided itself naturally into two parts, the western end, which was quickly depopulated, and the eastern part near Santo Domingo, which was occupied by at least forty sugar plantations, each using above fifty black slaves. Unfortunately, Santo Domingo was replaced by Havana as a port of call, was sacked by the English fleet in 1586, and began to lose its population to Mexico. Settlements in the north and west were abandoned to French, Dutch and English pirates. Those remaining in the eastern parts took up cattle-ranching in the hills and on the plains until the beginning of the nine-teenth century. Meat was exported to Saint Domingue. Once again sugar began to displace cattle on the southern plains, supplemented by mixed farming around Vega Real. By 1800 there were less than 130,000 inhabitants, though a considerable influx of refugees after the slave revolution in Saint Domingue added to the total.

Puerto Rico, a long-neglected colony but useful as a timber source, remained a community of subsistence smallholders until 1815, when the Spanish govern-ment rescinded most of its trade restrictions. Thereafter large sugar-cane hold-ings developed in the plains and broad alluvial valleys, while the smallholders retreated to the mountain slopes to grow coffee. This was the nineteenth-century landscape upon which the United States laid its protective hand in 1897.

Both Jamaica and Trinidad were for some period in their history occupied by Spain. Jamaica, closer to the great trade routes on the way back from El Dorado, benefited from its role as a supplier of salt beef, hides and tallow which it exchanged for European foods. When the English took the island over in 1655, it was not regarded as an attractive place for settlement, but rather as the home of the Port Royal buccaneers. Only after 1664 did colonists from the eastern Caribbean begin to flow in and take up the generous land grants offered them. By 1670 sugar, indigo, pimento, cocoa, cattle and salt were established products. Within 30 years sugar and slaves had turned the island to monoculture.

Trinidad might have proved a stepping-stone to South America and a base for exploring the Orinoco River. Apart from the occasional visiting ship it was neglected by Spain for more than two centuries. Cocoa cultivation, introduced in 1700, brought slaves and brief prosperity, and with the relaxing of immigration laws after 1783 Europeans, and in particular refugees from the French Revolution, flocked to the island. Captured by a British force in 1797, it was ceded in 1802 and became a British colony.

English settlements

The highly sophisticated Spanish systems of colonization and land development were in no wise imitated by the earliest settlers from England, France and Holland who sailed for the Lesser Antilles to take up their forms of colonial life. What particularly distinguishes the English method of colonization is its quite haphazard approach. The Proprietary System, evolving in the royal courts of sixteenth and seventeenth-century England (120 years after the Spanish had made their first permanent settlements in the western hemisphere), consisted in the granting, by royal favour, of blocks of overseas territory to groups of wealthy, influential persons intent on making vast fortunes from these lands. In this way Barbados and St Christopher (St Kitts) in 1625 became the sole property of James Hay, the Earl of Carlisle, against strong opposition from the Earl of Pembroke. Jamaica, despite its prior occupation in 1512 by the Spaniards, was similarly exploited under the Proprietary System.

Colonization in these 'English' West Indian islands was thus partly conditioned by earlier experience gained along the eastern Atlantic seaboard of North America and in the Guianas. Clearing the forest, building houses and forts, planting various crops and coming to terms with the aboriginal communities were generally first priorities in these areas. Absence of any form of freehold tenure under such proprietary patents, and even of the most elementary surveys, permitted a chaotic pioneer frontier to move gradually westward into the interior. Tobacco and cotton planted on the small, cleared patches of virgin soil added further to the disordered nature of the settlement. Accordingly, those who led their parties ashore at St Christopher in 1624 and Barbados in 1627 were men familiar with such empirical techniques, which they could apply immediately.

It was not long, however, before sugar cultivation dominated the landscape. In 1634 Courteen had brought the crop to Barbados, and by 1655 most of the island's smallholdings had been consolidated into plantations with slave labour. It would be difficult within this text to delineate the sequential changes in land use and ownership for all the English possessions in the Lesser Antilles, but most underwent similar processes in the seventeenth and eighteenth centuries. Jamaica became English in 1644 and rapidly turned to cash crops such as sugar and coffee.

As the system of proprietary grants faded, in concert with the decline of monarchical absolutism in the seventeenth century, new systems of controlling overseas territories evolved, in order to ensure some continuity of administration and security for investment. This was gradually accomplished in London through an officially appointed body known eventually as the Lords Commissioners for Trade and Plantations, derived from a sub-committee of the Privy Council. These officials recommended to the Crown suitable governors and military commanders to administer each colony under Common Law, certain English Acts and various mercantilist laws. A legislature modelled on Westminster, both elected (sometimes by dubious means) and nominated, helped the Governor to serve the island in the best possible way, generally in favour of the sugar interests.

From the mid-seventeenth century onwards a closer network developed within the English colonies of the Caribbean, involving regular inter-island transfers of officials and settlers. Sporadic outbreaks of hostilities among the metropolitan powers leading to naval engagements, especially in the waters off the Lesser Antilles, and subsequently the signing of a peace treaty assigning one or more of the islands to another nation, were part of common experience. It seldom made any difference to the planter, his slaves or his crop.

After the Treaty of Paris in 1763 the English and French colonists became very wealthy by growing sugar and high-grade cotton in virgin soils, and selling them in Europe. Delegation of authority to attorneys and managers, natural disasters, emancipation and changing economic conditions in Europe turned these plantation owners into bankrupts by 1850, and dragged the islands down to penury. Faced with such problems, the Colonial Office, a direct descendant of the Commissioners for Trade and Plantations, was constantly engaged throughout the nineteenth century in trying to integrate the Leewards, the Windwards and combinations of each into more economical 'units'. Some succeeded for a while, but the vexed problem of divergent interests was always present in small communities which had already developed different collective personalities. This is a facet of English West Indian life that nobody has been able to reconcile with closer union in the Caribbean Basin.

The Dutch traders

The Dutch followed belatedly on England's sixteenth-century buccaneers in the Caribbean. Having been denied entry to the Mediterranean by Spain, they penetrated her *mare clausum* in order to acquire raw materials such as salt for their herring fleets, hides, cocoa and wood in preference to settling on any particular island. From the Canaries and Cape Verde Islands (the 'salt' islands), armed merchantmen sailed westward to Brazil and the Wild Coast (between the Amazon and the Orinoco), and clung to the southern Caribbean rim. Margarita Island, therefore, became a primary centre for gathering cargoes from Cuba, Hispañola and *tierra firme*. Vast quantities of salt were taken from Punta de Araya on the Venezuelan mainland, south of Margarita, at great risk from punitive Spanish government forces which drove the Dutch further west to Venezuelan Tortuga and therefrom to Curaçao, Bonaire and Aruba in search of salt.

By the turn of the century Dutch privateers were active throughout the Greater Antilles, and ceased their depredations only during the Twelve Year Truce (1609–21). In view of the success of the Dutch East India Company in south-east Asia, the authorities in Amsterdam decided to create in 1620 a similar trading organization known as the Dutch West India Company. From the outset it was considered an 'instrument of war' against Spain and its empire, with the right to capture the Spanish treasure fleets, and to raid ports and cities while trading illegally with all and sundry. By impoverishing Spain, the Dutch United Provinces freed themselves from its yoke, and accumulated wealth enough to make them the world's leading trading nation. The company itself lasted but 28 years.

While busy destroying Spanish naval power, the Dutch failed to settle any of the islands of the Lesser Antilles ('las islas inutiles'), except for Saba, St Eustasius, and half of St Martin, and thus lost the opportunity seized upon by the English and French to develop virgin soils for sugar cane and cotton. Instead, their eyes were fixed on the Guianas and northern Brazil, making Curaçao, with its fine port, the centre for co-ordinating all Dutch maritime activities. Bonaire and Aruba close by were also claimed by the Dutch. Attempts to settle Spanish Trinidad proved fruitless, while sporadic settlements on Tobago, from 1628 to 1678, brought conflict with the Caribs, Spain, the Courlanders (from Latvia), the English and finally the French. Naval and land engagements in 1672 and 1673, ostensibly for the control of Tobago, brought the Dutch to their knees and compelled the signing of the Treaty of Nijmegen in 1678. Thereafter, a nation which had at great profit traded in slaves, bought and bartered tropical crops and created capital for others to use, faded from the Caribbean. Only those minuscule islands, known today as the Netherlands Antilles, remained under the Dutch flag as a bleak reminder of a meteoric, short-lived empire.

French colonization

When turned away from piracy along the West African coast in the sixteenth century, French corsairs chose to harass, from new bases in the Bahamas and Tortuga, the Spanish fleets coming through the Mona Passage, the Middle Passage and the Florida Straits. A settlement in 1625 on St Kitts, already occupied by the English, aroused enough interest for Cardinal Richelieu to create a chartered company to exploit the island. In 1635 he organized a larger company, the Compagnie des Isles d'Amérique, to occupy Martinique, Guadeloupe and several small islands nearby. Early attempts in 1639 to grow sugar on Martinique were followed by more successful efforts in 1647.

It was Colbert's Compagnie des Isles Occidentales, formed in 1664, with its built-in mercantilist monopolies, which made French colonization in the Caribbean into a successful venture. Its first official act was the occupation in 1665 of western Hispañola (St Domingue) which rapidly developed into the richest territory of the Caribbean with more than 100,000 slaves and 30,000 whites. Now that the Dutch had faded from the scene and by the Treaty of Rijswijk (1697) the Spanish had recognized French and English territorial claims in the *mare clausum*, there followed more than a century of naval rivalry between these two nations, leading to bewildering exchanges of various islands in the Lesser Antilles by capture and treaty. With the permanent loss of St Domingue to rebel slaves in 1791 and the reversion in 1815 to the English of all the islands they had taken, the French had to be content with Martinique, Guadeloupe, some small neighbouring islands and half of St Martin and Cayenne. Despite such territorial losses, islands occupied at any time by France retained indelible features of French language and culture.

The chief relict upon the French Caribbean scene, however, was an orderly landscape, dominated not merely by sugar monoculture, but including mixed cropping of tobacco, cocoa, coffee, indigo, ginger, cotton and various ground provisions, together with animal husbandry. Accurate cadastral surveys were common in the first half of the eighteenth century, and planners took great care to vary the size of holdings. Before they fell into neglect and collapsed, the irrigation systems watering the plains of St Domingue provided the high yields of cane for which it was renowned. In 1764 the French Governor of Grenada was able to produce a detailed map and description of every acre of his densely populated island, and show how intensively it was cultivated. Organization was Colbert's gift to the Caribbean, and this is what distinguishes the French islands even today.

Sugar and slavery

In the end all Caribbean islands with their widely varying forms of settlement and exploitation fell under the domination of sugar monoculture. Four phases may be discerned.

The early period, 1638–1763
During this period St Kitts, Barbados, Jamaica, Guadeloupe and Martinique received the sugar cane plant, grew it on smallholdings, and began the process of consolidation of holdings essential to greater efficiency of land and slave labour.

The middle period, 1763–1870
During this period the Ceded Islands, Tobago, Grenada, the Grenadines, St Vincent and Dominica, together with Martinique, Guadeloupe, Antigua and St Kitts, grew a major part of Europe's sugar requirements in virgin soils. This was the period of highest slave importation, leading to the protracted struggle for abolition, and of plantocracy's period of greatest affluence, followed by technical innovations in sugar manufacture, which placed most absentee landlords in the bankruptcy courts.

The industrial period, 1870–1948
During this period Cuba, Jamaica, Trinidad, Puerto Rico and the Dominican Republic were subjected to massive capital investment, land purchase and consolidation, centralization of mills and factories and cartelization without the field machinery yet to mechanize growing and harvesting techniques.

The modern radical period, 1948 to the present day
This has involved decolonization, nationalization of land and mills, mechanization and the disappearance of uneconomic sugar cultivation in small islands.

Few remains of the early period are still to be seen in the cultural landscape except the rare ruins of an old building. It is the middle period that has contributed so much to the relict Caribbean landscape. On every island the successive techniques of sugar manufacture are demonstrated by abandoned windmills, water wheels and steam boilers, and the great plantation houses sited on high points to catch the breeze, with the slave barracks below them. On some islands, such as Barbados and Martinique, the fields survive almost in their original shapes of two centuries ago. It is difficult on the other hand to discern such patterns in many parts of Haiti, Jamaica or Puerto Rico, where population has overwhelmed field boundaries.

By contrast, the modern *central* or factory of Trinidad, Puerto Rico, Cuba or the Dominican Republic, each with its highly organized transport system, its bulk loading warehouse, its *bagasse* and paper mills alongside, represents the only possible way of staving off major competition from the sugar beet. Most of the smaller islands such as Tobago, St Vincent, St Lucia, Dominica, Montserrat and Nevis, have abandoned sugar plantations during the past 50 years.

The presence of black peoples everywhere is an even stronger reminder that labour from the canefields came not from an indigenous people, but from a continent that yielded a myriad of souls to the western hemisphere.

Slavery

The aboriginal peoples of Hispañola and Cuba, decimated by diseases soon after the Spanish conquest, were replaced by African slaves, the first batch of whom probably arrived in 1517. Once the gold and silver myth had been exploded, they became the key element in the plantation of several tropical crops for export to Europe. This trade cycle, part of the Atlantic triangular trade, thrived for some two and a half centuries, with consequences yet to be measured. Of some 10 million Africans transported to the western hemisphere, about 4 million reached the islands (*Table 3.1*). Few ever returned to Africa, and today their descendants represent all but a slight fraction of the island's peoples, many of them carrying varying proportions of the former slave-owners' genes.

Table 3.1 Slave imports into the Caribbean islands during the period of slavery

Islands	Total ('000)
St Domingue (Haiti)	864
Jamaica	748
Cuba	702
Barbados	387
Martinique	366
Leeward Islands	346
Guadeloupe	291
Puerto Rico	77
St Vincent, St Lucia, Tobago, Dominica, the Ceded Islands	70
Grenada	67
Dominican Republic	30
Danish W. Indies (US Virgins)	28
Trinidad	22
Dutch Antilles	20
Bahamas	10
British Virgin Islands	7
Total	4035

Source: Curtin, 1969, Table 24.

Data for the slave trade are obviously suspect. None the less, scholarly efforts, such as Philip Curtin's *The Atlantic Slave Trade*, have made it possible to assemble more precise factual material. What is certain is that it made the plantation possible, thus establishing sugar, coffee, cocoa and other items as indispensables in Europe's dietary, while introducing salt pork, cod, African tubers and breadfruit to the islands as food for the slaves.

Emancipation was only gradually achieved in the islands during the nineteenth century after the ocean trade had been abolished. Haiti's slaves had already taken matters into their own hands in 1791 through a bloody revolution, but elsewhere there was a wide gap between the cessation of trading in slaves and their actual release. Although the Danes abolished the slave trade in 1805, their Caribbean possessions did not release their slaves until 1821. A French revolutionary government abolished slavery in 1794, but it was not until 1848 that all the French colonies freed their slaves. Clandestine trading, particularly into Cuba and Puerto Rico, made it difficult for the Spanish government to impose its ban on these islands, and despite many official declarations, it was not until the Spanish American War (1898) that emancipation became universal in the Caribbean.

Left behind were the Caribbean communities of free black people whose forebears had been torn from many parts of West and East Africa. Apart from distinctly negroid characteristics, they must have also retained important cultural traits from Africa. A number of writers have demonstrated that African slaves came from a large array of ethnic groups, tribes and other identifiable groupings. Curtin has gone so far as to map the origins of many cargoes of slaves from historic data. Others have shown that buyers preferred certain tribes for the physical and mental qualities they were reputed to possess.

Within Cuba itself, 'los negros bozales' (those blacks who wish to stress their African origin) readily identify themselves with certain areas of the continent such as Guinea. The Yoruba (S.W. Nigeria) known locally as 'Lucumí', are the most widespread and the largest group. The second largest group, the Carabalies (from Calabar in S.E. Nigeria) associate themselves with the Efik and Ibibio tribes. The Araras (Dahomey), the Congos or Bantu (the lower Congo or Zaire basin) make up the ethnic groups of the island. Each group is located in a specific region of Cuba, and their presence is acknowledged today through the teaching of Yoruba and several Bantu languages in a number of schools. In Trinidad, Yoruba is also a recognized grouping, although those who belong are descended from free nineteenth-century merchant traders rather than slaves. Many similar examples of ethnic identity can be found throughout the area.

Indigenous languages, quickly suppressed in favour of some European language such as English, Spanish, Dutch or French, failed in large measure to survive the Atlantic crossing, although the various forms of 'creole' or 'patois', particularly in Jamaica, Haiti, British Virgin Islands, St Lucia, Trinidad and Tobago, clearly reflect an African linguistic origin. In Cuba and the Dominican Republic many Yoruba, Efik and Fon phrases and words are in use to this day in speech, song and religious cults.

Most Caribbean slaves soon discarded their overt African religious beliefs in favour of Christian conversion, while retaining much of the mystical and magical aspects within their barracks. The Obeah (Obi) man or woman, a caster of spells, is still hired to effect changes for better or worse on behalf of a client. Jumbies and Duppies (ancestor spirits) have to be placated when disturbed, and Animism is reflected in folk tales and legends. It is in Cuba (and to a lesser degree in Haiti and parts of Trinidad) that African religious cults have survived best, through admixture with Roman Catholic ritual. Santería is a faith of the Lucumí or Yoruba, while Naniguismo, the secret cult of the Abakua from Eastern Nigeria, is a form of ancestor worship founded in 1836. Other Afro-Cuban cults are Regla de Palo, Regla Congo and Mayombería, generally associated with more obscure forms of sorcery. Similarly, Haitian Voudon (Voodoo) derives much of its ritual from Roman Catholic practices. Many of the rites are intimately associated with drums, particularly those of the Yoruba (batas).

Secular music expressed through dance and song throughout the Caribbean owes much of its rhythms to West Africa. Modern sculpture and woodcarving are only indirectly inspired through Europe's twentieth-century discovery of the African plastic arts. Many food dishes share a common heritage with those of Africa. It would, however, be dangerous to extend the principle of 'single centre' cultural diffusion in matters of land use (slash-and-burn cultivation, use of the moon's phases to control planting, customary tenure, polyculture and herbal medicine), all of which may well have evolved independently of African influence, especially in post-emancipation years.

Modern Caribbean island societies have naturally been strongly motivated by the Black Consciousness movement in the United States (which had in fact been initiated by West Indians who had migrated to the mainland) and so have sought in Africa the roots they left behind as slaves. African revivalism has become the predominant cult, much to the chagrin of those whose ancestry was long ago mixed with whites, Chinese and East Indians, as well as of those who have avowed a non-racialist stance. Over-emphasizing Africa in such variegated communities may well prove disruptive and unrewarding in the end.

Effects of emancipation on the plantation economy

Once Great Britain had in 1807 abolished the slave trade, the lot of the slave in the West Indies began to improve in some ways. He became a valuable commodity not easily replaced except through his progeny. Emancipation in the British colonies in 1834, followed by 4 years of apprenticeship, was regarded by many as the ideal solution to the problem.

Within the British islands the Apprenticeship Act, creating in essence a 4-year labour contract between ex-slave and master, failed almost without exception. From Jamaica to Trinidad, most freed slaves abandoned their plantations for an independent subsistence life, often as far from their ex-owners as possible. Only in Antigua could it be said that the population passed from slavery to wage-labour without much difficulty.

Although many English plantations were inactive by 1834 as a result of bankruptcy, natural disasters or neglect, there was usually a manager or attorney to maintain authority. His principal task in the years following emancipation was to prevent squatting by ex-slaves on the best plantation lands. Instead, plots were offered for sale on the marginal fringes, usually steeply sloped and useless for sugar cultivation. Where no authority existed to prevent them, ex-slaves set up villages in remote areas, sometimes grouping their houses about a church. In many instances Moravian missionaries were the guiding lights in such settlements, and the names adopted by the settlements are a reminder of this fact. From such villages men went out to grow sugar, corn, roots and tropical vegetables on the steep slopes of the local hillside. An examination of remote villages in West Indian islands shows that people were not forced to seek such isolation, but usually chose it.

The consequences for the West Indies are manifest. In all but a few instances, land-title deeds did not exist for such holdings until a few years ago. As a result, the process of inheritance was resolved by customary tenure. Litigation among heirs was, and continues to be commonplace, exacerbated by the absence, through emigration, of the beneficiaries. Although it was possible in the early years to derive enough food from the soil, lack of conservation, loss of topsoil and many other factors reduced productivity to the minimal level that is encountered today. It is not difficult to visualize the consequences: malnutrition, outmigration to the plantation life again or to town.

The gap left in plantation life by emancipation was serious enough to reduce many to bankruptcy. Where it was possible, *métayage* (sharecropping) was introduced by local managers and owners, and lands exhausted after a century of exploitation went further downhill under the system, which persisted on islands such as Tobago into the twentieth century. British colonial records contain many references to negotiations between governors bent on importing the 'best' type of labour. This process went on well into the second half of the twentieth century among plantation owners who sought Grenadians above all others as overseers on their estates. Barbadians, who in the first half of the nineteenth century were regarded as scoundrels and thieves when exported to other islands, have since developed a reputation as the finest engineers, agricultural officers and economic planners throughout the British Caribbean, and are among the most reliable bus drivers and conductors in the London Transport Authority.

Unwillingness on the part of black West Indians to enter plantation life as paid labourers led directly to the introduction of East Indians to the Caribbean. Between 1838 and 1917 300,000 of these indentured men and their families came to British Guiana, Trinidad, Jamaica, the Lesser Antilles and the French islands. Their effect was demographically significant in British Guiana and Trinidad only (*Table 3.2*).

Table 3.2 Introduction of indentured East Indians into the Caribbean islands 1838–1917

Trinidad	143,900	St Lucia	4,400
Guadeloupe and Martinique	78,600	St Vincent	2,500
Jamaica	36,400	St Croix	351
Grenada	5,900	St Kitts	300

Another source of indentured labour was found among liberated Africans freed from captured slavers in the Atlantic Ocean. These men, unable to return to their homeland, were set ashore in Sierra Leone and St Helena, and then dispatched to the West Indies as paid workers for a period of 3 years, followed by a grant of free land (*Table 3.3*). The scheme freed the British government from embarrassment, but failed to increase the labour supplies of the West Indian islands to any marked degree.

Table 3.3 Released African slaves indentured in the British West Indies 1849–53

Jamaica	10,000	St Lucia	730
Trinidad	8,390	Tobago	650
Grenada	1,540	St Kitts	460
St Vincent	1,040		

Population and demographic changes

There are today probably 26 million people living in the Caribbean islands, which comprise some 250,000 km, and the three territories of Cuba, Haiti and the Dominican Republic, with 85 per cent of the total area, contained in 1976 between 16 and 18 million of them. The remaining 7–8 million were distributed among the many islands and islets that ring the Caribbean Sea (*Table 3.4*).

Except for the more obvious cases it would be difficult to break down the total population into categories by race or colour. In the Hispanic and Dutch islands many more white persons are found than black. In the British and Francophone islands, black populations predominate and probably less than 5 per cent of these populations would be considered white. In all the islands, however, there is abundant evidence of considerable intermixture, giving rise to so-called mulatto or creole communities of various shades.

Differentiation of the islands' populations in terms of race is significant only because it serves to distinguish certain marked demographic differences that emerge between one island and another. Experts assert that the 'British' islands owe their relatively low growth-rate to the various facets of black slavery and what it implied in terms of reproductive capacity. Cuba and Puerto Rico, largely Spanish in origin, showed growth-rates that increased five or sixfold between 1834 and 1955, whereas the British islands showed only a fourfold increase.

The most pathetic assemblages by race in the modern Caribbean are those who, by rejecting contact with black communities, maintained their 'whiteness' in isolation through exclusive inbreeding. Many of these groups have a long lineage, and can boast that their remote ancestors were among the first white men in the area. Thus the 'Redlegs' of Barbados, Grenada and St Kitts are still found today in scattered communities, sadly degenerate, living in poverty, degraded in intelligence and wits, without prospect of improvement unless a bolder member breaks away to marry 'outside', often to his and his children's great advantage. There are many such relict communities – the fishing folk of Petit Martinique, the Dutch farmers of Saba and St Martin, the French on St Thomas, the English of the Grand Caymans, and the German peasants of Seaford, Jamaica.

Marked changes in individual patterns of growth are also noteworthy where there was immigration of alien groups such as the East Indian indentured labourers. These increased their numbers very rapidly because the laws covering their entry to Trinidad insisted that women should be included in the parties of new arrivals. That this was not always possible is reflected in the marked sex-ratio imbalance to be seen on occasion, but for Trinidad it brought about a tenfold expansion in total population between 1844 and 1955. The gap between East Indian and black is thus rapidly narrowing.

Immigration on a massive scale from the Mediterranean countries of Europe (particularly Spain) has also been responsible for the continued predominance of white people in Cuba and Puerto Rico. Whereas Cuba in 1841 had some 590,000 blacks and 420,000 whites, it had 1.4 million whites in 1907, 3.55 million in 1943 and an estimated 6.5 million in 1965 (out of a total of 7.8 million). Similarly Puerto Rico, with only a slight fraction of its present population of black origin, has shown a remarkable capacity for demographic growth.

Table 3.4 Population characteristics of the Caribbean islands

	1	2	3	4	5	6	7	8	9
Cuba	9,824	8,569.1	114,524	86	1.2	18	6	25 (77)	70 (77)
Dominican Rep.	5,551	4,284.0	48,734	114	2.5	36	9	96 (65–70)	55 (65–70)
Haiti	5,670	4,330.0*	27,750	204	2.4	42	16	150 (65–70)	46 (65–70)
Puerto Rico	3,395	2,712.0	8,897	382	1.1	23	6	20 (77)	74 (76)
Jamaica	2,215	1,938.0	10,991	201	1.2	27	6	17 (78)	67 (70)
Trinidad and Tobago	1,150	1,027.0	5,128	224	1.5	25	7	29 (78)	67 (78)
Guadeloupe	312	324.5*	1,779	175	−1.3	18	7	29 (73–78)	65 (63–67)
Martinique	310	324.8*	1,100	282	−1.3	16	7	25 (73–78)	65 (63–67)
Barbados	279	238.7	431	647	2.3	16	7	27 (78)	69 (69–71)
Neths Antilles	240	223.0*	961	250	0.9	29	7	24 (73)	62 (66–70)
Bahamas	236	168.8	13,935	17	3.6	25	5	25 (76)	66 (69–71)
St Lucia	121	100.9	616	196	1.6	32	7	27 (76)	57 (59–61)
St Vincent	111	87.3	388	287	2.9	33	10	56 (77)	59 (59–61)
Grenada	105	93.9	344	307	1.1	25	8	29 (78)	63 (59–61)
US Virgin Islands	99	62.5	344	287	2.9	26	4	27 (76)	68 (70)
Dominica	78	70.5	751	104	0.7	22	6	20 (78)	58 (59–62)
Antigua	74	65.5	442	168	1.3	20	7	24 (77)	62 (59–61)
St Kitts–Nevis–Anguilla	57	45.6	357	160	0.8	24	10	42 (77)	64 (69–71)
Cayman Islands	17	16.7*	259	64	3.9	16	3	18 (74)	n.a.
Br. Virgin Islands	12	9.8	153	80	1.7	18	6	38 (68–72)	70 (68–72)
Montserrat	11	11.7	98	110	−1.1	18	13	47 (78)	68 (68–72)
Turks and Caicos	7	5.6	430	15	1.9	26	9	43 (70–75)	n.a.

Identity of columns:
1 Estimated population, '000, mid-1979.
2 Population at latest census, '000; adjusted population given where appropriate. Dates are for 1970 with the following exceptions: Neths Antilles and Haiti, 1971; Guadeloupe and Martinique, 1974; Cayman Islands, 1979.
3 Area, km².
4 Density of population/km².
5 Annual rate of growth, 1978.
6 Crude birth-rate per 1000, 1978.
7 Crude death-rate per 1000, 1978.
8 Infant mortality-rate per 1000, date given.
9 Expectancy of life at birth in years, date given.

Source: World Population, 1979, US Department of Commerce.
Note: *See note for column 2.

Generally, the Caribbean island populations have increased according to the conventional demographic processes. They have reacted quickly to health and sanitary measures which were introduced by the United States into the areas taken over at the end of the nineteenth century, and by the British and French colonial authorities in the twentieth century. Malaria, yellow fever and other diseases were reduced well below danger level, with consequent beneficial effects on infant mortality, death-rates and, in some instances, the birth-rate.

Though food intake is considerably higher than in most undeveloped countries of the world, malnutrition remains everywhere a barrier to totally healthy Caribbean populations. Furthermore, the ever-present risk of resurgent diseases in such densely populated islands is very great. The pernicious spread of dengue fever in rural Puerto Rico is indeed serious, as are the occasional outbreaks of malaria in islands that disregard preventive programmes. Water-borne schistosomiasis has also been allowed to spread from one island to another.

Distribution of population

Any map of settlement, however inadequate, shows that most islands have a widely scattered distribution pattern not always easily explained. Such patterns may generally be classified as follows: (a) Along spurs and ridges in the interior, above the oppressive heat of the deep valleys also associated with malarial infection. (b) Along the sea coast, strung out in long, discontinuous settlements. (c) In nucleated villages well away from old plantations. (d) Near or on plantation lands and sugar factories. (e) In towns, varying considerably in size and function.

Urbanization

In the islands, as elsewhere in the world, it is the town that has attracted large segments of the population in the twentieth century. No Caribbean island is without its primate city. Usually sited on the leeward coast, safe from the brisk trade winds and the full force of the hurricane, the city nestles snugly along the shores of a deep indented bay, protected on either side by steeply sloping hillsides. In the days of sailing ships these were ideal places for a port or anchorage for unloading cargoes and loading sugar, rum and spices, and where men-of-war could lie unseen by enemy squadrons at sea. The commanding hills above were generally occupied by forts and gunsites in case the enemy came too close. Much of the central part of such a town, usually the island's capital, was occupied by government offices.

It is apparent to the observer that such primate cities or capital towns suffer great disadvantages in the twentieth century. Growth in the capital automatically means greater concentration on the coastal flats. Expansion of the city core is possible thereafter only by an uphill climb or a series of paternoster satellites along the coast. Industry must find its way to alluvial flats well outside the town, and international airports are perforce located at great distances from the city, usually reached only after a long taxi ride across the mountain divide. Thus, the capital cities of Tobago (Scarborough), Grenada (St Georges), St Vincent (Kingstown), St Lucia (Castries) and Dominica (Roseau) are today confined by historical circumstances. Even the secondary towns established two centuries ago to provide an outport, maintain contact with the capital by sea rather than by time-consuming and unreliable roads carved out of steep, coastal hillsides. In most of the islands it is quite apparent that new capitals or primate cities should be established in order to meet the requirements of a modern city which include accessibility, the opportunity to expand and the ability to absorb a burgeoning population.

The many towns of the Hispanic Caribbean were deliberately created on the assumption that the landscapes were merely an extension of Spain. Because each town is fixed in its position by historical circumstances, many of the natural changes that might have occurred were inhibited. What saved them was the capital design embodied in the original plans, allowing for continuous outward expansion and growth. It is interesting to note that the seven cities established in Cuba are still ranked in the same order as they were in the sixteenth century.

Regardless of their origins, all Caribbean towns and cities have suffered the common fate of the world's urban areas: growth, industrial expansion, unrestricted immigration, shanty towns and, ultimately, submergence beneath a flood of humanity. Little is known of the structural changes that occurred in the earlier periods of urban growth in the islands. Most of the coastal port cities remained static after they had been established, wharfside and warehouse giving way to government offices, shops and trading houses. Such towns were not desirable areas for residence, except in the Hispanic islands where most of the original settlers were townsmen from Spain adapting themselves very quickly to tropical conditions.

Largely built of timber, most colonial Caribbean towns suffered periodic fires which swept through street after street. Until recent years no attempt was made to reduce such risks. Even a destructive earthquake in 1907 was not enough to re-create Kingston, Jamaica, although valiant attempts were made to carry through a commendable 'town plan'. Such plans exist in every island today, but there are few signs that much has been or can be accomplished in the way of modifying the haphazard patterns of historical growth. Castries (St Lucia) seen from above, is divided into two major sections, the old and the new. Only financial aid from the British government following devastating fires in 1948 and 1949 made a rebuilding programme possible.

It would also be difficult to say at what stage urbanization as a demographic process began to make itself felt. The early town life of most Caribbean islands revolved round the slave market, the trading shops and the port. Emancipation brought few changes, because ex-slaves opted for isolation and moved as far away from the town as possible into small villages, which in themselves were but nucleations of subsistence farmers. During the nineteenth century there was little in the town to interest most Caribbean people except the port, which offered the chance to emigrate from one island to another, to the United States, South America or Panama. Urbanization is thus a twentieth-century phenomenon, and its motivation is as little documented with certainty as similar movements in the rest of the world.

Disillusionment with 'peasant' farming, rural isolation and sheer poverty on the one hand, combined with the lure of bright lights and the possibilities of some kind of industrial work on the other – the so-called 'push-pull' effects mentioned by Clarke (1964) – all these are sufficient to explain the rapid rural depopulation that has occurred on every Caribbean island from the largest to the smallest. Cuba, well endowed with towns and cities, best illustrates the trend (*Table 3.5*).

Whereas in 1931 44 per cent of Cuba's population was to be found in urban concentrations of 2000 people or more, this had increased in 1953 to 51.4 per cent, and to 58 per cent in 1963. At present some 60 per cent of the Cuban population is considered to be urban. In 1963 there were twenty-six towns with more than 20,000 inhabitants. Towns of 10,000 to 20,000 in size increased in number from thirteen in 1931 to nineteen in 1953. The primate city, Havana, including its main suburbs, contained 1,861,500 in 1975, but its rate of growth has slowed. Among the demographic aims of the present government is the deliberate diversion of rural populations from the primate cities to decentralized growth areas, with all the attractions of the city but few of its disadvantages.

Similar urban patterns may be observed in most of the remaining islands. The population of Kingston, Jamaica, the largest city of the British Caribbean (480,000), has increased by at least 90 per cent since 1943. Some 25 per cent of the island's people now live in the capital. All the usual symptoms of rapid growth are present in this city whose present-day layout reflects the half-completed plans for its reconstruction after the 1907 earthquake. Its central business district is sadly cramped and quite inadequate, while at its western end a vast shanty town has mushroomed over many hundreds of square kilometres of valuable industrial/commercial land. Despite a remarkable public and private building programme in new suburbs and residential estates, there seems little prospect of stemming the tide of rural immigrants who take over shacks abandoned by those moving to a new house. Such rural folk account for about 50 per cent of the city's growth, the other 50 per cent being attributable to natural growth.

Since 1943 there has also been an increase in the number of Jamaican settlements with more than 1000 inhabitants, from nineteen to thirty-six, reflecting once more the marked shift of population from rural areas into nucleated communities. In some instances such moves have resulted from deliberate governmental planning or from growth based on industry and mineral extraction. Urban areas in Jamaica are growing at the cost of the rural economy, and in the present state of industrialization in towns such as Kingston, there is little prospect of absorbing more than a small fraction of the immigrants – a feature that characterizes all Latin America.

Despite its remarkable growth, Trinidad does not suffer from population pressures. Sugar and petroleum provided an early outlet for surplus labour, while farming land was available for all who sought it. None the less, the island's larger towns, Port-of-Spain, San Fernando and Arima, have all increased their populations rapidly enough to suggest that incoming rural migrants are responsible for a large proportion of urban growth. Since Tobago is a ward of Trinidad, there was also for many years a strong flow of black families from this island 'paradise' to the oilfields, United States military bases and the factories. As a consequence there are more Tobagonians in Trinidad than in Tobago itself.

A racial imbalance is also created by the fact that the East Indian section of the community is stabilized in its small towns and villages close to the flat ricefields, sugarlands and horticultural lots, resulting in the virtual exclusion of most blacks

Table 3.5 Growth in population of Cuban cities, 1899–1968

	Province	1899	1919	1943	1953	1964	1968
Havana (metropolitan)	Havana	247,000	408,700	868,400	1,139,500	1,517,700	1,577,300
Santiago de Cuba	Oriente	43,000	62,000	118,200	189,200	231,000	259,000
Camaguey (Puerto Principe)	Camaguey	25,000	42,000	80,500	129,500	153,100	178,600
Guantanamo	Oriente	7,100	14,800	42,400	76,700	122,400	135,100
Santa Clara	Las Villas	13,800	21,700	53,900	83,200	120,600	137,700
Cienfuegos	Las Villas	30,000	37,300	52,900	62,700	78,700	91,800
Manzanillo	Oriente	14,500	22,300	46,300	51,100	78,000	91,200
Holguin	Oricate	6,000	13,800	35,900	68,300	77,700	100,500
Matanzas	Matanzas	34,400	41,600	54,800	72,900	75,500	84,100
Pinar del Rio	Pinar del Rio	8,900	13,800	26,200	43,000	66,700	67,600
Cardenas	Matanzas	22,000	27,500	37,000	41,200	57,200	67,400
Sancti Spiritus	Las Villas	12,700	23,600	28,200	44,900	55,400	62,500
Ciego de Avila	Camaguey	3,000	16,400	23,800	40,700	51,000	54,700

who fail to meet the high farming standards of the East Indian smallholder. With plantation or estate labour as his only alternative, the black Trinidadian prefers the uncertainty of urban poverty to rural helotry, and soon becomes a landless resident of Port-of-Spain.

Apart from several persistent patches of shanty town on the outskirts, Port-of-Spain (pop: 67,000) has not suffered the indignities of overcrowding, mainly because its Hispanic ancestry provided the original, generously laid-out open expanses that exist today. Furthermore, residential and industrial expansion has taken place eastward along the foot of the northern range, creating a coalescing, linear conurbation of smaller towns. San Fernando has found expansion possible along the central ridge of the island. Generally speaking, careful planning, adequate water and energy supplies, together with a fair degree of industrialization, have made Trinidad's problems seem slight compared with those of other Caribbean cities.

Similarly, such towns as Willemstad in Curaçao and Orangestad in Aruba appear to be tropical copies of small seaports in Holland, under complete planning control and developing in an orderly fashion, thus disguising their phenomenal expansion in recent years.

Few people care to define the geographical limits of Bridgetown, Barbados. With an estimated island population of 248,000, a widespread network of roads, houses and schools makes it difficult to discern where town begins and country ends. The observer can merely reflect upon a contradictory situation by comparing the astonishing degree of overcrowding with the obvious socio-political stability of the island.

Of all Caribbean cities none is more remarkable for its growth-rate than San Juan, Puerto Rico. Neglected by the Spanish authorities until the nineteenth century, it retained within its defensive walls a small community of some 35,000 (1900). United States intervention in 1897 brought few changes to the city itself, although along the banks of the rivers and lagoons to the south the first of the shack-dwellers began to congregate. The Puerto Rican sugar industry tied most rural workers to the land during much of the first half of the century, and only with internal, political and social change did rural depopulation begin.

For a variety of reasons San Juan was by 1950 one of the fastest growing cities in the Caribbean. Its primacy on the island was undisputed. Its boundaries expanded rapidly, absorbing many neighbouring suburbs and small towns (e.g. Bayamon, Rio Piedras) until metropolitan San Juan occupied some 445 km² with a population in 1976 of 750,000, representing 38 per cent of the island's inhabitants. This conurbation has extended as far as Caguas, some 40 km to the south. An astounding industrial expansion and tourist trade, a steady rise in living standards, and immigration, have all led to a characteristically North American type of suburban sprawl, which has enveloped large areas of old wooden houses and shanties in low-lying parts of the city itself. These are occupied by families who earn fair incomes, but seem content to remain with such communities. If vacated and left standing, these shacks are immediately occupied by incoming rural migrants.

There are several other cities in the island, of which Ponce (160,000) on the southern coast is second in rank. A decision has been made to divert industry and port facilities to this very attractive town in order to remove unhealthy pressures on San Juan. Most other towns, scattered across the island, are today little affected by the rural immigrants who once used them as a temporary staging post on their way to San Juan or New York. Most of these prefer nowadays to make the journey directly, even by-passing the capital. Such towns with market and transport functions, have lately shown a decline, though they were at first stimulated by factories set up under a programme designed to industrialize the island, known as 'Operation Bootstrap' (see pp. 126–7).

Migration

Total destruction of the Amerindian tribes of the Caribbean islands meant that all succeeding populations were immigrants from Europe, North America and Asia on the one hand and from Africa on the other. The former came voluntarily, while the latter were imported as slaves. From the earliest period of settlement onward, European colonists exercised great mobility. For most sixteenth- and seventeenth-century Spanish immigrants, the three Hispanic islands were but staging points on the way to the Central American mainland, and permanent settlements were sparsely populated until the eighteenth century.

Under the Proprietary Patents of the seventeenth century, British islands such as Barbados and St Kitts attracted freemen and indentured servants from England in large numbers, but with the growth of the sugar plantation system and consequent consolidation of land, some 30,000 white smallholders left Barbados for North America or other English islands. They moved to any island that was advertising vacant land. Thus St Kitts became the focus of attention in 1714, while the British Ceded Islands attracted so many men away from large estates in Jamaica and Antigua between 1764 and 1775 that complaints were registered in the British parliament by the plantocracy. Similarly, Frenchmen moved between islands such as St Kitts, Martinique, Guadeloupe and Grenada, though barred from English islands until 1763 if they were Catholics. In an attempt to win over the French Catholic populations of Grenada, St Vincent and Dominica, the English government, by the Treaty of Paris in 1763, adopted a policy of religious toleration in these islands.

European wars occasionally impinged upon inter-island migration, when whole populations were suddenly transferred by treaty to a rival nation. Evacuation programmes seldom lasted long enough before the onset of another war, and in the latter part of the eighteenth century the Lesser Antillean islands switched loyalties on a number of occasions, so that the local communities learned to disregard these external pressures.

European emigration to the Caribbean declined sharply in the early years of the nineteenth century when sugar-cane farming ceased to yield quick profits, slaves became expensive and English colonies in other parts of the world were beginning to attract investment, capital and potential settlers. White populations in the British islands, already static, fell away sharply in the nineteenth and twentieth centuries. Those who remain today are traders, shipping agents, bankers, government officials and some retired people.

The Hispanic islands, in strong contrast, received their major population boost throughout the nineteenth century, principally from the mother country and the Mediterranean fringe. Cuba, the Dominican Republic and Puerto Rico therefore surged ahead of their black populations, and are today dominantly European. The Dominican Republic has since 1938 encouraged colonization of its western frontier by central European Jewish refugees, Hungarians and Japanese.

Cuba, the largest of the Caribbean islands, presents an excellent example of the variety of influences that affected population growth. In the closing years of the eighteenth century and the beginning of the nineteenth the island was in a fair position to receive any immigrants who cared to come. Thus, 30,000 persons made their way from Santo Domingo as a result of the instability following the negro insurrection there. Large numbers of Spanish people also left New Orleans when Louisiana was transferred to France in 1803.

Recognizing the need for increased European immigration, the Spanish government in 1817 issued a decree encouraging foreigners to own land in Cuba. Steps were taken to provide transport costs and a monthly stipend for 6 months for anyone wishing to settle in newly founded towns in Cuba. Cienfuegos was thus established in 1819 on the basis of a French immigrant group. The towns of Nuevitas, Guantanamo, Nipe Banes and Santo Domingo were all founded about this time. Cuba was also fortunate in its ability to accept a large influx of Spanish-speaking peoples from the South American mainland, fleeing from the consequences of various independence movements that were sweeping the continent.

On the other hand the slave trade, largely concentrated in Cuba in the first half of the nineteenth century, resulted in an influx of some 387,000 black slaves between 1800 and 1865, despite the abolition laws introduced by the Spanish government in 1845. These regulations sought also to encourage the migration of a non-black population to offset the effects of so large a black influx. Yucatán Indians were deliberately invited to take up residence in Cuba, and a campaign to encourage Chinese to enter the island resulted in a total influx between 1847 and 1860 of some 48,000 Chinese immigrants (excluding about 8000 who died en route). This campaign was brought to a halt in 1860 by Governor Francisco Serrano in view of the rapid local population growth, and only a sporadic trickle of Spaniards was allowed during the next 40 years. In all, some 1,260,000 immigrants entered Cuba between 1902 and 1931.

Table 3.6 Immigration into Cuba 1902–19

Europe		America	
Spain	436,005	Jamaica	50,368
Denmark	6,372	Haiti	39,906
France	3,213	USA	44,054
Turkey	1,219	Antilles (non-Spanish)	24,976
England	1,013	Mexico	19,621
Germany	636	Puerto Rico	13,631
Italy	760	Central America	4,240
Portugal	108	Panama	4,154
Other countries	838	Other countries	7,420
Totals	450,164		208,370

In the 10 years after the First World War the sugar companies sought cheap labour from Jamaica, Haiti and the Lesser Antilles. But immigration to Cuba virtually came to an end in 1933 when, as a result of the world depression and a fall in sugar prices, the so-called '50 per cent Law' required at least half of any payroll to be Cuban. Black West Indian labourers from neighbouring islands returned home, and many Spanish families sailed back to Spain or migrated further afield to Venezuela or the United States. But there were still some 15,000 immigrants from Haiti in 1959.

Movement to and from the United States was a comparatively simple matter throughout the next 25 years, but with the Socialist Revolution of 1959 migrants became refugees who by various means had fled the island. By 1968 at least 350,000 Cubans had found clandestine or official haven in Miami, where they have prospered and prepared themselves for an improbable 'Day of Judgement'.

No more dramatic instance of the influence of migration upon a socio-economic situation exists than that of modern Puerto Rico. This island, dominated by sugar cane, was about to receive substantial devolution from Spain in 1897 when it was seized by force of arms and became a United States territory. An important consequence of this event in the evolution of the Commonwealth of Puerto Rico was that the island became part of the domestic United States, permitting free movement of peoples to the mainland. Puerto Rican migrant farm workers, who played an important role in the harvesting of various crops in all parts of the United States, were able to move freely about the country, and increasing shortages of labour in the service and textile industries of New York enabled them to take up permanent employment there in the late 1930s. Without this safety valve, Puerto Rico would long since have had to face a population explosion.

Benefiting from United States health services and improvement of local conditions, the island saw its infant mortality-rate fall remarkably, its death-rate decline, and its life-expectancy rise to a value close to that of New York, within the brief space of 30 years. The 'Operation Bootstrap' programme was able to restrict the unemployment rate only to 14 to 18 per cent between 1952 and 1978. Exporting large fractions of its population to New York and other United States cities over the same period, has prevented a serious socio-economic crisis from developing in Puerto Rico. Improved education facilities in the island, together with social security pensions provided for workers by the United States, have brought about a reverse flow of immigration among both the older generation and the young people. The outward movement of young adults, nevertheless, goes on, providing valuable technical experience for those who will eventually return. Despite continuous industrial programmes designed to absorb the natural population increase and the incoming rural immigrants, it would appear that accelerated emigration to the United States must continue for many years to come, varying according to economic conditions.

For those who did not have the protective mantle of the United States there were other alternatives. Although West Indian ex-slaves might have rejected the indignities of the plantation, they were quick to accept the idea of manual labour in other parts of the Caribbean. During 8 years of canal construction in the isthmus of Panama from 1881 to 1888, the French employed 25,000 workers from Jamaica alone, despite high mortality-rates on the site. Most returned home to their island when the French company gave up for lack of funds. A second phase of Jamaican and Barbadian migrations occurred between 1905 and 1913 when the United States took up this abandoned project. Some 35,000 Jamaican workers were always present, though individuals constantly replaced themselves with relatives from Jamaica. The Canal Zone remained a popular work place for West Indians as late as the Second World War. Consequently there are today in this territory some 65,000 West Indians who keep their British citizenship and preserve many of their West Indian customs.

As early as 1887 Latin American countries such as Colombia, Nicaragua, Honduras, Guatemala and Costa Rica used Jamaicans and Barbadians for their vast banana and sugar plantations, until political pressures resulted in their exclusion in favour of local labour. Cuba and Haiti also accepted Jamaican labourers.

The opening of the Venezuelan oilfields in 1916, inaccessible across the shallows of the Gulf of Maracaibo to all but small tankers, resulted in the establishment of off-shore refineries in Curaçao (1916) and later in Aruba (1925). Some 10,000 West Indians from Trinidad, Barbados and Curaçao itself were regularly employed on the Venezuelan oilfields until 1929, when restrictions were once more applied after local objections to foreign workers had been voiced. Both Curaçao and Aruba employed some 10,000 workers coming from as far afield as the Windwards (St Lucia) and the Leewards (St Kitts). This figure declined after the Second World War, and with the introduction of increased automation in the 1960s not only has the total labour force on Curaçao and Aruba declined, but West Indians have been required to return to their homes.

The United States has received large numbers of British West Indians who were able to enter on unfilled British immigration quotas until 1952, and many of these have been absorbed into continental black communities. Many have become prominent in the professions and in politics. During the Second World War about 120,000 West Indians were allowed to enter the United States to remedy a manpower shortage in agriculture and industry. Jamaican workers continued to flow in during the 1940s and early 1950s, but have now been severely restricted in numbers. Some 9000 continue to cut the sugar-cane harvest in south Florida. Since the new quota laws, most British West Indians by-pass the United States and take up residence in Canada, where their citizenship is not a barrier. Other United States possessions, such as St Thomas and St Croix (where important oil and aluminium plants are being erected), are the target not only of contracted migrant labour from St Kitts and the British Virgin Islands, but of men and women seeking illegal entry.

Increasing restrictions on the permanent movement of people in all islands of the Caribbean has become the norm since the Second World War. Trinidad introduced such legislation as early as 1942 when the island became an important military base, attracting large numbers of migrants from the Lesser Antilles. Permanent entry to any island, from Trinidad northwards, in order to seek employment, is virtually impossible today.

The remaining outlet for British West Indians was, for those who could endure the climatic transition, the British Isles. The cost of a sea passage was the only barrier to free movement. After the Second World War many West Indians with experience of continuous employment in many areas sought economic alternatives to poverty at home by migrating to the United Kingdom, where thousands of unskilled jobs were available.

From 1950 onwards migrants from Jamaica, Trinidad and the Lesser Antilles began to arrive in ever-increasing numbers, carried in Italian and Spanish ships that offered cheap passages to European ports after they had unloaded their cargoes of European emigrants to Hispanic America. Charter flights, particularly from Jamaica, accelerated this flow.

For various reasons accurate statistics of in- and out-migration were not kept in sufficient detail for the United Kingdom authorities to measure this West Indian immigration pattern. Census records of 1951 reveal some 15,301 UK residents who were born in the British West Indian islands. The 1961 census recorded 171,796 in the same category. This was a measure of the rapid increase that took place in 10 years, all but a tiny fraction being black rather than white and estimates of the present West Indian population vary considerably (*Table 3.7*).

Political pressures within the United Kingdom brought about the 1962 Commonwealth Immigration Act, severely restricting all in-migration from the British Commonwealth. Yet some immigrants are highly desirable in the eyes of the United Kingdom. Barbados, with its carefully regulated, long-standing emigration programme, continues to supply transport workers and nurses to London and other cities.

Table 3.7 Immigration of British West Indians into the UK 1951–61

Year	Number	Year	Number
1951	1,000 (approx.)	1957	22,473
1952	2,000	1958	16,511
1954	10,000	1959	20,397
1955	24,473	1960	45,706
1956	26,441	1961	66,260

The classical Caribbean preference for metropolitan Europe was displaced by a surge of refugees to the United States in the 1970s on the grounds of political and economic oppression. Between 1972 and 1981 an estimated 40,000 to 51,000 illegal immigrants from Haiti were added to 41,670 legal entrants, most pleading 'economic oppression'. Only in 1981 did the United States decide to reduce the flow through air and sea surveillance. It is not unlikely that South Florida has today become the home of some 100,000 ex-Haitians.

A short-lived period of free migration from Cuba to the United States in 1965 was unexpectedly repeated in 1980, when the Cuban government allowed exiled relatives to collect from the port of Mariel (west of Havana) any Cubans 'ideologically in disagreement with the revolution and socialism'. Between 21 April and 26 September 1980 some 123,000 islanders were carried in small craft to South Florida to be absorbed among the 500,000 ex-Cubans already living there. Without the deliberate intervention of the United States government in June 1980, the total entry could well have doubled.

Population control

For the islands the imposition of immigration restrictions has forced government leaders to re-examine their population problem as one that can no longer be exported, but must be solved locally. One solution lies in family planning, and birth-control campaigns continue to reduce the birth-rate, which has remained fairly constant throughout the Caribbean, while commendable reductions in death-rates and infant mortality-rates have made even a marked decline in the birth-rate unimportant. Religious attitudes on the one hand, particularly in the Hispanic islands, do not favour such control programmes, while on the other hand *machismo* acts as a deterrent to such programmes throughout the Caribbean. A further factor is the custom of permissive cohabitation, probably a relic of slavery, leading to an illegitimacy-rate as high as 40 per cent in some communities.

The Puerto Rican authorities, faced with overwhelming population growth, have undertaken an oral contraceptive programme, resulting in an undramatic but steady decline of the birth-rate from 39 per thousand in 1940 to 23.5 per thousand by 1976. Voluntary sterilization operations for men and women are also carried out as a matter of routine. The island's population, however, continues to grow alarmingly despite large-scale emigration to the United States, and cannot lead to any optimistic long-term view of the problem in this part of the Caribbean.

Barbados introduced a family-planning programme in May 1955, following a long soul-searching among its people, religious leaders and legislators. It has been made gradually acceptable throughout the island, and appears to all but a few to be at least a modest contribution to the solution of a major problem. It is impossible to visualize a similarly successful campaign and result on an island such as St Lucia, where literacy, education and motivation are so far below those of Barbados. A modest programme has also been introduced in Grenada.

Trinidad, with a nominal Roman Catholic majority and a reputation for being underpopulated, in 1967 reversed its policies, seeing the high birth-rate as a 'threat to economic and social well-being'. The programme was designed to reduce the birth-rate from 38 to an unrealistic 19 per thousand in 10 years, a decline that could be rivalled by very few countries in the developing world. In 1976 it had fallen to 23 per thousand.

In his perceptive study of Jamaica's population Eyre (1972) listed the following options applicable to all Caribbean islands seeking to control population growth: (a) increased primary production, (b) extension of utilized land, (c) increased non-primary production, (d) external migration, (e) fertility control and (f) reduction in *per capita* levels of living. All have been tried and found inadequate.

Prospects and projections

No matter what political philosophies or governments prevail in the Caribbean, it is clear that the larger island communities such as Cuba, Haiti, the Dominican Republic, Jamaica and Puerto Rico can offer no ideal solution to the problems associated with population growth. Overpopulation, as it is understood in some Latin American or Asian countries, is absent, i.e. there is land enough (except in Haiti and possibly in Jamaica) for those who wish to grow their own food, though without any prospect of raising standards of living. For Haiti, only the most optimistic can envisage anything but a totally Malthusian solution in an island lacking any kind of resource, including soil productivity.

The small islands of the Caribbean, with so much of their land consisting of steeply sloping hillsides, or lacking vital water supplies, appear to have reached a point of no return. Unless their inhabitants can miraculously become terrace-agriculturists, such lands cannot support expanding populations. It is, of course, possible to advocate a policy of robber-economy during the next generation, by making available all forest reserves and unalienated lands for cultivation, and encouraging the total destruction of each island landscape – much as Haiti has been reduced to its bare bones. This would lead inevitably to a complete break-down in the social structure and economic life of each and every island.

Yet throughout the islands, except in present-day Cuba, the trend is one of drift to the town after the abandonment of plantations, farms and small-holdings. Manufacturing and industrial expansion has already fallen behind in the task of finding jobs for the present generation of immigrants. The alternative – sending people back to their rural plots – has rarely succeeded in any country, and, furthermore, requires a degree of compulsion which the Caribbean would not tolerate. By 1990 the present total population of these islands will have doubled itself, even under the most optimistic assumptions of slower growth. Unless a dramatic change in attitude towards population control occurs, there can be little to comfort a student of the Caribbean.

Agriculture and mineral resources

Cash crops
Socio-economic patterns in the Caribbean islands, as elsewhere in the tropics, have always been closely linked with cash crops produced for the temperate world. Beginning with tobacco in the seventeenth century, eventually cultivated to excess, there followed a succession of plantation crops whose popularity in Europe waxed and waned in accordance with the potential of other new countries discovered and exploited as colonies.

Tobacco remains a staple crop of Cuba and Puerto Rico in particular, mainly cigar leaf. Other islands have tried repeatedly to grow a cigarette tobacco commercially (where it might have been an import substitute) in order to supply a local factory, but there appears to be little enthusiasm for a crop requiring delicate handling throughout its growing life.

In the two centuries before the American South dominated the European market, Sea Island cotton became an important crop in the Caribbean islands wherever it would grow, and remained a vital crop there with a very high sales value in London as late as 1795. Only a few cotton areas survive, specifically in Antigua where about 500 tonnes are produced annually from a local ginnery. Other islands such as Barbuda, St Kitts, Montserrat, St Vincent and Haiti are known to have grown cotton, but extensive droughts have tilted the balance and caused growers to abandon the crop. The destruction by fire of St Vincent's cotton ginnery in 1959 brought cultivation to an immediate standstill, and as no attempt has been made to rebuild the ginnery, cotton has ceased to be a cash crop.

A sharp transition to the monoculture of sugar cane in the mid-seventeenth century drove most other cash crops out of the islands. It is not difficult to follow the effects: consolidation of landholdings, slavery, wholly inefficient production and eventually bankruptcy in the nineteenth century, for most of the sugar planters. Yet a major resurgence of the industry in Cuba, Puerto Rico, the Dominican Republic, Guadeloupe, Jamaica and Trinidad in the latter half of the century, and the persistence of the crop into the present day in Barbados and St Kitts, are a measure of the power of capital investment in tropical production. Elsewhere, as in Puerto Rico, St Lucia, Grenada and Antigua, commercial sugar cane has disappeared or is in process of disintegrating owing to rising labour costs. In other islands the sugar-cane plantation has been kept solvent by subsidy, guaranteed price arrangements and an international quota system devised by the United States. Without these aids much of the present-day industry would collapse in the face of sugar-beet production, as in Martinique, where local producers cannot compete with their competitors in the European Economic Community. Sugar cane has probably outstayed its usefulness on the Caribbean scene; it is certainly not a smallholder's crop, as many theorists would like it to be.

The vicissitudes of sugar-cane cultivation, together with its sensitivity to world prices, rising labour costs, hurricanes and drought, were already well known to the nineteenth-century plantation owner. Encouraged by enthusiastic governors, he planted various cash crops from one era to another in order to meet world demands. Sugar cane gave way to coconut groves and copra extraction. Wet hillsides were given over to cacao, long a popular crop in the French and Spanish islands, and coffee, well established in Cuba, Haiti and Puerto Rico, spread as far south as Trinidad. Dominica became a lime-growing and processing centre; Grenada and others cultivated the nutmeg tree; and St Vincent mono-polized the world's arrowroot production. Aruba contributed the bitter aloe to the world's pharmacopoeia until it was displaced by synthetics. Citrus and pine-apple have appeared in Jamaica, Trinidad, Puerto Rico and Martinique in recent years.

News of successes in one or other of these many crops spread from one island to another in colonial dispatches during the nineteenth century, and in the twentieth many agricultural departments and plantation owners turned from one to another when world markets turned fickle or plant diseases devastated an optimist's new and carefully established crop.

One of the most important crop innovations in the Caribbean islands was the banana (*Musa sapientium*). Introduced into Jamaica in the early days of commercial exploitation in the late nineteenth century, it has played a vital role in many island economies, both as a plantation and as a smallholding crop, and in the Windward Islands could be called the only viable cash crop of consequence. Despite such disasters as Panama disease, Sigatoka (leaf spot), frequent hurricanes, rains and droughts, the banana has been bred in size and resistance to suit the markets of western Europe and North America. Transportation by sea, rail and road has been superbly organized by several companies, and marketing arrangements are carefully managed from the moment of harvest to purchase in the shop. Here at least is one tropical crop that has succeeded in providing a reasonable living for all those who handle it, and will continue to do so until economic depression makes it a luxury once more, and until the ravaged hillsides of the islands can no longer support its roots.

Food crops

For those pioneers who landed upon Hispañola and Cuba with Columbus, the Caribbean islands were not a particularly promising source of food. Accustomed to Mediterranean variety, they were discouraged by their inability to transfer wheat, temperate fruit and vegetables to the tropics. Such native foods as cassava failed to satisfy the colonists, who were forced to import their own requirements. This situation persisted until a time in the Second World War, during the German submarine campaign, when the entire island fringe faced starvation because so many Allied supply ships were being sunk – a measure of the increasing dependence of the islands on sophisticated European foods.

From the earliest days of slavery plantation-owners allowed the cultivation of sugar 'banks' for food crops and the care of livestock near the slave barracks, expecting not only enough for the plantation, but a surplus for sale in the local town market. Such privileges were well guarded and continue today to be part of the payment in kind for workers who commonly receive a plot of land on the less fertile part of the plantation, and have the right to tether and graze cows and goats on the owner's property. The chicken and pig remain an immediately available food source throughout the Caribbean.

Such local husbandry did not suffice, however. Salted cod, imported from the eastern seaboard of North America, and salt pork became the basic protein food for slaves (who seldom looked to the sea for their own fresh-fish supplies). Salted fish continues to be an important item in the average Caribbean diet, but must still be imported from the temperate north, at inflated prices.

With the African slaves came novel foods brought in by sea captains who observed what their human cargo was given to eat while awaiting transfer from West African stockades. Most were starchy tubers such as the many varieties of yam. Others were legumes such as the pigeon pea, millets and sorghums. Fruits such as ackee made cod dishes more palatable, while the mango offered an easy alternative to starvation in hard times, and continues to do so in modern Haiti. Other vegetables, such as okra, are now a standard item in island diets. Of the exotics introduced into the Caribbean, especially designed to reduce slave-food costs, the most important plant proved to be the breadfruit. Although introduced in 1776, it failed to impress the slaves immediately as a possible food, and was mostly fed to animals. None the less its spread throughout the Caribbean landscape, and the great demand for fruit in local markets, made the tree a vital part of any food garden.

Increasing rural depopulation throughout the area has automatically reduced food surpluses which could be sold in the marketplace, and has led to serious food deficiencies especially in the Lesser Antilles. Some enterprising islanders have developed a brisk inter-island trade by transporting breadfruit, plantains, yams and other food crops in schooners from Montserrat and Dominica to Antigua, the Virgin Islands, Trinidad and Barbados. Similarly, Venezuelan farmers bring boatloads of fruit and vegetables to Curaçao and Aruba, both islands being quite incapable of growing their own requirements despite a good hydroponics garden on Aruba.

Tourists on holiday in the Caribbean seldom eat the local foods. Every island seeking this trade must therefore allow considerable quantities of frozen, tinned or fresh food and Scotch whisky to be imported. Since these are also displayed in the local retail shop, they become part of an islander's regular purchases. Not all the vast increase in imports of flour and processed foods can be blamed on the increasing tourist trade. Tastes have actually changed in all the islands. Puerto Rico, for example, has almost abandoned its 'native' foods in favour of those being imported in ever-increasing quantities. These include fresh meats, dairy products, chicken, pork, fish, fruit and vegetables, including tropical tubers such as *malanga*, now grown near Miami. In the matter of chicken and beef production, aggressive competitors from the mainland have shown that it is possible to undercut the local market without any difficulty.

There is in fact no more vexing problem in the modern Caribbean than that of food production. Every island government since the mid-nineteenth century has entertained schemes for larger food harvests, assuming that peasants, given sufficient encouragement, would remain on their lands. Empty lands were divided into 2 ha- and 4 ha-holdings, and granted to the landless on condition that a certain proportion of all crops should be food. Barbados introduced a strict law requiring 15 per cent of all lands to be devoted to food crops. In the post-Second World War years economic devices such as marketing boards and subsidized or guaranteed minimum prices, accompanied by wide propaganda, have had mixed success.

Settlement schemes
Of greater interest to the geographer have been the innumerable settlement schemes initiated mainly by government departments of agriculture in order to overcome agrarian discontent, to raise local food production or to satisfy political promises. Others on a grander scale have aimed at the total rehabilitation of communities. Some 300 such schemes are known for the British islands alone. Most of these are found in Jamaica, an island suffering intense land hunger and population pressures. The Yallahs Valley Authority, now in its third year, was designed to restore eroded lands in Jamaica, to relocate communities and bring about an economic revival in peasant fortunes. That it has achieved some of its objectives is a tribute to the Authority and the inhabitants of the valley. Few others have succeeded except temporarily in the first flush of enthusiasm. Lack of capital, preparation and education are generally to blame when ultimate failure is analysed.

Land reforms followed by planned food production have been the chief preoccupation of the Cuban government since 1959. Forced to arrange its internal food supplies very carefully, including the rationing of all major items such as beef, milk, sweet potato and maize, the Cubans have placed greater emphasis on food farming. Egg production more than trebled between 1965 and 1978. Since Cuban farms vary from state farms based on the Russian *sovkhozes* to cooperatives and individual smallholdings, it is difficult to discern any patterns or trends that would indicate a greater success on the part of the Cubans in handling their food problems than anybody else in the Caribbean. For all their land reforms it is still the small (private) farmer who produces the bulk of the island's food.

Fishing

Since salt cod plays such an important role in all island diets, fresh fish should also be considered as part of the food cycle. Barbados is renowned for its 'flying fish' harvest, but elsewhere, until recently, the rowing boat, a seine net and co-operative hand-hauls within 450 m of the shore were the only evidence of a fishing industry. Pathetically small catches of undersized fish and a quick sale to local bystanders remain the current pattern. Training facilities initiated by the FAO and other organizations, together with improved equipment, have in St Lucia done little to increase fish supplies. It has been left to Japanese trawlers to station themselves in Trinidad and Puerto Rico and to set up a tuna-canning industry. Barbados and Trinidad have similarly become shrimping centres for Venezuelan concerns, and only in the Dominican Republic is there any sign of an active fishing industry. Cuba has, with the help of the Soviet Union, established large-scale trawling fleets, but these have been hampered by local consumer resistance to fish as a substitute for meat.

Livestock

Early Hispanic settlement involved a flourishing cattle industry, well adapted to the Caribbean climate and pastures. Most were scrub cattle and have remained so until recently, but they provided crudely butchered meat and small supplies of milk for local sale. Succeeding generations of veterinarians introduced their favourite breeds from India, Africa and Great Britain, resulting in a wide range of animal breeds and cross-breeds. A deliberate breeding programme carried out in recent years in Jamaica yielded two important types of animal suited to the Caribbean, and 'Jamaica Hope', a dairy cow, and 'Jamaica Red', a beef stock animal, are represented in all large herds today. At the same time Friesian dairy animals have been successfully adapted to Puerto Rico, Jamaica, Trinidad and Tobago and to Cuba.

More important to the smallholder are sheep and goats, which can survive under most difficult conditions. Large numbers of black-bellied Barbados sheep have been successfully exported to several islands. A wide variety of goats exists throughout the Caribbean and all are used for milk and flesh.

Mineral resources

The Spanish conquistadores in Hispañola thought for one brief moment in time that they had discovered a sure source of gold. With a few major exceptions, however, the lack of minerals of any kind throughout the Caribbean has meant that each community has been thrown back on its only resources – the land and the people.

There are, of course, a few deposits that offer a restricted profit. Limestone in abundance provides the raw material for cement, while in many islands certain clays offer a good base for a ceramics industry. Trinidad has a rich iron-ore lode in its northern range. Copper deposits in Puerto Rico and the Dominican Republic are now being worked experimentally.

Only the Greater Antilles contain minerals of any value to the outside world. Cuba, for example, is undoubtedly rich in minerals. There were some 287 mines in operation in Cuba in 1958: 68 manganese, 9 copper, 12 chrome, 6 iron, 4 pyrites, 2 lead, 1 silver, 1 tungsten, 3 zinc, 2 nickel, 1 cobalt, 4 barytes, 3 gold, 8 gypsum, 6 kaolin and 1 lignite. At that time 98 per cent of the island's produc-tion went to the United States, where technical skills made exploitation of some difficult ores possible. It has been estimated that some 3000 million tonnes of iron ore (consisting of limonite and haematite of varying grades) await exploita-tion in Cuba. Similarly, manganese, chrome, nickel, tungsten and cobalt – all regarded as strategic minerals in the Free World – are available in that island. Few are exploited at present, however, owing to lack of a suitable market. The rich iron ores of the Dominican Republic on the other hand, are exported to West Germany, while its nickel and bauxite go to the United States.

The most generously endowed of the Caribbean islands today is Jamaica, whose bauxite resources were among the greatest in the world. Five major aluminium companies are engaged not only in exploiting large areas of open-cast deposits, but in establishing concentration plants within the island. The govern-ment has now required all bauxite reserve lands to be handed over, and has acquired a major interest in all processing plants.

Petroleum in Trinidad is the only other exploitable mineral of any interest in the Caribbean, though small deposits of it have been developed in southern Cuba. The south-western corner of Trinidad was known for its asphalt lake as early as 1572, when Sir Walter Raleigh visited the island. This mineral has been extracted for many years from the same lake. From 1908 onwards a number of oil wells by the lake, as well as a number of later off-shore wells, have provided Trinidad with 50 per cent of its national income through three refineries set up to process the crude oil. The strategic value of these refineries has led to plant expansion far beyond local production capacity, resulting in the importation of crude to keep the refineries in full production.

Trinidad's oil industry not only provided a source of taxable revenue, but has enabled many thousands of workers to acquire a variety of mechanical skills invaluable in the island's industrialization programme. Since 1974 the Trinidad and Tobago government has acquired majority rights in all the local petroleum companies, and has not only established a wide-ranging petrochemical industry, but has extended its off-shore fields to major proportions. Natural and waste gases also provide thermal energy for power production and domestic uses. At the moment the only other island with oil pretensions is Barbados, where petro-leum has been located at great depths which are at present uneconomic. Natural gas has, however, been tapped, and provides a valuable source of heating on a fuel-hungry island.

Twentieth-century economic development

Economic theorists have commonly postulated that industrialization should be able to absorb the rural migrants and the growing urban populations of the Caribbean. Such a process is thought capable of bringing about a reduction in population pressure through a lowering of the birth-rate, associated with rising levels of living. It is natural, therefore, to expect island politicians, planners and economists to support industrial programmes, and to offer every incentive to manufacturers to establish themselves. Industrialization as a panacea for com-munity ills has been best developed in Puerto Rico, a model, too, for many other similar programmes in the Caribbean.

In Puerto Rico, in view of the lack of local resources, it was necessary to concentrate on labour-intensive industries producing goods that could be processed from imported raw materials, and exported to world markets. A wide range of textile manufactures, including men's shirts and women's underwear and corsetry, remain among the island's chief exports, but with a general rise in local technical skills, electronic components of all kinds are now being made for delivery to many parts of the United States, mainly by air freight. The most important developments in the past decade have been in the field of petro-chemicals, based on the by-products of two very large petroleum refineries erected in the island.

Most islands have been infected by industrialization fever. Throughout the Lesser Antilles and Virgin Islands, pioneer industrial legislation has been introduced and incentives, including the 'free port' principle, offered to those who are prepared to manufacture import-substitutes. The smaller islands can thus boast a brewery, a cold-drink plant, a cigarette factory, a margarine or coconut-oil plant, a mattress factory and a tyre-retreading plant. All have extended their incentives to the hotel and tourist industry, usually on generous terms. Antigua and St Croix are headquarters of petrol refineries and crude-oil bunker stores, while the latter island now has an aluminium plant. St Thomas is today the focus of a number of minor industries including watch-assembly lines. The ABC islands of the Netherlands Antilles, now associate members of the European Common Market, are likely under new regulations to become important centres of aluminium processing and certain other industries.

The French Antilles, administered as *départements* of France, have also sought to develop pioneer industries. Major interest is centred on Martinique, the more sophisticated island. Incentive industrial legislation is partly inhibited by metropolitan self-interest and partly by local preference for better-paid, white-collar jobs. Too many investors also wish to avoid the losses that followed the collapse of French hegemony in south-east Asia, and have in any event found more lucrative openings in francophone West Africa.

It is difficult to sum up the value of development programmes involving industrialization. They offer prestige and self-respect to local populations, who begin to feel themselves in the van of progress. They involve relatively few workers compared with the total population, but do cause more money to circulate. Industrialization as the culminating stage of an economy cycle to which mineral resources have contributed little or nothing, is at best precarious. For most Caribbean islands it is an act of defiance in the face of population pressures and poverty. Without massive emigration, a renewed interest in agriculture and the stimulus provided by packaged tourism, industrialization has very little to offer most islands.

Tourism

Tourism is today of vital importance to Caribbean island economies, if only because it brings in foreign exchange. Before 1950 the visitors were mainly members of the wealthy, closed society which could afford to leave home during the northern winter to enjoy sunshine in Cuba, north Jamaica, Antigua, the Virgin Islands and the Grenadines. This so-called 'jet set' gradually moved away to new and more distant paradises when the international hotel chains and the airline corporations, together with the travel agencies, popularized the Caribbean among American, Canadian and European middle-class holiday-makers in search of a glamorous, exotic island.

In those early days tourism was looked upon as part of the local development programme bringing in foreign capital, giving work to local people, improving or creating infrastructures and services such as airports, highways and reliable water and electricity supplies. Encouraged and supported by various inter-national agencies, each island offered tax and other pioneer incentives to any foreign investor with plans to build large hotel complexes.

The trade grew rapidly in Jamaica, Puerto Rico, Cuba (until 1959), Antigua, Barbados and the Bahamas, so that by 1977 some 4 million visitors (excluding cruise-ship transients) were recorded throughout the islands. It has taken some time for the local population to realize that these visitors are no longer the extra-vagant 'jet set', but a new cost-conscious generation bent on getting value for money.

The benefits of tourism as part of the island economy can be shown to be sub-stantial. In the Dominican Republic some US$3 million were added to the local economy in 1973 as a result of building one thousand hotel rooms. Generally from five to eight jobs are created for every hotel room completed. Airports, highways and other services have improved out of all recognition.

Despite the huge investments made during the past two decades, most governments have discovered that economic returns remain low. Hoteliers import most of their furniture and other capital equipment under non-taxable incentive laws. Managerial and executive positions are filled by expatriates. Most profits have leaked back to the United States, Canada, Sweden and the United Kingdom through multi-national corporations. Local farms can harvest neither the types nor the amount of produce required by the large hotels. More than 90 per cent of all food eaten by the Caribbean tourist, including fish, therefore, comes from Jacksonville and Miami by container ship or freight plane.

Amid a growing radicalism in the Third World the more negative aspects of tourism have become a political issue. Low economic returns, declines in local agriculture (through loss of manpower to the tourist centres), a rapid rise in the value and price of land often resulting from a spin-off growth of expatriate retire-ment or second-home settlements, and the absence of local executives in the tourist trade are all cited as inimical to the country's best interests. However, it is 'relative deprivation' which is most resented in such communities where poverty is widespread. The local citizen compares his lot with that of the tourist, and either seeks redress through chauvinism and political activism, or simply becomes a beggar in the streets. Islands visited by cruise ships, now a significant fraction of the tourist trade, are especially prone to such activities.

In response to such complaints Jamaica, Puerto Rico, Barbados and Trinidad and Tobago have arranged for training facilities, locally as well as overseas in spe-cialist countries such as Switzerland, in order to create a cadre of hotel executives and staff. Planning authorities have belatedly intervened with moderate success when tourism has threatened the bio-physical environment on beaches, coral reefs and coastal vegetation, which are the most vulnerable. It is clear, though, that not enough has been done to weave tourism into the fabric of economic and agricultural life in the islands.

Government policy, particularly in the smaller islands, has lately turned towards the financing of guest-house accommodation rather than large hotels, in the hope that visitors will get to know the local people. Since much of the Caribbean tourist trade is oriented towards large conferences held at all times of year, the large hotel must remain an essential part of the landscape. Regardless of local feelings, it is the city travel agent who, in arranging package tours to the islands, chooses those places which seem best to fit the seductive advertising posters, and which are best equipped to cater to their clients' preference for high-quality accommodation, service and food. Not to do so would be to invite bankruptcy.

Cuba, once the most boosted tourist area, ceased to be so in 1959. Now, once again, the Republic has cautiously opened its doors to Canadian, European and finally, to United States visitors. Those who go there will enjoy the advantage, not only of unspoilt beaches, but of being able, from their newly built modern hotels, to view this socialist paradise without fear of assault by the indigenous population.

Caribbean integration: closer relationships

The mainstream of world trade and shipping routes has in this century shifted away from the Caribbean area, and all the islands, including those which are still metropolitan-controlled by either France or the Netherlands, share the effects of this relative isolation, albeit more noticeably in the former English colonies. Changing technologies, revised geopolitical viewpoints, shifts in the world balance of power and other imponderables have left the islands stranded economically and politically. Most would now call themselves Third World or developing countries, lacking the resources, capital and skills to improve their lot. Several are endowed with vital minerals which have been thoroughly exploited by foreign interests according to the classical forms of capitalist *Raubwirtschaft* (robber economy). Some have preserved parts of the plantation system in order to export tropical products such as sugar, bananas, citrus, cocoa and coffee to European and American consumers. Others have experimented with various types of agrarian reform in order to overcome the greatest problem of all – feeding a rapidly growing population. Despite protests to the contrary, land is available, though much of the soil has been eroded away as a result of poor farming practices. Driven by rural poverty, people have moved to towns where jobs are scarce. How best to stabilize and then develop the islands has occupied the minds of both local and metropolitan governments for many years.

The anglophone islands

A brief historical review of constitutional changes needed to create governable units in the anglophone islands since the seventeenth century, has shown continual re-grouping of the Windwards, Leewards, and Barbados by the British Colonial Office, while Jamaica and even Trinidad were divorced from the process by distance. Strategically vital in the days of sail, each small island played its part during extended European wars, but in peacetime they were collectively a nuisance except for the sugar and cotton crops they produced. The disadvantages of being peripheral, small and resourceless, and subject to the caprices of world prices, were partly offset by their membership of an imperial trading system which sought protection through Commonwealth Agreements. By 1939, when a Royal Commission prepared its report, it was clear that the entire British Caribbean was in dire economic and social straits.

Post-Second World War demands for a better world accelerated the recommendations of the Royal Commission, through Commonwealth funding and investment, so that by 1955 oil from Trinidad and bauxite from Jamaica and what was then British Guiana, were adding to local revenues. Growing political demands for decolonization through constitutional reforms brought about the first attempt in modern times to integrate a number of disparate units into a political federation. After long debate Port of Spain, Trinidad, was designated the legislative capital, where in May 1958 the Federal Parliament convened. By 1962, when it was clear that free trade and free inter-island migration could not be sustained, Jamaica withdrew, followed by Trinidad, leaving the small islands (the 'little eight') to fight for themselves once more. The British Government, in one more constitutional manoeuvre, granted six of these 'little eight' associated statehood, while Barbados, Guyana, Jamaica and Trinidad and Tobago became independent states within the Commonwealth. Trinidad and Tobago and Guyana have both adopted a republican form of government.

Since the missing link in the West Indies Federation had been economic integration, steps were then taken to create a Caribbean Free Trade Area (CARIFTA), which came into being in May 1968. It comprised eleven members, four of which (Trinidad and Tobago, Jamaica, Guyana and Barbados) were More Developed Countries (MDC), and seven (Grenada, St Lucia, St Vincent, Dominica, St Kitts, Nevis and Montserrat) were Less Developed Countries (LDC). A twelfth member, Belize, joined in 1971. Barbados having departed, within CARIFTA the 'little seven' were allowed to form their own 'Common Market'. A Regional Development Bank was established in Barbados, and many other regional associations dealing with transport, communications, universities and other activities also came into being.

That CARIFTA could not provide total integration was made clear when Barbados, Guyana and eventually Trinidad and Tobago, saw fit to join the Organization of American States (OAS). At the same time the European Economic Community (EEC) was gradually drawing the United Kingdom into its orbit, thus weakening economic links with the Commonwealth. By August 1973 the next step towards a Caribbean Community and Common Market (CARICOM), together with the Caribbean Development Bank, had been taken by the anglophone islands and two mainland countries, Guyana and Belize, representing together some 4.6 million people.

CARICOM faces a very uncertain future in view of the recent world recession and the severe trade imbalances between Trinidad and Tobago on the one hand and Guyana and Jamaica on the other. Trinidad, being an OPEC member, is now comparatively rich, while Guyana and Jamaica suffered from the fall in bauxite prices as well as over-extending their financial resources. Guyana and Jamaica are now seeking bilateral connections outside CARICOM with Eastern European countries (COMECON) in order to relieve their economic distress, and are also restricting imports from other CARICOM members. Both opted for a socialist model, while Trinidad and Tobago have chosen joint partnership with private enterprise. The small islands have not derived much benefit from membership of CARICOM, although United States AID (Agency for International Development) funds have recently been channelled through the Caribbean Development Bank for distribution to small farmers. Guyana and Belize are receiving big grants to establish large food farms in order to reduce food imports into the CARICOM community. Pilot projects with a corn-soya bean combination have proved most unsatisfactory.

One alternative suggested by West Indians as a counter to political and economic isolation, was a closer political association with Canada. Dried codfish and timber in exchange for molasses and rum had long been the trading staples between the two areas, but in the twentieth century Commonwealth preference gave Canada considerable advantages over the United States, and widened the range of goods sold to the British West Indies. Canadian banks, insurance companies and shipping lines have dominated commerce for many years, while large numbers of Canadian citizens in post-war years have moved into the area to retire or to build and occupy second homes.

Canada has granted technical and financial aid in many fields of endeavour, including education, geological surveys, the fishing industry and agriculture. West Indian students have poured into Canadian universities to acquire badly needed skills. Permanent immigration, sharply restricted until 1966, has been made easier for islanders. Discussions directed towards closer union have been fitfully conducted between the two sides during the past two decades, but local xenophobic fears of foreign investments on the one hand and Canada's painful internal problems on the other have made further negotiations most unlikely.

The French islands

While anglophone countries have, through the Lomé Convention, derived certain benefits as sugar producers, they are not as privileged as the French islands Martinique and Guadeloupe, both of which have been administered since 1946 as integral parts of metropolitan France (overseas *départements*). Their export crops such as bananas and pineapples, sugar and rum have direct access to all EEC countries. Consequently neither island has any direct external relationships in the Caribbean area.

The Dutch islands

Similarly, the Dutch possessions of Aruba, Bonaire and Curaçao (Leewards) and St Martin, St Eustatius and Saba (Windwards), producing no export crops, have had associated status with the EEC for petroleum products, chemicals and manufactures. Their peoples have since 1954 formed an integral part of the Netherlands and have migrated in large numbers to the metropole.

The Hispanic islands

Neither Haiti nor the Dominican Republic has sought any economic or political tie with any other nation in the Caribbean outside the OAS since they were released from United States control in 1922. In this they contrasted strongly with Puerto Rico, whose association with the United States has remained all-embracing since 1898. While in the earlier part of the twentieth century Puerto Rico was dominated by a sugar-plantation economy, the mid-century period has seen the rapid growth of industry and manufacturing, as well as considerable modernization. The island has always sought its markets in the continental United States, and for the present would not be interested in a Common Market.

The Republic of Cuba, occupied in 1898 by United States forces and made politically independent in 1902, ceased to be an economic colony in 1960 when all large landholdings and industries were confiscated by the Castro regime. The island's linkages were switched from the United States to the USSR and Eastern Europe. A United States embargo on trade with Cuba has allowed the United Kingdom and Canada, among others, to fill the trade gap. Cuba has largely retained its agrarian mould, and continues to export sugar to COMECON countries. Its ideologies and political structure do not encourage membership of CARICOM, although certain countries in that community sympathize with the Cubans for the way they have confronted their internal problems.

Post-independence development of the anglophone islands

If the Federation, CARIFTA and CARICOM appear to have followed too slavishly the European development models, each has moved its islands towards independence within, and in several instances, outside the British Commonwealth. Subsidies through annual grants from the United Kingdom have prevented a truly Caribbean solution from emerging, but once they disappear, the islands individually or collectively have to examine the realities of population growth, food production, industrial development and tourism as part of the development process.

Trinidad and Tobago and Jamaica are strikingly contrasted examples of the consequences of independence. From 1962 onwards the aim of both was the control of the economy from within and the elimination of monopolistic capitalism as practised by multi-national companies. They serve to illustrate the different methods by which each country has attempted to do this.

Throughout the twentieth century Trinidad and Tobago was dominated by the international sugar and petroleum companies employing a major part of its labour force. Latterly, skilful parliamentary leadership and 30 years of stable government have encouraged import-substitution industries and diversification through incentive legislation. Under trade union pressures for land reform, sugar-company lands have been purchased for re-distribution to cane-farming co-operatives, while the largest of the present operating companies sold its entire equity to the government. Productivity, however, has at the same time fallen alarmingly.

By 1978 the central government also held minority interests in nine companies, majority interests of 51 per cent or more in eleven, and wholly owned twelve. Petroleum corporations have been encouraged to diversify into liquid natural-gas production, ammonia, fertilizer plants and other petrochemicals, while an iron-and-steel foundry and mill is being built at Point Lisas. Automobile assembly, food processing, clothing manufacture and other consumer industries have added to an impressive list of commodities to be exported throughout CARICOM and beyond. With foreign assets totalling more than $300 million in 1978, Trinidad and Tobago has little difficulty in raising international loans and financing the exploration and exploitation of off-shore oil and gas fields. By buying into private enterprise the country has recovered its self-respect and commands its own resources. Tobago, incidentally, plays no part in Trinidad's industrial revolution except by supplying additional labour. It has been designated a tourist haven.

From the early 1950s Jamaica, with its population of more than 2 million, sought economic salvation through industrial estates, tax-incentive legislation and import substitution, while allowing foreign corporations to dominate not only the production of sugar, bananas and bauxite, but also their transport and disposal abroad. Politically the island was governed in large part by the so-called middle class, while the poor, unable to find work, congregated increasingly in the capital city, Kingston, and other towns. Desperate efforts by the central government to accommodate the landless through well-founded land settlement schemes, failed to stem the flow. Mechanization of sugar cultivation was strongly resisted by field workers. The bauxite companies spread work by hiring workers for three-month spells, so that four men could partially benefit from one annual wage. By 1962 the safety valve of large-scale emigration to the United Kingdom had been shut off, thus compounding an already dangerous population problem. Violence and gang warfare in West Kingston slums, together with the abuse and molesting of visitors, quickly reduced the tourist trade and the volume of foreign investment.

A change of government in early 1972 brought into power a party espousing democratic socialism, non-alignment, the nationalization of all resources and the elimination of a stratified class society. Powerful fiscal restrictions were introduced in an attempt to balance the national economy. Increased violence and anarchy during the next 3 years forced the government to declare a state of emergency on 19 June 1976, and for a short time martial law in a section of Kingston. Yet another general election in December 1976 returned the government with a renewed mandate to move toward democratic socialism.

From 1974 onward, in addition to paying higher levies and royalties, the multinational aluminium companies were asked to accept Jamaican participation in their bauxite-alumina operations and to return to the state their rights to all bauxite reserves. By 1979 this process was virtually complete. Some bauxite and alumina were now being transported to the United States in state-registered ore carriers, thus involving Jamaica in every sector of the aluminium industry, but North American companies have since moved to cheaper sources of bauxite in Australia.

Similarly, much of the sugar industry has come under the control of National Sugar, the government agency, while in June 1977, under the Emergency Production Plan (EPP), 16,200 ha of arable land leased from former estates had been granted to 24,000 new farmers in order to reduce a food import bill of $200 million in 1976 to $70 million in 1977. A further 15,400 ha have since gone to 10,000 more food farmers. Food imports were brought under control of a state company, Jamaica Nutrition Holdings, one of seven basic organizations which co-ordinated all Jamaica's imports through a state trading corporation. In 1977 an export trading company, the Jamaica National Export Corporation, was established to conduct aggressive export-sales programmes.

In 1978 the government took over the only remaining independent radio station, thus bringing under control all media except the opposition newspapers. Efforts to have those nationalized did not succeed.

Declining revenues from bananas and sugar, together with a fall in world demand for bauxite and for tourist delights leading to trade imbalances with the MDCs in CARICOM, forced Jamaica to borrow extensively from various United States and international monetary bodies, the United Kingdom, Norway and even Trinidad and Tobago, at the cost of three devaluations of its dollar. A technical co-operative agreement with the Soviet Union in late 1977, together with attempts at closer trade links with the COMECON countries, Cuba and other non-aligned countries, pointed to new roles for a Jamaica seeking practical solutions to its unenviable problems – population growth, poverty and food supplies!

A general election on 30 October 1980 saw the overwhelming defeat of the People's National Party (PNP) under Manley by the Jamaica Labour Party (JLP) under its leader Edward Seaga. The latter immediately severed previously close relations with Cuba, and set about trying to repair the severe financial damage inflicted on the economy as a result of State Socialism. Restoring the bauxite and tourist trade with the United States, regaining the confidence of the international money markets have been immediate priorities, but nothing can conceal the innate poverty of this heavily populated island.

By 1985 the last of the smaller anglophone islands of the Lesser Antilles will have discarded their remaining ties with the United Kingdom and formed, whenever possible, new economic ties within the Caribbean Basin. A thriving, prosperous Barbados, despite its teeming population, has found a western democratic way of handling its economic problems without violence. For those remaining – Grenada, St Vincent, St Lucia, Dominica, Montserrat, Antigua, St Kitts-Nevis, Anguilla and the Virgin Islands – sporadic rioting, local political and trade union confrontations are part of everyday life.

Every executive, parliamentary and advisory function is replicated in each island. All suffer from budget deficits against which British governments have always made grants-in-aid. Each island, with its incentive legislation, has attracted small-scale consumer manufactures and modest hotels to its shores without ameliorating the unemployment rate in any way. Some of the Windward Islands have found partial salvation in bananas grown by smallholders and exported to the United Kingdom. St Vincent discovered that the computer industry requires its arrowroot for making print-out paper. Some cry out for the return of a revitalized sugar industry in islands where, for very sound econo-political reasons, it has long ceased to exist.

The political leadership in the small islands has, up to now, derived largely from a trade union movement inspired by TUC educational programmes shortly after the Second World War. This was reflected in widespread syndicalist tendencies throughout a critical period of development when foreign investors were seeking new outlets. Many of the younger élite, freed from monarchical loyalties and educated by radical teachers in the United States, Canada and Britain, have fled the islands whenever possible, seeing little evidence that they could successfully resolve their countries' problems without drastic socio-political changes. A planned communal settlement in Dominica is still at the experimental stage, but experience suggests that individual landownership, a rugged independence of mind and the open 'culture of poverty' are strongly opposed to the re-ordering of the social fabric. None the less, Marxist cells are active throughout the Lesser Antilles, ready to convince each island that through self-sacrifice it can achieve its destiny, and an increasingly youthful unemployed population may well accept such a proposition.

The extent to which such radical influences had penetrated in the 1970s is borne out by the almost bloodless coup executed by the New Jewel Movement (NJM) and led by Maurice Bishop, on 13 March 1979 on Grenada. The Cuban government quickly moved into action providing weaponry and technical aid as well as building a strategically vital international airfield. The new government adopted a non-aligned stance, but its proximity to such key points as Barbados, Trinidad and Tobago as well as Venezuela has resulted in an increased nervousness about future actions in the area.

An extensive group of small islands, the Bahamas, has through its proximity to Florida, escaped some of the consequences of being small by becoming not only the focus of American package tours and cruises, but also a major off-shore 'free port' for oil refining and other major industries. Vast areas of Grand Bahama, Eleuthera, Andros and other Out-Islands were subdivided for sale to North Americans between 1950 and 1970. Some recession in the United States, and the new status of the Bahamas as an independent country within the British Commonwealth, were factors which combined to reduce commercial investment there by those who seek speculative land deals and tax havens, resulting in a shift to the Caymans in recent years. Having lost their post-independence xenophobia, the Bahamian leaders are now becoming reconciled to the fact that the islands lie somewhat within the sphere of influence of the United States, although they regard themselves as an extension of the Caribbean Basin and continuously seek to strengthen their links with that area.

Developments in the Hispanic islands

Under the protection of the British, French and Dutch governments and friendly alliances with the United States, the anglophone, francophone and Dutch-speaking islands were able to free themselves from colonial bondage through a tidy succession of constitutional (in the case of Haiti, revolutionary) changes towards some form of autonomy or independence.

For the Spanish-speaking islands it has been a very different story. The direct control exerted by metropolitan Spain upon its overseas empire was so intolerable that by the mid-nineteenth century its colonies in South America had fought their way to freedom, but the Caribbean islands did not succeed in overthrowing their Spanish masters for more than a year at a time. Despite their revolutionary fervour, Puerto Rico, Cuba and Santo Domingo (the eastern part of Hispañola) remained in the grip of Spain for much of the century, enduring internal disturbances, vigorous colonial repression and economic systems dominated by a mercantilist philosophy. Only the growing power of the United States, recovering from its own internal war and determined to apply the Monroe

Doctrine to the Caribbean Basin, made it possible for these territories eventually to acquire autonomy, investment capital, technical aid, some of the 'American way of life' and their own home-grown dictators. The ways in which each country is coping with its problems cannot altogether disregard the 'yanqui' factor, for whatever happens in the Greater Antilles must concern policy makers in the United States, ever wary of a vulnerable Panama Canal.

Hispañola

The contrast between the Republic of Haiti and the Dominican Republic is best observed from the air along their international boundary running north-south across this mountain island. The Haitian side is bare, stripped of forest, heavily eroded and densely populated. Most of the stream beds are heavily silted up. Apart from small patches of coffee trees, there is no evidence of cultivation beyond the traditional slash-and-burn. To the east, the lands of the Dominican Republic are forested and relatively thinly populated, except where planned settlement schemes have been developed. There is no evidence of population pressure on the land itself, and from time to time the government has in fact summarily driven out all Haitians invading the area in search of new land.

Haiti

Every aspect of the economic life of Haiti reflects nearly 180 years of turmoil and strife since the country became independent. Even in the period 1915–34 when the United States occupied the country, very little changed beyond the building of roads, health care and some attempts at vocational education (the latter strongly resented by a small mulatto élite which saw its status endangered by a rapidly increasing black population). Although some sugar, coffee and sisal are still produced under plantation conditions, most of the people remain in the countryside at subsistence level, despite many efforts to introduce modern farming techniques and conservation measures. Irrigation, once the foundation of Haiti's agricultural wealth in its wide alluvial valleys, and very necessary in a country liable to long droughts, is seldom practised, and the Peligre Dam, completed in 1956, is now so heavily silted that it can supply only one third of its hydroelectric potential to Port-au-Prince because afforestation and removal of populations from the Artibonite watershed were not part of the total plan.

The capital, Port-au-Prince, and its purlieus have attracted large numbers of rural migrants, a tiny fraction of whom can be absorbed into small component-assembly plants built to lure US industrialists seeking low-wage labour. With the passing of the regime of Duvalier the elder, tourist hotels and cruise ships have made Haiti a popular stop on the Caribbean run, especially for the purchase of local folk art and for visits to lively markets and historic sites in the almost inaccessible north.

Few of Haiti's educated élite who fled the country in the past have returned to take their place in society. Most of them easily found employment with international organizations requiring competent French linguists. Innumerable foreign medical and church missions, trying valiantly to improve the lot of an impoverished nation by adding a few calories to an inadequate diet and to reduce its population growth, have taken their place. There can be no more depressing experience for a social scientist than to ponder the past, present and future fate of this sad country, where the problems appear insoluble even to the most sanguine, even though the USA has opted to pour vast amounts of money into the economy.

The Dominican Republic

Santo Domingo, as the country was once called, was among the first Caribbean countries to achieve independence, if only momentarily in 1821. Occupied thereafter for 22 years by Haitian invaders, it became an independent republic in 1844. Four years of re-occupation by Spanish troops between 1861 and 1865 were the prelude to some 50 unstable years in which there were twenty-eight revolutions and thirty-five governments of every kind. The United States, once more the guardian of the Antilles, occupied the republic from 1916 to 1924. In 1930 a military *coup d'état* brought Trujillo Molinos to power, establishing a regime whose stability was unquestioned, though at a terrible cost to human dignity and rights, and which lasted some 31 years. Another brief occupation by the United States armed forces between April 1965 and September 1966, later to be broadened into an occupation by the OAS, resulting from internal rivalries in part fomented from outside, was strongly resented by the circum-Caribbean nations, but it brought peace and relative stability to the country for the first time in its history. The latest election held in 1977 allowed the elected winner to take his rightful place as a civilian president, in spite of threats of a military *coup d'état.*

With so violent an historical background, the Dominican Republic cannot avoid being an under-developed nation in the twentieth century when investment capital goes only to those who are politically stable. It is true that the Trujillo regime brought much in the way of prosperity and investment, but only at a price. With a population of little more than 4.8 million, it need not suffer land shortage. Smallholding agriculture predominates, permitting the sales of surpluses in rural markets. Until recently there has always been room for the large, private, sugar and banana plantations employing large numbers of rural labourers. The largest sugar plantation in the country, as well as twelve of the sixteen mills are, however, state-owned. Coffee, cocoa, rice and tobacco are secondary cash crops. Cattle have increased from 0.75 million in 1960 to 2 million in 1978. Idle land owned by very large United States corporations, especially on the southern coastal plain, is now being taken back into government hands for re-distribution to the landless.

Among the many changes introduced in the past decade is the concept of regional planning, largely in the hands of well-qualified technocrats. The tourist industry, managed by big multi-national corporations, has developed under careful environmental controls. Among the government's priorities is the completion of as many as 600 irrigation dams to overcome frequent droughts and water shortages. Several of the largest have hydroelectric barrages for supplying much-needed power to the island. Ever since the sanitizing of the Haitian borderlands with the aid of planned settlements, the Republic has used these to improve productivity, and has even encouraged foreign settlers from Eastern European countries.

Bauxite deposits in the south-west and ferro-nickel ores at Bonao in the central interior are exploited under licence by North American companies. It is now considered that at the appropriate moment, large deposits of copper, zinc and mercury could be developed. Otherwise the Republic cannot boast of any major industrial development beyond its Free Zones in Santo Domingo, La Romana, San Padro de Marcovis and Santiago, if only because power and water are not yet readily available for large-scale expansion. Instead, the country looks to Western Europe as a market for its tropical crops, and, apart from its OAS connection, has steered clear of treaty obligations and of other ties in the Caribbean. It is in fact a largely agrarian society with old, long-established cities, a large near-white and white middle class, and with abundant land and food for its population. Given a stable political climate, the Republic apparently wishes to use nothing more than pragmatiç measures to achieve its economic goals.

Puerto Rico

Puerto Rico, occupied and ceded to the United States in 1898, was the sole remnant of the Spanish colonial empire to be retained for its strategic value as an organized but unincorporated territory. For the next half-century the island was administered by the United States Congress until 1952 when it became the Commonwealth of Puerto Rico or a 'free associated state'. Except for federal controls over customs and excise, defence and immigration, the island government is autonomous. In 1967 an island-wide referendum confirmed the status quo, despite the demand for independence by a small minority and a strongly supported campaign for statehood.

Nineteenth-century local and American investment in the sugar industry brought Puerto Rico out of obscurity, allowing the island to develop a classical plantation economy in which landholdings grew larger through corporate control, and smallholders were reduced to landless vassals (*jíbaros*). By 1939 nearly 80 per cent of the island's economy w̨as connected with sugar production. The federal government disregarded this agrarian imbalance, yet at the same time introduced big changes through health-care programmes, roads and education. Malaria and yellow fever were eliminated, so that birth-rates steadily rose while death- and infant mortality-rates fell rapidly. Only in later years was a family-planning programme to have a profound impact on the birth-rate by reducing it from 45 to 23 per thousand. Part of the subsequent population surplus was absorbed into the continental United States, since Puerto Ricans could migrate freely to and fro. By 1939, however, the island was an overcrowded rural slum.

Under New Deal policies attempts had been made to reverse the trend towards over-large holdings by using a long-ignored '500-acre Law', and the creation of land-reform programmes. These were reinforced in 1940 when the Popular Democratic Party (PDP) swept the polls. In 1948 the first Puerto Rican governor was appointed, and in 1952 the Commonwealth came into being.

Earlier efforts by the PDP to industrialize had been based on the sole use of local natural resources such as leather (footwear), clay (ceramics), sands (glass) and limestone (cement) in factories controlled by the government, but this policy soon gave way to 'Operation Bootstrap', an intensive programme of feasibility studies, tax-incentive legislation, factory building and worker-training directed at United States manufacturers seeking a large, lower-paid labour pool.

Beginning with textiles (women's underwear) and small electronic components, many diversified industries from petrochemicals to pharmaceuticals, as well as tourist facilities, have been attracted to the island. Through the Economic Development Administration (EDA), a national investment bank (Banco de Fomento), working with the Puerto Rico Industrial Development Corporation (PRIDCO), has been able, until recently, to apply capitalistic solutions to development problems in the island. Hindered by local bureaucratic delays in making decisions, foreign investors are today venturing further afield. None the less, 'Operation Bootstrap' has proved so successful that the various incentives are now being modified and reduced in order to bring Puerto Rico closer to the realities of 'statehood'. A referendum on this delicate issue in 1980 favoured the status quo.

Until 1963 out-migration to the mainland partially solved the island's population pressures, particularly in the younger, economically active sector, but official unemployment rates have never fallen below 14 per cent. They have risen as high as 22 per cent during periods of recession in the United States, and the unofficial figure is undoubtedly higher. In spite of economic theory, an expensive industrialization programme has not cured a basic ill. Furthermore, the past decade has seen the return of large numbers of Puerto Ricans from the mainland's northern cities, in search of a safe education for their children. Others, now qualified to receive social-security pensions, are retiring to peri-urban San Juan and other cities. Some 80 per cent of the island's families are now included in the Federal Food Stamp Programme. The contribution from all federal sources is more than US$2 billion per annum and is likely to increase. The budget for medical welfare and education was more than doubled in 1979. Larger grants for infrastructural improvements to roads and housing are providing work for those willing to abandon their unemployment benefits. Such federal financial aid is designed to show that a US$14-billion investment by North American capitalists has not been in vain.

A majority (60 per cent) of the Puerto Rican electorate have clearly voted in favour of remaining an anomaly in the Caribbean Basin in their relationships with the United States. Their current *per capita* income ($2025) places them well below that of the poorest mainland state Mississippi ($2550). Furthermore, statehood implies federal taxation and many other administrative burdens.

The 'Independistas', comprising the island's intellectual and cultural leaders and the young radicals from the universities, are a small but vociferous minority (5 per cent) who envisage Puerto Rico's future as linked with a Third World community and with social democratic systems as the driving force. An island largely bereft of its sugar, coffee and pineapple crops, with most of its farmers urbanized, and 90 per cent of all foods except dairy products imported from the mainland, would undoubtedly suffer intensely under independence. The United Nations Decolonization Committee insists, however, that Puerto Rico's relationship with the United States is that of an inferior colonial client and that it merits its freedom. Feeding a population of 3.5 million people, however, remains the fundamental dilemma.

Cuba

It was inevitable that Cuba, the first and last Caribbean possession of the Spanish empire, should attract the attention of the United States. As the largest of the islands, close to the mainland, it had a complex and turbulent record of foreign invasion, insurrection, civil war and piracy until occupied by American forces from 1898 to 1902, when it became a republic. Freedom from Spain made surprisingly little difference to the political life of the island, compelling the United States government, under the so-called 'Platt Amendment' to intervene or to mediate frequently in the republic's affairs.

Once the Platt Amendment was abrogated in 1934, there followed until 1958 a period of *caudillismo* (military dictatorship), during which the country was dominated by Fulgencio Batista both in the role of *eminence grise* and as president. His flight in 1959 after 3 years of rural and urban guerilla warfare led by Fidel Castro, was precipitated when the Cuban army withdrew its support from Batista.

No event in the Caribbean did more to change the face of history than the appointment of Castro as prime minister, coming to power with the support of all but the *Batistianos*. A brief euphoric period of democratic government yielded to a full-blown Marxist socialism within a year of the revolution, offering yet another method of solving the problems of the Caribbean.

Nineteenth-century Cuba had made itself the world's premier sugar producer by consolidating landholdings to create ever larger units of production, and by introducing modern technologies. The island's chief client was the United States, which bought not only 75 per cent of the annual sugar crop, but also large quantities of tobacco, coffee and tropical fruits. Between 1911 and 1925 the sugar crop increased from 1.4 million tonnes to more than 5 million tonnes, partly to replace beet sugar destroyed during the First World War. Vast areas of forest and pasture were converted to cane fields. American investment and ownership rapidly increased, particularly in 1921 when United States banks took over sugar companies rendered bankrupt by a sudden collapse in the price of sugar.

Ever closer economic ties between Cuba and the United States were effected through all forms of commercial undertaking, tourism, gambling and the evasion of Prohibition laws, without much hindrance from either government, so that by 1952 it could be said that Cuba was just as much an economic captive as Puerto Rico and another victim of 'yanqui imperialismo'.

From 1960 onwards a radicalized Cuba undertook extensive land re-distribution programmes and the virtual confiscation of US-owned businesses valued at more than a billion dollars, leading to the withdrawal of Cuban sugar quotas by the United States and a break in diplomatic as well as trade relations. Cuba was quick to turn to the Soviet Union, which soon replaced the United States as principal sugar buyer, trading partner and protector.

The first popularly supported Agrarian Reform Law, passed in 1959, limiting the size of holdings and re-distributing confiscated lands to tenant farmers and squatters, was but an intermediate stage in developing the centralized state farm, together with government control over the land. A second Agrarian Reform Law in 1963 brought more than 80 per cent of farmlands under state control, through the National Institute of Agrarian Reform (INRA).

With the flight to the mainland of at least 350,000 Cubans, the island lost much of its technological and managerial talent. By 1961, following a policy of reducing its role in the national economy, more than 80,000 ha of sugar-cane land had been ploughed out, and increasing emphasis placed on food production. At the same time rapid industrialization to provide import substitutes was given national priority. Factories purchased *en bloc* from Eastern European countries were set up in Cuba to manufacture spare parts for machinery, most of which was North American in origin. High costs, lack of technicians and bureaucratic inefficiency brought the industrial phase of Cuba's revolution to an end in 1963. Five-year plans since 1965 have restored sugar to its original importance. Some 120,000 professional cane-cutters (*macheteros*) have replaced enthusiastic foreign amateurs and part-time labour from the city, and in 1978 some 30 per cent of the cane crop was mechanically harvested with Soviet-designed machines. The long-vaunted 10-million tonne target has given way to more realistic estimates of between 7 and 8 million tonnes, about 5 million of which are destined for the Soviet Union. Irrigation and chemical fertilizers have also done much to raise output, but poor weather conditions are still capable of reducing it drastically.

Cattle ranching, pig and poultry farms have all been collectivized, but it is still some 200,000 small or 'private' farmers owning about 30 per cent of the arable land, who produce 60 per cent of the vegetables, 50 per cent of the citrus and *viandas* (ground tubers, etc.), 80 per cent of the tobacco and coffee and 30 per cent of the beef cattle. Despite large rice harvests and imported foods from COMECON, food rationing continues to be used to maintain a daily intake of 8820 J per person. A large fleet of modern trawlers, with Soviet technical help, has made Cuba the largest fish- and shrimp-catching nation in the Caribbean, thus providing foreign exchange as well as valuable protein for internal consumption.

Cuba is fortunate in having large exploitable deposits of nickel in association with iron, chrome and cobalt – all difficult to separate in the production process. Manganese and copper in small but workable quantities cannot compete with other world producers. Domestic petroleum and natural gas, though significant, provide less than 5 per cent of the island's thermal power and petrol needs, so that high-sulphur oil from the Soviet Union must be imported to supply the refineries originally constructed by United States companies.

Although industrialization *per se* no longer plays a major role in Cuba's 5-year plans, it nevertheless remains a part of the island's economic life. Steel-fabricating plants, rice and flour mills, textile factories and many others help to supply the local market, but consumer goods are difficult to obtain, even though people have the money to buy them.

The seven major cities of Cuba have, however, benefited little from the revolution. Except for the creation of several 'new' towns and ports, capital expenditures and infrastructural improvements have been channelled instead to secondary and tertiary centres in order to attract rural populations away from the cities. Within a comprehensive educational network students of all ages are being trained to view the land enthusiastically as an integral part of their national and personal heritage so that they will be members of a decentralized, semi-rural landscape rather than cyphers in the squalor of an urban slum. No facet of the

Cuban revolution is more remarkable and more worthy of a trial by other Caribbean nations than this solution to the major problem of rural depopulation.

Cuba's avowed role as an exporter of revolutionary Marxist thought has already exerted its seductive influence on Puerto Rico and Jamaica, and on many of the younger leaders in the smaller islands. A poverty-stricken, broken-down country such as Haiti, is fertile ground for yet another violent revolution. Nevertheless, Cuba survives on vast subsidies (in the form of high prices for sugar, for instance) from the Soviet Union, while at the same time a near neighbour, Puerto Rico, relies on the United States to keep its head above water. Meanwhile Cuba's world stature among the nations of the Third World rests on an ideology which has apparently made some 9 million people content with their relatively meagre lot despite a considerable loss of individual liberties. They are no longer starving, are satisfactorily housed, literate and enjoying adequate health care. Would that this could be said of all the other Caribbean islands!

Faced with increased Cuban-backed radicalism and declining standards of living, the United States, Canada, Mexico and Venezuela held bilateral and multilateral talks throughout 1981 and 1982, in order to co-ordinate development aid to the mainland states of Central America and to the mainly island states of the 'wider' Caribbean. On 24 February 1982, the President of the United States announced details of a 'Caribbean Basin Initiative' (CBI), designed to improve the economic stability of the region.

Free trade, preferences (for all goods except textiles and apparel), tax incentives, financial aid, technical assistance and international co-ordination are among the far-reaching elements of the plan, now being debated by a US Congress, reluctant to expend large sums on new commitments to a region so perilously placed, economically and politically.

Bibliography

ACADEMIA DE CIENCIAS DE CUBA (1970) *Atlas Nacional de Cuba*, Havana, Inst. de Geografia y Cartografica.

ANDERSON, J.R. (1972) *World Atlas of Agriculture Vol. 3 Americas*, Novara, Instituto Geografico de Agostini.

ANGLADE, G. (1974) *L'espace Haitien*, Quebec.

ANTONINI, G.A. (1975) *Urban and Regional Planning in the Caribbean*, Kingston, Assn Caribbean Universities and Research Institutes.

ANTONINI, G.A. *et al.* (1976) *Population and Energy: a Systems Analysis of Resource Utilization in the Dominican Republic*, Gainesville.

BLUTSTEIN, H.I. *et al.* (1971) *Area Handbook for Cuba*, Washington, DC. Am. Univ. Foreign Areas Studies Division.

BLUME, H. (1969) *The Caribbean Islands*, trans. J. Maczewski and A. Norton, 1974, London.

BOORSTEIN, E. (1968) *The Economic Transformation of Cuba*, New York.

BRUNNER, H. (1977) *Cuban Sugar Policy from 1963 to 1970*, Pittsburgh, Pa.

BARRATT, P.J.H. (1972) *Grand Bahama*, Newton Abbot.

BURNS, A.C. (1954) *History of the British West Indies*, London (2nd rev. edn, 1965).

CLARKE, C.G. (1964) 'Population pressure in Kingston, Jamaica: a study of unemployment and overcrowding', *Transactions, Inst. Brit. Geog.*, 38, 165–82.

CLARKE, C.G. (1976) *Kingston Jamaica: Urban Development and Social Change 1692-1962*, Berkeley.

COMITAS, L. (1977) *The Complete Caribbean 1900-1975, A Bibliographic Guide to the Scholarly Literature*, 3 vols, New York.

CRATON, M. *et al.* (1978) *Slavery Abolition and Emancipation. Black Slaves of the British Empire*, London.

CUBAN ECONOMIC RESEARCH PROJECT (1965) *A Study on Cuba*, Coral Gables.

CURTIN, P.D. (1969) *The Atlantic Slave Trade. A Census*, Madison.

DAVISON, R.B. (1962) *West Indian Migrants: Social and Economic Facts of Migration from the West Indies*, London.

DEERR, N. (1950) *The History of Sugar*, 2 vols, London.

DEMAS, W.G. (1965) *The Economics of Development in Small Countries with Special Reference to the Caribbean*, Montreal.

DEMAS, W.G. (1971) *Carifta and the New Caribbean*, Georgetown, Commonwealth Caribbean Regional Secretariat.

EDEL, M. (1962) 'Land Reform in Puerto Rico 1941-1959', *Carib. Stud.*, 23, 26-60; 24, 28-50.

EHRLICH, A.S. (1971) 'History, Ecology and Demography in the British Caribbean: An analysis of East Indian Ethnicity', *South West J. Anth.*, 27, 166-80.

EYRE, L.A. (1972) *Geographic Aspects of Population Dynamics in Jamaica*, Boca Raton, Florida.

GOSLINGA, C.C. (1971) *The Dutch in the Caribbean and on the Wild Coast 1580-1680*, Gainesville.

HILLS, T.L. (1965) 'Land Settlement Schemes: lessons from the British Caribbean', *Revta Geogr.*, 63, 67-82.

HILLS, T.L. (ed.) (1972) *Resource Development in the Caribbean*, Montreal.

JOUANDET BERNADAT, R. (1967) 'L'économie des Antilles françaises', *Carib. Stud.*, 7, 3-22.

KAPLAN, I. *et al.* (1976) *Area Handbook for Jamaica*, Washington, DC, Am. Univ. Foreign Areas Studies Division.

KLASS, M. (1962) *East Indians in Trinidad: A Study of Cultural Persistence*, New York.

LASSERRE, G. (1961) *La Guadeloupe: étude géographique*, Bordeaux, Union Franç. d'Impression.

LEWIS, V.A. (1975) *Size, Self-Determination and International Relations: The Caribbean*, Mona, University of the West Indies Institute of Social and Economic Research.

LOWENTHAL, D. (1972) *West Indian Societies*, New York.

McDONALD, V.R. (ed.) (1972) *The Caribbean Economies*, New York, MSS Information Corp.

McFARLANE, D. (1964) *A Comparative Study of Incentive Legislation in the Leeward Islands, Windward Islands and Jamaica*, Mona, Jamaican Institute of Social and Economic Research.

MANIGAT, L.F. (ed.) (1976) *The Caribbean Yearbook of International Relations 1975 and 1976*, Groningen.

MESA LAGO, C. (1974) *Cuba in the 1970s: Pragmatism and Institutionalization*, Albuquerque.

MINTZ, S.W. (1974) *Caribbean Transformations*, Chicago.

NELSON, L. (1972) *Cuba: the Measure of a Revolution*, Minneapolis.

NIDDRIE, D.L. (1961) *Land Use and Population in Tobago: an Environmental Study*, Bude.

NIDDRIE, D.L. (1966) 'Eighteenth-century settlement in the British Caribbean', *Transactions, Inst. Brit. Geog.*, 40, 67-80.

PAYER, C. (ed.) (1975) *Commodity Trade of the Third World*, London.

PEACH, C. (1968) *West Indian Migration to Britain. A Social Geography*, London.

PHILPOTT, S.B. (1973) *West Indian Migration: The Montserrat Case*, London.

PICO, R. (1974) *The Geography of Puerto Rico*, Chicago.

PROUDFOOT, M.J. (1950) *Population Movements in the Caribbean*, Port of Spain, Central Secretariat, Caribbean Commission.

RAGATZ, L.J. (1928) *The Fall of the Planter Class in the British Caribbean 1763–1833*, (repr. 1963) New York.

REINECKE, J.E. *et al.* (1975) *A Bibliography of Pidgin and Creole Languages*, Honolulu.

RICKARD, C. (ed.) (1977) *The Caribbean Yearbook 1977–78*, Toronto.

ROBERTS, C.P. (ed.) (1970) *Cuba 1968. Supplement to the Statistical Abstract of Latin America*, Los Angeles, University of California.

ROBERTS, G.W. (1954) 'Immigration of Africans into the British Caribbean', *Pop. Stud.*, 7, 235–62.

ROBERTS, T.E. *et al.* (1966) *Area Handbook for the Dominican Republic*, Washington, DC, Am. Univ. Foreign Areas Studies Division.

ROUSE, I. (1949) 'The West Indies: an introduction: the Arawak, the Carib', in STEWARD, J.H. (ed.) *Handbook of South American Indians*, Washington, DC, Bureau of Ethnology, vol. 4, pp. 495–565.

RUDDLE, K. and ODERMANN, D. (eds) (1972) *Statistical Abstract of Latin America 1971*, Los Angeles, Latin America Center, University of California.

SAUER, C.O. (1966) *The Early Spanish Main*, Berkeley.

SEERS, D. *et al.* (1964) *Cuba: The Economic and Social Revolution*, Durham, N. Carolina.

SELWYN, P. (1975) *Development Policy in Small Countries*, London.

SHEPPARD, J. (1978) *The 'Redlegs' of Barbados*, New York.

SHERLOCK, P.M. (1973) *West Indian Nations. A New History*, Kingston, Jamaica.

TANNEHILL, D. (1956) *Hurricanes*, 5th edn, Princeton.

TAYLOR, D. (1978) *Languages of the West Indies*, Baltimore.

THOMAS, M.E. (1974) *Jamaica and Voluntary Laborers from Africa 1840–1865*, Gainesville.

WEIL, T. *et al.* (1973) *Area Handbook for Haiti*, Washington, DC, Am. Univ. Foreign Areas Studies Division.

WEISBROD, B.A. *et al.* (1973) *Disease and Economic Development: The Impact of Parasitic Diseases on St Lucia*, Madison.

WEST, R.C. and AUGELLI, J.P. (1976) *Middle America, its Lands and Peoples*, Englewood Cliffs, NJ (2nd rev. edn, 1976).

WILHELMY, H. *et al.* (1977) *Die Industrialisierungs-bestrebungen auf den Westindischen Inseln*, Tubingen, Geographical Institute.

WOODRING, W.P. (1954) 'Caribbean land and sea through the Ages', *Bull. Geol. Soc. Am.*, 68, 719–32.

YOUNG, G. (1973) *Tourism – Blessing or Blight?*, Harmondsworth.

4 · Central America, including Panama

D.J. Fox

There are many ways of viewing Central America (*Figure 4.1*). An atlas map may suggest the isthmus as a bridge joining the two great Americas – a tenuous link between the land masses to the north and south. A marine chart proclaims it as that part of the western hemisphere where the two oceans virtually touch; man has continued where nature left off, and the narrow neck of land is now pierced to become a corridor housing one of the busiest shipping lanes in the world. A visitor may adopt yet another view: that Central America is, in truth if not in form, a peninsula, the tapering end of North America. This view is likely to be reinforced if the visitor has travelled overland from the north; he may enter by a variety of routes, but, as he continues deeper into the isthmus, his choice is whittled away until he finally reaches the end of the road in the forests and swamps of Darien, on the edge of South America. Beyond that point land travel is as yet unfeasible. It is a view shared by sailors. Most ships using the Panama Canal are trading in the northern hemisphere and are forced into latitudes more southerly than direct routing would demand; for them, rounding Central America has taken the place of rounding the Horn. It is a view substantiated in the commercial life of Central America: her ties are largely with the north, and particularly with the USA, and she has no comparable links with South America.

The viewpoints of the inhabitants of Central America are equally varied. A few have a supranational outlook; they see the isthmus as a unit and political, economic and social integration as a desirable goal. Most, however, view Central America as a chain of independent countries, perhaps geographically contiguous, sharing certain historical associations, and producing some of the same products but with the nation as the real concern, and their neighbours and the isthmus as foreign and hazy concepts. Many Central Americans have an even more restricted view: other parts of their own country may be as remote and unknown as the most distant of countries overseas, and an organized picture of the isthmus is quite outside their comprehension.

The area with which this chapter is concerned includes Central America proper – that is, the republics of Guatemala, El Salvador, Honduras, Nicaragua and Costa Rica, and the newly independent state of Belize, together with Panama, a country nominally and historically a part of South America. Unless otherwise specified, however, the phrase Central America as used in this chapter includes Panama.

Nature in Central America

Whatever one's view of Central America today, it is a fact that it has been

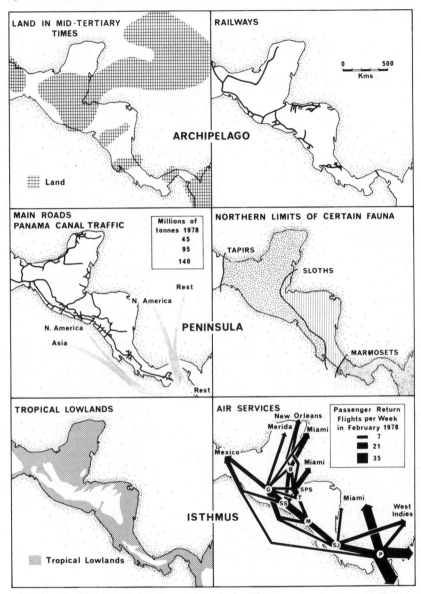

Figure 4.1   Central America

isthmus, peninsula and even archipelago in the not-so-distant, geological past. It has, therefore, been both a corridor and a barrier to landward movements, and it has been an area in which migrants have flourished, new forms have developed and new ways of life have evolved (West, 1964).

Central America became recognizable early in the known geological history of the area. In Palaeozoic times it was probably a peninsula of central Mexico, somewhat wider than it is now, and separated from Andean South America in Darien by a narrow but deepening gulf; it is from this epoch that the folded rocks of central Guatemala, Honduras and northern Nicaragua date. In early Mesozoic time the land connection with North America had become weaker and the isthmus firmly attached to South America as part of a Caribbean land mass, whose north coast ran along the backbone of Cuba, and whose south-west coast lay well beyond the modern Pacific coast of Central America. During Cretaceous times southern Mexico, northern Central America, Jamaica and Cuba formed a great arcuate land mass. This was separated from North America by a proto Gulf of Mexico, and from South America by straits that submerged southern Nicaragua and all of Costa Rica, but left eastern Panama attached to the mainland. The folds in the rocks exposed in the highlands of northern Honduras and southern Guatemala date from this period, as do the mineralized zones associated with the massive intrusions of igneous rocks that occurred in the four northern republics.

The southern straits remained in Early Tertiary times, but the Caribbean island shrank and became firmly attached to Mexico. Later, in Mid-Tertiary times, the Caribbean island again became detached when the isthmus of Tehuantepec foundered. At the same time three elongated islands arose to the south: one in middle Honduras, another across Costa Rica, and the third lying across central Panama. The straits between these islands proved no barrier to certain plants and animals, and the islands served as stepping stones for the invasion of South America by, for example, the primates. Great volcanic activity in the Mio-Pliocene age and general land uplift resulted in the linking of the islands and the emergence of a true isthmus, narrower in the north-west and wider in the south-east than it is today. The equids (horses) used it to enter South America, the opossums to invade North America. Local uplift (especially in Honduras) and depression (particularly in Panama) meant that by about a million years ago (Early Pleistocene times) the isthmus had clearly taken on the shape it has today.

The isthmus has suffered two important changes since that date. In the first place volcanic activity has continued unabated, and the beauty of the landscape has been enhanced, not only by the great volcanic cones which now form the high points in most of the countries, but also by the *calderas* and lakes created by volcanic collapse and the blocking of former drainage outlets. There are 109 large volcanoes, some over 4000 m high, and it is the most active volcanic zone in the Americas today. Secondly, the isthmus has been affected by the changing climatic circumstances of the Pleistocene. During the glacial periods of mid-latitudes, the higher reaches of Central America, for example the Cordillera de Talamanca in Costa Rica, carried glaciers, the climate was cooler and the sea-level fell. During interglacial periods the sea-level rose. The presence or absence of Pleistocene straits across the isthmus has been widely debated: the findings of marine biologists suggest that there has been no deep-water connection between the Caribbean and Pacific since the Miocene and that, although a shoal-water link may have persisted into the early Pleistocene, this formed no barrier to the movement of land animals, and had certainly ceased to exist well before man arrived on the scene.

It is in the context of this eventful geological history that the development of the flora and fauna of the area may best be viewed. A tropical situation in combination with a great variety of local relief and climate has not only allowed a large range of species to exist and multiply, but has also meant that refuge areas for species endangered by changes in environmental conditions have been widely available. The opportunity Central America has given for insulation of plants and animals – whether by sea or land – has resulted in the evolution of large numbers of endemic species. To these indigenous species must be added plants and animals spreading from the adjoining continents, creating a biota that in many important respects is richer than that of either North or South America.

The vegetation of the isthmus reflects the various factors that have been at work. The lowland rain-forests of Caribbean and Pacific Central America have, therefore, strong affinities with the *selva* of South America. From the botanical point of view the Caribbean rain-forests below about 1000 m form a distinctive assemblage of species in which the large number of palms, tree ferns, lianas and epiphytes emphasizes the humidity of the region. It ends in the Maya mountains of Belize, north of which the climate is drier. The tropical rain-forest of Panama and eastern Costa Rica gives way on the Pacific side to a dry, evergreen forest at lower altitudes; this remains essentially of South American composition. At elevations about 1000 to 1600 m species are fewer and the affinities with North America are stronger. The important trees in the forested highlands of northern Central America are pine and oak as in the highlands of Mexico. The pines do not extend further south than north-central Nicaragua, however, and in the Cordillera de Talamanca of Costa Rica the oaks are joined by conifers of South American provenance. Similarly, above the tree line in Guatemala, the bunchgrass assemblage has Mexican and even US associations, whereas in Costa Rica, above, say, 3100 m, the shrubby, open landscape has closer floristic affinities with the Andean *páramos*. The proportion of indigenous plants increases with height.

The fauna of Central America is essentially a diluted assemblage of nearby South American species, with a substantial minority of northern creatures, and two highland areas (one western Panama and eastern Costa Rica, the other the Guatemala–Honduras–northern Nicaragua region) that have served as local areas of evolution or endemism. The jaguar, important in mythology, the ocelot, the jaguarundi and the margay, all members of the cat family, are at home in the South American forests, and have spread throughout Central America. In contrast, the puma, like the gray fox and the coyote, is of North American origin; the coyote does not extend into southern Central America. The racoon from the north, which reaches South America, and the kinkajou from the south, which reaches Mexico, are similar, gentle animals that suggest the manner in which different geographical distributions overlap in Central America. There are also surprising gaps in the animal population of the isthmus: for example, although there are bears in Mexico and adjoining South America, there are none in Central America. Perhaps the most distinctive mammal of Central America is the aquatic, herbivorous manatee, which still survives in some of the more isolated lagoons and estuaries of the east, despite persecution for its flesh and blubber; it is larger than its Amazonian cousin.

The various families of land turtles found in Central America have both northern and southern relatives, although there are fewer species here than in either Mexico or in northern South America. The huge, fresh-water turtle, once distributed throughout Europe, Asia and North America, survives today as a living fossil only in Central America. The sea turtles are, of course, common to the eastern hemisphere as well. It is the green sea turtle that is most prized for its meat, although, as with other turtles, its eggs are also eaten; it is found off both coasts, but is economically more important in the Caribbean (Nietschmann, 1973, pp. 23–45).

Central America is extraordinarily rich in bird life, residents being augmented by annual migrants. Parrots are abundant, and the variety of species is very large in the southern lowlands. Humming-birds, too, are numerous and can be seen even in the higher mountains of Guatemala and in the Cordillera de Talamanca. Perhaps the *quetzal*, a trogon, is the best-known Central American bird; its brilliant tail feathers were prized by Mayans and Aztecs (and later the Victorians), and have been adopted as a national emblem by Guatemala. At home in the damp cloud-forests from southern Mexico to Panama, it has now become a rarity and is confined to the less accessible parts. The toucan family is essentially from the northern part of South America, but toucans, with their outrageous beaks, do penetrate the isthmus and some even reach central Mexico. Large numbers of North American species winter in Central America, their flight paths converging as the land tapers southwards.

Central America in prehistory

Man was added to the fauna of the Americas relatively late, perhaps about 60,000 years ago during the last major glacial period. By 9000 years ago, he had spread from the Bering Straits to southern South America, presumably by way of the Central American corridor. As yet, no date can be usefully offered for the beginning of human occupation of the isthmus. Archaeological sites abound, but many have not been scientifically investigated, though it is clear that Central America was very close to the centre of the Neolithic revolution, which introduced agriculture to the Americas.

There are two schools of thought on the origins of agriculture in the New World. Sauer's (1952) attractive hypothesis places the agricultural hearth just to the south of Central America in, perhaps, northern Colombia. He believes cultivation began with the planting of roots, such as manioc, by sedentary river folk, and the idea then spread southwards along the Andes and northwards through the isthmus, becoming modified by new techniques and new plants as conditions changed. Given the necessary geographical and botanical environment, security and leisure, the intellectual step required to move from gathering to planting appears seductively slight. It is interesting to note that archaeologists are now placing the American origins of pottery and ceramics in northern South America, and consider that a planting culture of the type Sauer envisaged was in existence in Venezuela 4000 years ago. Unfortunately, it is in the nature of the evidence that few relics of this period should have survived. It is perhaps partly for this reason that most archaeologists incline to an alternative but more demanding hypothesis, which gives pride of place to the invention of seasonal seed agriculture and regards root-planting as a derivative. The earliest evidence of agriculture of this type comes from northern and central Mexico. Amongst the earliest domesticated plants recorded there are various species of *cucurbitacea*, (pumpkins, squashes) and beans, which grow wild in Central America, and it is a noteworthy fact that many of the other plants that became cultigens are indigenous to the isthmus. One of the most interesting and suggestive of recent excavations in Middle America was made on the Pacific coast of Guatemala, at Salinas La Blanca, near Ocós, where a kind of non-hybridized maize was grown and a system of agriculture practised, which, it is claimed, were both more advanced than those extant at Tehuacán at the same time (2800 to 3000 years ago). Analysis of the middens shows very clearly that, although vast numbers of shellfish and, especially, crabs from the adjoining estuaries supplemented the agricultural diet, hunting of the locally abundant animals played no part in the economy. Coe and Flannery (1967) believe that it was in an amphibious situation rather than in the dry highlands of Mexico that a properly settled agricultural way of life first became a reality, and that it was made possible by the introduction of highland maize some 3500 years ago. In this context the work done by MacNeish and others in Belize in 1980 is significant because it suggests that a sedentary agricultural life existed in the colony 5500 years ago, and that it was preceded by an earlier economy which combined seed and shellfish gathering and fishing; the earliest artifacts associated with these sites go back to 11,000 years BP. This evidence supports Willey's long-held view that village life existed in Central America before the existence of cereal agriculture and that the southward spread of maize cultivation (which reached Peru 2700 years ago or earlier) was preceded by a type of forest-planting culture akin to that envisaged by Sauer.

Whatever early role Central America played in the transformation of man's position in relation to his environment, the archaeological record makes it clear that later developments followed two divergent paths. Northern Central America shared the remarkable culture that blossomed in Yucatán and southern Mexico; in contrast, southern Central America vegetated and attracts archaeological interest today largely because of its curiously negative quality, separating, as it does, the two areas of high civilization, Meso America and the Andes. The southern boundary of the high culture area ran southwards approximately from what is now Tela on the Gulf of Honduras, up the Ulua river almost to the border of El Salvador, and then turned south-eastwards some 40 or 50 km inland until it reached the sea by the Gulf of Nicoya in Costa Rica.

The areas in northern Central America shared a number of advanced cultural traits in late prehistoric times. One of the most important, as well as the most easily appreciated today, was the development of a monumental stone architecture. Planned ceremonial centres began to appear in a landscape that formerly mustered no more than mud-walled houses loosely grouped together into hamlets and villages. Such centres frequently had an essentially rectilinear plan, which combined stepped and truncated pyramids, platforms and spacious courtyards, perhaps flanked by tiered stands or columnated arcades, in an open and organized design. The earliest of such centres in Central America were in the Guatemalan highlands; the site of Kaminaljuyu, which is now being engulfed by the suburbs of Guatemala City, is not only one of the most accessible but also one of the oldest. Here, a ceremonial centre was in existence as early as 3000 years ago; at its peak, the centre had at least a hundred major structures and 50,000 people were tributary to it in a type of association that has been compared with sixth-century BC Athens. The buildings showed very strong links with central Mexico, but other evidence shows these links weakened in the first millennium AD and by its end defensive sites comparable to European hill-forts had replaced the open cities of earlier days. In the centuries immediately before the arrival of the Spaniards many highland sites took on the character of medieval European castle towns used as religious, commercial and administrative centres, and as a refuge for outlying peoples. Zaculeu was an example of such a centre that had long been occupied, and was able to put up a spirited and damaging resistance to the Spanish when attacked in 1525.

It was in the lowlands of the Yucatán peninsula, however, rather than in the Guatemalan highlands, that the architectural qualities of Meso-American civilization reached their zenith. It was about the time of Christ that Tikál and Uaxactún first rose above the tropical forests of the Petén; 150 years later, the building of Copán in Honduras was in full swing, soon to spawn its daughter settlement at Quiriguá. Such well-known sites are extremely large and impressive, and are frequently uncluttered by more recent buildings; some, including Copán, have dramatic settings. Sites are sufficiently numerous for the adventurous to share with such earlier visitors as Stephens and Catherwood the thrill of discovery of major ruins, perhaps all but blanketed by the tropical forest (Stephens, 1963). Tikál, in the north-eastern Petén, is the largest site. Here the central cluster of ceremonial constructions, many linked by causeways, occupy 3 km², 'suburban' Tikál occupies another 13 km², and there are many detached but associated buildings. In many lowland sites the underlying, gentle topography was remoulded to fit into the orderly concepts of the planner; in the lower Usumacinta sites (for example, Piedras Negras) the terrain is too accidented and the plans were bent to take advantage of the hills. In addition to the major sites there are innumerable house-mounds and lesser ruins testifying to a substantial population in the first millennium AD. The Usumacinta river and its tributary, the Pasión, attracted settlements as did the middle and upper Belize. These, together with the drier north-eastern Petén and the Copán–Quiriguá region, appear to have been the most densely populated parts of the isthmian lowlands at that time. In general, the settlement pattern of the first millennium AD appears to have been a dispersed one: the peasants lived in hamlets and periodically visited the ceremonial sites. These sites served not only as centres of religious experience and, presumably, political control, but also as regional and interregional markets; the market as an institution remains one of the lasting traits of northern Central America.

The cause of the collapse of the system, witnessed by the gradual abandonment and neglect of the centres from late classic times (perhaps about 800 AD) onwards, remains a matter for speculation. Appeal has been made to such cataclysmic events as earthquakes; epidemics may have scourged the area, but, again, proof is lacking; a critical change in man–land relationships, brought about, perhaps, by overpopulation, soil exhaustion and a failing water supply, has also been a popular hypothesis, although difficult to accept as anything more than a local explanation. Internal political changes and pressures from outside seem more likely causes, if difficult to demonstrate conclusively. Mayan civilization proved more tenacious in the highlands (perhaps because it was more urbanized and the centres were not so limited in their functions); it enjoyed a renaissance in the more northerly part of the peninsula under Toltec influence, but even in pre-Columbian days northern Central America had fallen into the role of a buffer zone separating the higher cultural area of Mexico from that of the southern isthmus.

While monumental architecture, mathematics, writing, astronomy and human sacrifices to propitiate the gods were occupying the minds of large numbers in the north, the southern isthmus was a part of a less flamboyant cultural area that included the Caribbean fringes of South America and coastal Ecuador (Helms, 1975, pp. 111–19). There are no signs of large, organized communities or of substantial architectural remains; the most interesting archaeological finds relate to pottery and metal-workings. The art of gold-working, derived from South America, was practised in the Azuero peninsula of Panama and in the highlands of Costa Rica for, perhaps, a thousand years before the Spanish Conquest, and both these areas were more advanced than the rest of the isthmus. Gold, and cacao from coastal Guatemala, formed items of trade in an area of mainly self-sufficient, thrifty, root-crop cultivators, and this trade continued into Aztec times.

Between southern Central America and the Maya lands, there lay a fringe area where Maya cultural traits have faded from the archaeological record. El Salvador forms part of this zone, and Mayan sites are much more numerous in the intermontane basins of the west than they are in the flatter lands east of the Río Lempa. Oases of high culture existed in the area bordering the Gulf of Fonseca, but links were stronger all through its history with peripheral sites in central Honduras (for example, Comayagua) than with western Salvador. In late prehistoric times trade in pottery from the Nicoya peninsula of Costa Rica and intervening coastal Nicaragua brought these areas into the Meso-American cultural area.

In summary, the prehistory of the isthmus shows that it was at or near the heart of many of the truly remarkable cultural advances made in the Americas prior to the arrival of the Spaniards. To what extent it owed this to the great variety of opportunities the area offered and to the merging of experience that may have happened as the routes of land migrants converged in the isthmus, it is impossible to know. In broad terms, it is clear that Central America had no distinctive cultural unity in the two millennia preceding Columbus; rather, northern Central America shared many of the attributes of central and southern Mexico, while southern Central America was much more akin to contiguous South America. There was no isthmian unity in 1492.

The conquest and colonial integration

The divisions in the isthmus before the arrival of the Spaniards continued to some extent during colonial times; they are reflected even in the contemporary situation. The early colonial history of Central America involved two spheres of activity – the one based upon Darien and Castilla de Oro in Panama, and the other based upon Mexico City (Sauer, 1966). It was the lower part of the isthmus that was first brought into European knowledge during the last voyage of Columbus in 1502: he made landfall in the Bay Islands of the Gulf of Honduras and then turned east and south around Cape Gracias a Dios past the Mosquito coast to the gold-bearing areas of Veragua. Had he turned the other way, he might have discovered the high civilization of Yucatán. Loot (especially gold) and slaves drew adventurers to Darien, and their numbers were reinforced after Balboa's discovery of the Pacific in 1513. Barbaric treatment and epidemics of diseases carried from Europe soon decimated the native population and antagonized those who survived. Panama City was established in 1519 and became the base for exploration. An expedition under Hernandez de Córdoba led to the founding of both León and Granada in 1524 as centres from which to tap a new source of slaves, some of whom were sent to work the local gold deposits of the Nicaraguan interior, while others were shipped south to Peru.

In that same year, 1524, 5 years after Cortés first entered Mexico City, two expeditions were mounted from Mexico that were to form the springboards for the Spanish empire in northern Central America. On the one hand Pedro de Alvarado fought his way southwards through the Guatemalan highlands to the lowlands of the Gulf of Fonseca, following established Aztec tribute paths but finding few amenable subjects and little gold; on the other a sea-borne force was sent by Cortés to the northern Honduran coast where Puerto Caballos (now Puerto Cortés), San Pedro Sula and Trujillo were founded to exploit the placer gold and high-grade cacao of the Ulua and Aguán valleys. Thus the Mexican-based and Panama-based Spaniards met in the same frontier zone that had once served as the major cultural separation in pre-Spanish days.

But, although the early centres of exploration in Central America often survive as political centres today, the major division of early Spanish days was soon reduced. The conquest of Peru by Pizarro in 1532-3 turned Panama into a mere gateway for transients to and from the south; the pacification of Yucatán and the first of the great silver strikes in northern Mexico in the 1540s turned Mexican eyes away from Central America. In 1543 the Captaincy-General of Guatemala, answerable to the Viceroy of New Spain (Mexico), was created, with jurisdiction from the Isthmus of Tehuantepec to the empty lands of Costa Rica; Panama, in view of its new function, was incorporated in New Granada and was answerable to the Viceroy of Peru. It was thus in the very early days of the Spanish empire that certain of the prevailing characteristics of the area were established: the political separation of Panama from Central America, the establishment of a core area around the two lakes in Nicaragua, the dualism of the north coast and the interior in Honduras, the emphasis on the Pacific side of the isthmus, the political supremacy of Guatemala and the late emergence of Costa Rica.

In Costa Rica, the core area of today, the cool Meseta Central, was an area unattractive to settlers in the sixteenth century; it proved to have neither gold nor tractable Indians. The small number of white settlers who moved southwards from Nicaragua from the 1560s onwards had to take up subsistence agriculture and to manage on their own. Later, exports of wheat and tobacco placed the colonial economy on a sounder economic basis, and encouraged the intensive settlement that characterizes the Meseta Central today. Nicoya and Guanacasti on the Pacific side offered an easy overland route from Nicaragua to Panama and a malleable Indian population, and were administered quite separately in colonial times from the rest of present-day Costa Rica. They fell within the Nicaraguan sphere of influence and the cattle-ranching economy and the more traditional society that arose then persists today. Unlike Costa Rica, the interior of Honduras did contain precious metals, even if the Indians were difficult to subdue, and in the 1570s the silver deposits of Tegucigalpa and Comayagua were discovered. Interior Honduras was to Central America what northern Mexico became to New Spain, and mining and ranching sustained the economy for the next three centuries; the landscape of central Honduras today, with its red-tiled roofs and white-washed walls, scattered groves of dark-leafed pines, dusty tracks and open views, could easily be mistaken for, say, parts of Guanajuato in Mexico or even parts of Andalusia.

The situation in Guatemala was different. Here there was no mineral wealth, but there was a sizeable Indian population. Difficult to enslave, the rural Quiché and Cakchiquel were of little interest to any save the Catholic Church, and their pacification was left to the Dominican priests. Today Indians survive in large numbers; the Spaniards never formed a numerically significant proportion of the population, and mixed-bloods and *ladinos* (mestizos) represent a much smaller element in Guatemala than they do anywhere else in the isthmus, except in Belize and Costa Rica. In the other territories the experience of, for example, El Salvador is more typical (Browning, 1971, pp. 31–138). Here the Indian was more timid: he was required to adapt to the attitudes of the newcomers and was incorporated into post-conquest society. In earlier days the land not only gave sustenance, if properly propitiated, but also offered a permanent home to the souls of dead ancestors and of living plants and trees: it was the continuum in which the transitory life of the individual Indian had its due place. After the conquest his land became one of the spoils of war and different values were placed upon it by the new masters. The Indians were organized and new villages established in places most convenient to house the labour for the new haciendas and estancias; epidemics swept through the new congregations and helped decimate the Indian population. Spanish overseers were installed in the villages, the early distinction between Indian and Spaniard soon became blurred, and a rather confused ladino society emerged. A new and more varied landscape of conquest came into being with the creation of commercial plantations (of cacao and indigo in El Salvador) and cattle ranches, but the Crown made sure that not all the communal lands were usurped, and the new rural landscape was in practice an amalgamation of both the traditional and the introduced.

After the initial impetus given by the discovery, the Spanish Main lapsed into being a backwater of the empire. The largely unincorporated Caribbean coast provided a safe haven to buccaneers and smugglers, and by the middle of the seventeenth century the English had secured a number of footholds on what was nominally Spanish land (Helms, 1975, pp. 210–16). The off-shore islands, such as Providencia, San Andrés and the Bay Islands, the Mosquito coast of eastern Honduras and Nicaragua, and the Yucatán coast of the Bay of Honduras, fell under unofficial English sway. Seventeenth-century pirate strongholds became eighteenth-century shipping points for logwood (from which inferior dyestuffs were extracted) and later mahogany, and English interests extended round the Yucatán peninsula into Campeche. Ephemeral agricultural colonies were started along the Mosquito coast together with more permanent trading centres like Bluefields and Greytown (Nietschmann, 1973, pp. 23–45).

The weakness of the Captaincy-General, and the extent to which the Caribbean side of the isthmus was beyond the effective control of the government in Guatemala City in colonial days, is further suggested by the Spanish decision of 1803 to switch nominal jurisdiction over the Mosquito coast to the Viceroyalty of Bogotá (Parker, 1964, p. 234); even today the English-speaking islands of Providencia and San Andrés off the Nicaraguan coast are Colombian territory. Some three centuries of English associations and of neglect by the proper authorities have created a very different cultural milieu on the Caribbean underside of Central America from that on the Pacific side. Amongst the more obvious signs to the contemporary visitor is the English, or an English patois, used as the common language of large areas, the weakness or non-existence of Spanish institutions and the high proportion of Negroes in the population. In fact, it was not until the middle of the nineteenth century that the political situation of the Caribbean rimland of Central America was regularized: Britain gradually withdrew her protection from the Mosquito coast, but confirmed (although not to Guatemala's satisfaction) British Honduras (now Belize) as a Crown Colony.

The illusion of Central America's colonial unity was weakened in the waning stages of the Spanish empire as interest in, and the ability to maintain, the rigid administrative structure declined. Central America, like the rest of its territories, was viewed from Spain in the light of the contribution it could make to the economy of the mother country. Towards this end trade commodities and routes were restricted, such artificial difficulties compounding those created by nature. Smuggling flourished, and the profits and convenience to be won by illicit trading further weakened central authority. Not that Central America offered much to the outside world at this time; the traditional cacao trade with Mexico persisted, indigo became a speciality crop of San Salvador during the eighteenth century, and the Honduran mines continued to yield silver, but the total trade was small and the interest of Spain correspondingly slight. Within Central America authority relied on an attenuated set of transport links: coasting vessels were supplemented by trackways, one of which ran the length of the isthmus on the Pacific side between Guatemala City and Panama while another linked the oceans across Panama and two fed the Gulf of Honduras ports.

Independence and political fragmentation

Independence of Central America from Spain in 1821 came on the coat-tails of Mexico's declaration earlier in the same year. It followed a decade of sporadic insurrections by disaffected regional interests (Parker, 1964, pp. 77–90). After the declaration effective power lay in the hands of the separate towns of the isthmus, and it took two years for a stable pattern of political alignment to emerge. All except Chiapas eventually rejected union with Mexico, and in 1823 the United Provinces of Central America was proclaimed with its capital in Guatemala City. The five provinces were created through the amalgamation of existing local government districts, and approximated very closely to the independent countries of today. They had a total population of about a million and a quarter, with the lion's share in the north (Guatemala around 500,000 and El Salvador around 250,000); the three southern provinces shared the remaining one-third (Honduras and Nicaragua about 180,000 and Costa Rica 60,000). The Federation lasted only 16 years. The liberal ideas that had helped to establish the

Federation also helped in its downfall; local quarrels turned into civic unrest and then to civil wars; the three southern provinces seceded first, followed by Guatemala, leaving El Salvador to remain true but only by default. Slowly the provinces declared themselves independent republics and, although the idea of a unified Central America was kept alive, none of the twenty or more attempts in the nineteenth and twentieth centuries to resurrect the Federation has succeeded.

Until recently both internal and external circumstances have been against unification and have worked in favour of the continued Balkanization of the isthmus. Political instability has been one such circumstance. There has never been a sufficiently strong popular mandate to allow unification a chance of success, while the periods of stability have usually coincided with the kind of *caudillismo* for which the isthmus is notorious, and few strong men have had territorial ambitions beyond their national boundaries. Large areas still remain, especially on the Caribbean side, where effective government is still not established; perhaps this task is a necessary alternative or preliminary to extra-national aspirations. It is a situation that reflects the continued small size and uneven distribution of the populations of Central America: the city regions remain, and it is only in recent years that transport both within and between the separate countries has been of a nature to encourage the growth of non-local links.

The post-colonial period of political fragmentation has suited outside powers. It suited the British with their *de facto* colonies on the Caribbean shore in the early nineteenth century, and it has suited the United States once the isthmus was brought within its sphere of influence. United States interest stemmed partly from appreciation of the role the isthmus could play in one of the manifest destinies of the United States – the provision of a water route between her east and west coasts. This role first became obvious when the 'forty-niners' retrod the route of the *conquistadores* across Panama and, like them, re-embarked on the Pacific on the way to their El Dorado. The forty-niners also followed the Río San Juan–Lake Nicaragua route. For the next half-century proposals for, and attempts at, canal-building across the isthmus were numerous. It was convenient for foreign negotiators to have alternative governments to bargain with as well as alternative routes to explore. The creation of the Republic of Panama itself is, of course, a striking measure of the manner in which political *minifundismo* was encouraged by United States actions; seen in this light, the creation of the separately administered Canal Zone was a natural corollary.

The manner of the negotiations for the Panama Canal in the early part of this century, the large scale of the enterprise, and the creation of an almost entirely foreign enclave to operate it, all have their counterparts in the other parts of the isthmus. The establishment of the great banana plantations in Costa Rica, Honduras and Guatemala involved similar massive injections of capital, the same abrogation of national rights to United States interests and the immigration of foreign labour, and identified similar poles for recent political agitation. The exploitation of areas so affected involved the building of roads and railways and ports, and the creation of other facilities that eventually revert to the local governments; it would have been impossible for the host country to have financed this infrastructure itself. The host countries gained, because areas that had previously been to all intents and purposes outside the national economy were made part of that economy. But the process was a reversible one: the interests of the company became more and more the interests of the country, which frequently had no other significant export to turn to. Institutional changes were made in response to this, sometimes unwilling, identification of interests, while the companies, by spreading their assets over several countries, placed themselves in very strong bargaining positions. These developments did little to improve communications between the republics: in fact, their interests tended to parallel rather than to complement each other. In contrast, links between the individual countries and the United States were strengthened, sometimes to the point of becoming bonds. The strength of these bonds may be judged from the several examples in this century of official American intervention in the internal affairs of the Central American republics – perhaps the presence of US Marines in Nicaragua from 1909 to 1933 is the best example – and the current difficult position in which the United States finds itself in trying to match its officially proclaimed ideals of social justice with the desire of US business interests for political stability in the isthmus.

Not all economic developments in the countries since the middle of the nineteenth century have worked at maintaining separatism. The emergence of coffee as a mainstay of the economies of Guatemala, El Salvador and Costa Rica has introduced a new wealthy class within which there have always been liberals drawn to the political concept of isthmian unity, even though the economic forces in the coffee trade tend to diverge from it. Of more practical importance, however, has been the emergence during the last 30 years of a number of circumstances that have not only made Central American integration more feasible but have also given it a much sounder economic *raison d'être*. An improved road system has helped bring the countries closer; a growing realization that the separate domestic markets are often too small by themselves to support efficient manufacturing industries has helped to promote the idea of a common market; the weakened position of the traditional export staples and the massive increase in population have required a reappraisal of public policy; and the support of the United States and of the United Nations for efforts designed to promote concerted Central American action in tackling some of the economic problems of the isthmus: all these have helped promote a climate in which rebirth of the United Provinces is a political possibility, although the possibility may still be remote. In 1951 the five republics established the Organization of Central American States (ODECA) to resolve local international disputes. During the 1950s several bilateral trade agreements were signed, as was a convention on so-called integration industries or those new industries attracted to Central America by the promise of a monopoly of the isthmian market; another covention equalized certain import duties. The General Treaty of Central American Economic Integration was ratified in 1963 giving effect to the formation of the Central American Common Market (CACM): a secretariat (SIECA), a bank (Banco Centro-Americano de Integración Económica (BCIE) and a stabilization fund came into existence. Most of the progress to date has been at the level of economic co-operation where many controversial matters are not solved, but sidetracked; the political hurdles to union remain high. The gaps between noble constitutional principles and political reality remain wide in some republics, and the normal, built-in difficulties of reassigning political responsibilities remain very strong in the highly diverse political regimes now in power. Movements towards Central American reunification are well publicized, but a closer look at some of the economic, social and political problems in Central America today may produce a more sober assessment of the weaknesses in these movements.

Contemporary Central America

Figure 4.2 may serve as an introduction to some of the features of life in Central America today (The World Bank, 1979). The three cartograms make several facts clear at once. First, Central America is dwarfed in area, population and economic production by Mexico in the north. The isthmus bears the same kind of areal relationship to Mexico that Mexico does to the contiguous United States, each approximately one-quarter the size of the other. In 1980 there were over 22 million Central Americans, less than one-third the number of Mexicans, and the population density of about $270/km^2$ is still very low in comparison with many parts of the world and in particular with the Caribbean islands. In economic terms Central America produces about one-quarter of the volume of goods that Mexico does; this means that, for example, the gross national product of Central America can be equated with that of Colombia or Ireland or with that of one of the minor states of the United States of America. In the second place, all sets of comparisons between Central America and Colombia are to the advantage of the latter, although the differences are less than those between the isthmus and Mexico. Although less than half the size of Colombia, Central America, if treated as a unit, would rank between Colombia and Peru, in any demographic or economic ordering of Latin America.

Figure 4.2 shows that there are some notable differences within Central America. The disparity between area and population is greatest in the cases of El Salvador, the only Central American country without a Caribbean flank, and Belize, the only country without a Pacific backbone. But for large-scale emigration, the population density in El Salvador, already the highest of all the countries of Latin America, would be even higher; illegal settlements in neighbouring countries have been a source of political friction in recent years. Other difficulties, notably those of leading an independent existence in the modern world, face Belize which, with only about 135,000 people in 1980, was by far the least populous of any country on the mainland of the western hemisphere. All the other countries of Central America have a density of population that is above the average for Latin America, although that of Nicaragua is very close to this figure. A more refined measure of man–land relationships is given by the amount of agricultural land available per head of population, especially since the majority of Central Americans still live directly off the land. It is clear that there is relatively less agricultural land available in the three northern countries, albeit for different reasons in each, than there is in the south; the diagram emphasizes the temptation to cross the border offered by the disparity between Honduras and El Salvador. The general relationship between population and economic production is suggested by the similarities in the two relevant cartograms: Guatemala and El Salvador, although with only one-quarter the area, have just over half the population, and account for half the production of the isthmus. Nevertheless, a comparison of GNP on a *per capita* basis shows that the average

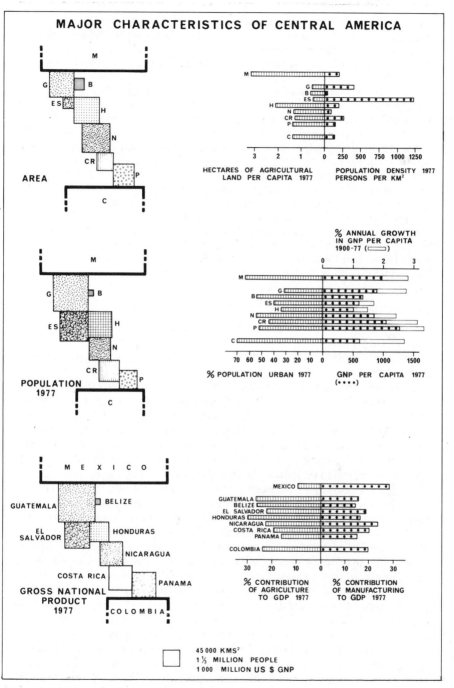

MAJOR CHARACTERISTICS OF CENTRAL AMERICA

AREA

HECTARES OF AGRICULTURAL
LAND PER CAPITA 1977

POPULATION DENSITY 1977
PERSONS PER KM²

POPULATION
1977

% ANNUAL GROWTH
IN GNP PER CAPITA
1900-77 (☐)

% POPULATION URBAN 1977

GNP PER CAPITA 1977
(••••)

GROSS NATIONAL
PRODUCT
1977

% CONTRIBUTION
OF AGRICULTURE
TO GDP 1977

% CONTRIBUTION
OF MANUFACTURING
TO GDP 1977

45 000 KMS²
1⅓ MILLION PEOPLE
1 000 MILLION US $ GNP

Figure 4.2 The major characteristics of Central America
Sources: Inter-American Development Bank, *Economic and Social Progress in Latin America, 1977 Report*; Latin American Bureau, *The Belize Issue*.

Panamanian is three times as productive as the average Honduran, and that the average Costa Rican is better off than any of his northern neighbours. The gap between the wealthier countries and the poorer countries has widened since 1960: the standard of living has risen twice as rapidly in richer Costa Rica than in poorer El Salvador.

Such comparisons partially help to explain why Costa Rica was the most reluctant of the countries to join the CACM, and why Panama, in 1982, still remained outside. In general, those countries where agriculture plays the smallest role in the overall employment structure enjoy not only the highest *per capita* productivity, but also the most rapidly rising overall standard of living. This is a familiar pattern in the developing world. A less universal feature of such developing countries, and certainly not one of some others in Latin America, is the good record of economic growth during recent years. Despite an explosive growth in population, the lot of the average citizen has improved.

Significant though the relative national standings may be at a political level, and important though international boundaries remain in affecting economic developments, many of the features of Central American life today can best be viewed detached from the political map. Some phenomena are common to all the countries; others are regional or local in their impact.

Rural conditions

Central America is one of the decreasing number of regions in the world where country people remain in a majority. In 1980 two out of every three Central Americans were classed as rural, compared with only one in two South Americans and one in three North Americans, and many of those classed as urban were farmers living in towns that could perhaps be more properly described as agricultural villages. Only one in four lived in towns or cities with populations of over 50,000. It is a situation which is changing under the impact of active demographic factors. For example, Central America has the highest crude birth-rate in Latin America: the 1970–5 rate of 42.3 per thousand compares with that of 38.3 in tropical South America, with 32.8 per thousand for the Caribbean islands and 23.3 per thousand in temperate South America (Fox and Huguet, 1977, p. 37). With death-rates ranging between 5.1 per thousand in Costa Rica and 11.6 per thousand in Honduras and little loss or gain of people by migration, the overall rate of population growth in 1980 was about 3 per cent per annum. Most of this increase accrues in the countryside where, for example, the population rose by about 3.6 million between 1970 and 1980. This growing pressure on the countryside is often overlooked because of the more conspicuous changes in the towns. For example, in 1970 Central America held only a dozen towns with populations of over 50,000, and only six of these (all national capitals) exceeded 100,000 people; only one (Guatemala City) passed the half-million figure. By 1973 the isthmus could boast its first millionaire city, and in 1980 San Salvador was pressing hard to become the second. Panama City, Managua and San José each had over 500,000 inhabitants and Tegucigalpa over 400,000. Yet dramatic though these recent changes in the cities have been, changes in the rural populations have been greater. The rural population of both Guatemala and El Salvador rose by over one million in the 1970s, of Honduras by two-thirds of a million, Nicaragua and Costa Rica by 400,000 each and Panama by 160,000.

Population pressures and land reform

In general, the countryside remains uncrowded. There are a number of areas, however, in which the pressure of population on local resources is becoming difficult, if not impossible, to contain (*Figure 4.3*). This is the case in certain of the upland basins of western Guatemala, in the core area of El Salvador, in the Meseta Central of Costa Rica, and perhaps in the area between the two lakes in Nicaragua, in the more densely settled parts of rural Honduras and in western

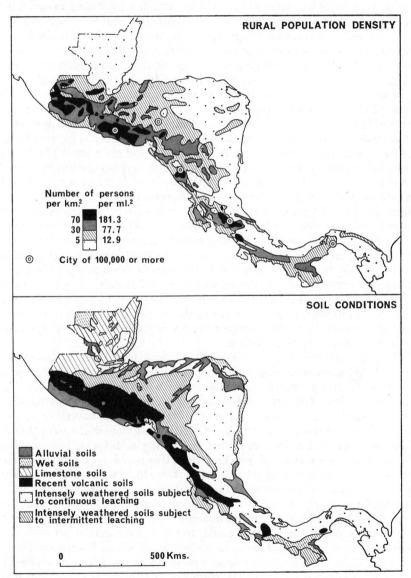

Figure 4.3 Central America: rural population densities and soil conditions

Panama. Rural population densities of over 200/km² exist in places in all the countries except Honduras and Belize, a figure in sharp contrast to the average rural density of below 24/km², and in even more striking contrast with the Caribbean half of the isthmus where rural densities almost nowhere exceed 5/km² (Nunley, 1967).

The *altiplano* of western Guatemala offers many overt signs of the impact of land pressure. It is a landscape in which steep slopes rim such lakes as Atitlán and the western volcanoes, while a blanket of volcanic dust has muted an earlier hill country to the north-east, and produced extensive flattish basins crevassed by deep *barrancas*. Substantial villages fringe the lakes and punctuate the basins, bearing witness to the heavy population; they may be overlooked by an old Maya temple reminding us of their long history of settlement. The remarkable feature of the region, however, is the almost complete cultivation of the land, no matter how steep the slope; it is particularly notable in a part of the world where, still, most of the land remains uncultivated. Apparently every hectare is carefully husbanded, producing a manicured landscape with nothing of the untidiness characteristic of most parts of tropical America. Slopes as steep as 30° are planted by hand, and terraces built to minimize the chances of soil erosion. On the flat land the tiny size of the individual plots accurately suggests the minimal size of most holdings, and helps explain the intensity of their cultivation; no land is wasted on field boundaries and few fossil structures are permitted to exist in the agrarian scene. This is Indian territory and, unlike most of Guatemala, has been largely protected since the Spanish Conquest from enclosure attempts. Under these circumstances the overwhelming concentration on food staples becomes readily explicable. In August the entire agricultural area is under ripening corn – usually maize, but with some wheat above 3000 m – usually grown alone, but sometimes in conjunction with beans and squash. The great range of corn grown – early and late, tall and short, hybrid and non-hybrid – ensures that the countryside, far from presenting the geometrical uniformity of mono-culture elsewhere, is a polychrome patchwork of greens and browns: a human-ized, not a mechanized, landscape.

Perhaps two-thirds of the families own land, but extremely few own more than 2 ha; the population has quadrupled since the beginning of the century, and in many *pueblos* the average landholding is now under 1 ha. Most of the holdings have been inherited, and the land owned by any one family may be scattered as a series of fragments. Land is sold, but only with the greatest of reluctance for it is a form of basic insurance; it is a measure of the dire straits into which many families have fallen that perhaps one-quarter of the existing holdings has been acquired by the present owners through purchase. Land that is sold is not normally sold to outsiders, and, since most of the *minifundistas* are Indians, the relatively homo-geneous patterns of landownership and rural practices are retained. This intro-spective attitude unfortunately makes innovations difficult to take root and capital for long-term improvements difficult to obtain. Little money is available for purchase of fertilizers or new seeds, and it is as difficult to persuade sub-sistence farmers to vary traditional habits here as elsewhere. Continuous mono-culture and the ever-increasing incentive to cultivate land which nature places at risk, have meant that average yields are probably declining, at a time when most yields elsewhere are rising. Yields of wheat and maize average only approximately half the figure on the Pacific Coast; a moderate application of capital and an effective agricultural advisory service could easily increase these figures. But a *per capita* income, including the value of home-grown food-stuffs, of under US $200 a year, a virtual lack of other funds and an understandably jaundiced view of outside bodies make improvements difficult to achieve. However, the emergence of market gardens in the eastern *altiplano* between Guatemala City and Lake Atitlán, where traditional methods of careful hand-tillage on tiny plots are combined with the use of modern fertilizers, insecticides, hybrid seeds, etc., to produce onions, potatoes, flowers and green vegetables, shows that it is possible to overcome these obstacles given appropriate market conditions.

The main agents of change in recent years have been the (Spanish) Catholic missionaries, allowed back into Guatemala after exclusion in 1956: they have been actively involved in the introduction of co-operative purchasing organizations and their influence helps to explain geographical patterns in the diffusion of the use of artificial fertilizers (Demyk, 1975, p. 320).

Few, even of the landowning families, can subsist on the products of their plots, nor need their cultivation absorb more than one-third of the working year. Forty per cent of the income of the *minifundistas* is now drawn from outside their holdings. The opportunities locally for supplementing income are small: the gathering of fuel and making of charcoal is one, but is becoming more arduous as an increasing demand means that more and more distant supplies are tapped; work in cottage industries is another. Public-works programmes are often run only on unpaid local labour. A few families have additional plots in other parts of the country, notably in the Costa, and some men work in the towns or as traders; but it is as migratory agricultural labour that most are able to improve their incomes. Between 200,000 and 250,000 families worked as migrant labour in Guatemala in 1965–6 and most of these were from the *altiplano*, although some came from Baja Verapaz and Jutiapa (Schmid, 1967, p. 1). During the 1970s and early 1980s the size of the coffee crops and the demand for seasonal labour have risen substantially. The coffee harvest of the lower Pacific slopes is the main draw, partly because it begins in November after the corn has been safely harvested at home. The coffee *finqueros* frequently work through contractors (*habilitadores*), who arrange for truckloads of Indians to report at the appropriate time. Sometimes the workers are able to pick over a coffee *finca* on the lower slopes (say about 500 m) in time to move to a higher *finca* in the main coffee zone, but most are away no more than 3 months. The newer cotton farms and sugar plantations on the Costa itself also absorb some of the migratory labour at harvest time, but, although wages are slightly higher, the harvests fall at less convenient times and working conditions are tougher. The complete lack of any shortage of agricultural labour on these commercial farms on the Pacific slope is a measure of the land hunger in the *altiplano*. Some move from the Costa to temporary employment in the towns or the forests before returning, such is the attachment of the Indian to his *tierra sagrada*.

A number of other escape valves have been opened up in recent years to supplement the temporary relief provided by seasonal work away from the *altiplano*. The improvement in the Inter-American Highway from Guatemala City via Huehuetenango to San Cristóbal de las Casas in Mexico has opened up the limestone soils of the western border to the Indian from the *altiplano*, and some have settled in adjoining Chiapas; such a move has changed only slightly the cultural milieu in which the Indians live. Some of the land in Guatemala was supplied from the National Finca, government lands taken from German coffee planters during the Second World War. The improved roads to Cobán and from the capital to Puerto Barrios on the Caribbean have also acted as catalysts to settlement, as will the road to Flores when it is secured. But the most important road to come into commission in recent years has been the Coastal Highway, a surfaced road running along the Pacific coastal plain from the Isthmus of Tehuantepec to the Gulf of Fonseca; it is along this road that the greatest expansion of cultivated land, and of commercial production, has occurred. The largely successful outcome of the campaign begun in the early 1950s to eradicate malaria from these lowlands has also played an important role in opening up this area, as has the world demand for cotton. Although much of the growth has been of large commercial farms raising sugar cane, cotton or cattle, there has been spontaneous and organized settlement by refugees from the *altiplano*, as well as by infiltrators from El Salvador. The best publicized of the organized settlements has been on lands between Tiquisate and Nueva Concepción, which were formerly owned by the United Fruit Company and grew bananas. In 1963, plots of 20 ha each were distributed to needy and favoured individuals, and the area now supports a population of about 5000 at a more-than-minimal standard of living. An equal amount of land has been distributed in other parts of the Pacific lowlands, although the land-reform programme in general lacks the impetus some would like to see it have.

Many of the characteristics of population pressure extant in the Guatemalan *altiplano* are duplicated in large parts of rural El Salvador and are most obvious in the east and north-east. Forty per cent of the farms in El Salvador are under 1 ha in area and grow subsistence crops; as in Guatemala, many farmers must supplement their incomes, and it is estimated that about 300,000 move to seek temporary work in the coffee *fincas* and in the cotton and cane fields at harvest time (West and Augelli, 1976, p. 424). But in other respects the situation differs in El Salvador. In the first place, it is a *ladino* population without the cohesion and traditionalism of the Guatemalan Indian. Although maize remains the main staple, sorghum is widely grown in the drier parts (intertilled with maize as an insurance against drier years), and the diet of the rural Salvadorean is more varied than that of his Indian neighbour. It is true that steeper slopes are cultivated by hoe and digging stick, but here the plough may be used on the flatter lands. Another important difference is that most of the smallholders rent their land from the large landowners, paying rent either in cash or kind, and only a minority holds title to the lands. Frequently leases are for a matter of months, and security of tenure for more than a year is rare. Such circumstances mean that the Salvadorean does not have the same attachment for the land as his Guatemalan counterpart and has shown himself more prepared to move: nowhere in El Salvador has population pressure built up to the same level as, for example, it has in the vicinity of Quetzaltenango. In the nineteenth century this mobility resulted in the widespread spawning of hamlets and *aldeas*, transforming the colonial landscape of compact villages and towns into one today characterized by a much more dispersed and even scatter of settlement than that found in any other area of comparable size in Central America. Even the poorer land in El Salvador is now in danger of over-exploitation, however, as the pressure mounts, and many Salvadoreans have found it expedient, if not mandatory, to seek a livelihood abroad. Some have taken temporary work in the banana plantations of the north coast of Honduras, and more have squatted on unused land across the Guatemalan and Honduran borders. A detailed map of rural population densities prepared in 1967 shows that there is, in fact, no break as the border is crossed (Nunley, 1967, p. 82). Rather, the level only dropped below 75/km² between 15 and 30 km deep into Guatemala, and it is no coincidence that the heaviest rural population density in Honduras, behind Choluteca, is only 30 km from the Salvadorean border. An estimated 200,000 Salvadoreans now live in Guatemala and 300,000 in Honduras, but, since illegal migration is a smouldering issue between El Salvador and her neighbours, reliable statistics are obviously not available. A measure of the scale of the phenomenon is given by the expulsion of about 130,000 Salvadoreans from Honduras in 1969, when the Hondurans decided to apply a dormant provision of their Agrarian Reform Law, which allows only native-born citizens to own land. These reverse migrants joined the growing army of unemployed in El Salvador, and many were drawn into the ranks of those who participated in the civil unrest of the early 1980s. With the exploitation of the new lands along the Pacific coastal highway settled in the last decade, there is little scope for extension of agriculture to uncultivated areas; the government is concentrating on encouraging the intensification of land use. The Lempa river is already harnessed as a source of power and of irrigation water, and further works are planned.

The only part of Nicaragua with a relatively high density of rural population is the neck of land separating the two lakes near the capital, Managua. The manner in which most people live in the countryside today follows patterns inherited from colonial days. Thus the majority work as labourers on what were, until 1980, large haciendas and now may be full collectives or co-operatives of family plots; most live in villages sited to serve the convenience of the haciendas. The improvement of the coastal road, to Tanque via Chinandega in one direction, and to Costa Rica in the other, and an anti-malarial campaign have opened up fresh lands and supplied some relief to labourers displaced by mechanization of farm work. The land bounding the Rama road, east of Lake Nicaragua, is potentially the most interesting prospect.

One-half of the population of Costa Rica lives, as in Nicaragua, within 80 km of the country's capital, and the populations of the two countries are approximately the same. Nevertheless, the rural population density around San José is four times as high as it is in the lake lowlands of Nicaragua, and appears to be reaching saturation point at current levels of management. The difference between the two areas is suggested by the fact that, whereas the proportion of the Nicaraguan population living in the lake lowlands is rising, that in the Meseta Central is dropping: all the municipalities around the Meseta have gained agricultural migrants for whom there is simply no room in the core region of the country. There are other differences. For example, the social structures of the populations are different. The large hacienda is foreign to the traditions of the Meseta Central and, although the average size of farm unit is rising, this is due to the operation of modern cost factors favouring larger enterprises and not to the existence of colonial or post-colonial land grants. The very small holdings are less common also, and three-quarters of all farms are owner-occupied. Family labour is more important than hired help. The economic and social ambience produces a more equitable system of rural exploitation, and the fabric of the rural landscape is of a finer, more even texture, reminiscent of certain parts of peasant Europe. Without the focus of the hacienda, the settlement pattern is irregular. Red-tiled and corrugated-roofed houses straggle along the close network of farm roads and, by their pattern and numbers, indicate the way in which pressure on the land has mounted. Some relief has been found by intensifying cultivation of coffee and, west of Alajuela and at lower levels, sugar cane; elsewhere in the higher, more temperate areas dairying is becoming more important in a mixed-farming economy that has been a feature of the Meseta since the end of the nineteenth century. The situation is a far cry from the very limited economy of the *altiplano* at similar altitudes in Guatemala.

Emigration from the Meseta Central has taken people in all directions (Sandner, 1962). Some have moved down on to the edge of the Caribbean low-lands, and at Siquirres, now accessible by road from the Meseta and long accessible by rail, white and negro smallholders come face to face, both assisted by their government through various incentives to be successful colonists. Former banana lands are being rehabilitated, and new villages, with social and other services supplied, have been completed. Other moderately successful colonies have been established on the northern side of the volcanic range. But the most attractive areas of spontaneous settlement have been on the Nicoya lowlands on the drier part of the Pacific coast, and on the alluvial soils of the Valle del General in the south. The Inter-American Highway first penetrated the Sierra de Talamanca in the 1930s, and San Isidro became the regional centre for the 60,000 people who now live here, 500–700 m above sea-level, within the *tierra caliente*. The dirt highway continues southwards, and has attracted settlement, aided by the stimulus of successful banana plantations along the coast. The border between Panama and Costa Rica is now quite densely settled, colonists from Italy as well as the Meseta Central being grafted on to the local population.

Rural population densities are low in Panama, but because shifting agriculture is still important, a figure of 40/km^2 for the Azuero peninsula creates pressures that other agricultural systems would not; new roads and the possibilities of irrigation should moderate a situation in which demographic growth has only just begun to be matched by economic changes.

Frequently linked to the question of rural population pressures is the question of land reform. It is argued that the traditional pattern of rural life that prevails over most parts of Central America is inimical to the release of those pressures. *Table 4.1* shows that in 1970 almost half (44.3 per cent) of the families relying upon farming for their livelihood in Central America (excluding Panama) either had no land at all, or possessed plots too small to supply even minimal subsistence; a further third (32.2 per cent) survived on tiny land holdings which averaged no more than 1.1 ha in area. In contrast, 4.4 per cent of the families owned almost three-quarters (73.2 per cent) of the land. Indeed, it can be cogently argued that the 'soccer war' of 1969 between El Salvador and Honduras was fuelled not so much by absolute pressure of peasants on the land as by the increased competition between the peasants and the large landholders on both sides of the border (Durham, 1979). In El Salvador the large commercial farmers successfully resisted the Agrarian Transformation Movement, and reduced the amount of land available to peasants to the degree that many were forced to migrate (Browning, 1971, pp. 271–303); within Honduras the impact of the migrants was felt most strongly by the larger landholders, rather than by the Honduran peasantry, and it was they that engineered the expulsions of 1969. In doing so the Hondurans capitalized on a popular national disenchantment with the results of membership of the CACM. The aftermath of the war in Honduras saw a strengthening of the peasant organizations and a more determined move towards agrarian reform. In El Salvador no accommodation was made towards the position of the peasants, the large landowners retrenched, and the battle lines separating the rich from the poor were more firmly drawn.

Table 4.1 Agricultural holdings in Central America (excluding Panama) in 1970

Size of family holding	Area '000 ha	% total	No. of families '000	% total
Landless families	0	0	476.0	27.7
Under 0.7 ha	85	0.6	285.0	16.6
0.7–4 ha	868	6.0	552.0	32.2
4–7 ha	583	4.0	126.5	7.4
7–35 ha	2,350	16.2	181.0	10.5
35–350 ha	5,121	35.2	68.8	4.0
Over 350 ha	5,535	38.0	7.0	0.4
Total (incl. others)	14,542	100.0	1,717.2	100.0

Source: Fox and Huguet, 1977, p. 17.

In most parts lip-service has been paid to the need for rural change, but little has been achieved. Guatemala has its Instituto Nacional de Transformación Agraria, which has redistributed the former German-owned coffee lands and the banana lands formerly owned by the United Fruit Company. Nevertheless, little has changed since the mid-1960s when only 1.6 per cent of the agricultural families owned 72 per cent of the land, and accounted for over 56 per cent of the value of agricultural production (Pearse, 1966, p. 62); one thousand large estates occupied over half of the land in cultivation, while 27 per cent of the families were without any land to their name. In El Salvador, as elsewhere, the contrast between the small and the large landholdings grew with time. In 1839 the top 10 per cent of the agricultural population owned 40 per cent of the land, in 1932 65 per cent, and in 1971 80 per cent (Durham, 1979, p. 45). Recent civil violence has overshadowed the impact of the US-inspired land reform decree of 1980; yet by mid-1982, 223,000 ha had been transferred to 178,000 beneficiaries. In Nicaragua, before the Sandinista Revolution of 1979, one-third of all the privately owned land was divided amongst only 350 estates and probably even fewer families; in contrast the poorer 50 per cent of rural families shared no more than 3.4 per cent of the agricultural land. Costa Rica has a different tradition of smallholdings, but even so there are parts (for example, Guanacaste) where the statistics on rural income distribution resemble those of the rest of Central America. Even in Honduras, where agrarian reform and land redistribution has gone further than in most of the republics, 5 per cent of the agricultural population received 40 per cent of the agricultural income in 1977, whereas those 65 per cent of families with holdings of under 1 ha shared only 23.6 per cent of the national income derived from agriculture: the disparity in family incomes was twentyfold.

It is foolish to place too much reliance on statistics on incomes in Central America, and a definitive cadastral survey has yet to be completed. Nevertheless, the data make it clear that two different social and economic classes characterize most of Central America; they are widely separated by a deep, and apparently widening gap. It is perhaps fortunate that dense rural populations are the exception rather than the rule. They are exceptions of long standing. A map of the pre-Spanish population of Central America would show broadly the same distributional pattern as now, although without the same degree of concentration in the Meseta Central of Costa Rica. It is notable that this long-standing pattern broadly corresponds with the qualities of soils found in Central America (*Figure 4.3*): the highest densities are associated with the young soils of recent volcanic origin subject to only intermittent leaching, while the older, intensely weathered soils carry only light populations; the heavier the rainfall, the more continuous the leaching of the soil and the lower the rural population density. Until a century ago the correspondence between the volcanic soils and the populous areas of Central America was even stronger than it is today, and the growth in importance of coffee as a commercial crop from the 1870s onwards did nothing to weaken the association. The moderate-to-heavy population densities found on many of the pockets and zones of alluvial soil in Central America are largely a recent phenomenon, and their settlement has normally awaited the application of considerable capital resources and technological expertise. Ports, railways and roads have made them accessible, and public health campaigns have made them habitable. Voluntary migration to these areas has been a much more important demographic fact than the organized land-settlement programmes of the various official agrarian reform and colonization agencies that have sprung up in recent years. Nevertheless, the large landholder has, in the nature of things, been more important and played a more positive role here than elsewhere in Central America. The incentive for the development of the alluvial lands has been the commercial profits to be won in the export markets, and it is the importance of the crops grown for foreign consumption that set these alluvial soils apart from most of the rest of the isthmus.

The export crops
The dependence of the economies of many of the Central American countries upon a small range of certain rural products is notorious. In typical years over half the total value of goods exported is contributed by only three products: coffee, bananas and cotton. The proportion of the total and the contribution of each item varies substantially from year to year as commodity prices fluctuate. In the 1960s coffee provided 33 per cent of the export income of the isthmus, and both bananas and cotton 15 per cent each; in the 1970s the figures were 30, 13 and 9 per cent. Soaring coffee prices in 1977 pushed the share of coffee up to 50 per cent from 34 per cent in 1976, and the contribution of the three crops to over 70 per cent. This was the same contribution as they made in 1967, another unpredictable boom year for the Central American economies.

Coffee   Coffee was the first of the staples to emerge in Central America; it was first grown there on a commercial scale in the late eighteenth century and had become securely established in Costa Rica in the 1830s. It spread to El Salvador and Nicaragua in the 1840s and to Guatemala in the 1860s; it has attained importance in Honduras only since about 1940. It is a highly profitable crop in Central America, and production has risen markedly during the last 20 years: annual coffee production in Central America is on a par with that of Colombia, and represents about 12 per cent of the world trade in the bean.

Several circumstances made coffee successful in Central America. In the first place the region offers excellent natural conditions for the coffee plant, which needs a seasonal, almost monsoonal climate: wet during the early part of the year, but followed by several dry months for harvesting the bean. Such a climate is typical of the Pacific slopes of the isthmus. Most commercial coffee is grown at altitudes between 300 and 1700 m, where the narrow annual temperature range falls within the appropriate limits for the plant; the best coffee (that is, mild coffee commanding the highest price on the world market) is grown near the upper altitudinal limits where the bean takes longest to mature.

Coffee grows best in well-drained, fertile soils and the shrubs can be planted safely on steep slopes without necessarily exposing them to erosion since most coffee is grown under shade trees. The soils of the volcanic slopes of the Pacific side of the isthmus are ideal. In some places the shade is from natural woodland, in the lowlands it may be from bananas and elsewhere from trees specially planted for the purpose. In El Salvador trees whose leaves hang pendant during the night are used, for these generate exactly the right local climatic conditions for the coffee below.

Successful coffee cultivation calls for generous supplies of cheap labour, and Central America offers this in abundance. About a million temporary workers help to bring the coffee harvest home each year. The workers include entire families, and they are paid partly in cash and partly in rations of food. Total labour costs on a typical Central American coffee *finca* would normally amount to no more than one-quarter of the f.o.b. price of coffee at the port. Further, reasonable accessibility to shipping points and to the main market, the United States, may be added to the advantages which Central America offers coffee producers.

There are some 150,000 coffee farmers in Central America. This statement in itself might be misleading; 66 per cent of the total area in coffee and between 80 and 90 per cent of coffee production is in the hands of the top 5 per cent of farmers. The concentration of production by a few individuals is more striking in Guatemala and El Salvador, which account for over half the coffee produced, than in the isthmus as a whole. The trend is towards an even higher degree of concentration. Many of the larger coffee plantations in these two northern republics are very well managed, and yields may reach 2000 kg/ha. By investing in improved varieties of coffee plant, double-planting the coffee rows, making increasing use of fertilizers and insecticides, and making economies, production costs per hectare can be kept low. Production costs commonly absorb only a minority of the price received for the coffee, giving the grower an income (on which he will, quite legitimately, pay little or no tax) which may be in excess of US $1000/ha. In El Salvador, where the pressure on land is greatest, the average yield is now about 1000 kg/ha, compared with a figure of 700 kg/ha for the isthmus as a whole.

Yields are highest in Costa Rica, where the natural conditions provided by the Meseta Central for the growth of high-quality coffee are almost ideal, and where pressure on the land has induced the adoption of the most modern and intensive methods of coffee cultivation. Costa Rica produces as much coffee as do Nicaragua and Honduras combined, and that is from only one-quarter of the area; in recent years production has risen, but the area in coffee has remained the same. One feature of Costa Rican modernization has been the amalgamation of many of the traditional small farms to form larger working units. In consequence, even though the structure of land tenure is less unbalanced than elsewhere, one-half of the coffee crop is now produced by fewer than 2 per cent of the producers. The point above which modern intensive production methods begin to lower production costs is measured by a yield of about 1000 kg/ha. This was the average yield first reached in Costa Rica in 1966; by 1976 it was 1260 kg/ha and rising.

Yields in Honduras and Nicaragua are between only 250 and 400 kg/ha. Natural circumstances are less propitious to coffee cultivation, and cultivation methods, although changing in Nicaragua, remain largely traditional. Nevertheless, coffee is a well-established crop in Nicaragua (Radell, 1964). Its first commercial success was due to the demands created by the forty-niners, who followed Vanderbilt's route across the country on their way to California; it remained the most valuable export of the country for almost a century. Half the coffee comes from the volcanic uplands to the south of Managua, but here periodic emissions of sulphurous fumes and cinders from the Masaya Caldera are a hazard to the crop, and the scarcity of water in the scorching dry season limits the processing of the coffee. Production is on large haciendas, 90 per cent of which were owned by absentee landlords in 1978, and a higher proportion of the labour is seasonally employed than is the case elsewhere. Most of the rest of Nicaragua's coffee comes from smaller plantations in the highlands behind Matagalpa, where the cultivation methods are extremely poor, but the coffee comes from above the 800 m contour line and its better quality commands a higher price than coffee from Masaya. Labour is in relatively short supply, and frequently part of the crop remains unharvested.

Coffee has only recently been introduced into Honduras. It is suggestive that this should have occurred only since immigration from El Salvador has reached substantial proportions. It is particularly in the areas adjoining the north-eastern border of El Salvador and the area behind Choluteca that coffee is grown. *Fincas* are small, traditional methods are employed, and yields are poor.

One of the interesting aspects of coffee production in Central America has been the stimulus it has provided for the development of improved transport routes (USAID, 1965). The first railway in Guatemala (1877) was built to link the coffee market of Escuintla with the sea at Puerto San José; by 1904 a line ran at the foot of the whole of the coffee zone to the west of Escuintla, and a link to Guatemala City had been extended across the country to Puerto Barrios, providing a direct Atlantic outlet for Guatemalan coffee. The Verapaz railway in eastern Guatemala was built in 1884 specifically to serve the coffee interests, and it remained open until 1963. In El Salvador the assembly, link by link, of the existing railway system in the late nineteenth century went hand in hand with the reorganization of the land-tenure system (which favoured the creation of large estates at the expense of the communal village lands or *ejidos*), and both were spurred on by the profits to be won from large-scale coffee production. In Nicaragua the fragmentary road, railway and river-transport systems have always been inadequate. Economic development in general has suffered; even today some of the coffee produced travels by mule train on its way to market (at an approximate cost of ten to twenty times the comparable rates by lorry or rail). The need for a Caribbean outlet for the coffee produced in the Meseta Central was the major factor behind the decision of the Costa Rican government in 1871 to com- mission the Northern Railway; it was from Minor Keith's association with this enterprise that there sprang the subsequent involvement in the isthmus of the United Fruit Company. Although most coffee now travels by the new roads, there is only one road competing with the Caribbean section of the Northern Railway, and Puerto Limón, to which the railway leads, still handles a proportion of the coffee exports of the country.

Bananas Good transport routes have been of even more critical importance in the development of the banana economy of the isthmus. This is in part because bananas are a perishable commodity and must be consumed within weeks of being cut, even if they are cut green: unlike most fruits, whose markets have been expanded by modern preserving methods, almost all bananas are eaten fresh. Further, bananas, on a weight-for-weight basis, are far less valuable than coffee and so are far less able to carry anything other than low transport costs. Fast and frequent shipments to market demand large-scale operations and heavy capital investments; were it not for the fact that bananas, unlike coffee, may be harvested throughout the year, such investments would not have been forth- coming. Capital was necessary because in the areas best suited to bananas – the virtually unpopulated alluvial lands of the wet Caribbean coast – not even a rudimentary transport system existed; in contrast, the coffee economy flourished in areas already long populated and under circumstances in which production could be expanded fairly easily as markets emerged.

The first commercial banana shipments of any importance were made from the Caribbean lowlands of Costa Rica at the time when the Northern Railway was under construction. The difficult engineering works, disease and other hazards made construction unexpectedly expensive, and banana shipments were begun as the Caribbean spur was extended to help recoup some of the high outlay. By the end of the nineteenth century three districts (the area tributary to Puerto Limón and the Northern Railway, the Bocas del Toro area across the border in Panama and the Bluefields region of Nicaragua) had all been transformed from chaotic tropical rain-forest into geometrical plantations of bananas. Many of the pioneers in bananas, as in coffee, came from outside Central America. But whereas some of the coffee-growers came to identify themselves (sometimes involuntarily) as Central Americans, the creation of the United Fruit Company in 1899 and then of the Standard Fruit Company in 1924 meant that the banana economy became identified with alien, United States interests, and this element of national life became a butt for local politicians.

Between 1900 and the peak year of Central American banana production in

1930, shipments increased fivefold. This was due largely to the taming of the alluvial valleys of the north coast of Honduras and of the lower Montagua valley of Guatemala. In 1930 one-third of the world's trade in bananas originated in Honduras. The narrow-gauge railway of the Standard Fruit Company was built to serve the eastern plantations of the northern Honduran coast, and the United Fruit (Tela) and Ferrocarril Nacional de Honduras system was built to serve the west; the restricted purpose of the railway is underlined by the fact that there is still no land link, by road or rail, with adjoining Guatemala. Railway construction, land clearance and drainage, the building of ports, towns and plantation facilities, all required the importation of labour; as with the building of the Panama Canal, much of this labour came from the British West Indies, particularly Jamaica, and many stayed on to man the plantations once installation works had been completed. Although some of the negro population of the Caribbean coast of Central America dates from long-standing British interests in the area, most of it is a phenomenon of the twentieth century. Today it is in process of dilution as access from the interior has improved, and mestizo labourers are finding their way to the coast.

During the 1930s, 1940s and 1950s the banana economy suffered a series of setbacks. Panama disease swept the plantations and the Sigatoka leaf blight made matters worse. There is no cure for Panama disease, and many plantations were abandoned. There was a wholesale shift of interest to the wetter parts of the Pacific coast, notably to the Chiriqui area of western Panama and southern Costa Rica, to the alluvial soils of the Golfo Dulce and Parrita in Costa Rica, and to the well-drained, sandy soils of the Tiquisate area of Guatemala; but costs were higher. The economic depression of the 1930s and the disruption of trade during the Second World War reduced production to the level of 1905, but post-war recovery was rapid, and production in both Costa Rica and Panama soon reached record levels. There was a shift to new varieties of banana much more resistant to Panama disease, less vulnerable to high winds and giving higher yields; they are, however, more fragile, and must be shipped in boxes rather than on the stem. The box is important. On the one hand it has created a demand for locally produced cardboard: box-making provides jobs for a quarter of the labour force on the plantations, and the box represents a quarter of the f.o.b. price of boxed bananas, thereby extending substantially the value of bananas to the economy of Central America. On the other hand boxing bananas has reduced handling costs and facilitated direct retailing in the United States market.

The position of Central America *vis-à-vis* the other major American producer of bananas, Ecuador, has improved. Although Ecuador is free from hurricanes and was developed as Panama disease swept Caribbean Central America, it, too, is now ravaged by the plague, and is less able to meet the standards of quality now demanded by the United States market; it is also twice the shipping distance from New Orleans than is Puerto Limón, and bananas from Ecuador have to bear the considerable cost of tolls through the Panama Canal.

Within Central America there have been a number of interesting changes in recent years, and many of these have been occasioned by changes in the ownership of the plantations and in the relationships between the companies and the individual states. No longer do the two traditional fruit companies dominate the banana economies of Central America. The hegemony of the United Fruit

Company and the Standard Fruit and Steamship Company was broken in 1967 when Del Monte, the Hawaiian-based fruit-canning company, now a subsidiary within the R.J. Reynolds tobacco empire, purchased some banana plantations in Costa Rica. At about the same time some European money was invested in Costa Rican banana production. In 1972 Del Monte extended its holdings in Central America with the purchase from United Fruit of the Bananera plantations in Caribbean Guatemala. These had been allowed to run down by United Fruit, who had been loath to invest in an area where not only were floods and hurricanes genuine hazards, but political unrest was rife. By 1980 Del Monte had transformed these plantations into reputedly the most productive in Central America, and the company could claim 10 per cent of the world trade in bananas. The move into growing bananas by Del Monte was against the more general trends of the 1970s as demonstrated by the two longer-established banana-producing companies. They had begun to withdraw from growing their own bananas and, particularly in Costa Rica and Nicaragua, were establishing themselves as marketing agents for independent producers. This shift in emphasis was prompted partly by the growing use of the companies and their plantations as convenient targets at which to direct the growing disquiet over the inequitable pattern of ownership of rural land. Land owned by foreign corporations became an increasingly emotive popular issue. The policy of divestment of land was endorsed in 1969 when United Brands, the large and diversified food concern, purchased the United Fruit Company. The policy has been carried furthest in Costa Rica, where a *modus vivendi* appears to have been established between the government and the companies. The government, as part of a policy to promote settlements outside the Meseta Central, has helped finance small producers; the companies have assisted with stocking materials, advice and long-term (10-year) purchasing contracts. The construction in the 1970s of new feeder railways, and the extension of the fragmentary road network on the Caribbean coastal plain have helped expand the volume of banana shipments through Puerto Limón. Yet in spite of a reduction in the involvement of the companies in the plantations the three continued to employ over 24,000 in Costa Rica in 1979.

In Honduras traditional arrangements persist, but there are signs of change. In 1978, in an attempt to relieve political and social tensions, the Honduran government expropriated 4500 ha of land belonging to the Tela Railroad Company (United Brands) and Standard Fruit. The tensions were partly caused by damage in the wake of the 1974 hurricane 'Fifi', which devastated 70 per cent of the banana plantations. The top-heavy banana plant is particularly vulnerable to damage by high winds, the low-lying plantations to flooding, and the north shore of Honduras particularly open to hurricanes. It took 5 years for Honduran banana output to return to the 1973 level, and for yields of 30,000 kg/ha to become normal again. The effect of the hurricane on production was exacerbated by the spread of Sigatoka disease into the Ulúa valley. In Panama during the same period the export of bananas was similarly cut, but the cause was a dispute between the government and United Brands. The dispute was resolved by the purchase by the government of the plantations in Chiriqui and their lease back to the company for a 5-year period. This offers another example of the apparent loosening of the physical ties of the multi-national fruit companies with Central

America. It would be a mistake, however, to underestimate the continuing importance of the companies to Central America. A striking indication of that importance was provided in 1980 when the United Nations Food and Agricultural Organization (FAO) revealed that two-thirds of the world's potential breeding stock of bananas was possessed by United Brands.

Cotton Large-scale cotton production is a much more recent development in Central America, and it is only since the Second World War that cotton has become significant. Production in the middle 1960s was ten times the volume of the early 1950s; in 1965 exports of cotton from the isthmus were marginally more valuable than were exports of bananas and accounted for over a third of her foreign-exchange earnings. In the 1970s the fraction fell to less than a tenth. This was less than in the 1950s, but three times as much cotton was baled in the 1970s as in the two decades before. In the late 1970s Central America supplied 6 per cent of the world trade in cotton, and only Brazil in Latin America produced more. This extraordinary development was a response to the rising world demand for cotton. One of the most important internal circumstances which permitted this reaction was the construction of roads in the Pacific coastal plains of the northern republics, which opened up large areas of land highly suited to cotton cultivation. The littoral highway in El Salvador was built in the early 1950s, and made the alluvial soils east of La Libertad to La Unión easily accessible to the capital; most of El Salvador's cotton comes today from this area. In Guatemala the newer Pacific coastal road to Mexico has similarly attracted plantations during the last decade. The coastal road from Managua to León was extended to Chinandega and Corinto in the late 1950s, and the rich volcanic soils of this new area have become the most important single area of cotton-growing in the isthmus. More recently in Nicaragua the Managua to San Benito section of the improved Inter-American Highway, before it leaves the alluvial soils of the Lake Managua lowlands behind, has become bordered by cotton fields. Land of cotton quality is in shorter supply in Honduras, and the Choluteca region is the only area of intensive cultivation, cultivation which awaited the surfacing of the Inter-American Highway in the early 1960s. In Costa Rica and Panama land which might grow cotton is better suited to rice, and rice is the crop preferred.

Nicaragua produces about 40 per cent of Central America's cotton, and the sale of cotton supplies about 25 per cent of her foreign exchange; in no other Central American country are these proportions as high. This situation is in part because coffee and banana cultivation is less advanced than in the other countries, but the major causes are the more favourable conditions for cotton. Average cotton yields were higher, production costs lower and profit margins wider in the 1970s in Nicaragua than in any of the other countries of the isthmus. It seems that the risk of adverse weather conditions in Nicaragua is less than in, for example, El Salvador; the cotton areas of El Salvador have suffered more from too dry conditions than have those of Nicaragua, where the climate appears to be more reliable. Nicaragua has been freer from insect pests and plagues than the other countries until very recently, and has been saved some of the cost of expensive insecticides and spraying. Cotton cultivation is more mechanized, cotton farms before the 1979 revolution were larger and tillage practices are more advanced; it is the only country in Central America where the greater part (60 per cent) of cotton is machine-picked, although at some loss of quality (Cruz and Hoadley, 1975, p. 457).

Guatemala vies with Nicaragua as a cotton producer. Yields in both countries have been consistently rising, and it is a remarkable fact that there have been returns of 2400 kg/ha (in 1977) from some of Guatemala's cotton land and an average of 2300 kg/ha (in 1972–3) from those of Nicaragua. Central America can claim some of the most productive noñ-irrigated cotton farms in the world. The cotton farms in Guatemala are large and highly mechanized in most phases of their operation, although they also employ large numbers of migrant pickers at harvest time. Ownership is restricted: in 1966–7, there were fewer than 400 cotton farmers in Guatemala as compared with over 3000 in El Salvador and over 4000 in Nicaragua. Prospects for the future expansion of cotton-growing are probably brightest in Guatemala, where large areas of the coastal plain seem physically suited to cotton production.

Most of the rest of the cotton produced comes from El Salvador, though here circumstances make it a less profitable crop. Yields are much lower, perhaps only one-quarter to one-third of those in Guatemala, and have fallen; costs are also higher. The fertility of the cotton soils of El Salvador, amongst the first in Central America to be planted in cotton, is dropping, and expenditure on fertilizer, which is mounting everywhere in Central America, is highest in El Salvador. The much smaller size of the individual cotton-farmer's holding and the fact that most cotton land here is rented, encourage a non-conservationist approach to farming. Monoculture, the multiplicity of farms and the longer period during which cotton has been cultivated, have all allowed plant plagues to establish themselves and spread more widely. Costs of buying and applying insecticides now represent fully one-third of total cotton-production costs. For these and other reasons the area under cotton has contracted sharply, and recently many marginal producers have gone bankrupt or have switched to other crops. Half the cotton fields have reverted to maize, and most of the rest have gone back into pasture. The very close correlation since 1950 between cotton production and the import of basic foodstuffs, particularly maize, has been maintained, but the trends in values have been reversed (Durham, 1979, p. 33).

Although cotton production is small in Honduras, yields are high, and Choluteca resembles many of the other towns of the cotton areas – flushed by the success of the cotton bonanza and alive with advertisements for various brands of fertilizers, insecticides, machinery and seeds, and for the cotton co-operative. Here, as in El Salvador, the smallholder is the typical cotton farmer, and in both countries he is backed by strong co-operative processing and marketing organizations.

Marketing Marketing is, of course, one of the crucial aspects of the cotton, coffee and banana economies of Central America. It is worth reiterating that one-half or approaching one-half of the export earnings of all and each of the Central American republics in recent years has been derived from the sale of one of these three agricultural products. A fall in world commodity prices quite outside the control of Central America may have crippling consequences upon the local economies.

The degree of vulnerability to price changes of the commodities and the countries varies. Banana prices are probably the most stable. Until recently most of the production and all the marketing of Central American bananas was controlled by the United Fruit Company and the Standard Fruit Company: almost all the bananas exported were destined for the United States, where the two companies account for a large majority of the market for bananas. The addition of Del Monte to the banana companies in the 1970s did nothing to change the nature of the trade nor the dependency of the banana economies on the multi-national fruit companies. It is a dependency which the Central American countries resent and have begun to challenge. For example, since 1972 an export tax has been levied on bananas in an effort to retain more of the value of the product in Central America; by 1978 the proportion of the value retained in Central America through the application of such taxes and the payment of wages and other local purchases had risen to 17 per cent of the final retail price, but still remained relatively small. Some Central American countries have argued that the proportion is low because of a deliberate under-valuation of the bananas at the stage when they are exported from Central America to the marketing area of the companies overseas. In part to counter this possibility and in part to try to extend the market for Central American bananas beyond the United States, Spain and Western Europe, the Costa Rican, Panamanian and Colombian governments established in 1977 their own marketing body, Comunbana. In 1978 the two state-owned banana companies in Panama supplied the first fruit to be exported through the new venture. But to date the quantities of bananas supplied to the new outlet have been small, the market has remained neither an elastic nor an expanding one, and the advantages of longer-term contracts at stable prices with the traditional companies are not lost on the banana growers.

Japan is the main market for Central American cotton. It was the economic resurrection of Japan during the 1950s which was the key stimulus in creating a minor cotton boom: between 1958 and 1965 the cotton trade between the two areas quadrupled. The Japanese stake in Central American cotton has gone beyond the mere commercial: Japanese expertise, advice and credit have also been made available, while quality standards have established a degree of uniformity in the product which may not be altogether to the advantage of Central America. World cotton prices have generally subsided since the boom years of the 1960s, although untoward circumstances outside the control of Central America have led (as in 1973–4) to a short-term doubling and a subsequent halving of the price. Most cotton farmers in Central America have been able to survive on the reduced margins of recent years; the only notable exceptions have been in El Salvador. One of the lessons being learnt in Central America is that there is room for a better co-ordination of the flow of information on international market conditions amongst the growers and for an improvement in domestic marketing arrangements. The application of these lessons would allow a higher proportion of the value of the cotton crop to remain in Central America, and so provide a cushion against any further deterioration in the terms of the trade in cotton, a change which Central America is almost powerless to influence (Cruz and Hoadley, 1975, pp. 453–89).

The coffee-marketing situation is more critical, partly because of the greater importance of coffee in the economy and partly because of the volatile nature of coffee prices. As an illustration of both characteristics, coffee exports from Central America in 1976 were worth about US $1025 million and represented 31 per cent of the total trade of the isthmus; in 1977 they were worth almost double (US $1940 million), boosting the total export earnings of Central America by fully one-third, and the share of coffee to 44 per cent. The price of coffee reflects world supply and demand, with Brazilian production and United States consumption being the major contributors to the equation. The 1977 bonanza to the coffee growers of Central America was due to a disastrous frost in São Paulo and Paraná in July 1975 which destroyed immense quantities of market-ready coffee beans, and reduced the 1976-7 contribution of Brazil to the world's coffee markets by three-quarters. The effect on the income of Central America was dramatic, and was magnified by the unusually low price for coffee which had prevailed in the conditions of over-supply of the early 1970s. Between 1973 and 1975 the average coffee price ranged between 62 and 72 US cents/lb: these quotations were on a par with those of 20 years before, but twice those of, say, 1963. The cold wave in Brazil in 1975 meant that the average daily price of coffee on the exchanges in April 1977 reached US $3.15/lb. By July 1978 it had fallen to US $1.30/lb at which level, allowing for inflation and the weak performance of the dollar, it remained until the beginning of the 1980s. Valuable though such unexpected windfalls are to Central America, the inevitable reaction brings hardship in its wake. It is partly to mitigate against the effects of such enormous swings in the price of coffee that the first International Coffee Agreement was brought into being in 1962. The effect of this Agreement was to allocate to each producing country within the Agreement a quota of the anticipated market of the major consuming countries, all of whom are signatories to the Agreement, and so maintain prices between agreed levels. International co-operation has been intensified in recent years, and the Central American countries have joined other Latin American coffee producing countries to co-ordinate marketing policies and, in particular, to use their buffer stocks more effectively to balance supply and demand. Nevertheless, the economies of Central America remain particularly prone to the effects of fluctuations in the world prices of coffee (Struckmeyer, 1977).

Other economic products From this review of the three key export commodities of Central America, it may be evident that there is little room for a sustained increase in the income to be won in this field in the near future; a modest expansion in production may be anticipated, but a decrease in the real unit price seems to be more likely than an increase. A similar assessment and a reluctance to be so dependent upon such a limited range of products has encouraged Central American governments to promote a number of policies designed to strengthen and diversify their economies.

These policies have taken two directions. One is towards diversification of production for both the domestic and export markets; and the other is towards greater international co-operation leading to certain market advantages of scale not open to the countries of Central America individually. Most attention in the matter of diversification appears to have been devoted to measures designed to encourage the development of manufacturing industries; the success of this endeavour is considered below. It may, however, be argued that diversification of the agricultural sector offers a greater chance of long-term gain. The concentration of agricultural production on coffee, cotton and bananas for the export market and on maize and beans for domestic consumption is paralleled in a geographical sense by the concentrated exploitation of very restricted parts of Central America to the virtual exclusion of other, much more extensive, parts of the isthmus. Diversification policies in the field of agriculture have been aimed, on the one hand, at attempts to reduce the degree of concentration on certain products in areas already settled and well established and, on the other hand, at the exploiting of lands whose resources remain virtually untapped.

Agricultural diversification

It is not easy to find a satisfactory substitute for coffee either in the farming economy or in the national commerce. Much coffee grows on slopes too steep to serve as ploughland, and the profits and privileges to be won from coffee cultivation would be difficult to match in another crop. Further, it appears that, within the coffee-growing regions, returns are governed by the quality of farming enterprise rather than by natural conditions: high yields mean adequate profits and no incentive to diversify; poor yields can normally be equated with a low level of modern farm-management skills, and this is an unpromising situation for any would-be innovator hopeful of commercial success. It has been argued that the taxation on coffee and coffee lands could be increased; this not only might make alternatives to coffee relatively more attractive, but might also raise the funds available to help the diversification programme. But, even at the present low levels of taxation, there is widespread evasion of payment in, for example, El Salvador and Guatemala, and it seems politically unrealistic to anticipate the kind of drastic change in this direction which might substantially help a diversifying programme. Locally, alternative crops have replaced coffee. The expanding urban markets for fruit and vegetables have led to some substitution near the capital cities, while tobacco cultivation has received a fillip, notably in Nicaragua, from Cuban refugees. Elsewhere normal market forces have led to the failure of marginal producers and to a less intensive use of the land. In the context of substitute products for export, it must be remembered that Mexico is in a better position to supply the off-season demands of the United States market, and that her marketing arrangements are often superior.

The situation in the banana and cotton lands is more open; here the policy is not so much one of diversification through substitution as diversification through addition (Stouse, 1967). Cacao, rubber and oil palms are the commercial companions most frequently proposed for bananas and all have been tried. For example, Guatemala supplies one-third of her own rubber requirements from domestic sources. Yet there are serious problems in such substitution: for example, the proposal by United Brands in the late 1970s to replant all their remaining banana plantations in Costa Rica with African oil palms was resisted by the government on the grounds that unemployment would result. Change is more difficult for the ordinary Central American peasant to contemplate because of the absence of any return during the first years of most new products, and because of his greater uncertainty about prospective markets. The proven commercial alternatives to cotton are sugar, maize and rice. Land withdrawn from cotton cultivation in El Salvador is most frequently turned over to commercial crops of maize; in Costa Rica and Panama rice is more favoured, while in Guatemala sugar cane is a more likely contender. All these alternatives are destined mainly for the local consumer market, and the prices received by the grower are higher than those obtained on the world market. Sugar is the only one of these products which has an export market, and this is a protected one: the key factor is the United States sugar quota, and this may be further increased. Sugar exports still account for less than 3 per cent of the total value of Central American exports, and even an increase in income of one-third would have less effect on the economy of Central America than a rise of only 1 US cent/lb in the f.o.b. price of coffee. Citronella and lemon grass are two other grasses of minor commercial interest; oils for perfumes and vitamin preparations are distilled from them, and their cultivation, which is being extended, is a speciality of the Mazatenango–Tiquisate area in Guatemala.

Pastoralism Some agricultural economists see the most hopeful future for grass in Central America as the basis of a beef and dairy products industry. Cattle-raising is no novelty in Central America, of course, and many of the traditional features of the colonial cattle ranch still survive, especially in parts of western Panama, Guanacaste, interior Honduras, El Salvador and Pacific Guatemala. Cowboys still solve the problems of a lack of herbage during the dry season by rounding up the cattle and trailing or trucking them up to the high pastures or down to the coast. For several centuries the Pacific coastal plains of Guatemala, El Salvador and Honduras have fattened cattle brought from the interior prior to slaughter, and today cotton seed, molasses and other crop by-products supplement these superior pastures. A much more intensive form of cattle-raising is evolving on the plains; this is taking advantage both of the considerable work done on developing improved strains of tropical grasses and improved breeds of cattle, and of the growing demand, locally and in the United States, for beef (Miller, 1975; Spielman, 1972). Meat-packing plants have come into existence, and chilled and processed beef is shipped by sea and air to the United States; its quality, however, remains so poor that much of it is mixed with low-grade domestic meat and eaten in hamburgers. The former Canal Zone, and the shipping traffic using the Canal, supplies the main market for beef raised in western Panama. Fresh meat was third to cotton and coffee in the pre-1979 export lists of Nicaragua (more was sent abroad than eaten at home, was the rather ambiguous claim), third to coffee and bananas in Costa Rica, and is now more valuable to Honduras than silver exports. Central America is free from foot-and-mouth disease, an important attribute from the standpoint of the United States market, and there is still considerable room for applying a wide variety of well-tried improvements to take advantage of the rising demand for meat.

The cattle economy has drawn impetus from two other directions. The banana companies, to meet their own requirements, have introduced quality stock to the Atlantic lowlands and their example has been followed by many smallholders; when such veterinary problems as tick control have been solved, there is no reason why the sloping lands above the plantations should not carry a much larger cattle population. The second impetus has come from the rising demand for fresh dairy products from the Central American towns. Intensive dairy-farming is in its infancy, but Guatemala City and San Salvador are well supplied, often from herds fed on irrigated feedstuffs, and San José draws on the dairy farms above the coffee *fincas* of the Meseta Central, and urban Panama from the western departments. One interesting recent development is the construction of a dried-milk plant near Matagalpa in Nicaragua, drawing its supplies from organized herds in the vicinity; but the generally low fat-content of Central American milk and the relatively abundant alternative sources of cheap dried milk cast doubt upon the economies of this project.

Many Central Americans are aware that large sections of the isthmus at present unexploited could be most profitably used under cattle, but some balk at the cost of making such lands accessible. It has been shown in a part of Guanacaste that the return per hectare from cattle farms is only between one-quarter and one-third of the return from agricultural farms. Such disparities in returns give fuel to arguments concerning the best use of the scarce funds available for promoting new developments. Should new lands be opened up or should the money be spent on improving the existing, inadequate infrastructure in these parts of the isthmus, notably those with volcanic or alluvial soils, already occupied and of proven productivity? It may be argued that more attention should be focused on the poorer agriculturalists to help them become more productive so that, by entering the commercial life of the country, they would more fully expand the national economy as consumers not only of primary foodstuffs, but also of services and manufactured goods as well. In practice, political issues, the sources and availability of external funds and other relevant factors rarely allow questions of alternatives to be reduced to such simple terms.

Forestry   One further topic deserves mention in connection with diversification of the export economies and wider utilization of the empty lands of Central America. This is the commercial role of the forests. About one-half of Central America is forested and, like many other parts of tropical America, the first commercial product of many of its Caribbean shores came directly from the forest. Dyewoods and then mahogany, chicle gum and pine timber formed the backbone of the economy of British Honduras (Belize) for over two centuries; until about 1960, forest products supplied over half of that country's exports. Mahogany grows best and in purer stands on the drier, limestone soils of the lower Petén peninsula, on the edge of the tropical rain-forest belt; and it was the accessibility that the Belize and Sarstoon rivers gave to these forests, fashionable demand, and the advantages of a stable government, that combined to give the colony its economic *raison d'être* in the nineteenth century. During the first half of the present century, chicle, a gum tapped from the chicle zapote tree and used as the base for chewing gum, became another staple, until it was partially superseded by artificial substitutes. Further, with a change in taste, the availability in the United Kingdom of alternative furniture woods, and the removal of most of the accessible mahogany, the timber men switched their attention to cedar and to the stands of Caribbean pine that grow on the leached soils of the interior. Mahogany and cedar drew British timber companies to the Mosquito coast and other parts of Caribbean America during the nineteenth century in much the same way that they were drawn to Belize.

It is the pine forests and not the tropical rain forests that are the more important commercial forests in Central America today. The most extensive coniferous forests are on the heavily, if intermittently, weathered soils (*Figure 4.3*) of the interior of Guatemala and Honduras, and on the siliceous soils of the wet lowlands behind Cape Gracias a Dios. It is possible that in neither area is the forest 'natural', but rather a response to long-standing human interference through burning: many areas now under pine are capable of maintaining tropical rain forest. The number of important commercial species declines from five in Guatemala, through three in Honduras, to two in Nicaragua; only *Pinus oocarpa* and *Pinus caribaea* are common to all areas.

Over half the coniferous forests are in Honduras, which supplies half the timber cut in Central America today; timber comes after bananas and coffee in the list of Honduran exports. Most of the timber comes from the interior, where haphazard exploitation, frequently in conjunction with shifting agriculture, has given the pinewoods an open and decimated appearance. In the middle 1960s one-third of the pine timber of Honduras was destroyed by the spread of the *Dendroctonus* beetle. The best remaining stands and the most hopeful prospects for the future lie in the remote, less populous, national forests of the eastern province of Olancho. In 1977 the Industrial Forestry Corporation of Olancho (Corfino) began its operations by implementing a large pulp and paper project after more than a decade of planning. Initially the native pines are being cut for timber; later emphasis will shift to less profitable kraft-paper production. This is one of three such projects under way or proposed in Central America. Another is the private Istmo project to produce pulp wood from the pinewoods now being opened up by roads in eastern Guatemala; the third is a Nicaraguan project to use the north-eastern pinewoods tributary to Puerto Cabeza, woods that have been logged-over in the past, to feed a kraft-paper mill in the port. All these schemes face difficulties. Large-scale capital outlays are needed to provide the essential infrastructure on which the successful commercial exploitation of these Caribbean backwoods will depend. Private capital is only forthcoming if timber rights can be convincingly guaranteed for a long period – not an easy thing to do in present circumstances.

Manufacturing industries

It is apparent from any survey of the commercial prospects for the products and regions of rural Central America that possibilities for diversification, extension and improvement of the isthmian economy exist. With positive guidance these possibilities could be realized, but Central America will still have to rely heavily upon market decisions made abroad and on foreign sources of investment capital for the success of new developments. In order to lessen the vulnerable position that most Central American countries feel themselves to be in, they have resorted to a variety of programmes in recent years to expand domestic manufacturing industry. With few raw materials and only small local markets, governments have been obliged to use all the normal devices of import substitution programmes to encourage growth. High external tariffs have been erected to protect domestic industries and, since 1957, industrial development laws have given new (and some existing) manufacturers certain fiscal exemptions, some extending over periods of up to 10 years. In the 1960s, national incentive schemes for promoting industry were supplemented by Central American Common Market legislation which, by abolishing many tariffs between members (in 1980 only Belize and Panama were not members) and by erecting a common external tariff wall, has offered the advantages of a larger market, and sometimes a monopoly of it, to manufacturers of a widening range of products. In addition, the overall growth in GNPs, the widening of the effective markets in the populations of Central America, and the improvements in the transport arrangement of the countries of the isthmus have all made conditions better for the successful development of new manufacturing activity.

There have been certain advances in the field of import substitution; domestic production now supplies two-thirds of the internal demand of the CACM countries (Willmore, 1976; The World Bank, 1979). It is claimed that, perhaps, between one-fifth and one-quarter of the total expansion in manufacturing activity during the first decade of the market was due to CACM legislation. In visible terms, large factories manufacturing detergents, paints, tyres, paper and cardboard articles, certain fertilizers and insecticides exist today where none existed 20 years ago; and these supply all or part of a market that itself may not have existed before or, if it did exist, was served from abroad. Such growth is registered statistically in actual and proportional terms: since 1961 the expansion in the manufacturing sectors of the economies of the CACM countries has been greater by a third than the expansion in the economies as a whole. In 1960 manufacturing contributed 14.3 per cent to the GNP of the isthmus; by 1980 the proportion had risen to 18 per cent. Although these figures are substantially below the average (25.8 per cent in 1980) of Latin America, they show that the growth in the value added by manufacturing in the CACM countries during the period was at a rate above that typical for the continent at large.

But, in spite of improvements, manufacturing still plays a relatively small part in the economic life of the isthmus, and there is little to suggest that industrialization is going to transform the essentially agricultural nature of the Central American economies even in the more remote future; they are underdeveloped countries, far from the 'take-off' stage of self-sustained economic growth. The figures largely speak for themselves. In 1980, the proportion of the GNP derived from manufacturing ranged from 14.2 per cent in Panama to 24.3 per cent in Nicaragua, and the proportion from agriculture, forestry and fishing from about 14.3 per cent (Panama) to 28.1 per cent (Honduras). By comparison, agriculture yielded only 10.4 per cent of Latin America's gross product and 8.7 per cent of adjoining Mexico's. The relative insignificance of manufacturing is emphasized by employment data. In 1968, there were only about 450,000 employed in the industries of Central America or perhaps 8 per cent of the economically active population, and no more than 3 per cent of the population at large. Even these low figures were arrived at after taking a generous view of what constitutes 'industry'; if 'cottage industries' are excluded and the phrase 'manufacturing employment' restricted to workers in establishments employing five or more people, only about 250,000 were so employed in the isthmus in 1968. Nor is the figure rising rapidly. Ten years later manufacturing still provided work for only one Central American worker in nine, and the potential labour force was growing more rapidly than the ability of industry to provide jobs.

The scope for the expansion of manufacturing industry in Central America is limited. Hydroelectricity is the only domestic source of power; local industrial raw materials are restricted to certain agricultural products, wood and a small output of mineral ores; and the market for manufactured goods is largely restricted to the commercial farmers and the town-dwellers, who together form a minority of a population already small and with a low purchasing power. The existing structure of Central American industry reflects these factors. The typical industrial unit is small. As recently as 1962 there were as many people working in cottage industries as in factories, and only one in seven worked in a large factory (one employing 100 or more); in Mexico at the same time only one-sixth of industrial employment was in cottage industries, and over half was in large factories. Large factories are usually more efficient and Central America is no exception: 70 per cent of the value added in manufacturing in 1962 came from factories, and in all the countries factories were more productive than were the cottage industries, even though it was only in Costa Rica that cottage workers were outnumbered by factory workers. Since 1962 the trend has favoured the factories of Central America, although artisans still practise their skills to a degree unusual in most parts of the Americas today. The dichotomy in the structure of Central America's manufacturing sector is well illustrated by data for Panama. In 1972 there were about 40,000 employed in Panama in manufacturing. Of these, two-thirds (26,229) worked in establishments or factories employing five or more people. Most factories were small: 78.7 per cent (536) employed fewer than 50 people, and these accounted for 36 per cent of factory employment but only 31 per cent of wages paid, 27 per cent of the value added in manufacturing and 24 per cent of the return on fixed capital assets in manufacturing. In contrast, only eighteen factories employed over 200 people, yet these factories employing a fifth of the total employed in manufacturing met a quarter of the wage bill, cornered a third of the national income from sales and of the value added in manufacturing, and yielded half of the return on capital invested in Panama in manufacturing.

The preponderance of small enterprises is reflected in the great importance of traditional industries: over 75 per cent (85 per cent in 1962) of industrial production in the isthmus is of this type. The food, drink and tobacco industries alone account for over half the total value added in manufacturing, employ about 40 per cent of the industrial labour force, and occupy about one-third of all factories employing more than four workers. A further fifth of the goods produced comes from textile, clothes and shoe manufacturers giving employment to about a quarter of the industrial working population. Very little capital is required to reach a relatively efficient level of production in these fields. The wood-working industries, including furniture-making, and the leather and printing industries make up the rest of the traditional industries. The participation of cottage workers and the smaller factories in the newer industries is much smaller. Typical of such industries and accounting for over one-third of this type of activity is the chemical industry, perhaps the most dynamic element since the 1950s, and stimulated by the rising demand for fertilizers and insecticides. Oil-refining is of major importance in Panama and of lesser importance in the other republics; it is a symbol and makes little direct impact on the employment structure of the isthmus. The Guatemalan refinery receives a little oil from the Rubelsanto field in the south-western Petén, a token of the national hope that the Mexican fields will prove to extend deeper into Guatemalan territory, but in 1980 the refinery remained dependent upon imported crude.

Metal-working, including the assembly of motor cars, is another growth sector of the economy, but employment statistics are inflated by the inclusion of essentially repair establishments. Even if, as anticipated, the newer industries expand production at twice the rate of the traditional ones, the latter are going to dominate the industrial scene for a long time to come.

The location of industrial activity in Central America follows a fairly simple pattern. In the first place, there are industries such as sugar mills, coffee *beneficios*, cotton gins, lumber mills, fish-processing plants and cement works, which are located close to their raw materials, frequently in the nearest small town. Secondly, those based on imported raw materials may be in or near the ports: the oil refineries or fertilizer plants of, for example, the new national port of Matias de Gálvez near Puerto Barrios in Guatemala, Acajutla in El Salvador, Puntarenas and Puerto Limón in Costa Rica, and Colón and Panama City, are representative of this category. Finally, the great majority of industrial employment is in the large towns and particularly the capital cities. Two-thirds of all factory employment in Central America is in the ten most populous towns, and the larger the town the greater its share of industry. Guatemala City can claim one-fifth of all industrial establishments and employment in the isthmus, followed (in order but at a distance) by San José, San Salvador, Managua and Panama City. The inflated commercial role of Panama City helps to account for its disproportionately low rank amongst industrial towns of the isthmus. The dominance exercised by these cities over the national industrial structure is often very strong: three-quarters of Costa Rica's industry is in the San José metropolitan area, two-thirds of Guatemala's industry is in Guatemala City, Managua accounts for half of Nicaragua's industry, and even in El Salvador and Honduras, where the capital cities are least dominant, they each accommodate one-third of the national total. Only in Guatemala has there been any deliberate governmental attempt to disperse industry more widely in the country (for example, to Amatitlán) but with little apparent effect: the market advantages, the better public utilities and facilities and the easier access to sources of finance and influence are powerful centralizing factors.

The distribution of industry within the towns is not remarkable. Detailed mapping substantiates casual observation that smaller workshops and factories are scattered through the fabric of the towns, although they are less frequently met in the better-class housing areas (López Toledo, 1965, p. 9). Larger factories are frequently related to transport facilities: such is the case in the older industrial areas around the railway terminals in San Salvador, Guatemala City and San José, and the newer factories along the exit roads to Escuintla and the west in Guatemala City. The roads leading to the airports from the other capitals have special attractions; they are invariably the best-quality roads, and conditions suitable for airport construction are similar to those required for modern industrial sites.

Between one country and another there are differences in the degree and pace of industrialization. It is interesting to view these in the light of one of the stated aims, variously interpreted, of the CACM countries to promote a balanced industrial development in the isthmus. In absolute terms Guatemala accounts for just over 36 per cent, and with El Salvador, 50 per cent of the industrial production of Central America; Panama, Costa Rica, Honduras and Nicaragua follow in that order, with Belize of negligible importance. In relative terms industrial production is of greatest importance in Nicaragua and Costa Rica (*Figure 4.2*). Nicaragua, having been the least industrialized country in 1960, had become the most industrialized in 1978, partly because reconstruction after the 1972 earthquake, which destroyed much of the industrial capacity of the smaller factories, stimulated an exceptional demand for building materials and for consumer goods. Much of the money made available to the Central Bank for relief was channelled into the manufacturing sector – particularly into the footwear, food-processing and chemicals branches.

El Salvador has harnessed a larger proportion of her hydroelectric potential than any other Latin American country, with the exception of Uruguay; the Cerro Grande hydroelectric plant on the River Lempa, inaugurated in 1977, is the biggest plant of its kind in Central America, though with a relatively modest capacity of 270 Mw. El Salvador is also developing her geothermal resources. In the form of reduced distributional costs El Salvador reaps some industrial advantage from her small size and relatively good transport system. The similar advantages of the Meseta Central in Costa Rica are given more tangible support by comparing the *per capita* contribution (in US $) of industrial production in 1980 in Costa Rica ($339)·with that of Nicaragua ($218), El Salvador ($109), Guatemala ($200) and Honduras ($108). Costa Rican industry benefits from a larger installed capacity for generating electricity than any other Central American country. Further, it is hoped that the El Arenal project will be completed in the mid-1980s, thanks, in part, to money provided by the International Development Bank, which has supported many of the other hydroelectric schemes in Central America. The Boruca Project remains at the planning stage, but if completed would double the installed hydroelectric capacity of Costa Rica and yield more power than current total demands; a new aluminium smelter is part of the project. If the average Costa Rican derives most from industrial developments, the average Honduran gains least. Honduras suffers from poor roads, has few raw materials and only a small and unusually dispersed national market. San Pedro Sula is almost as important as Tegucigalpa, but both offer a market more comparable with that of, say, Santa Ana, the second city of El Salvador, than that of, say, the capital, San Salvador.

Panamanian industry is less protected from external competition but does have easier access to investment capital, and has some stimulus from the passing needs of canal traffic; these factors help explain the presence there of the only steel mill, even if it is a small one, and the largest oil refinery of the isthmus. Panama is not alone in possessing an oil refinery. It is an interesting reflection on the industrial policy of the CACM countries that four of the five now possess a domestic refinery (Guatemala has two: one at Galvez, the other at Escuintla) and all except the refinery at Puerto Cortés in Honduras were built in the 1960s. One refinery, or at most two larger ones to meet the needs of the CACM would have been more economical. In this way, oil-refining would have been treated as an integration industry under the terms of the 1958 industrial regulations of the Common Market Treaty. These regulations, which have since been strengthened, were designed to encourage the establishment of industries of a scale to serve the whole Central American market; their products were to be given free entry to the national markets and protection from outside competition. Up to 1968, however, only three integration industries had been designated: tyre plants in Guatemala and Costa Rica (making different sizes of tyres) and a caustic soda and insecticide plant in Nicaragua. This poor response arose in part from the success of another aim of CACM policy – the reduction and abolition of industrial tariffs between the countries of the Common Market – and in part from the fact that better terms, apart from those incentives specific to CACM regulations, may often be obtained from individual countries. In the particular case of the oil refineries, no doubt the symbolic status of a national oil refinery played a part in countering the generally stated objective of industrial development.

All the oil refineries are subsidiaries of American companies, and their presence reflects one aim of the CACM authorities: to attract foreign investors to Central America. They can claim some success, and draw attention to the tripling of direct foreign investment in the isthmus between 1959 and 1976: it now totals over US $1000 million, of which 80 per cent comes from the United States. Manufacturing has particularly benefited from these overseas funds, its share rising from only 4 per cent in 1959 to over 30 per cent in 1976. The greater part of goods manufactured as a result of such investment is destined for the regional export market: in 1976 90 per cent of such exports were to other CACM countries. In general, CACM legislation has encouraged local trade, particularly in manufactured goods. The member countries shelter behind a higher common-commodity tariff than they did before the treaty became effective, and this helps explain why intra-regional trade now accounts for about 20 per cent of the total trade (as compared with only 6 per cent as recently as 1960). Almost all (95 per cent) of this declared trade is in manufactured goods, particularly clothing and chemicals. International co-operation in other spheres has also helped stimulate trade amongst neighbouring countries in Central America. Perhaps the most significant has been the road-building programme, which has vastly improved their physical accessibility. Paved roads now link all the capitals of the CACM countries, and trailer trucks and scheduled bus services run almost the length of the isthmus, despite anachronistic customs arrangements. The benefits of this new accessibility and increased trade have not been spread evenly. Guatemala is the only CACM country with a positive balance of intra-zonal trade, thus continuing its traditional role as the principal exporter of the region. Economists calculate that Guatemala, Nicaragua and El Salvador have benefited from the shift from importing lower-cost overseas produce to goods manufactured at home. In contrast the benefits to Honduran and Costa Rican consumers have been from a shift away from high-cost domestic products to lower-cost products imported from their partner countries within the Common Market (Willmore, 1976, 411). The unequal nature of these benefits suggests why Costa Rica and Honduras have been the least enthusiastic members of the CACM: indeed, strained political relations with El Salvador since the 'soccer war' of 1969 have led Honduras effectively to withdraw from the CACM, and thwarted some but not all local attempts at further integration. For example, it did not prevent the linking in 1976 of the electricity-supply systems of Honduras and Nicaragua, thereby giving both countries a larger reserve capacity than either would have on their own. Negotiations are under way to link the supplies of Costa Rica and

Nicaragua, Guatemala and El Salvador, and Guatemala and Honduras. These developments are symptomatic of the high priority that a unified regional distribution system of electricity has, not only in the plans of the CACM, but also in the allocation of funds from international development banks, which would not necessarily be available to the Central American countries without the CACM umbrella.

Panama and Belize remain outside the Common Market, and remain physically and spiritually detached from their isthmian neighbours. The Inter-American Highway leading to Panama in southern Costa Rica is tortuous and remains unpaved; it does not exist in Darien. The peculiar circumstances of the Canal give Panama a different perspective of the world. Commercial interests are more important than industrial interests, and the benefits won from joining the CACM less clear. Belize, too, would not fit easily into an enlarged common market. She is linked with the Mexican road system and not with that of Guatemala. More important, her cultural and commercial ties are with Britain and the Caribbean rather than with Latin Central America. Both Panama and Belize are anomalies when viewed from a Central American vantage point, and their peculiarities deserve some special comment.

Belize and political independence
Belize, called British Honduras until 1973, achieved independence in 1981, having had nominal self-government since 1964. It is by far the least populous of the Central American countries; its population of about 150,000 in 1980 was less than one-tenth of the population of any of the republics. Belmopan, the new inland capital, is safe from the hurricane-induced tidal surges of the coast, but mustered a population of only 4000 in 1980. Even Belize town, the former capital and largest city, housed only 42,000. The majority of the citizens are negro, and English is the common language. The three staple export crops of Central America – coffee, bananas and cotton – play almost no part in the economy of the country. In people, outlook, history and past connections, Belize is closer to the British West Indian islands than to the other Central American countries. Peripheral to Central America and on the rim of the Caribbean, Belize seems to experience only the disadvantages of such a position. The problems of accommodating the country to modern life may not be insoluble, but they are certainly very difficult. The country is, at the moment, too small, too empty and too poor to stand completely alone on her own feet as an independent country, yet she rejected association with the abortive West Indian Federation, and is too divergent in sympathies to relish political attachment to, or absorption into, her Latin American neighbours.

History has given Belize, as it has given Panama, an outlook at variance with that of the other countries in the isthmus (Waddell, 1961). Overseas contacts have been of great importance here. Belize may be viewed as a beach-head set up by buccaneers and smugglers, loggers and slaves; the Belize river was a main route inland to the forests of the interior, and Belize town grew up on the mangrove swamps at its mouth. It is significant that Belize had the only capital in Central America proper that was at the same time a port. Successful lumber operations demanded freedom to scour the forests for marketable species; they did not allow the rigid master–slave relationships of, for example, the British

West Indian sugar islands. There was initially no real clash of interest between the logger and the indigenous Mayan Indian. The Indian practised shifting agriculture, and valued the trees only for the land they occupied; the logger had a completely opposite attitude. The negro logger based himself on Belize or on one of the logging camps from which he worked the forests seasonally; the Indians accepted the incursions of the loggers and were left almost unbothered by the British colonial administration. The urbanized negro lumberjack regarded agriculture as effeminate; even the gathering of chicle was left to the Indians. It is possibly for this reason that only about 5 per cent of the country is farmed, and an even smaller proportion is actually under cultivation at any given time (Weight, 1959).

It may also be significant that the Belize Estate and Produce Company (BEC), formed in 1875 and recently absorbed into the United Kingdom-based conglomerate International Timber Corporation, has long been a very strongly established force in the colony. Although the company has sold some of its holdings, it still owned about 405,000 ha in 1971, and many of the ten largest land companies who together own two-thirds of the freehold land in Belize had acquired their land from the BEC. A number of the owners are American speculators, a trend which has been discouraged by legislation since 1973 (Latin American Bureau, 1978a; Comercio Exterior, 1975, 1231–8).

In an easy social milieu and with no lack of unexploited agricultural land, the indigenous Indians have retained their traditional manners, and a large number of immigrant groups have come to the country. This has given a remarkable variety to the small population. It has also added to the present-day problems of emergent independence. The largest single element until the late 1960s was the English-speaking negro or creole; his main focus is Belize. Another negro group is centred on Stann Creek, the second largest town of the country, but with a population of only slightly over 8000 in 1980. This group is Carib-speaking. They are descendants of those escaped negro slaves who had absorbed the culture of the Carib Indians in St Vincent and, after deportation to Ruatán island, spread to the shores of the Gulf of Honduras in the early nineteenth century. A third group is Latin American. The Indian Wars of Yucatán in the 1850s led some Mexicans to move south and settle in northern Belize, and modern Mexicans and Guatemalans have followed in their footsteps: one-third of the country's population today is Spanish-speaking. Other white refugees came from the Confederate States after the American Civil War and settled on the south coast around Toledo. Since 1957 some 3500 German-speaking Mennonites have established several thriving agricultural colonies in the Cayo and Orange Walk areas, where they have been promised a greater degree of autonomy than they found in Mexico, Canada or the United States (Sawatsky, 1969). Sprinklings of British administrators and expatriates, Lebanese entrepreneurs and Chinese businessmen add further spice to the immigrant mixture. Even the indigenous Indians are not homogeneous. In the northern interior they are Mayan-speaking and many are of mixed, and some are of very mixed, blood; in the south, away from the main mahogany stands and less exposed to outside influences, many live on the Kekchi reservation and speak no Spanish or English. The net result is a 'colony of colonies', a potpourri of peoples speaking several different languages, in which the task of creating a unified national outlook seems near-impossible (Furley, 1968).

The portents for true, economic independence seem no more clear. Belize has run a balance of trade deficit for most of the years of this century: in 1977, as in 1967, it amounted to about a quarter of the country's GNP. It has been an apparent drain on the United Kingdom economy for many years, and the subsidy rose with self-government. There has been a marked diversification of the economy. The old and ruthlessly exploited staple, timber, is in decline and sugar from the Corozal area now provides half the country's exports; tinned citrus fruits from Stann Creek and fresh and frozen lobster and crayfish for the American market are of increasing significance. Two recent economic developments are intriguing: the first has been the revival in 1974 of a long-defunct trade in bananas, which by the late 1970s was helping to swing Belize's trade with Britain towards the black; the second is the revival of interest in the oil potential of the country following the recent oil discoveries in Mexico. In 1979 the Texan company, Spartan, held leases on 3200 km^2 of territory, most of it close to the Yucatán border and including 500 km^2 off-shore; the hope in Belize is that Spartan's drillings will prove more fortunate than Exxon, Shell and Philips have been in the past. Meanwhile the most tangible asset is relatively abundant agricultural land whose capabilities have been subjected to detailed scientific assessment by the Land Resources Division of the British Ministry of Overseas Development (LRD, 1970, 1973, 1976).

The present population of Belize is too small to provide a market for large-scale domestic industrial development, let alone to support the administrative superstructure of an independent country. There is agricultural land available in Belize and surplus labour in the English-speaking Caribbean islands. It has long been argued that the densely settled, English-speaking Caribbean islands could solve some of their population problems by decanting their surplus to the empty lands of English-speaking Central America. But it is difficult to harmonize these two elements. Few West Indian emigrants seem to relish the role of a pioneer farmer in Central America as an alternative to life in an English, Canadian or American city; and even the limitations of the United Kingdom, Canadian and US immigrant quotas have produced no increase in migration to Belize – a country, after all, almost as remote for West Indians and as expensive to reach as is the United Kingdom. Nor has Belize paid much more than lip-service to the need to encourage immigrants.

Her land neighbours are no great help to the concept of independence. Guatemala has reiterated the claim that 'Belice es nuestro' and, for a time, political factions in Belize played on this claim. The claim rests upon Guatemala's assumption that she inherited from Spain the lands of the Captaincy-General of Guatemala and that these included Belize. This is disputed by Mexico, who has a dormant claim to the northern part of the colony, which may have been part of the Captaincy-General of Yucatán. Both claims ignore the fact that the United Kingdom had *de facto* possession of Belize since before the disintegration of the Spanish empire. Whatever the merits or demerits of these claims, it is difficult to imagine Belize deriving any long-term benefit from becoming, in fact, the department of Belice, for so long claimed on the maps of the Guatemalans. A marriage – willing or unwilling – with Guatemala would run into difficulties. The political instability of Caribbean Guatemala might spread northwards; the majority of Belizeans would find themselves a coloured minority speaking an alien language, governed from a capital at present inaccessible by land, and within an economic system isolating them from their traditional markets. Belize might reap some economic advantage as an entrepôt in the event of the Petén being developed. A union of Belize with Mexico, not attractive in itself, might be less unattractive than union with Guatemala. Meanwhile her political and economic interests seem better served by her membership of the Caribbean Community.

In 1982 Belize safeguarded her new independence, retaining a British military force while fostering her ties with Central America and a more varied economy shored up by financial aid from Britain and the United States. The British government's view remained (in contrast to that of the Belizean) that the cession of a strip of land widening Guatemala's corridor to the Caribbean would be a price worth paying for the removal of one of the last of her post-colonial encumbrances in the New World.

Panama, the Canal Zone and the Panama Canal

Panama is a corridor as Belize is a beach-head, and just as the imprint of Britain is clearly visible in the landscape of that part of northern Central America, so that of the United States is unmistakable in the Canal Zone. As Britain relinquished its sovereign rights in Belize, so the United States has revised its attitude towards its undeclared colony in Panama. The Canal Zone and the Panama Canal are indelible marks of that colonial status. Less obvious is the web of international links which focus on Panama and increasingly identify her as an off-shore financial centre and as a node in the strategic defence system of the western hemisphere. Whatever her formal status Panama is becoming more, rather than less dependent upon decisions taken outside her boundaries, decisions most likely to be taken in the United States (Comercio Exterior, 1974, pp. 1052–7; 1975, pp. 793–806; 1976, pp. 268–77).

The United States first became officially involved in Panama in 1846, when it guaranteed the neutrality of the isthmus in return for freedom of movement across it for American citizens. The American West was just beginning to be opened up. In 1849 the Panama Railroad Company of New York acquired a concession from the government of New Granada (Colombia) to build a railway or canal across the isthmus. A Committee of the US House of Representatives supported the project, but recommended a canal as a better long-term proposition. No financial aid was forthcoming, however, and had it not been for the sudden influx of the forty-niners and their successors en route for California, no railway would have been built. That railway was in use well before its completion in 1855; although expensive to build, it had a virtual monopoly of traffic over the isthmus and was highly profitable. In 1866, the US Senate ordered a survey of all feasible routes for a ship canal through the isthmus; nineteen were considered and a route through Darien favoured. The authoritative First United States Interoceanic Canal Commission in 1876 favoured a route through Nicaragua, a decision which coloured United States opinion until the beginning of this century. Meanwhile, a French company headed by Ferdinand de Lesseps, builder of the Suez Canal (and President of the Geographical Society of Paris), actually began the first serious attempt to join the oceans by a sea-level canal across Panama. Two-fifths of the work had been completed when, in spite of a late switch to a lock-canal design, the De Lesseps Company collapsed in 1889. By 1903 the United States government was ready to build a canal. An official investigation reported that there was little to choose between a route through Nicaragua or one through Panama, and the former was authorized. It was only then that the French company holding the concession rights through Panama came to heel. This was followed by the unilateral declaration of independence from Colombia by the Panama Republic. The situation changed in favour of the Panamanian route; the extraordinarily generous treaty terms offered by the fledgeling republic clinched the choice of the new route. A very narrow majority decision in the US Senate favoured a cheaper lock canal over a sea-level cut, and the construction of the present canal began. The route chosen closely followed the railway, and the railway in its turn closely followed the sixteenth-century Camino Real, over which much of the gold of Peru and silver of Bolivia had made its way to Europe; all these routes recognized the fortunate coincidence of the lowest point in the Continental Divide with the narrowest portion of the isthmus. In so doing, the importance of Panama City as the fulcrum of the country was reinforced.

The Canal, planned over 70 years ago and little modified since, is still capable of meeting most of the demands placed upon it today. Almost each year has brought record traffic, and the rise from about 30 million tons of cargo carried in the early 1950s to over 140 million tons in the late 1970s was dramatic. But when the Canal was built it could accommodate all ships then afloat; today a growing number of ships are too big to use the Canal. The size of the lock chambers closes the Canal, in effect, to fully laden ships of over 80,000 tonnes and to many below this size. Of the average 38 daily transits being made by vessels in 1980, 45 per cent were of ships with beams of at least 24 m across; the peak daily capacity was 45 transits of which no more than 75 per cent could be vessels with beams of over 24 m. Meanwhile, the world trend to larger ships continues, with vessels of over 300,000 tons currently on order. The limits placed by the locks on the use of the Canal is being increasingly felt, and longer shipping routes, avoiding the Canal, are becoming genuine alternatives. The time element in passing through the triple-stepped Gatun locks at the Caribbean end of the Canal imposes restrictions on the flow of vessels. There is a limit to the supply of water available to replenish the canal after each lockage. The demand for lockages through the canal is naturally uneven; sailing schedules can rarely be governed solely by considerations of Canal convenience. In 1952 there were about 7200 ship transits through the Canal and the number has been rising; in 1978 it was over 12,600 but 2000 below the 1971 figure. The depressed level of the world economy in the late 1970s took some of the pressure off the Canal, and the prospect receded of its reaching a saturation rate of 20,000 annual transits in the early 1980s. This was the expectation of the Atlantic-Pacific Interoceanic Canal Study Commission, which exhaustively reviewed the operations of the Canal in the late 1960s; but it is a prospect which now seems unlikely to be realized until early in the next century. Nevertheless, the potential inadequacy of the Canal remains a live issue and one with very serious implications for Panama (Fox, 1964).

Suggestions for relieving future pressure have followed four lines. The first is to increase the capacity of the present canal. If the bottleneck at Gatun were broken by the construction of an additional, third, flight of locks, more vessels could use the Canal; in fact, this was begun, and $75 million were spent until work was suspended in 1942. To permit larger vessels to use the Canal would require sets of larger lock chambers on both sides of the divide and the widening and deepening of existing canal in many places. New water storage capacity would be needed. A second alternative approach demands the construction of an entirely new canal as a replacement or a supplement to the existing canal. All the traditional choices of routes have been reconsidered, and the earlier controversy over the relative merits and costs of a lock canal and a sea-level channel has been revived. The advantages of a sea-level channel, recognized in 1905, remain valid today: it would be cheaper to operate and passage time would be shorter. It has been proposed that nuclear explosions be used in the excavations. A variety of estimates suggests that a canal so built would be considerably cheaper and quicker in construction and wider, deeper and easier to maintain than one built by conventional engineering methods. The most attractive site for a seaway so constructed would be the Sasardi–Morti route from Caledonia Bay to the Gulf of San Miguel, 160 km east of the present canal. But the ability to limit radio-active fall-out to acceptable levels and areas, and the wider political repercussions of employing nuclear devices have proved hurdles more difficult to overcome than was anticipated when the proposals were first broached. In consequence, it was a sea-level canal cut along the line of the existing canal and built by conventional engineering methods that the Canal Commission recommended in the early 1970s. At that time the cost would have been about $3000 million; in 1980 the figure had risen to about $20,000 million. The Commission further recommended that if it were impossible to reach agreement with Panama, a Colombian canal, built by conventional methods, would be the next most economical alternative. The consideration given to a Nicaraguan route appeared to be for diplomatic reasons only.

The third and fourth approaches to the potential plight of the Canal are aimed at deflecting traffic; it seems that whatever other decisions are taken some such steps will be required by the 1980s. The bulkiest cargo using the Canal is petroleum and tonnage is likely to increase as more Alaskan oil is shipped to the East Coast of the United States: the creation on either side of the isthmus of deep-water terminals linked by oil pipeline might remove some of the pressure on the Canal. The other approach is a fiscal one (CTRS, 1967); it would manipulate the toll structure to make it more expensive to use the Canal and other routes more attractive. The terms of reference for the Panama Canal Company required it only to cover running costs. The use of increased tolls is obviously a cheap and easy way to ensure that the existing canal does not become choked with traffic, but such a measure would raise shipping costs and so be to the disadvantage of current users. It is to be noted that two-thirds of all cargo using the Canal originated in, was destined for or was part of the intercoastal trade of the United States. The whole concept of deliberately killing growing traffic is alien to United States philosophy, and would probably accentuate the demand for a new canal.

Any approach to the future of the Canal would have to take account of the two 1979 Canal Treaties which have reduced and more closely defined the involvement of the United States within Panama. The first treaty abolished the Canal Zone and transferred sovereignty to the Republic. Over 300 km² (58 per cent of the zone) was immediately absorbed into Panama: the rest was transferred to a new commission. During the 1980s, the commission is required to have a majority of Americans on its board, while in the 1990s Panamanians will be in the majority; and in the year 2000 the Zone, including the Canal, will come entirely within the jurisdiction of the Republic. The treaty requires the two nations jointly to prepare another feasibility study for a sea-level canal. It also establishes that the United States will have the first option on the construction of any Panama Seaway. Cynics claim that this option means that Panama could find herself in a few years heir to an obsolete canal whose former traffic would have been syphoned off to a United States controlled Seaway — a Seaway whose operation would require a staff of only 1000 compared with the 14,000 employed by the present canal. Others believe that the United States — many of whose ports cannot accommodate deep-draft vessels — might be prepared to internationalize such a construction or, perhaps, cede the right to build to Japan. The heavy dependence of Japan upon such raw materials as Brazilian iron ore, Venezuelan and Middle Eastern oil makes her particularly aware of the advantages to be gained by increasing the capacity of the present routeway. The treaty provides for an increase in the income to Panama from the operation of the Canal, and the new arrangements gave the Republic a gross income of about $70 million in 1980, a figure much larger than the $2 million received in the past. It also provided for the removal of any privileges previously enjoyed by the 'zonians' and the provision of training schemes to allow Panamanians access to jobs still performed by expatriates.

The retreat of the United States in Panama heralded by the latest treaties is cushioned by some important provisos in the agreements and by recent changes in legislation governing financial activity in the Republic. The treaties allow the United States to retain her military bases in the former Canal Zone. This is a very substantial proviso: the 14 bases form the headquarters of the US Southern Command, and together with the School of the Americas (where Latin American and US personnel are trained in the arts of tropical warfare) occupy over one-third of the former Zone. Panama, together with Guatanamo Bay in Cuba and Puerto Rico, forms one of the pivots of a defensive triangle protecting America's 'Mediterranean'. The military significance of the former Zone is suggested by the level of military investment in the Zone, which was twice that for civilian purposes, and by the fact that the military normally exceeded the civilian population. This substantial American enclave will remain.

Another territorial enclave to remain is the Colón Free Trade Zone. This is a small (37 ha) site at the Atlantic end of the canal which since 1953 has acted as a tax-free, warehousing, processing and redistributing entrepôt serving much of Latin America. The unrivalled transport facilities of the Zone — served by sixty shipping lines and, via Tocumen airport, by thirty airlines — have made it one of the most important free-trade zones in the world. The Zone employed about 3500 people in 1976, and it contributed $50 million to the Panamanian economy in 1974; in 1978 the 600 companies in the zone had a turnover of about $500 million. The success of the Colón Free Trade Zone has prompted the suggestion that a second such zone be created at Tocumen international airport. It has also prompted other Central American countries, although lacking most of the advantages of Colón, to establish or contemplate establishing similar zones (at Limón in Costa Rica, Santo Tomás de Castilla in Guatemala, Puerto Cortés in Honduras, and the Corn Islands of Honduras).

But Panama has capitalized on her worldwide contacts and the cosmopolitan and entrepreneurial attitudes induced by the presence of the canal to extend her services to world trade well beyond the confines of the free-trade zone. Panama, has popularized, like Liberia, the advantages to world shipowners of registering their ships overseas. Five per cent of the world's merchant marine tonnage now sails under Panama's flag of convenience, at some benefit to the Republic. The Panama Bank Act of 1970 extended the services provided within Panama to the world by legitimizing 'off-shore' banking activities (McCarthy, 1979). The effect of the Act is to provide multi-national companies with a capital market and Latin America with a financial entrepôt through which overseas funds can be transferred without undue regulation. The presence in Panama of the US military bases and the parity of the Panamanian balboa with the US dollar have helped secure a transformation in the national banking system. In 1960 Panama had five banks and $125 million of funds on deposit; 70 per cent of the funds were of local provenance, and 96 per cent of the credit given was to local enterprises. In 1976 the number of banks had grown to seventy-four, $10,000 million was on deposit, 90 per cent of this money was drawn from overseas, and over 75 per cent of the credits granted went overseas. Such banking activities directly and indirectly boosted employment in the service sector of the economy, yielded direct benefits to Panama worth about $6 million, but did little for agriculture and industry. Additional legislation which has made companies registered in Panama not subject to taxation on profits won overseas, has attracted 50,000 new companies to the country; such companies may be used by parent institutions in, for example, the United States (where there is a tax liability on all profits no matter where earned) to reduce the overall tax paid by the joint enterprise at the cost of a modest contribution to the Panamanian exchequer (Latin American Bureau, 1978b).

It will have been appreciated that Panama owes much to the Canal. Its polyglot, many-hued population, its relatively open and fluid society, its skin-deep industrial development, its economy finely tuned to the requirements of world commerce, its ambivalent relationship with the United States: all are aspects of the national character which have their origins in the cutting of the Canal. The geographical expression of these characteristics is best seen in the vicinity of the Canal. Half the population of the country lives within 20 km of the Canal: this same strip is responsible for 80 per cent of the country's national product. Away from the Canal the style of life becomes more typically Central American, and exotic influences are confined to a few enclaves. The Chiriqui banana plantations are the oldest (1899) economic enclave in the country; the nearby low-grade copper mines, of Cerro Colorado and elsewhere, are the most recent. In the plantations the Panamanians have recently been sharply reminded of their continued dependency upon the marketing power of United Brands, and of the importance of American, Japanese and British metallurgical skills and capital. Elsewhere the peasants live lives of rural poverty subservient to the limitations of the land and the will of the landowner. Even more distant from the affluent 'pro-consuls of American capitalism' in Panama City are the small groups of Indians that survive in the forest of Darien oblivious of the larger world outside. It is a world which has impinged more dramatically on Panama than upon the rest of Central America in the twentieth century.

Concluding remarks

Belize and Panama have their peculiar characteristics and problems, but share with the rest of Central America many others common to developing countries elsewhere. For example, the towns are growing rapidly and strains are being placed on municipal resources and abilities. So far modern architecture is little in evidence away from the capital cities, and the townscapes of Central America retain their traditional lines. But the population pressures are already present, and substantial changes in urban functions and form can be anticipated. In the countryside new roads have transformed life in certain limited regions yet large areas remain little used and isolated. An anachronistic land-tenure structure mitigates against fuller exploitation of the agricultural potential of the isthmus, and provides a focus for widespread rural discontent. Economic and social inequalities in the countryside and in the towns have played their parts in the recent weakening of authoritarian rule in Central America. The civil war of 1978–9 which finally ended the Somoza family fiefdom in Nicaragua, cost more lives (35,000) and left more damage ($500 million) than did the Managua earthquake of 1972. In the early 1980s the events in Nicaragua have been followed by violent uprisings and reactions against the oligarchic governments of El Salvador, Guatemala and Honduras. In contrast, democratic Costa Rica and outward-looking Panama were less threatened by the changes in Nicaragua. The growing disparity between the political stances of the countries of Central America combined with the traditionally centripetal policies they follow make it difficult to anticipate any sustained resumption of the moves towards political and economic integration. Rather it may be in the very diversity of national approaches towards the search for solutions to common problems that the brightest hope for the future of Central America lies.

Bibliography

BOLSA (Bank of London and South America) *Review* (monthly 1967–80; quarterly 1980–).

BROWNING, D. (1971) *El Salvador: Landscape and Society*, Oxford.

COE, M.D. and FLANNERY, K.V. (1967) *Early Cultures and Human Ecology in South Coastal Guatemala*, Washington, DC, Smithsonian Contributions to Anthropology, vol. 3.

COMERCIO EXTERIOR (1975) Banco Nacional de Comercio Exterior, SA (monthly), Mexico.

CRUZ, E. and HOADLEY, K.L. (1975) 'The effect of government trade policy on private sector exports: the case of Nicaraguan cotton', in Inter–American Development Bank (1975) pp. 453–89.

CTRS (1967) *Canal Tolls and Route Studies*. Hearings before the Sub-committee on Panama Canal of the Committee on Merchant Marine and Fisheries, House of Representatives, 90th Cong., 1st Sess. HR 6791, 9 and 18 May 1967.

DEMYK, N. (1975) 'Marchés et Minifundio en pays quiche (Guatemala)', *Annales de Géog.*, 84, 463, 318–50.

DURHAM, W.H. (1979) 'Scarcity and survival in Central America', *Ecological Origins of the Soccer War*, Stanford.

FOX, D.J. (1962) 'Recent work on British Honduras', *Geogr. Rev.*, 52, 112–17.

FOX, D.J. (1964) 'Prospects for the Panama Canal', *Tijds. Econ. Soc. Geografie*, 55, 86–101.

FOX, R.W. and HUGUET, J.W. (1977) *Population and Urban Trends in Central America*, Washington, DC, Inter-American Development Bank.

FURLEY, P. (1968) 'The University of Edinburgh British Honduras–Yucatán Expedition', *Geog. J.*, 134, 38–54.

HELMS, M.W. (1975) *Middle America: A Cultural History of Heartland and Frontiers*, Englewood Cliffs, NJ.

INTER-AMERICAN DEVELOPMENT BANK (1975) *Agricultural Policy: a Limiting Factor in the Development Process*, Washington, DC.

INTER-AMERICAN DEVELOPMENT BANK (1981) *Economic and Social Progress in Latin America*, Washington, DC.

LAND RESOURCE DIVISION (1970, 1973, 1976) 'A forest inventory of part of the Mountain Pine Ridge, Belize', *Land Resource Study* 13 (1970); 'An inventory of the southern coastal plain pine forests, Belize', *Land Resource Study* 15 (1973); 'The agricultural development potential of the Belize Valley, Belize', *Land Resource Study* 24 (1976). London, Ministry of Overseas Development.

LATIN AMERICAN BUREAU (1978a) *The Belize Issue*, London.

LATIN AMERICAN BUREAU (1978b) *Panama and the Canal Treaty*, London.

LOPEZ TOLEDO, J. (1965) *Informe sobre la colonia San Diego*, Guatemala City, Estudios Geográficos, Dir. Gen. de Obras Publicas.

McCARTHY, I. (1979) 'Offshore banking centres: benefits and costs', *Finance and Development*, 16, 4, 45–8.

MILLER, E.E. (1975) 'The raising and marketing of beef in Central America and Panama', *J. Trop. Geogr.*, 41, 59–69.

NIETSCHMANN, B. (1973) *Between Land and Water: The Subsistence Ecology of the Miskito Indians, Eastern Nicaragua*, New York and London.

NUNLEY, R.E. (1967) 'Population densities using a new approach', *Revista Geográfica*, 66, 55–93.

PARKER, F.D. (1964) *The Central American Republics*, London.

PEARSE, A. (1966) 'Agrarian change trends in Latin America', *Lat. Am. Res. Rev.*, 1, 3, 45–77.

RADELL, D.R. (1964) *Coffee and Transportation in Nicaragua*, Berkeley, University of California, Dept of Geography.

SANDNER, G. (1962) *Colonización agrícola de Costa Rica*, 2 vols, San José, Instituto Geográfico de Costa Rica.

SAUER, C.O. (1952) *Agricultural Origins and Dispersals*, New York.

SAUER, C.O. (1966) *The Early Spanish Main*, London.

SAWATSKY, H.L. (1969) *Mennonite Settlements in British Honduras*, Berkeley, University of California, Dept of Geography.

SCHMID, L. (1967) *The Role of Migratory Labor in the Economic Development of Guatemala*, Madison, University of Wisconsin, Land Tenure Center.

SPIELMAN, H.O. (1972) 'La expansion ganadera en Costa Rica: problemas de desarrollo agropecuario', *Informe Semestral*, San José, Instituto Geográfico de Costa Rica, 2, 33–57.

STEPHENS, J.L. (1963) *Incidents of Travel in Yucatán*, 2 vols, New York.

STOUSE, P.A.D. (1967) *Cambios en el uso de la tierra en regiones ex-bananeras de Costa*

Rica, San José, Instituto Geográfico de Costa Rica.

STRUCKMEYER, H.J. (1977) 'Coffee prices and Central America', *Finance and Development*, 14, 3, 28–40.

USAID (US Agency for International Development) (1965) *Central American Transport Study Summary Report*, Guatemala City, Regional Office, Central America and Panama Affairs.

WADDELL, D.A.G. (1961) *British Honduras*, London.

WEIGHT, A.C.S. *et al.* (1959) *Land in British Honduras*, London, Colonial Office research pub. 24.

WEST, R.C. (ed.) (1964) *Natural Environment and Early Cultures*, Austin, Texas, Handbook of Middle American Indians, vol. 1.

WEST, R.C. and AUGELLI, J.P. (1976) *Middle America, its Lands and Peoples*, 2nd edn, Englewood Cliffs, NJ.

WILLMORE, L.N. (1976) 'Trade creation, trade diversion and effective protection in the Central American Common Market', *J. Devel. Stud.*, 12, 396–414.

WORLD BANK (1979) 'A model of agricultural production and trade in Central America', *World Bank Reprint Series*, 82. Reprinted from CLINE, W.R. and DELGADO, E. (eds) (1978) *Economic Integration in Central America*, Washington, DC, Brookings Institution.

5 · Colombia and Venezuela

David Robinson and Alan Gilbert

Complexity in diversity

Few areas in Latin America provide so great a challenge as the territories of Colombia and Venezuela to those who seek to describe, understand or explain the dynamic relationship between man and nature, culture and landscape. All writers on these countries stress the diversity and complexity of both natural and human phenomena. Viewed together, Colombia and Venezuela stand at an intermediate location in the Americas, apart from but part of both Caribbean and Andean regions, intimately linked by both natural and cultural history to Meso-America as well as to South America. It was southwards, across the late-glacial Central American isthmus, that man spread into southern America; across Colombia and Venezuela historic migrations took place, migrations not only of man, but of plants and animals. Though much more needs to be learnt of earlier habitats, it is now well established that by the fifteenth century AD the area of present-day Colombia and Venezuela was inhabited by a diverse mixture of agriculturalists, fishermen, hunters and gatherers. None of these groups, except perhaps for the Chibchas, had reached the same cultural levels that distinguished Central or Meso-America.

Within the pre-Hispanic period, one can identify a wide range of adaptation of cultures to ecology and examples of the interaction of the two. The distinction between bitter-manioc vegeculturalists and maize horticulturalists may have been as significant as that between swidden farmers of the forest and hunters of the grassland plains, though much more needs to be known of the rhythms of life, of agricultural implements, of social structures and of the range of domesticated plants and animals before it will be possible to attempt to distinguish meaningful regions within the northern part of South America before the arrival of the Europeans.

Their coming, however, marked a turning point in the historical geography of the area, yet despite the ubiquitous activity of Spanish authority, diversity, from a cultural viewpoint, was increased rather than diminished. By complex processes of assimilation, adaptation and introduction, people, products, plants, animals and a new technological age were spread through the area. The maritime margins became colonial contact zones. Hispanic modes of urban and rural settlement spread first inland and thence overland to and through all parts of Colombia and Venezuela. The objectives of the Spanish colonists demanded a reappraisal of indigenous cultures and peoples, ranging from their potential as suppliers of precious minerals, or as a labour force, or as new citizens of Spain, to their role as enemies of civilization and obstacles to the colonial experiment.

Upon the indigenous patterns, then, were imprinted Hispanic frameworks: road networks, trading organization, rhythms of activity, value systems, agricultural practices, urban living. The processes of contact, diffusion and acculturation were themselves differentiated throughout Colombia and Venezuela, no less dependent upon routes of penetration and personal ambitions and decisions as upon the distribution of aboriginal population and mineral and agricultural resources. New settlements were founded, older ones deserted; in some areas forest was cleared, while elsewhere grassy savannas became more woody; turf-covered flat land was ploughed, whilst terraced hill slopes were abandoned. Here, dispersed native population was in part congregated; elsewhere, driven further towards the forested margins of colonial control. Where densely settled Indians died of introduced diseases, they were sometimes replaced by imported African slaves. Skin colour added to the distinction between highland and lowland, forest and grassland, coast and interior. The development of the colonial economic system engendered the growth of isolated, independent city regions. Their numbers, size and interrelationships resulted from many factors. Within the latitudinal ecological grain of Venezuela a small but increasingly significant core of activity developed, centred on the coastal and highland zone of the north. In Colombia the more complicated physical structure encouraged greater regional dispersion and diversity.

Once more, as in pre-Colombian times, Colombia and Venezuela found themselves in an intermediate position. Venezuela, administered first from Santo Domingo, later fell within the jurisdiction of Santa Fé de Bogotá, a colonial centre that was never to attain the status of Mexico or Lima. As far as population, agriculture and mining were concerned, the area was of secondary significance to the Spanish crown; Colombia and Venezuela were sub-centres of colonial activities, whose margins included some of the most peripheral parts of the empire.

It is interesting to note that when a political challenge to Spanish rule appeared in the early nineteenth century, the foothold established in just such a forgotten corner of the empire, the Venezuelan Guayana, was of critical importance. From this base, with links to allies in Caribbean islands, the independence movement spread territorially across the *llanos*, to take the colonial administrative centre of Caracas from the rear. Progress thence lay in moving from one former colonial centre to another. Politically independent, the new nations, whose adolescent attempt at political integration within 'Gran Colombia' had failed miserably, faced the task of adjusting to new conditions. They had to come to terms with industrializing north-western Europe, and with the economic forces that this process was unleashing throughout the world. New crops or products were developed to satisfy the demands of new markets. In Venezuela gold, then oil, replaced agricultural products as major contributors to the national exchequer. In Colombia, upon a more diversified base, coffee cultivation eventually assumed paramount importance. New landscapes were created; regular plantations of coffee trees and lines of oil derricks spoke of management and capital investment. New settlements were founded, older agricultural areas declined, population migrated to the nearest town in search of the urban El Dorado. Gradually, Colombia and Venezuela were to experience the benefits and difficulties of twentieth-century development. The processes dictating change were similar: industrialization, commercialization of agriculture, greater integration into the world economy and population growth. These processes were to make Colombia and Venezuela less diverse. Urbanization produced unequal cities. Built with international technology and through the self-help of the poor, prosperity grew side by side with poverty.

And yet, despite these increasing similarities, a vital feature of Colombia and Venezuela remains the level of contrasts in cultures, landscapes and peoples. Faced with tropical forest and Andean snow, sprawling metropolitan centres and aboriginal longhouses, microwave radio links and dirt roads, blast furnaces and wooden ploughs, the visitor cannot fail to become curious as to why and where such contrasts exist, or for how long they have existed and may continue to exist. Within an area endowed by nature with a remarkable richness of variety, it is possible to trace, albeit still in a qualitative and fragmentary manner, the process of cultural development that helps one to understand the juxtaposition of landscapes, societies and economies of the twentieth century alongside those of the eighteenth or sixteenth, close by others nearer to the millennia before Christ.

The physical framework

Altitude above sea-level induces variations within the physical geography of the area that have been of significance to man since the remote past. Entering the south-west corner of Colombian territory, the high Andes, less than 200 km wide and comprising two parallel ranges separated by a high plateau, diverge as they pass northwards into three main ranges, the Western, Central and Eastern Cordillera (*Figure 5.1*). The central range, a massive crystalline block topped by volcanic peaks over 5500 m, is separated from its eastern and western counterparts by the Magdalena and Cauca rivers respectively. The Serranía de Baudó, adjacent to the Pacific coast north of Buenaventura, forms a structural outlier of the Cordillera Occidental, and is noted for the roughness of its surface caused by intense differential erosion of uptilted strata. The Cordillera Occidental is separated from this coastal upland by a lowland strip picked out by the courses of the rivers Atrato and San Juan. Eastwards the land rises steeply up the flanks of the Cordillera Central, whose northern extremity fans out before plunging beneath the sediments of the Caribbean coast, dissected by the rivers Sinú, San Juan, Porce and Cauca, the latter stream having traversed northwards through spectacular gorge sections from its upper structural trench section south of Cartago. Like the Cauca, but in a different mode, the Magdalena valley reflects its origin in orogenic faulting and folding; its valley is deep as far south as Girardot, with rapids at Honda and south of Neiva.

From around Pasto the Cordillera Oriental proceeds north-eastwards, its physique characterized by discontinuous high mountain crests that stand above the series of intermontane basins in which are situated Bogotá, Tunja and Sogamoso, and the rugged relief features of its periphery, particularly on the south-eastern margin. Crossing into Venezuela, the Cordillera de Mérida still attains altitudes of 4900 m, but the width of the range decreases consistently as it trends ever more parallel to the southern Caribbean shore, providing little space for the intermontane basins such as those around Mérida, Valera and Trujillo.

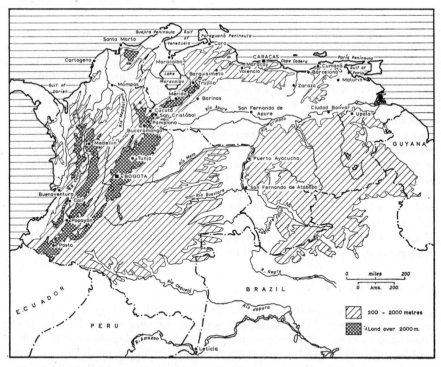

Figure 5.1   Colombia and Venezuela: relief and drainage

This elongated arm of the Andes continues eastwards through the central highland zone of northern Venezuela, in which are found small depressions occupied by Lake Valencia and the river Tuy, until east of Cumaná they leave Venezuela along the París peninsula for Trinidad.

North of the Andean ranges are found several important upland areas. In Colombia, west of the Gulf of Urabá, rise mountains that continue into Panama; east of Barranquilla the abrupt crystalline slopes of the Sierra Nevada de Santa Marta form Colombia's highest peak. In Venezuela the rocky headlands of the peninsulas of Guajira and Paraguaná stand out from the lowlands; east of Lake Maracaibo a tract of plateau country is dissected on its eastern edge by the waters of the Yaracuy and Tocuyo.

If the densely populated regions of northern Venezuela and north-western Colombia lack relatively level land, there is an abundance in the southern and south-eastern portions of the two countries. On the *llanos* one moves within a horizontal rather than vertical world; the only relief is between low, tabular interfluves and shallow, meandering, braided river courses. Although the terrain may vary markedly from one area to another, the overriding feature of this area that extends from the Amazonian tributaries to the Orinoco delta is its similarities. If, in the north and west orogeny and erosion are keys to understanding the land forms, in the south and east it is deposition. To the east and south the sedimentary trough abuts against the old structures of the Guiana Shield that in

Venezuela form the distinctive region of Guayana. The planation surface steps rise from the Orinocan fluviatile terraces around Ciudad Bolívar, across the lateritic inclusions of the lower Caroní to reach their maximum altitude in the Gran Sabana's southern margin, the plateau of Roraima.

Within and between such major relief components there exists a complexity of variation that defies brief description. Rivers are found shallow with deltas, or deep with estuaries; coasts high and rocky, elsewhere flat and sandy. Ruggedness is influenced not only by rock type and dip but also by past and present erosional or depositional activity. Whereas in one area the economic benefits of geology accrue from deposition, in another they derive from erosion and exposure (Vila, 1960).

Such degrees of diversity of altitude and relief are reflected through the climatic patterns of Venezuela and Colombia. A distinction can be made between sharp altitudinal variation and the general seasonal climatic rhythm. To the Spanish colonists, as in their Mediterranean homeland, time and weather were synonymous, and the seasonal droughts and rains formed the basis for their division of the year. In the *llanos* the seasonal change of precipitation, of cloud cover, of direction of winds and rates of evapotranspiration, affects both the ecology and the economy. In the mountains, on the other hand, altitude induces great variety in local climate in spite of low seasonal ranges of temperature. Sharp differences of rainfall between small areas are common, and a vertical zonation of average annual temperatures is commonly recognized, extending from the 'hot land' below 1000 m to the 'cold land' above 2000 m (*Figure 5.1*). Such approximate divisions based on altitude are better set aside in favour of the more subtle indices of vegetation and crops and the rhythms of economic activities. Climate has always meant more than mere rainfall and sunshine in Colombia and Venezuela; it has continuously and variously affected the pattern of vegetation, both before and during its alteration by man and it has promoted, permitted or prevented the spread of diseases, crops and human population. It may be said that in certain places man himself, on both the localized scale of urban building and the more general scale associated with forest clearance and reservoir construction, is affecting the micro and macro-climatic patterns of the area. In the past the southern Caribbean shore's longitudinal succession of arid and humid conditions was of more than local significance, as was the altitudinal seasonality of Colombia. Today man's impact on the balance of nature, his reduction of natural variety by cultural demands, has affected both the short and long-term importance of climate. The critical thresholds that initiate soil erosion or that prevent plant regeneration have in many areas been crossed.

It is impossible to describe the patterns of soil and vegetation without reference to the seasonality of the climate, climatic change, the variation of parent material and the rates and types of erosional and depositional processes at work. It was not without good reason that aboriginal Indians shifted their *conucos* (plots), that the colonial plantation-owner put a premium on the flat coastal, lacustrine or riverine zones. It is relevant to note that within the vegetational sequences of Colombia and Venezuela one can find problems that still challenge the biogeographer: the status of the so-called virgin tropical rain forest; the significance of riparian vegetation; the origin of the tropical grass and woodland savannas; the introduction and spread of alien plants. The variation and richness of the flora has caused comment since the early colonial accounts of Gumilla and Castellanos, long before Linnaeus sent Löfling to collect and classify the plants of eastern Venezuela and Guayana; and comment continued through the work of Humboldt and Depons, Codazzi, Ernst, Hettner and Pittier in the eighteenth and nineteenth centuries.

If the patterns of contrasts and variety were the corner-stones of early work in the physical geography of the area, in the more recent period greater, and justifiable, emphasis has been placed upon processes of change. On all sides one can find evidence for change in the physical framework. Rivers appear to be 'misfits' in their valleys, and anastomosing channels speak eloquently of climatic change. From the Caroní to the Sinú evidence is available in the shape of depositional and erosional features: river fall-lines, *ciénegas,* ox-bow lakes, raised marine and fluviatile terraces. Geological processes of volcanic activity have produced ash-filled valleys, and fault belts have briefly added a new dimension to natural instability. As relief and drainage have changed, so have vegetation, soils and climate, all five in some complicated relationship. New techniques make it possible to reconstruct the vegetational succession of the past, to date deposits and the surfaces upon which they rest.

Pre-Hispanic cultures

Developments to AD 1500
At least as long ago as 12,000 BC man was to be found in Venezuela, and probably also in Colombia (Rouse and Cruxent, 1963, p. 27). Both countries endured different climatic conditions from those of today. Average temperatures were lower and precipitation totals were higher, both caused, like the lower level of the sea, by the presence of glacial ice sheets within North America. The present complexity of lagoons, inlets, promontories and islands of the southern Caribbean shore was probably absent, replaced by a wide coastal margin that extended westwards to a wider Central American isthmus. Relatively large ice masses were present on the highlands of the Sierra Nevada de Santa Marta, around Caracas and Mérida, and over considerable portions of the three Colombian ranges. The pattern of rivers was different, too – in Venezuela, for example, the Orinoco entered the Atlantic by the extreme southern edge of its present delta; a wider Caroní entered the Orinoco further east than its present junction, and Lake Valencia's level stood higher and had an outlet southwards to the Orinoco. Although little is known of the late Pleistocene and Holocene landscape of Colombia, recent palynological studies have shown that in the Andes the tree-line fluctuated by some 1600 m from its present position, and that during the fluctuation of pluvial and interpluvial periods the margin between woodland and grassland varied considerably (Reichel-Dolmatoff, 1956, p. 41). During the so-called Palaeo-Indian epoch (15,000–5000 BC) man had moved into and through northern South America along a variety of possible routes: the southern Caribbean coastal zone; the lowland valleys of Magdalena and Cauca that led south to the Andean ranges and in turn gave way to the Orinoco and Amazonian river networks; and the Pacific coastal route south from the isthmus of Darien to the latitude of Buenaventura or Tumaco, whose short rivers led east

to the highland core. In Venezuela south from the Caribbean it would have been relatively easy to cross the narrow coastal ranges or avoid them by passing through the former exit of the Orinoco, around the mouth of the modern river Unare (*Figure 5.1*).

Large mammals roamed the area, and were hunted by men who also lived on the fruits of the forest. At Canaima in the Venezuelan Guayana and in Tolima in Colombia remains have been located of hunting-kill sites or stone spear-making localities. Settlements were widely dispersed and temporary in character, and population densities were exceedingly low.

By about 5000 BC the climate had improved and conditions were quite similar to those of the present. The large mammals had become extinct, possibly because of shortage of forage in the Boreal and pre-Boreal drier periods, or else through the increasing success of specialized hunting. Whatever the cause of the disappearance of the mammals, the beginning of the Holocene presented man in this area with a major problem of ecological adaptation. With smaller game unable to provide his dietary requirements, man began slowly to adapt to coastal fishing and forest agriculture. A series of coastal shell middens, stretching from the Paria peninsula to Barranquilla, are eloquent expressions of such adaptation. Other Meso-Indians who did not turn to fishing and sailing found it possible to obtain sufficient nourishment from fruits and seeds they collected from the forests and grasslands. Yet others, who in Colombia had moved south up-valley from the coast, and in Venezuela inhabited the forested mountains and basins of the north, began the process of plant domestication. At Rancho Peludo in Venezuela and Malambo in Colombia non-ceramic artefacts suggest bitter-manioc (*Manihot esculenta*) cultivation as early as 1000 BC. The striking similarity of associated pottery from this Colombian site to that of a Venezuelan series near Barrancas formerly supported the postulated origin of manioc cultivation in eastern Venezuela and its spread westwards into Colombia. In Colombia especially, sedentary village life still had a riparian orientation. Rich resources of fish, amphibians and bird life never made agriculture so necessary as elsewhere in the Americas.

Nevertheless, agriculture was adopted as a means of subsistence, not only manioc cultivation but also the domesticated grain, maize. At Rancho Peludo near Maracaibo and at Momil on the lower Sinú (*Figure 5.2*) evidence suggests that at some time around 1000 BC a change was made from a root-crop staple to the new seed crop. Though evidence is tentative, based upon the identification of manioc griddles and maize grinders, it seems clear that the use of maize was spreading into Colombia and western Venezuela from its domestication centre in Meso-America. Its arrival involved the learning of new agricultural techniques and the reappraisal and utilization of formerly under-used land; it included the possibility of larger and more durable surpluses. The distinction between the maize cultivation to the west and the manioc cultivation to the east of northern South America was to be of enduring significance. Hunting, fishing and gathering gradually became secondary to agriculture as a means of subsistence, except probably in the isolated interior areas of Guayana, the southern *llanos* and parts of the highland basins of Colombia. For Venezuela, it is possible to distinguish three geographical regions during the time of the Neo-Indians (1000 BC to AD 1500), but in Colombia the cultural diversity makes such generalized divisions

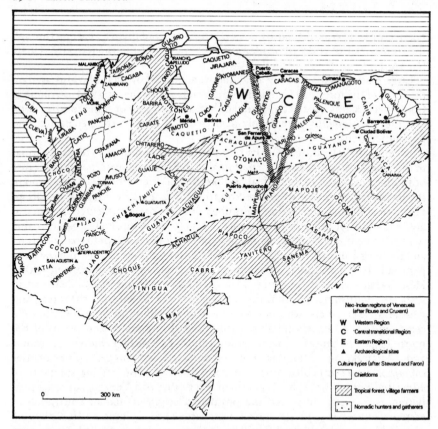

Figure 5.2   Colombia and Venezuela: cultural regions on the eve of conquest

worthless. Western Venezuela (*Figure 5.2*) was culturally closely affiliated with Colombia. Its maize and, in the highlands, potato cultivation, its pottery styles, its modes of burial and its earthworks distinguished the west from the east. If a nuclear area had to be chosen for western Venezuela, it would probably be the basin of Maracaibo. In the eastern region the lower Orinoco basin was its counterpart. From about 700 BC a series of Indian migrations down the valley of the Orinoco forced other groups east and northwards, driving them to the northern coast where they either migrated west or out into the Antilles, or else east and southwards into the delta of the Orinoco. This movement resulted in a displacement of pre-existing groups out of the lower Orinoco area, either north to the coast and thence along it westwards, or out into the Antilles, or alternatively east into the delta zone and south-east along the Atlantic coastal margin.

Between east and west there was a zone of transition (*Figure 5.2*), characterized by flows of peoples and crops and probably agricultural techniques and other cultural traits. Within this transitional region at least two centres were of more than local significance. One such centre was the locality around present-day Puerto Cabello. Here a number of routes converged: the coastal Caribbean passage; the Maracaibo–Yaracuy valley route; the Andean highland route out of modern Colombia to the Caribbean coast; and the Valencia–Tuy valley and upper *llanos* routes from the east. Further south the junction of the rivers Ápure and Portuguesa represented a convergence of routes from the lower Orinoco basin, the Amazonian–Orinoco headwater region, the Andean ranges of Mérida and the central highlands of northern Venezuela.

In Colombia it is more difficult to identify meaningful, large cultural regions in the period between 1000 BC and the eve of Spanish Conquest. With their new crop, maize, lowland horticulturalists gradually spread southwards into the valleys and on to the flanks of the Andean ranges, heralding a period of regional diversification – of agriculture, settlement forms and social and religious structures. The spread of settlement also involved isolation and localization of groups of Indians: river valleys, mountain flanks, upland plateaux – each saw the gradual development of their particular group. The influence of Meso-American culture, felt significantly around 1000 BC with the introduction of maize, was felt again around 500 BC with the landings of aboriginal colonists along the Colombian Pacific shore around Tumaco. From these footholds their influence spread eastwards into the Andean core area and thence downstream along the major river valleys. Gradually, at centres such as San Agustín, Tierradentro, Calima and Quimbaya intensive maize agriculture, metallurgy and a form of social stratification appear to have developed. Elsewhere, on the northern lowlands the earlier patterns and rhythms of land and life continued as before, influenced by coastal contacts from Panama at Cupica, and north-western Venezuela. Cultural variation was superimposed upon, and reflected, physical diversity. Access and contact were enhanced by the relatively dense network of waterways navigable by canoe. As in Venezuela, historic centres of contact can be identified, as exemplified by Zambrano where links existed between coast, Sierra Nevada, the Sinú valley and lands to the south. From the first century AD until the arrival of the Spanish, aboriginal developments in Colombia and Venezuela were probably characterized by an increasing density of sedentary agriculturalists in the northern and western highlands and fringe areas, whilst to the south and east and along the coastal margins Indians were migrating to and fro, adapting to and being influenced by the ecological opportunities of savanna, forest, marsh and lagoon. The interaction of cultural and ecological evolution resulted in the complex variations in styles of dress, food habits, form of dwellings, types of agricultural systems and range of implements, and in the differentiation of social institutions, which was the first feature to attract the attention of the Spanish as they made contact with this zone (Sauer, 1965).

Cultural patterns around AD 1500
A simplified and generalized sub-division of the cultural types to be found in the area around the Spanish contact period has been attempted in *Figure 5.2*. Necessarily, boundaries between cultures are as approximate as are the locations of many of the aboriginal groups. The key feature, however, is the tripartite division into chiefdoms, tropical-forest village farmers, and nomadic hunters and gatherers.

The chiefdoms were a cultural type characterized by small, class-structured states, which have variously been designated 'federations', 'realms' and 'kingdoms'. They have been viewed (Steward and Faron, 1959) as products of surplus agricultural production, increasing population density and intensification of military and religious practices. Whatever their origins, a useful distinction may be made between the militaristic chiefdoms of the Colombian area and their theocratic counterparts in northern Venezuela. For the former, warfare was not only necessary for co-operative efforts towards territorial expansion, but remained of paramount importance in the provision of human victims for temple rites. In the theocratic chiefdoms, however, warfare was far less important than religion as a means of community integration. It is of interest to note that the chiefdoms were by no means restricted to the highland areas, but extended downslope to include large areas of lowland savanna though not of tropical forest. In most chiefdom groups intensive farming provided the basis of economic life, and hunting, fishing and gathering never attained the significance they held for the tropical-forest farmers. In several areas the land available for agriculture was extended by means of terracing, and in the case of the Chibcha of Colombia and the Timoto–Cuica of Venezuela, irrigation systems were established (Eidt, 1959). Crop diversity was another characteristic of the chiefdom zone, and provided not only more abundant food, but, equally important, a well-balanced diet. Millennia of plant introductions and experimentation had produced a long list of domesticates: maize, sweet potatoes, sweet and bitter manioc, beans, peanuts, pineapples and avocado in the lowlands; at higher altitudes *quinoa, achira (Canna edulis)*, potatoes and *ullucos (Ullucus tuberosus)*. The variety of domesticated plants contrasts sharply with the paucity of domesticated animals. In the absence of large mammals, guinea pigs, muscovy ducks, mute dogs and bees provided alternatives to the vegetable harvests. Evidence also suggests that in most chiefdoms a very important protein source was that provided by riverine or coastal fishing, or, in the special case of the Gorrón in Colombia, by means of fish-breeding. A wide range of fibres, especially cotton, had engendered an interest and proficiency in weaving textiles, many of which in the highland areas were clearly related to central Andean practices. Pearls, gold and emeralds were but three of the highly valued commodities that entered into the trading economies of the chiefdom peoples. Their settlements were of two basic types; much of northern Venezuela and Colombia were characterized by relatively large wood-palisaded townships with populations of over 1000 inhabitants; elsewhere in the highland zones defensively sited villages were common. The peoples of these chiefdoms, the Cumanagotos, Caquetíos and Cuevans, were some of the first peoples of Tierra Firme to come into contact with European man (Sauer, 1965). It was only later that their more culturally advanced counterparts, the Timotos and Chibchas, were encountered.

The second major cultural division was the region occupied by tropical-forest farming communities, some of whose Indian groups, such as the Motilones west of Lake Maracaibo, still resist contact and assimilation. For decades the oral histories and linguistic affiliations of many of the Indian groups of these areas have posed problems of interpretation. Hypotheses relating to migrations, plant domestication and relationships between the forest groups and non-forest groups abound. Swidden appears to have formed the basis of the subsistence agriculture of at least a part of the tropical forest zone, though in south-eastern areas the staples of bitter and sweet manioc, beans and peanuts, which replace the maize-beans-squash complex more typical of north-western Colombia, are known to have been cultivated on fixed plots. The *montones*, or mound plots of the vegecultural *conuco* system of Venezuela, could be contrasted with the seed-based farming systems of the more complex Colombian areas. In all of the forest areas fish and small game provided vital sources of protein to balance the dietary qualities of the starch-rich crop foodstuffs. Though of the forest, the Indian settlements were dominantly riparian in location, the river networks providing excellent channels of intercommunication. Canoe construction, together with basket manufacture and heddle-loom weaving, characterized the diversified economic base of many of these Indians. The extensive range of forest products utilized included *anatto* (*Bixa orellana*) and *genipa* dyes, *barbasco* fish stupefiers and calabash utensils. Though their political organization was rudimentary compared with those of the chiefdom peoples, the tropical-forest Indians were structured upon a kinship base, and sex-differentiated occupational structures and shamanism were common elements. In certain groups cannibalistic practices were evidently of long standing. Perhaps the most significant features of this major region were the low densities of population, usually between a tenth or twentieth of the chiefdom area, and the lack of well-developed social structures and economic specialization.

The area of hunters and gatherers constituted the third major division of pre-Hispanic northern South America. In this zone small bands of nomadic Indians, within a more limited environmental setting, gathered fruits or hunted small game, peccaries, deer and armadillos in the riparian forests or extensive grassland savanna. Fire was an essential tool in their delicate relationship with the savanna grasslands, the marshes and swamps, and the narrow bands of woodland. The absence of agriculture as a mainstay of the economy was as fundamental a feature in setting them apart from their neighbours as was their relatively simple social organization. Although field evidence is continually being discovered of earlier, more intensive use of many of the areas of savanna grassland, the Orinoco plains, away from regular air routes, may yet provide more such data; the fact is that by the time of the arrival of the Spanish this zone was occupied by dispersed and exceedingly low densities of population.

This, then, was the general situation in Colombia and Venezuela on the eve of the Spanish Conquest; the impact of the intrusive Spanish culture on the diverse and distinctive cultures of the area was to be affected by the choice, or chance, of penetration routes, the types of contact made, and the extent to which the Indians were incorporated within first colonial and later national economies and social systems. The fact that in the present century neither Colombia nor Venezuela have significant Indian populations is as much related to the situation at contact as it is to later developments in acculturation and assimilation.

Colonial contrasts and similarities

Colombia and Venezuela were both peripheral and second-rank Spanish colonies. The Viceroyalty of New Granada never achieved the status of its northern and southern counterparts of Lima and Mexico. Nevertheless, the differentiation of colonial developments within the region of modern Venezuela and Colombia, and their location at the junction of Caribbean, Central and South America, provide an instructive basis for interpreting the processes of change after the removal of colonial political controls. In mineral and agricultural development, rural and urban settlement patterns, demographic change, the establishment and development of trade and communication networks and the evolution of administrative frameworks, the region emphasized more than anything else the complex nature of the adaptations, extension and effectiveness of colonial institutions, practices and modes of life.

Fundamental to any understanding of Hispanic colonialism is the role of the urban foundation as the symbol and instrument of colonization. Towns became the foci of authority, the centres from which economic enterprise extended, the cores of cultural contact, the stepping stones in the effective control of territory (Morse, 1962). During the sixteenth century colonization was characterized by its ephemeral nature. From early coastal bases, such as Cumaná and Coro, colonization proceeded slowly and haltingly, hampered continually by hostile Indians, unsuccessful land settlement and disorganized administration. With the flow of gold objects along the Coquibacao–Curiana route interrupted by Spanish activities in northern Colombia, the mineral potential of Venezuela was soon seen to be extremely limited, and it was only after the middle of the sixteenth century that the number of urban settlements increased in harmony with agricultural development (*Figure 5.3*). By 1600 a network of settlements had been laid down predominantly on or near to the Caribbean coast and its mountain hinterland; only Santo Tomé on the Orinoco lay outside this zone. In Colombia, likewise, primary coastal centres at Santa Marta, Cartagena and on the Urabá gulf preceded a penetrative phase of foundations that included the settlements of Cali, Popayán, Santa Fé de Bogotá and Tunja. Naturally enough, the chronology and location of such settlements reflected the various enterprises of Spanish *conquistadores*: in the Colombian area Federmann moving south-west from Coro; Quesada and Vadillo penetrating south from Santa Marta and Cartagena; Belalcázar moving north from Quito to Popayán and Cali. In Venezuela Berrío and Ordáz led exploration parties into the Orinoco basin, while Ehinger and others colonized the lands of the western region (Morón, 1954). Despite the popular preoccupation of many authors with the niceties of the exact dates of foundation of colonial urban centres, of far greater significance was the ubiquity of settlement shifts. A great variety of factors led to settlements being moved. Some, like Venezuelan Angostura (previously Santo Tomé), were moved for strategic reasons; others, such as Remedios in Colombia, were moved owing to the exhaustion of the local resource base. Fires, insect plagues, health hazards and Indian attacks were common factors. The process was by no means restricted to the early phase of colonial rule. As the settlement frontier expanded and new situations arose, so there were many settlement casualties (Vila, 1966; Houston, 1968). Even in the seventeenth and eighteenth centuries, through the amalgamation of dispersed small settlements, abandonment and removal were commonplace (Martínez, 1967).

Perhaps the commonest denominator of the urban settlements was their form. The basis of their plan was a rectilinear grid of streets dividing the built-up area into blocks. The *plaza mayor* and the location of church, *cabildo* and other functionally specialized buildings gave most of the settlements a formal, planned

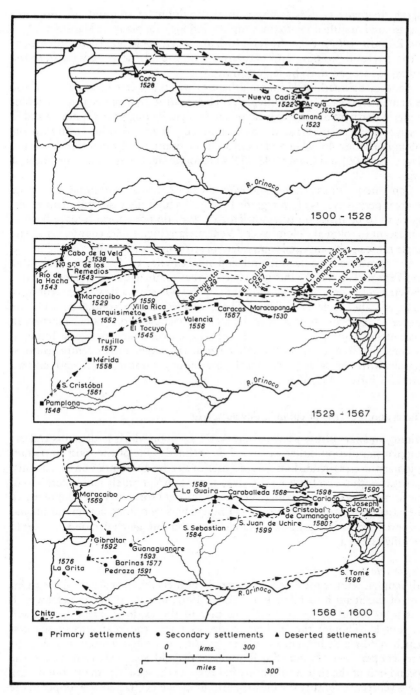

Figure 5.3 The spread of settlement in sixteenth-century Venezuela (after Vila, 1965)

aspect. In many cases, of course, local conditions of site and circumstances necessitated modification to this simplified model (for plans see Martínez, 1967, and Ricardo, 1967). Even the term 'urban', when applied to such settlements, needs careful definition, and barbarism was probably never very far from the central square, even in the largest towns.

Another hallmark of Hispanic urbanism was the hierarchical nature of the settlements. As early as 1600 there could be identified one or two primate settlements in northern South America; in size Santa Fé de Bogotá in Colombia and El Tocuyo in Venezuela stood apart from all others. Despite readjustments throughout the colonial period, even at the beginning of the nineteenth century Bogotá (25,000) and Caracas (48,000) were considerably larger than their nearest rivals.

Compared with the *ciudad* and *villa*, the *campaña* was as unimportant in the colonial period as it is often regarded today. Though there are still many investigations to be made on rural settlements, they do not appear to have been as significant in colonial Colombia and Venezuela as their urban counterparts. The single most significant rural component of settlement during the colonial period would appear to have been the mission village (Vila, 1965). In both Colombia and Venezuela it was the missionary orders that colonized, congregated Indians, and created distinctive rural landscapes. Over wide areas a new settlement form emerged, the nucleated village. These villages can be contrasted with the nucleated form and functions of the arable haciendas and livestock ranches. In Venezuela these agricultural settlements often contained over one hundred persons. In Colombia, on the other hand, the larger urban centres contrast sharply with the widespread small gold-mining camps, *minas* and *ranchos* (Armas Chitty, 1961; West, 1952).

The economic basis of colonial development

Mining and agriculture provided the dual base upon which settlements and the colonial population either expanded or declined. One of the most important colonial processes in northern South America was the differentiation of economic activities within and between the areas of modern Venezuela and Colombia. In the former agriculture provided the most significant stimulus to development; in the latter it was mineral resources – although neither mining nor agriculture was ever exclusive to each area. Local variations within this generalization have yet to be fully assessed.

Venezuela
Although it was the pearl beds of eastern Venezuela that first attracted the attention of the Spanish to Tierra Firme (Ojer, 1967; Morón, 1954), mineral wealth was soon found to be singularly lacking in the explored areas of northern Venezuela. Though the Welsers in the west and the Spaniards in the east searched diligently for vein and placer deposits, only small amounts of gold, and later copper, were found. Small amounts of gold were extracted during the second half of the sixteenth century in valleys to the west of Caracas, but during the seventeenth century the only significant mineral extracted was copper in the Cocorote mines (Brito-Figueroa, 1963, p. 82). Such mining necessitated the

importation of negro slaves to replace indigenous labour. Though colonial records mention eighteenth-century mining in places as far apart as Guayana (iron ore) and the Andean areas of the west, it never became an important element in the economic progress of the region. This paucity of mineral wealth undoubtedly had several fundamental effects on the development of the region. It meant first that it was to be largely by-passed during the sixteenth and seventeenth centuries as colonists were drawn to the riches of New Spain and Peru. It also involved the development of an agricultural colony that, compared with other areas, was to prove highly successful – at least from the point of view of the Spanish crown and the entrepreneurs involved. This agricultural base meant that for the most part the distribution of population and its changing structure, associated forms of settlement and concentration of wealth were controlled by the interaction of ecological and economic factors.

The basic feature of agricultural development in colonial Venezuela was the regionalization of activities and the establishment of dual systems of enterprise. The first was subsistence production based on small-scale units and related to aboriginal methods of cultivation; the second was commercial production on large, extensively used holdings. The extension of private ownership of land in the hands of a small but important group of families, and the concentration of the wealth derived from commercial agriculture in restricted zones, principally in the centre-north, were also significant processes (Brito-Figueroa, 1960; Arcila Farías, 1946).

As previously mentioned, the initiation of colonial agriculture was centred upon early urban foundations. In most parts of Venezuela the establishment of a township involved not only the distribution of urban *solares* between members of the community, but also the allotment of Indians, who were initially viewed as the means of cultivating the land or tending livestock. If one maps the location of *encomiendas* and the principal residences of their *encomenderos* (*Figure 5.4*), one can see the overwhelming significance of the north and western zones of upland Venezuela (Arcila Farías, 1946). As a result of recent researches it is now possible to trace the evolution of both arable and livestock farming in central Venezuela from the late sixteenth century through to the nineteenth century (Brito-Figueroa, 1961, 1963; Vila, 1965).

In the province of Caracas private agricultural land was acquired principally through *mercedes de tierras*, grants of land given to persons intimately connected with the foundation of the principal civil settlements. By 1600 almost 13,000 ha of land had been alienated from Indian groups in the northern valleys of Tuy, Caracas, Aragua and Barlovento. Of this land, some 50 per cent was controlled by twelve owners. The formation of large estates proceeded at an ever-quickening pace throughout the next two centuries. During the seventeenth century almost 1 million ha of agricultural land were included within private estates, almost 45 per cent of the total area of these valleys (Brito-Figueroa, 1963, p. 157). In the decade 1736–46 alone, a further 500,000 ha were incorporated within 190 estates. Holdings of over 2500 ha were relatively common. By the end of the eighteenth century, an estimated 18 per cent of the total area of the province of Caracas was in private ownership. Two associated features of estate formation were the control of most of agricultural production by a diminishing number of owners, and the low intensity of land use within the estates. By 1750,

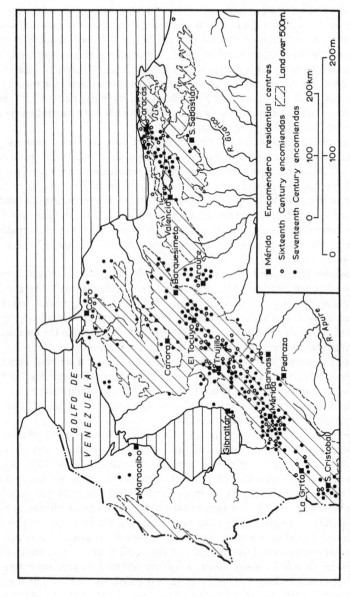

Figure 5.4 Distribution of *encomiendas* in north-west Venezuela

for example, some fifty owners controlled approximately 73 per cent of the agricultural land of the province of Caracas. As little as 4 per cent of the total land in estates was actually cultivated. The absorption of Indians into the labour pool of these estates, and the usurpation of their land by private owners, account for the widespread desertion of Indian settlements during the period 1600–1800. Caracas province alone lost at least sixty such settlements. Not only were large estates being created, but another significant trend during the colonial period was the reduction in the total number of commercial agricultural holdings, principally by purchase or inheritance. The number of cacao plantations, for example, was reduced from around 440 in 1746 to a mere 160 by 1800. Wealth was being concentrated in fewer hands, and in a decreasing number of centres.

Cacao was the most important arable crop during this period, providing the basis for plantation agriculture in many parts of northern Venezuela. By 1750 over 5 million cacao trees were under cultivation in the province of Caracas. The fifteen plantations of a single family, the Pontes, accounted for over 250,000 trees at that date. Cacao cultivation was located in the fertile soils of the valleys of northern Venezuela, extending from Cumaná in the east to Trujillo in the west. The two most important centres of production were the northern part of the province of Caracas, and around the margins of the Maracaibo basin, especially to the east of the lake. By 1800 almost 1000 plantations were in operation, involving some 13 million trees and 29,000 ha of land. During the whole of the colonial period Venezuelan agricultural production was dominated by cacao. Considering the inadequacy of the indigenous labour force (numbers were relatively small at contact and decreased through disease), it was not surprising that, as elsewhere in the Americas, the plantation owners imported large numbers of negro slaves (Brito-Figueroa, 1963, pp. 93–112). A high correlation is to be noted between areas of commercial plantation agriculture and negroid elements in the population. However, the Venezuelan total of 120,000 slaves legally imported during the colonial period accounts for a mere 12 per cent of the total entering Spanish American colonies.

Besides cacao, other commercial crops such as tobacco, indigo, cotton, coffee, wheat and sugar cane were cultivated within Venezuela. Tobacco, cultivated by the Indians of the Tocuyo district on the arrival of the Spanish, suffered from serious setbacks during the entire colonial period through the interference of the metropolitan monopolists. Though it was an important crop in the Valencia, Barquisimeto and Barinas regions, it never became an important item of export. Wheat was another crop widely cultivated during the sixteenth and seventeenth centuries, but it lost ground to the increasingly profitable cacao in the eighteenth. In the late colonial period the focus of wheat cultivation was in the Andean zone centred on Mérida. Sugar cane was another ubiquitous crop, which was provided principally for the Venezuelan market. Indigo, elsewhere in the colonies an important agricultural product, was left to grow wild in the northern central valleys until the 1760s, when price increases made its laborious cultivation economically viable. In the Aragua valleys, west of Caracas, it attained great local significance during the last two decades of the eighteenth century. Cotton was also grown widely in Venezuela, and from the middle Yuruari basin to Tocuyo it provided the basis for local textile industries.

It must be remembered, however, that these crops, accounting for less than 10 per cent of the area of private agricultural holdings, took the form of oases of intensive agriculture amidst a general scene of continuing aboriginal practices. Beyond, and probably within, the boundaries of private land lay large areas in which the aboriginal staples of manioc, maize, beans and plantains were cultivated by various methods of swidden. Though the newly introduced ploughs, water-wheels, stone aqueducts and the like affected great changes in certain localities, their innovative effects on agriculture were limited. Over large areas mestizos and negroes appreciated the suitability of indigenous agricultural techniques. What colonial crop agriculture did was to create a new, highly contrasted, agricultural economy and landscape.

The impact of arable plantation agriculture within the northern areas of Venezuela can have been no greater than that of colonial livestock ranching in the vast plains of the Orinoco basin. It is now known that, prior to the development of extensive ranching over the *llanos*, livestock farming was intimately associated with arable cultivation in the northern valleys. In the latter part of the sixteenth century, indeed, cattle were the most significant elements in agriculture from the basin of Valencia to Tocuyo. Only through the competition afforded by cacao and other crops was ranching displaced and extended southwards, downslope towards the River Orinoco. Gradually, during the seventeenth and eighteenth centuries, cattle ranches (*hatos*) were established in the *llanos*. Penetration of ranching practices into the plainlands came from three principal directions: from the Andean area south-east into Barinas and Apure; from the central northern valleys south via San Sebastián into Guárico; and from the eastern coastal lowlands around Barcelona-Cumaná, south to Zaraza and the Orinoco. The complicated chronology of advance cannot be discussed here. Suffice to say that, as in the arable areas, large estates (*latifundia*) emerged as the dominant unit of organization and production. By 1740, for example, there were more than seventy *hatos* in the valleys of Guárico and Apure, and more than 500,000 ha of open range had been alienated within the preceding 50 years. Some 300,000 head of cattle were grazed on the forty largest *hatos*, providing employment for almost 4000 persons, negro slaves accounting for 10 per cent of the total. Some of the largest *hatos* had herds of over 50,000 head (Brito-Figueroa, 1963, p. 217). Within the *llanos* zone two distinct ranching sub-regions can be identified. The first, the core area of ranching, extended around the southern fringe of the uplands, where free-range grazing was possible throughout the year. The second area was one of seasonal transhumance on the northern bank of the lower Orinoco, cattle being moved north and southwards as seasonal variations in precipitation presented problems of flooding and drought. It was in the upper *llanos* area that the introduced cattle multiplied rapidly. With no fences, the ranching economy and landscape was influenced by river lines, water-holes and scattered patches of woodland. Outside these two zones many secondary centres of livestock agriculture persisted until the end of the colonial period. The development of cattle-ranching under the syndicate system of the Capuchins in the Caroní–Yuruari area, the specialization of the Mérida district in sheep-rearing, and the north-east in goat-herding, provide good examples of localized developments during the eighteenth century.

Though we have summarized the salient features of Venezuelan colonial agriculture, space precludes any analysis of the local variation that was so prevalent. Individual valleys differed in their crop combinations; agricultural practices and tools were localized; production, productivity and efficiency were related to ecological circumstances, economic pressures and personal whims (Quijada, 1968). In widespread localities fishing provided a valuable addition to food supplies. In some areas aboriginal terraced land was abandoned, elsewhere irrigation systems were introduced and extended. To the *conuquero* were added the *hacendado* and the *llanero*, new Venezuelan folk figures (Mendoza, 1947).

Although the colonial period had witnessed an expansion and diversification of agriculture, in Venezuela the principal benefactors had concentrated with their wealth and families primarily in and around Caracas, or at least within the central northern zone, close to the coast. Though earlier the towns had spread economic activities into rural areas, as the colony flourished wealth and influence gravitated towards those same towns. Cattle estates and arable plantations were controlled by absentee landlords residing in the Caracas area. The mode of agricultural development had reinforced, rather than modified, the distribution of population. Colonial Venezuela had become differentiated: the northern regions contrasted with those of the south, the countryside had become equated with work, the town with wealth.

Colombia

It has been said that 'gold was for New Granada, up to a point, what hides, cacao and indigo were for Venezuela' (Bushnell, 1954, p. 3). This generalization has much to commend it. Colombia's most important economic resource during the colonial period was its mineral deposits, the exploitation of which brought about a pattern of settlement, a demographic history and a location of wealth and power quite different from those of Venezuela. Within the limits of modern Colombia diversity was a recurrent theme between AD 1500 and 1800, as it was before and has been since. Though mining was of paramount importance, agriculture also developed along lines similar to those in Venezuela, but to a lesser degree.

The three principal areas of gold-mining (both placer and vein) were within the Cauca valley, the upper and middle reaches of the Magdalena, and along the Pacific coast (*Figure 5.5*). Perhaps fortunately for Colombia, no one area completely dominated production or development at any one period. The Spanish entry into modern Colombian territory was at least in part a result of reports of gold deposits that had been heard of from the beginning of the sixteenth century. By the 1530s parties from the Caribbean coast and Quito had entered the area to find abundant evidence in both graves and functioning mines of the presence of gold and emeralds. The complexities of exploration and early settlement history have been traced in some detail by Henao (West, 1952). Gold-mining expanded rapidly in many localities. In the middle Cauca valley around Anserma and Cartago vein-mining was especially important, though the depletion of ore bodies necessitated the repeated removal of settlements. In the upper Cauca valley, around Popayán, placer deposits included within dissected Tertiary gravel trains provided rich but decreasing amounts of gold from the late sixteenth century.

Similar in morphology, but of Plio-Pleistocene age, were the flights of auriferous gravel terraces on the western flanks of the Cordillera Occidental. Between

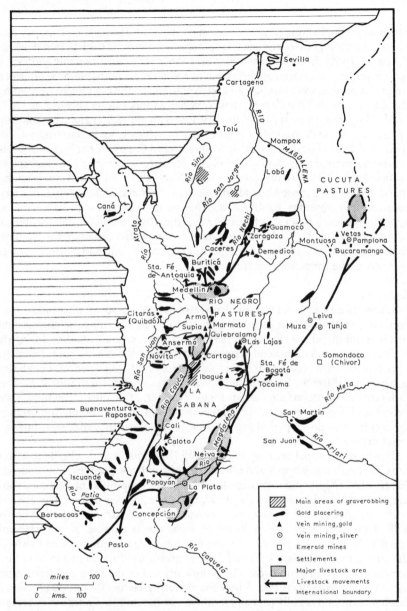

Figure 5.5 Patterns of economic activity in colonial Colombia (after West, 1952)

Buenaventura and Barbacoas practically every river basin provided ample returns. North along the littoral the Chocó district was rich not only in gold deposits but also in hostile Indians (West, 1957). Of special importance during the colonial period was the mining that went on around and within the Antioquia batholith at the northern end of the Cordillera Central. At Buritica the Spaniards had simply to persuade the Indian miners to continue their work, so productive were their methods. Although the constant shifts of the mining camps (*minas*) makes it exceedingly difficult to reconstruct the exact sequence of mining in the area, it appears that during the colonial period a gradual shift was in progress from Buritica northwards. This included the establishment of settlements at Cáceres and Zaragoza (*Figure 5.5*). Only with the decline of gold production in these northern areas in the second half of the seventeenth century was attention turned to the dissected high-altitude gravel spreads (*cerros*) within the Antioquia region north of Medellín (West, 1952, p. 27).

Another important mining district was that on the west bank of the Magdalena, centred on Ibagué. Activity spread northwards from this area to include the new and *ambulante* settlement of Remedios. In sharp contrast to the abundance of deposits in the central and western Cordillera was the low mineralization of the Cordillera Oriental. With the exception of the introduction of iron for tipping tools and water-powered stamp-mills, there were few technical innovations in colonial gold-mining in Colombia. Aboriginal methods of stream and pit placering, and panning, were extended over a much wider area. Far more important were the effects that mining had on agriculture and the demographic and settlement history of north-west Colombia.

Two processes characterized colonial agricultural change: first the development in and adjacent to the mining areas of subsistence arable cultivation; and secondly, the establishment and expansion of much larger scale commercial farming units, predominantly engaged in livestock-ranching, in the higher altitudes of the main mountain ranges, especially in the level *altiplanos* of the Cordillera Oriental (West, 1952; Marciales, 1948). Within each of the forested lowland mining districts were to be found plots in which first Indians and later negroes cultivated maize, beans and sweet manioc. As the supply of maize previously provided by tribute gradually diminished, its role, especially along the Pacific littoral, was taken over by the introduced plantain. In some areas the agriculturally unfavourable location of the gold camps made it necessary to utilize more distant fertile soils. Sugar cane was also widely grown within the lowland zone, providing a variety of strong liquors that did something to mitigate the hard labour. The most significant protein source available for the slave gangs and others employed in the goldfields, far outweighing local sources such as fish and *manatí*, was meat, produced on pastures as close to the populated areas as ecology and economy allowed. The flat, unforested floors of Andean river valleys and montane basins were ideal sites. In the middle and upper Cauca and the upper Magdalena (*Figure 5.5*) extensive areas were allotted to cattle farms (*estancias de ganado*). In the savannas around and to the north-east of Bogotá abandoned aboriginal agricultural mounds had been grassed over, and provided pastures for large herds of unfenced stock (Broadbent, 1968). In Cauca, Boyacá, Cúcuta and Medellín alike colonial documents record the establishment of the cattle industry (Crist, 1952, p. 83; Fals Borda, 1957, p. 143; Parsons, 1949). Unsuccessful attempts were made to clear forestland for cattle (West, 1957, p. 147). Meat, either fresh, dried or salted, became a permanent feature of the weekly food rations given to the workers. To be added to the complex pattern of arable and pastoral agriculture was the altitudinal differentiation of grain varieties. Dibbled maize plots of the northern humid lowlands gave way southwards to ploughed wheat fields that provided flour for the expanding mining communities. Compared with Venezuela, therefore, the colonial economic geography of Colombia was considerably more complex. A combination of mineral and agricultural exploitation, influenced by the variety of topographic and climatic conditions, produced a differentiation that was not to be found in agricultural Venezuela.

Demographic processes and patterns

To date few scholars have concerned themselves with the demographic character of colonial Colombia and Venezuela. Compared with the statistical analyses that have been made for Mexico and Peru, relatively little is known of this marginal area. If Eidt's (1959, p. 380) figure of 600,000 is accepted as the Chibchan population around AD 1500, then Colombian territory may well have contained as many as 1 million Indians on the arrival of the Spanish. For Venezuela it is unlikely that the more densely populated northern highlands held more than 250,000 Indians, which, added to an estimated figure for the remainder of the territory, means that Venezuela may have contained between 350,000 and 450,000 Indians by 1500. By the end of the eighteenth century Colombia's population was estimated at 1.5 million, that of Venezuela 900,000. Such totals, however, say little of the intervening years of demographic change, change that included not only periodic fluctuations in total population, but also major structural modifications to the ethnic and social character, and spatial distribution of the inhabitants of the area.

One of the most significant processes that affected and characterized the colonial period was the steady reduction of the Indian component in the population and its replacement by either Iberian colonists or their African negro slaves. While miscegenation between all these three groups gradually blurred the ethnic qualities of individuals, social status became all the more important. The aboriginal inhabitants of the area suffered grievously during the sixteenth century. After the psychological shock of the Spanish Conquest, which itself produced many instances of mass suicide, alien European diseases such as smallpox, tuberculosis, measles, typhoid and influenza began to take their toll. Of equal, if not greater, importance for the increased severity of these diseases was the continual intermixture and concentration of Indians. Whether by the practices of *mita* labour in the Venezuelan Andes, or the use of Indians in the placer mines of Colombia, decimation rapidly ensued. Some of the epidemics, such as the smallpox one in 1588, appear to have affected almost the entire area of Spanish contact. Within specific areas, however, the incidence and impact of disease was probably not unrelated to the general conditions of work endured by the Indians. From West's (1952) researches on the mining communities of Colombia, it can be seen that in many cases disease merely completed a task that arduous work and poor diets had begun. In Venezuela, too, the occupations of pearl-diving, mining and *encomienda* agricultural labour were often fatal for Indians. Not that attempts were not made to prevent these processes. Much more positive than the Utopian ideas of Las Casas in Cumaná were the many reservation tracts (*resguardos*) set aside in Colombia from the late sixteenth century for Indian communities (Parsons, 1949; Fals Borda, 1969). However, by that date much of the damage had already been done, and throughout the seventeenth century and in widely separated areas the migration of Indian groups out of reach of the Spanish can be noted. Only toponymic traces remained of many Indian groups by the year 1700. Elsewhere Indians were less harshly treated by the newcomers. In the widely dispersed mission villages only hard work was the price of spiritual salvation. In these communities Indians lived long enough to accept short-cropped hair and the clothes, language and music of their 'civilizers'. The Indians, for their part, demonstrated their native skills and adaptability. Other Indian groups remained, throughout the entire colonial period, outside either the knowledge or control of the new authorities. Their contact with other cultures had to await the nineteenth and twentieth centuries.

The demographic counterbalance to Indian decline was secured by the importation of negro African slaves. Though the influential white population gradually increased in numbers, it was this forced mass migration of labour that allowed the colonial system to function (Escalante, 1964). Very quickly it was perceived that what Indians could not, or would not, do, slaves could be forced to do. In Colombia the distribution of negroid elements soon reflected the patterns of gold-mining (West, 1952). In Venezuela their counterparts were to be found within the cacao plantations of the north. Although negroes appeared less affected by hard labour in the tropical lowlands and less susceptible to European diseases, the combination of unbalanced and inadequate diet with their own imported diseases (especially yaws) resulted in high mortality-rates (Vila, 1965; West, 1952). Intense economic activity, high mortality-rates and population replacement went hand in hand. Naturally, negroes were not exclusively to be found in the agricultural plantations and mining camps. Every township had its negroid population, engaged in household service and other menial tasks. Wherever work had to be done, there negroes were to be found, be it on the river boats, in cattle ranches, sugar mills or cathedral choirs. Exceptionally, negroes escaped their allotted role by fleeing to the more remote areas. More significant, especially during the eighteenth century, was the legal attainment of their freedom. Some 50,000 such free negroes formed a most important element in the agricultural economy of northern central Venezuela during the 1780s (Brito-Figueroa, 1963). In Colombia such negroes spread outside the mining zones to settle pioneer areas during the same period (Parsons, 1949). The integration process that negroes underwent, which so effectively adapted them to the habitat, economy and society of the colonies, is still little researched, let alone understood. Locally, where numbers or special circumstances permitted, they made many significant contributions to the cultural milieu, not least in music and agriculture (Fals Borda, 1969; West, 1957).

In Colombia and Venezuela Indians and negroes were instructed, ruled and manipulated by the third ethnic element, the whites and mestizos. Ethnic origin and social class, if not caste, were closely related. The whites' employment, income, residence, status and influence set them apart from all others. They stood at the top of the social scale as an aristocratic élite, an urban power group whose wealth and importance flowed from their hold over the organization of economic enterprise, and their manipulation of prices, interest rates and markets. Below them, and ever mindful of their relative position, came the group of mestizo artisans whose role in the functioning colonial economy is still little understood. One step further down stood the poor whites and mestizos, scorned by non-whites for being white and by other whites for being poor. The social stability of three centuries of colonial rule hinged on the acceptance by the majority of their social station.

The following figures reflect the numerical significance of each of the ethnic groups in Colombia and Venezuela at the end of the eighteenth century: Indians 18 per cent; negroes 60 per cent; whites and mestizos 22 per cent (Brito-Figueroa, 1961; Parsons, 1949; Vila, 1965). Of perhaps greater geographical significance was the spatial distribution of the influential white and mestizo class. Their locale was the urban settlement, not as the site of factories, but as the node of administrative and trading networks and symbol of culture. In Colombia this meant primarily the group of highland towns centred on Santa Fé de Bogotá. What the viceregal capital boasted could be found, albeit on a smaller scale, in many of the regional urban centres such as Cali, Popayán, Cartagena and Medellín. Only of necessity, and then often temporarily, did whites live in what were considered the unhealthy tropical northern lowlands (Parsons, 1949; Scott, 1968). In Venezuela, likewise, population was concentrated within the northern highland zone; and even though in the 1780s only 37 per cent of the population of the province of Caracas was urban, that sector contained the most notable entrepreneurs. The dominance of urban settlement in Venezuela is reflected in the fact that by 1810 out of a total of some 500 settlements, fifty towns accounted for one-third of the total population.

As a result of agricultural development (principally the cacao export crop) during the seventeenth and eighteenth centuries, the foci of commercial and administrative activities had been stabilized. Caracas dominated the central northern region, while to the west Barquisimeto and Maracaibo, and to the east Cumaná, Barcelona and, to a lesser extent, Angostura, each had their own distinctive hinterlands (Gormsen, 1966). The Caracas region contained within it by far the greatest concentration of financial capital in the whole of colonial Venezuela. In both Colombia and Venezuela the steady accumulation of secular wealth had been accepted and reinforced by the attitude and activities of the Catholic Church, whose increasing revenues from tithes and estates made it one of the most influential owners of land and financial resources (Brito-Figueroa, 1961; Fals Borda, 1969). The key feature of the late colonial period was the agglomeration of wealth in and around the principal urban settlements; in Venezuela this meant almost entirely the Caracas region, in Colombia, though perhaps less dominant, the Bogotá district. The benefits and profits from colonial economic enterprise were slowly but surely siphoned from cattle ranch, plantation and gold mine to the urban centres. The consequences of an urban-based colonization process had begun to appear; centripetal forces had replaced the centrifugal.

Patterns of trade and communications

The significance of the various settlements and zones of economic activity during the colonial period can best be judged by the type and state of the communications that linked them to the oceanic routes. For Colombia and Venezuela the critical interface between hinterland and foreland was the southern Caribbean shore, dominated by the ports of Cartagena and La Guaira. The scattering of smaller ports stretching from the Orinoco delta to the river Patía in south-western Colombia merely served as foci for restricted trading areas. Second only to the ports were the communication nodes formed by the administrative centres. It was between these and to the coastal settlements that the relatively rare royal highways (*caminos reales*) ran. In northern South America the most important of the *caminos reales* entered Colombia from the south, passing by way of Pasto, Popayán and Bogotá and thence into Venezuela via the Andean aboriginal route from Mérida and Trujillo on to Caracas. This was the northernmost extension of the Lima–Caribbean route. From and to this unpaved trail ran numerous pack trails, reflecting the significance of that ubiquitous colonial animal, the mule. In certain sections of Colombia the terrain was seasonally too difficult even for this sure-footed beast, and Indians carried heavy burdens over precipitous slopes (West, 1952). Traffic between settlements in such physically isolated areas as central Colombia put a premium on the availability of low passes. Even without a carting network present elsewhere in the New World, altitude often became a critical factor in determining the type of produce and the regularity of trade within colonial Colombia and Venezuela. In south-eastern Colombia and the major portion of Venezuela the overland trails were of critical importance. In Venezuela they connected La Guaira to Caracas, and Caracas to the agricultural valleys along and between the southern flanks of the coastal ranges. Along such tracks cacao, tobacco, cotton and hides moved out to the ports for export. In return came brandy and textiles and the necessary manufactured goods from across the Atlantic (Arcila Farías, 1946). In Colombia nature had bestowed upon the area a ready-made pattern of communications. Between the grazed mountain ranges ran slow-moving and navigable rivers. In no other Spanish colony were water transport links so important as in Colombia, Goods trans-shipped at Cartagena could penetrate far inland to the mining areas before the need for time-consuming and expensive mule trains or human porterage. In Venezuela, on the other hand, except for localized traffic, the colony turned its back to one of the continent's most extensive river networks. The Orinoco and its north bank tributaries remained on the margin of the unknown, rather than providing lifelines of commerce. As the heavy, cedar dug-out canoes gave way southwards to lighter rafts, so the seasonal regime of the longitudinal rivers affected the economic rhythms. In the wet seasons, when water was superabundant for sluicing out gold in the mountains, the rivers raced so fast as to make it impossible to pole upstream to collect or deliver products. Only in the dry season could trade be continued, and the building of river boats was a local industry in northern Colombia. While the Pacific rivers also had their own trading hinterlands they were never large enough to compete with those of central and northern Colombia (Parsons, 1949). Implicit in the use of mules, and in special cases ox-drawn carts, was the provision of pasturage along the main overland trails. When extensive trade in livestock on the hoof was concerned, this became of even greater significance. Some of the routes of livestock movements are shown in *Figure 5.5*.

Although the outline of the pattern of communications is known for both Colombia and Venezuela during the colonial period, relatively little is understood of the trading system used within those areas. West's (1952, 1957) researches in Colombia were the first to identify the activities of the itinerant merchant who appears to have been a key figure in the operation of the gold-mining industry. These merchants, who travelled on regular circuits within the goldfields, not only sold large amounts of imported produce, which were later delivered to the purchasers from warehouses in the port of Cartagena or the larger settlements of the interior, but they also carried gold dust back to the regional smelting centres on behalf of mining camp operators. Illegal practices were rife in this trading system, as was often the case in the Indies, since the merchants could equally well perform the function of smugglers – either bringing illicit goods into the mining zone to sell direct to the negroes or Indians, or taking gold out to the eager hands of French or English traders. The itinerant merchant was thus a middleman operating between the rich entrepreneurs who traded produce in the principal settlements and the dispersed population that characterized the northern region of Colombia. In Venezuela such a system does not appear to have operated so widely, though the significance of the trader class in both Caracas and Barquisimeto during the colonial period has been noted (Gormsen, 1966). Rather than itinerant merchants, there appear to have been periodic markets in most of the larger urban settlements. Thus the annual tobacco fair at Barinas attracted traders from over a wide area; cattle could likewise be purchased from the upper *llanos* settlements of Calabozo and San Fernando de Apure. In the north Caracas, Maracaibo and Cumaná each had its market for the sale and purchase of produce. Only in peripheral areas such as the south-east did merchants wander amongst the mission villages bartering hides, tobacco and cacao in exchange for metal wares and other imported products.

The pattern of trade routes reflected the geography of economic development in both Colombia and Venezuela. In the former they were orientated north–south (except the overland route to Venezuela), providing links between the urban centres of the interior and the Caribbean ports. Transverse links across the grain of the topography were less important and acted as mere feeders to the main routes. In Venezuela, with the topographic controls trending east–west parallel to the Caribbean shore, trade routes likewise shifted their direction. In Venezuela, however, distance between the principal agricultural centres of pro-duction and the Caribbean ports was never as great as in Colombia. Such proximity may well have made Venezuela more outward-looking during the colonial period. News soon reached Caracas from the Caribbean islands. Another distinguishing feature of Venezuelan trading was the use by Philip V of the monopolistic trading company to attempt to increase the productivity of the pro-vince of Venezuela during the eighteenth century. From 1728 until 1781 the Royal Guipúzcoa Company was ensured of monopolistic control over all produc-tion and trade (Hussey, 1934). Although some of the plans of the Company, such as the extension of wheat acreages in the Andean area, came to fruition, much resentment was caused amongst the influential merchants and landowners of Caracas. Eventually the monopoly was rescinded in favour of previous practices. The Barcelona Company, concentrating its efforts on the stimulation of agriculture in north-eastern Venezuela, appears to have been modelled along lines similar to those of the Guipúzcoa. After a brief spell of success it also ceased to have any effective power.

The end of an era

It might be argued that by the end of the eighteenth century the economic forces that had initiated and modified the process of Spanish colonial rule in northern South America had been practically spent. The reforms of the late eighteenth century created the *audiencia* of Caracas, and therefore retained the Maracaibo basin for what was later to become Venezuela. They freed commerce from some of the more exasperating controls that had prevented the development of inter-colonial trade, and they reorganized the administrative hierarchy. But there had been few economic changes. Agricultural production was still based upon serving the demands of Spain. Absentee landlords cared little about soil erosion or over-use, or about declining productivity. They were not permitted, let alone encouraged, to plan for new products or markets, and enterprise and initiative had to be set aside in favour of royal edicts. And all this while news was continually entering Venezuelan and Colombian territory from Dutch, French and English sources of the events that were gradually transforming western Europe. What was needed was a reappraisal of the situation by the men who had formed the colonial tradition, Spanish Americans.

The beginning of transition

In the same way as the conquest of northern South America by the Spanish in the sixteenth century initiated the establishment of a traditional order, or, as Fals Borda (1969) would prefer, a 'seignorial order', which evolved slowly during some 300 years of Spanish colonial rule, so the early nineteenth century marked the beginning of yet another formative period in the history of Colombia and Venezuela. Political independence from Spain heralded the beginning of a transitional phase in the development of the area. Whereas in the sixteenth century the complex regional variation of aboriginal patterns of economy and society provided the backcloth to Spanish endeavours, in the nineteenth it was the colonial system and patterns that were to be selectively retained, modified or elsewhere completely replaced by new elements. It was a process linked to developments in northern industrializing Europe. The pace and direction of change from the colonial models of Colombia and Venezuela were influenced by the interaction between capital, techniques, personnel and ideas of non-Latins, and the newly established economic, social and political power groups within those territories. Both countries have experienced a number of common changes. Each country has had, however, its own distinctive response.

The initial impact of political independence

Perhaps the most notable feature of the political independence gained by Colombia and Venezuela in the early nineteenth century was its insignificant effect upon the social structure and order within the old colonies. The former overseas centres of control were replaced with indigenous counterparts; in the main the individuals, families and strata within the society of the late eighteenth century took over the task of ruling the new republics (Picon-Salas, 1962). For the majority of the inhabitants of the two countries independence meant new masters rather than new power or influence, although there were exceptions to this generalization. The progress through the nineteenth century of insurrection, war and economic change naturally produced new members for the élite. Charismatic warlords and enterprising businessmen added their names to the roll of the strong.

That the initial impact of the independence movement was demographically, socially and economically disruptive, in certain regions almost of catastrophic proportions, there can be little doubt. Perhaps, like the comparable impact in the sixteenth century, it was a small price to pay for 'progress'. Though statistics of population in the early nineteenth century are notoriously unreliable, it is clear that a significant loss of population was experienced during the first two decades of that century. Population reduction was, of course, selective. Those settlements that lay on or near to what became historic routes of the independence armies were severely affected by the conscription of their male population. At least one-fifth of the Venezuelan population is estimated to have been lost in the cause of freedom, through death on the battlefields or through disease or hunger. More difficult to estimate, but undoubtedly of great significance, was the economic effect of the reduction in the males able to work the land on the farms of Venezuela, or in the mines in Colombia. In the labour-intensive systems then operative population and production were in delicate balance, and the sudden reductions caused grave hardship in certain regions (Bushnell, 1954). In Venezuela south of the Orinoco, in the bridgehead area of the Caroní plains, the impact of war, disease and economic disruption was exceptionally severe, and the majority of the villages lost more than 80 per cent of their populations. Westwards, in the *llanos* and foothills of the mountains villages had been deserted, the population fleeing to the forest or savannas to escape the notice of those organizing their freedom and seeking their services (Carrera Damas, 1964). In addition to the hunger and disease and the relentless cost in human lives of the wars themselves, natural disasters occasionally added to the toll, as in the 1812 earthquake, which cost a further 10,000 lives in Caracas. The demographic history of the period is still to be written. In the same way economic disaster followed hard on the heels of death and destruction; farms were deserted and crops were destroyed by plundering troops or invading weeds. Colonial irrigation systems were allowed to fall into disrepair, as had their aboriginal counterparts in the sixteenth century. The vast herds of cattle and horses in the savannas provided the perfect means to pay for the arms, ammunition and all the paraphernalia of armies that had to be imported and paid for by the liberators. As in Africa in the twentieth century European mercenaries were an expensive item in the budget of the new republics. Of equal but longer-term significance was the drawing off of the scarce operating capital of the business communities in both Caracas and Bogotá to repay the debts incurred during the war period. By 1820 the economic balance sheet in both countries looked little short of disastrous, but to the idealistic heroes of the period it was a small price to pay for political freedom and the right to make one's own mistakes.

Reconstruction, renovation and innovations

Given the essentially political nature of the independence movement in Colombia and Venezuela, it was to be expected that the first of the many tasks facing the new authorities was that of reorganizing the system of government and administration. Inspired by Bolívar, the notion of a single unit based upon the three former colonial units of Quito, Nueva Granada and Venezuela was accepted at Angostura in 1819. The new territory was to be called the 'Republic of Colombia', known better by the title given it by historians, 'Gran Colombia'. That such a large unit could have emerged by 1820 says much for the enthusiasm and personal qualities of leadership found in Simon Bolívar, all the more so since the period between 1810 and 1816 had been fraught with intercity and interprovincial rivalries and petty jurisdictional squabbles. However, there remained the question as to whether it should be a federation of independent states or a centralized union. The federalists' strongest support came from within the heartland of old Nueva Granada, with Caracas favouring the idea of union. The most significant defect of the grand design lay in the geographical variation that was to be found within its borders, a diversity of interests and uneven distribution of power and wealth, and economic advantages that were not adequately reflected in the administrative units proposed. Even in the stable days of the colonial period, the three colonies had enjoyed little more than peripheral contact with each other; given the heady days of almost continual change and adaptation, it was highly unlikely that they could be welded together. Even within all three colonies interregional disputes and rivalries over representation at the federal level continued after 1820. It was also decided that the three former colonies would be divided into departments, provinces and cantons, each approximately the same size, both to rationalize the democratic representation of the population, and also, by the necessary division and amalgamation of former units, to destroy some of the outmoded loyalties and traditions of the colonial past. The fact of the matter was that with population, and thus power and influence, so unevenly distributed, areal size of the administrative units was one of the least equitable factors to be taken into account. Why, it was argued by Caracas, should the newly created department of Apure have equal rights with Caracas, when it was so relatively insignificant? There were many cogent arguments put to the federal authorities. Gran Colombia was just too great, and what purpose was served by paper symmetry of representation when the decentralization of power was hardly ever discussed (Bushnell, 1954)? Like so many of the republican proposals, the administrative hierarchy of *cabildos*, provincial governors and departmental intendants beneath the federal authorities was an excellent intention, but proved quite impractical in operation.

The product of personal inspiration and a common enemy, Gran Colombia could not survive the secession of Venezuela. Perhaps the greatest enemy to success was sheer distance, or rather the cost and time of travel. With communications still affected by an almost total standstill in certain seasons, the movement of people, products and opinions could not take place from one area to another quickly enough to make so large a unit practicable. Other factors mitigated against success. In spite of the small-scale trade between the departments of Venezuela and Cundinamarca, and the family connections that tied the western Andean population of Venezuela closer to Colombian settlements than to others in Venezuela, the territories were too alike in their agricultural self-sufficiency and their range of imports and exports to make trade a possible catalyst of interdependence. But of still more significance was the deep-seated rivalry between Caracas and Bogotá. Caracas, the 'cradle of liberty', thought itself far more important than the 'Athens of South America'. There was a distinct reluctance in Caracas, as the centre of revolution, to accept the departmental status of some of its own former satellite regions, and to relinquish ultimate authority to a remote federal centre allegedly dominated by 'backward mountaineers' (Bushnell, 1954, p. 289). Maracaibo remained quite favourable to the idea of the large federal unit, but regional problems made the whole notion of Gran Colombia meaningless east of Caracas and to the south on the bandit-infested *llanos*. Support for the idea decreased in proportion to distance from Bogotá. Perhaps the war with Spain had ended too quickly, and technological changes had not made sufficient advances. It is difficult to explain the dissolution of the first major post-colonial administrative framework. After 1830 the independent nations, with minor adjustments to their boundaries, were left to pursue their separate destinies.

Just as in the sixteenth century the primary zone of cultural change had been along the southern shore of the Caribbean sea, so, too, in the early nineteenth century the Caribbean islands and the northern ports of Venezuela and Colombia were critical staging posts in the transference of ideas, products and personnel. From Trinidad, Jamaica and Barbados, and from other islands, came the foreigners who were to organize the early phase of economic renovation in Colombia and Venezuela. It was to the ports and later to the inland cities of Colombia and Venezuela that they brought their capital, influence and skills. Angostura, formerly a stagnating landing place on the south bank of the Orinoco, was completely transformed by the arrival of immigrant merchants, sailors and entrepreneurs. Very rapidly the tonnage of shipping using the port had increased by more than 200 per cent (Nichols, 1954). From overseas came a range of goods that could not be manufactured under Spanish colonial rule: textiles, ironware and machinery. Even more important were the imported foodstuffs, which were cheap enough to dominate domestic markets. Through the expanded ports were exported the products in demand in Europe and North America: hides, beans, coffee, gold, timber and the like.

If, however, trade was to prosper within either Colombia or Venezuela, then it was clearly vital that foreign traders should be able to obtain the agricultural or mineral resources they required. This involved both an extension of existing resources and an improvement in the system of communications within each of the countries. This, it was hoped, would mean more and cheaper products for the ports to export, and a better method of distributing imports internally.

During the nineteenth century very little improvement or extension was made to the network of roads in Colombia and Venezuela. For the most part they remained dirt tracks, seasonally impassable, and providing the minimal conditions for wheeled traffic. The horse and mule remained the principal mode of passenger transport until the arrival of the train and the motor car, and the inadequacy of the road system was felt most in terms of commercial developments. The basic pattern of roads remained as it had been at the end of the colonial period; the only extensions to it in either country were prolongations from the existing network to new areas of economic activity or to new entrepôts of trade on coast or river. In Colombia the mining industrial developments of the 1830s, but more especially the expansion of settlement in and around

Antioquia, involved an elaborate process of network extension (Parsons, 1949). Especially significant was the development of improved linkages between Antioquia and Cartago (*Figure 5.5*). Counter to the principal direction of Antioqueño colonization, efforts were made, most unsuccessfully, during the 1840s to open a route to the sea via the Urabá panhandle. A complex history of concessions, fruitless schemes and much endeavour has recently been revealed by Parsons (1967). In Venezuela the most important roads were concentrated in the northern third of the country, reflecting the intensity of economic activity in that area. The two penetration roads leading southwards to San Fernando de Apure and Angostura demonstrate the increasing significance of those two areas after 1850 – the former concerned with the extension of ranching on the *llanos*, the latter reaching great prominence between 1860 and 1885 with developments in the Caratal goldfield. It can also be noted that the roads connect riverine or coastal port settlements. All improved roads led to exports. Beyond the limits of profitable carriage, roads decayed rapidly into mere tracks, difficult to pick out from one season to the next.

If account is taken of the bulky nature of most of the exports of Colombia and Venezuela, it is hardly surprising that during the nineteenth century the advances made in water transportation were of far greater significance. This was one of the principal sectors of private and public capital investment during the century, linked, as were comparable port installations and improvements, with the profitable import–export trade (Parsons, 1949). In Colombia the neglected colonial route northwards from Bogotá and the central Colombian settlements became the most important channel of commerce in the entire country (Harrison, 1952). The river Magdalena now began to assert its real influence over the traffic in goods between the Caribbean ports and the stores of the capital. It was in relation to the Magdalena river that the first monopolistic contract was signed to permit Juan Elbers to introduce steamships. The contract signed in 1823 stipulated that he would be assured of the sole rights of operating steamships on the river for 20 years, in return for which he had to carry mails free of charge; not exceed certain maximum freight charges; build canals from the Magdalena to both Cartagena and Santa Marta; and connect the upstream terminus to Bogotá by means of a new road. Though the canals and road were never completed, the project appears to have been a stimulus to river trade (Gilmore and Harrison, 1948). A similar venture was proposed for the vast water network of the Orinoco in Venezuela (Gray, 1945), but the ambitious proposals to link the Orinoco and Amazon by steamboat service via the Casiquiare canal failed through lack of financial security, problems of navigability and the difficulties of stimulating sufficient trade on the Orinoco above the confluence of the Apure to make the enterprise economically viable. In western Venezuela the river Zulia and Lake Maracaibo also attracted a proposal for a German-controlled steamship service, but the enterprise failed owing to the restricted and not very prosperous hinterland. In north-western Colombia riverine trade increased with the road links established from the thriving communities of Medellín and Antioquia (Parsons, 1967). Water transport was one of the first media to benefit, albeit in a somewhat haphazard fashion, from technological inventions emanating from north-west Europe. It is clear from several studies of the evolution of the transportation networks in Colombia and Venezuela (Parsons,

1949, 1967; West, 1957), that the development of the waterways depended upon the economic potential of their hinterlands. Thus in western Colombia the rivers of the Chocó, and even the short navigable streams west of the Cauca valley, never developed to the same extent as the Magdalena. Not only was investment capital in short supply, but other links proved more attractive to the entrepreneurs involved. There is also much evidence to suggest that the improvements in navigation on the principal rivers of Colombia were economically worthwhile only because the principal staple for trade, coffee, was sufficiently highly priced to withstand the transport costs, and make investments profitable (Harrison, 1952).

Urrutía (1969) has noted that through the steady reduction in transportation costs on the Magdalena during the nineteenth century, British textiles could threaten the artisan trade in the Colombian textile centres. Until the advent of steamships, distance and topography had protected the home industry more efficiently than any tariffs. Reliable bulk shipping, unaffected by seasonal regime or shifts of wind, had brought north and south closer together. Coastal shipping and the river trade were certainly profitable, and helped to spread wealth by way of the ports and landing stages involved in the traffic.

In a similar manner the initiation and extension of railways in Colombia, and less so in Venezuela, also hinged on economic incentives (Rippy, 1943; McGreevey, 1971). Whereas in the difficult terrain of Colombia coffee paid for the reduction in grades, in the vast open plainlands of the Venezuelan *llanos* not a single line was ever laid. Only in the Yuruari savanna zone of the extreme south-east of Venezuela was a line proposed, and that, dependent upon the gold boom of the 1870s, never got further than a pile of rusting metal on the wharf at Ciudad Bolívar.

Developments in agriculture

Stagnation was the chief feature of agriculture in Colombia and Venezuela during the first half of the nineteenth century. Codazzi's surveys of the 1830s had shown that in Venezuela only 0.5 per cent of the national territory was under cultivation, approximately the area of the island of Margarita. In both countries the incessant troubles of the domestic political scene made agriculture a precarious enterprise. Uprisings had periodically removed the labour supply, damaged farm property and cropland, and devastated livestock herds. It is true that some of the earliest legislation ratified by the federal authorities of Gran Colombia had involved measures to prevent the entailment of estates, to reduce clerical mortmain and *ejido* land areas, and to sub-divide the communal lands of the Indians (henceforth to be known as *indígenas*) to permit them the dubious pleasure of private ownership. But most proposals were little more than paper plans. It has also to be remembered that relatively large proportions of the agricultural land remained effectively outside the national market until late in the century. Included in this category was church land (approximately a third of the cultivated land in Colombia), Indian commune land, *resguardos*, and town *ejidos* (Urrutía, 1969). When in 1850 Colombia lifted the restrictions on the scale of *resguardo* land, the rapid enclosure movement that was precipitated benefited local capitalists; Indians fell into the grips of usurers. With new, large estates (*latifundia*) being created, the Indians left the land, either to seek their fortunes in mining or to work as labourers in the expanding coffee zones; others remained as tenants. Cheap labour was just what was needed in the new boom areas. So it happened that the 'proletarianization of the workers benefited plantation agriculture rather than industry' (Urrutía, 1969, p. 28).

If it proved almost impossible to reform radically the existing agricultural systems of tenure and cultivation (Carrera Damas, 1964), this did not mean that it was not possible to look forward to new agricultural ventures. Indeed, ambitious colonization schemes have a long history in northern South America. On paper they had much to commend them: they usually proposed the occupation of land previously under-utilized; they often involved foreign capital, labour supply and technological expertise; and they needed little but the blessing of the respective national governments. Unfortunately, another common characteristic was their almost total failure. In Venezuela, from the Orinoco to the Andean mountains, various colonies were attempted, but, except for the German 'Colonia Tovar' west of Caracas, all were to fail (Rasmussen, 1947). In Colombia, too, colonization schemes were never short of official support, whether they were in the tropical forests of the north (Parsons, 1967, p. 34) or the cooler uplands of the central area.

Efforts were made in both countries to encourage export production. In Venezuela these efforts were largely confined to the *llanos* where large-scale ranching produced cattle and leather for export. In Colombia the granting of a monopoly over tobacco production to Montoya, Saenz and Company was a turning point (McGreevey, 1971). Rapid growth from 1845 to 1857 turned tobacco into Colombia's principal export, a position it maintained until the catastrophic fall in prices in 1875.

This sudden decline was the most significant of a series of spectacular swings in Colombian export agriculture. Various products enjoyed brief prosperity before world competition, disease or a downward shift in demand undermined the business. Cotton during the American Civil War, indigo and cacao in the 1870s, quinine in the early 1880s and even more briefly sugar, rubber and vanilla all passed through brief export cycles (McGreevey, 1971).

There was one exception to this sad catalogue: the rise of coffee. Coffee cultivation spread into Colombia during the latter half of the eighteenth century from Andean Venezuela. It gradually developed as a major export in both countries and was channelled through British and US merchants based in Maracaibo. In the late 1870s coffee represented 7 per cent of Colombia's exports, but its annual export of 100,000 60-kilo bags was a mere sixth of Venezuela's export total (Palacios, 1980; Quero-Morales, 1978). Between 1870 and 1897 coffee production in Colombia increased fivefold, and in 1900 contributed 40 per cent of the country's export earnings; in Venezuela coffee exports doubled in volume between 1872 and 1913 to 65 million tonnes.

In Colombia coffee was to transform an economy dependent on gold, mules and tobacco before the 1870s to one dominated by coffee, railways and banks (Safford, 1965). Its expansion led, and was stimulated by, the development of railways; 564 km of track were in operation in 1904, most of which linked the coffee areas with the major rivers, ports and frontiers. Coffee was eventually to stimulate the beginnings of industrial development and to change the rural social structure. By contrast with other forms of export production, coffee cultivation in Colombia had historically been a smallholder enterprise. Although its beginnings were dominated by large estates in the old slave areas of Santander, an area which produced 60 per cent of Colombia's coffee in 1900, the association with small family farms was rapidly developing. This association was to become more marked as the cultivation of coffee spread westwards into the Antioqueño colonization areas, which were to dominate Colombian production in the twentieth century.

The Antioqueño colonization, and its relationship with coffee, is one of the most distinctive and fascinating aspects of Colombian history (Parsons, 1949).

> The society which emerged from it on the hillsides, river banks, and valley slopes situated between the Cauca river basin and the peaks of the central cordillera found integration and economic progress in the early years of the twentieth century through the cultivation, processing, packing, and transport of coffee. (Palacios, 1980, p. 161)

Coffee was not initially, however, the primary crop in the Colombian agricultural frontier complex. Coffee, with its 5-year maturity period, had to wait until communities had been established on the basis of maize, bean, manioc and banana subsistence production, and until improvements had been made in the mule tracks to the major markets. At first this frontier merely offered the possibility of survival to a population which in its Antioquia homeland was growing more rapidly than elsewhere in Colombia, and which was concentrated in an infertile and eroded terrain.

Undoubtedly, the Antiqueño frontier experience created a more egalitarian society than that in most other parts of Colombia; a society based on 'an ethos of axemanship, effort and achievement' (Palacios, 1980, p. 161). But it was not the ideal society it has sometimes been portrayed. From the start, colonization was dominated by the élites of Medellín who provided credit, supplies and political support. In addition there were constant struggles over land; the great majority of peasant holdings were not recorded, and therefore insecurity of title remained a source of conflict, crime and violence. The picture described by Parsons (1949, p. 98–9) of the new colonies 'as closely knit, fraternal, agrarian associations in which co-operative clearing, seeding and harvesting, and the sense of communal responsibilities were highly developed' mirrored only one part of the truth.

Undoubtedly, the seeds of change had been sown in the latter half of the nineteenth century. Colombia's free-trade policies from 1845 to 1890 had increased export production, raised land values and redistributed communal and church lands, but to the benefit of few Colombians. Antioqueño colonization and coffee were establishing the foundations of twentieth-century change. But in many other respects economy and society remained unchanged. Population totals had increased slightly, and there had been some shift in the margins of settlement, but the overriding characteristic was the stability of inactivity. Disease still presented insuperable obstacles to human occupation of the major portions of the national territories; malaria in Colombia was endemic below 1500 m, and in Venezuela only scattered islands of immunity stood above the ocean of infestation. Elsewhere hookworm, dysentery, typhoid and yellow fever took their annual toll. In the coffee *fincas* intestinal parasitism thrived in the damp soils trodden by shoeless labourers. For the most part settlements established during the colonial period slumbered on, unaffected by the industrial revolution in northern Europe and the United States. In Venezuelan Guayana it is true that the gold strike had brought in Hamburg lager by the crate, elaborate mining machinery, negro labourers from the Antilles, British, German and French capital, and all the necessities of a latter-day California. But the dance halls were soon empty and the population dispersed, and the forest closed back in on this temporary economic heartland of Venezuela.

The huge *llanos* zone, though still occupied by very large cattle ranches, suffered from permanent rustling. It was only in the north of Venezuela that change was taking place. It was to Caracas and La Guaira that the telegraph came in 1856, and in 1883 the telephone. Here in the north were to be seen the first attempts to put the new technology to work: tramways and steam trains; plans for elaborate copper works at Aroa; and, equally abortive and ingenious, Cochrane's mechanized pearl-fishing using a diving-bell. For the majority of the population, however, a livelihood was only to be gained through hard work in agriculture, 'cultivating crops that had been cultivated for generations. The further away one was from the capital, or the ports, and thus the influence of the exterior, the closer one was to the colonial tradition.

In Colombia the areas affected by change were undeniably more widely distributed. The distribution of population called for better communications and this in turn meant a wider spread of knowledge, products and ideas. But the Colombian situation did not yet permit the steady progress from a developing agricultural to a thriving industrial nation, which was seen by some as a logical progression. Like Venezuela's, Colombia's history had been punctuated by a succession of civil wars and armed insurrections. Conservative 'centralists' and Liberal 'federalists' spared the country no economic harm to win their way to power. In Venezuela, too, militarism became a sure means to an end, and the country repeatedly succumbed to civil strife (Gilmore, 1964). Petty municipal regulations maintained the façade of decentralized control, but when anything significant had to be decided it was to the capital that everybody turned. It was in the capital and the larger cities that the majority of the immigrants were to be found in 1900, traders who recognized the advantages of being close to the centre of political power and influence.

The most significant feature of the transitional phase throughout the nineteenth century was the replacement of the Spanish colonial system based on political control, by an economic system geared to producing raw materials for overseas markets in return for manufactured products.

However, the practice of securing maximum returns for minimal effort clearly benefited only a minority – the politically and economically powerful – while the masses found the struggle to survive no less pressing than their forebears had in previous centuries.

The twentieth century

The geography of twentieth-century Colombia and Venezuela has been shaped principally by four fundamental forces: the expansion of coffee and commercial agriculture, the exploitation of petroleum, the development of manufacturing industry and the population explosion. While all four were present in both countries, however, they were not to have an identical effect; after all, Colombia has remained a relatively poor country, whereas Venezuela has become Latin America's richest nation. The difference in their growth experiences lies in the vast petroleum wealth of Venezuela and the more limited discoveries made in Colombia. This vital difference apart, the developmental experiences of the two countries demonstrate many similarities. Both economies and societies have changed enormously. Essentially traditional, rural economies have been transformed into modern, capitalist, urban economies. Not only have most people ceased to gain their livelihood from agriculture and now earn their living in cities, but their numbers have increased many times: in 80 years, the populations of both countries have increased roughly seven times.

In Colombia these changes were already under way by the end of the nineteenth century. The planting of coffee brought a dramatic increase in the levels of domestic prosperity and of interlinkage with the world economy. Coffee was to sustain the Colombian economy virtually alone during much of the twentieth century. Industrial development had also achieved a foothold in the country. The manufacturing plants which had emerged in Medellín during the last two decades of the nineteenth century were the precursors of twentieth-century industrial growth. If both coffee and industry were eventually to disappoint many of the hopes placed in them, they were also undoubtedly to change the face of Colombia.

By contrast, Venezuela in 1900 had still not found the key to the treasure chest of the twentieth century. The petroleum which would transform Venezuelan economy and society was not to be exploited for another 20 years. As in Colombia the coffee industry provided the major source of foreign exchange from 1900 to 1920. Unlike Colombia, however, coffee revenues were to fade into relative insignificance beside the gush of funds from oil. Nor had the first stirrings of industrial development begun: Venezuela contained only artisan industry in 1900.

Agriculture

During the greater part of the twentieth century agriculture has been the largest single source of employment in both Colombia and Venezuela. In neither country, however, has it brought affluence to most farmers. Low productivity in the agricultural sector and the inequitable distribution of land have left an indelible mark on both societies. While recent years have seen a marked rise in the efficiency of agriculture, agrarian reform laws have not improved the situation of the small-scale producer. When combined with the dynamism of the urban sector, this situation has led to a relative decline in the rural population. In recent years that population has even fallen in absolute numbers.

If the inequality of landholding in the country has not been remedied, the agricultural economy has changed out of all recognition. At the turn of the century most rural workers were bound into some form of exploitative relationship on the large haciendas. Labourers received the right to work small plots of land in return for their labour on the estate or a share of their production. There were of course major variations in agrarian organization in different parts of the two countries. In the western coffee regions of Colombia, the distribution of land was much more equitable than in most of the rest of the country as a result of the massive colonization movement of the second half of the nineteenth century.

In fact it was the development of coffee in these areas that stimulated a major transformation in Colombian agriculture. The tremendous growth of coffee production between 1903 and 1929 forced change on the hacienda economy by providing alternative sources of income for rural labour. The old *latifundia* sought to prevent their workers being contracted by the coffee farms and by the companies building the roads and railways financed by the coffee boom. Gradually, however, they were forced to change. Even before the Second World War, and accelerating afterwards, the increase in industrial production, the demand for agricultural products and the demand for foodstuffs in the cities encouraged higher levels of efficiency and enterprise. Throughout the twentieth century, haciendas dedicated to cattle-rearing have gradually given way to large estates dedicated to sugar cane, cotton, rice, sorghum or dairying. In 1950 around 270,000 ha were in commercial agricultural production, and by 1977 such production occupied more than 2.7 million out of a total of 4.2 million ha of agricultural land (Kalmanovitz, 1978, p. 37). As a result Colombian agricultural production has increased dramatically during the 1970s. Today it contributes around one-quarter of the GDP and contributes three-quarters of Colombia's exports.

The trend towards commercialized agriculture was encouraged further in the 1960s and 1970s as the government sought to stimulate agricultural-export production through tax and credit incentives. The expansion of agricultural-export production has been very impressive: between 1959 and 1978 rice production increased from 421,000 to 1,714,000 tonnes, that of cotton from 142,000 to 330,000 tonnes and soya beans from 14,000 to 131,000 tonnes. Indeed, the value of these new agricultural exports rose from $159.6 million in 1972 to $373.8 million in 1978 without including by far the most dynamic new agricultural export of all, marihuana. While figures for this product are naturally less than reliable, a huge area of production has been opened up in the Guajira peninsula to supply the market in the United States. Despite some efforts by the government to control exports, mainly at the behest of the United States administration, the profits to be made are such that its production has proliferated, and has made fortunes for the more successful producers and traffickers. In the early 1970s, when the prices of coffee were low, it is probable that the value of marihuana and cocaine exports exceeded those of coffee.

Coffee, of course, has continued to be vitally important to the Colombian economy, even if producers have suffered from major fluctuations in world prices. In 1970 300,000 coffee farms covered nearly 1.1 million ha of land, and produced around 12 per cent of the world's coffee. Today coffee production accounts for around 15 per cent of all agricultural value added, and contributes around 4 per cent of GDP. In 1978 coffee production reached an all-time record of 11.7 million bags, roughly double what it had been in the 1940s, 1950s and 1960s. The expansion in production has been due both to recurrent frosts in Brazil, which have increased the market for Colombian exports, and a major campaign during the 1960s and 1970s to improve the efficiency and quality of coffee production. The essential change has been the introduction of the *caturra* tree which matures earlier, gives higher yields and can be produced in direct sunlight, although it has a shorter life and requires larger inputs of fertilizer and capital. The introduction of the new tree and the associated new technology have had major effects beyond the mere increase in production. In the major production areas there has been a decline in the number of family farms as commercial production has increased (Palacios, 1980, p. 246). With the new trees and higher prices the economically viable farm size has risen considerably.

In Venezuela, the agricultural economy has until recently shown little of the dynamism of that of Colombia. Indeed, even though the 1960s brought considerable change, and agriculture still employed around one-fifth of the labour force in 1976, it added only 6 per cent to the GNP (BCV, 1977). As in Colombia, however, the commercialization of agriculture is now rapid and the old-style latifundia have been largely eliminated. This has had the effect of stimulating agricultural production, which from 1950 to 1976 increased in value three and a half times. Particularly rapid has been the growth of milk, livestock, poultry and eggs: this sector expanding tenfold between 1945 and 1976.

If productivity has increased dramatically in many sectors of Colombian and Venezuelan agriculture, it has not benefited all rural groups equally. Indeed, one of the critical problems facing the agricultural sector for many years has been the inequality of income due primarily to the unequal form of land tenure (Smith, 1967). Of course, the forms of land tenure and agricultural production have changed markedly during the twentieth century, and in some ways for the better. The contractual servitude that existed under the great *latifundia* has largely disappeared as a result of the growing intrusion of capitalistic methods of production. Nevertheless, despite the changes, the fact remains that most people in the countryside have access to little land. In Venezuela the situation in 1971 was that 170,000 holdings (60 per cent of the total) occupied a mere 2.2 per cent of the land, and 4900 holdings over 1000 ha (1.7 per cent of the total) occupied 67 per cent of the land. Indeed, in both Colombia and Venezuela it is becoming increasingly clear that the agricultural labour force is declining not only because of the opportunities in the cities, but also due to the increasing pressures on smallholders to vacate their lands.

The plight of the smallholder has in some respects increased despite, and some writers would argue because of, attempts at agrarian reform. In Colombia the expansion of coffee and parallel changes in the national economy gave rise to an attempted reform in 1936. While the reform outlawed unfair labour relationships, this change often had the result of encouraging landowners to expel their renters and sharecroppers, rather than ceding them land as the reform intended. While the reform was clearly aimed at improving the situation of the poor, it brought few benefits because of poor implementation by later administrations. In Venezuela the first government of Betancourt (1945–8), the first democratically elected government in Venezuela's history, introduced a land redistribution and assigned around 150,000 ha to 75,000 members of peasant unions. Unfortunately, even this limited reform was short-lived. A further period of military rule soon removed the government and reallocated the land to its former owners.

It was not until the time of the Alliance for Progress that further attempts were made in both countries to redistribute land (Warriner, 1969; Hirschman, 1963). Land-reform laws were approved in Venezuela in 1960 and in Colombia the following year. Both laws were heralded as a new beginning; both have been a total disappointment to the majority of poor rural workers. In Venezuela IAN, the land-reform agency, acquired large amounts of land, but most was of very poor quality and little was distributed to possible beneficiaries. Indeed, the outcome of the reform process seems to have increased the number of minifundistas while promoting the fortunate to the status of medium-scale commercial farmers (Cox, 1978). *Table 5.1* shows that between 1961 and 1971 the average size of holding smaller than 10 ha has actually declined in size from 3.5 to 3.4 ha. If the very largest category of farm has declined in size and share of holding, the medium and large farms have increased their share of land. Most interesting is that the number of landholders with less than 50 ha declined from 283,000 in 1961 to 245,000 in 1971. As Cox (1978) has stated 'those with power in Venezuela have opted for an agrarian reform and agrarian structure that have marginalized hundreds of thousands of their fellow citizens and accentuated urban problems' (p. 56).

Table 5.1 Venezuela: distribution of agricultural landholdings (1961–71)

Size	Number of holdings	%	Area in hectares	%	Average size in hectares
1961					
Less than 10 ha	213,419	66.7	753,288	2.9	3.4
10–50 ha	69,987	21.9	1,323,905	5.1	18.9
50–100 ha	11,567	3.6	719,241	2.8	62.2
100–200 ha	7,332	2.3	942,641	3.6	128.6
200–500 ha	6,147	1.9	1,766,319	6.8	287.4
500–1000 ha	2,802	0.9	1,844,246	7.0	658.2
1000 ha and more	4,223	1.3	18,655,219	17.8	4,175.3
Without land	4,617	1.4	–	–	–
Total	320,094	100.0	26,004,861	100.0	81.2
1971					
Less than 10 ha	171,173	59.5	577,087	2.2	3.5
10–50 ha	73,772	25.6	1,429,543	5.4	19.4
50–100 ha	14,308	5.0	920,140	3.5	64.3
100–200 ha	8,340	2.9	1,050,662	4.0	126.0
200–500 ha	7,903	2.7	2,291,200	8.7	290.0
500–1000 ha	3,883	1.4	2,533,584	9.6	652.5
1000 ha and more	4,904	1.7	17,668,318	66.7	3,602.8
Without land	3,636	1.3	–	–	–
Total	287,919	100.0	26,470,134	100.0	91.9

Source: III and IV Censos Agropecuarios (Quero-Morales, 1978, p. 1117).

The Colombian reform has been equally ineffective in helping the poor, with the inactivity of the land reform agency, INCORA, contrasting markedly with the dynamism of those agencies attempting to increase agriculture production. It is true that in the period of major activity from 1962 to 1971 INCORA distributed 3.9 million ha of land to 130,000 families. Unfortunately, most of this land was in the form of colonization projects in the east of the country on land that was difficult for smallholders to cultivate (Eidt, 1968; Hegen, 1966). In terms of effective land redistribution only 14,000 families received the 240,000 ha of land which had been expropriated, purchased or otherwise acquired by INCORA. The rest of the land, some 3.7 million ha, was merely state land in the *llanos* which had been subdivided. Even this pretence at redistribution was soon to cease. Under the Pastrana government (1970–4) the Chicoral agreement was reached. In return for paying their taxes to the government, large landowners received an effective guarantee that they would be free from the danger of expropriation by the land-reform institute (Kalmanovitz, 1978). Like the Venezuelan state, that in Colombia has resolved to increase production at the expense of smallholders. Indeed, the only real help that smallholders have received in Colombia has come from the recently introduced Integrated Rural Development Programme (DRI). This seeks to provide credit for viable-sized farms (those with more than 3 ha of land), to set up co-operatives and to improve health, education and road services. While the programme may well help many smallholders, it will lead to the expulsion of the more marginal farmers, and will in no sense reduce the need for a genuine redistribution of land.

Mining

Oil has dominated Venezuela's history since the 1920s (Lieuwen, 1954). It has consistently contributed over 90 per cent of the country's foreign-exchange earnings; it has contributed the larger share of government revenues; and it has been the basis of industrial and urban development and a principal reason why, until recently, agricultural production has stagnated.

Despite this impressive record it is clear that petroleum has distorted the country's development. Under the Gómez dictatorship (1910–35) the foreign petrol companies were permitted to repatriate most of their profits. Apart from Gómez and his henchmen few Venezuelans actually benefited from the fact that from 1928 Venezuela was the world's second oil producer and largest exporter. After 1935 the first efforts were made to invest the petroleum revenues in productive enterprise, 'sowing the oil', and to extract a better deal from the oil companies. Such policies continued under the Medina dictatorship (1941–5) and under the brief interlude of democratic government between 1945 and 1948, but were then interrupted by the 10 years of the Pérez Jiménez dictatorship. Under Pérez attempts to spread the benefits of oil wealth among a wider group of Venezuelans faded, vast new concessions were granted to major oil companies, and foreign investment and earnings poured in. During this period iron ore began to be extracted, huge sums were invested in a motorway programme, and most spectacularly, around half of the government's oil-inflated budget was spent in Caracas in an attempt to turn the city into a spectacular modern metropolis. The return to democratic government in 1958 led to greater efforts to invest the wealth in productive activities such as industry and agriculture. Nevertheless, this policy has not favoured all poor Venezuelans. In part such an aim is complicated by the enclave nature of the oil industry and the fact that it employs few people. Between 1975 and 1978 oil-extraction generated around 20 per cent of the GNP, but never employed more than 1 per cent of the country's labour force (Oficina Central, 1978). Essentially, however, the limited benefits to the poor derive from inadequate Venezuelan government policy, the failure to hold down imports, to stimulate agriculture and to establish viable industrial enterprises.

In 1976 the Venezuelan government nationalized the oil industry. There had always been tension between the foreign oil companies and the government, and nationalization was a long-cherished ambition. The transition to Venezuelan control seems to have been relatively smooth, but has undoubtedly been eased by the revolution in world oil prices which brought huge benefits for Venezuela. In 1973 the price of Venezuelan crude was US $4.23, in 1974 it was $14.06. If this benefited all the world's petroleum producers, the OPEC-inspired price rises were especially helpful to Venezuela. In the 1960s Venezuela had become increasingly aware both that it was producing high-cost oil compared to that produced in Africa and the Middle East, and even worse, that its reserves of easily exploitable oil were rapidly becoming exhausted. OPEC policy not only slowed down extraction (*Table 5.2*), thereby extending the life of the Venezuelan reserves, it also made exploitation of the tar reserves of the Orinoco river commercially viable (*Figure 5.6*). The strip of tar reserves will certainly cost much more to exploit than ordinary deposits, but promises to extend Venezuela's life as an oil producer for many years. Whatever the problems, there is no doubt that oil has been the major source of Venezuela's spectacular rate of economic growth: between 1950 and 1976, the GNP grew by over 500 per cent in constant prices (BCV, 1977).

Table 5.2 Venezuelan and Colombian petroleum production, exports and imports (thousand barrels/day)

| Year | Venezuela | | Colombia | | |
	Production	Exports	Production	Exports	Imports
1970	3,708	3,469	219	86	–
1971	3,549	3,282	215	70	–
1972	3,220	3,064	169	41	–
1973	3,366	3,151	184	26	–
1974	2,976	2,752	168	1	–
1975	2,346	2,086	157	–	–
1976	2,294	2,132	146	–	18
1977	2,238	1,985	138	–	26
1978	2,166	1,896	131	–	24

Sources: BCV (1977, p. 68); Oficina Central, 1978; *Colombia Today*, 14, 10.

If Venezuela's relationship with oil has been one of affluence and frustration, recent years have not favoured Colombia at all. If the OPEC agreement benefited Venezuela, it harmed Colombia. The latter's problem seems to be that it has not discovered and exploited the oil that is almost certainly to be found in the country. Since most of Ecuador's and Venezuela's oil is close to Colombian territory, it is hard to believe that the oil beds stop at the customs posts. Nevertheless, the fact is that Colombia ceased to be a net exporter of petroleum in 1974 and became a large net importer. In 1970 the country produced 219,000 barrels per day, of which 86,000 barrels were exported, a total which fell to 126,000 barrels in 1979. In 1980 imports were around 22,000 barrels per day at an annual cost of $500 million. Colombia is trying hard to discover new deposits, and fortunately much of the country remains unexplored.

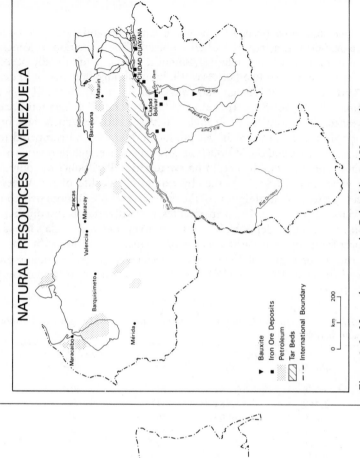

NATURAL RESOURCES IN COLOMBIA

Santa Marta
Barranquilla
Cartagena
Riohacha
Maicao
EL CERREJON
Valledupar
Monteria
Sincelejo
CERRO MATOSO
Medellin
Cucuta
Bucaramanga
Barranca-bermeja
BOGOTA
Buenaventura Cali
Rio Magdalena
Rio Cauca
Tumaco

★ Nickel
Natural gas field
Major coal producing areas
▲ Petroleum
--- International Boundary

0   km   200

NATURAL RESOURCES IN VENEZUELA

Maracaibo
Mérida
Barquisimeto
Valencia
Maracay
Caracas
Barcelona
Maturin
CIUDAD GUAYANA
Ciudad Bolivar
Guri Dam
Rio Caroni
Rio Paragua
Rio Churn
Rio Orinoco
Rio Orinoco

▼ Bauxite
■ Iron Ore Deposits
Petroleum
Tar Beds
--- International Boundary

0   km   200

Figure 5.6 Natural resources in Colombia and Venezuela

If oil is critical, both countries still have other important mineral resources. Venezuela has been a major producer of iron ore since 1950, with production reaching a peak of 26,000 tonnes in 1974. Colombia has discovered natural gas in two areas of the Caribbean coast. The Guajira gas fields have reserves equivalent to 521 billion barrels of crude oil, and in 1977 a gas pipeline was completed to link the Guajira fields with the port cities of Barranquilla and Cartagena. Offshore, new discoveries of gas are still being made. Coal offers one of the most promising areas for mining development in Colombia as the country has the largest coal reserves in Latin America. Although more than half the coal is located in the mountains, the Cerrejón deposits about 97 km from the Caribbean coast are well situated for export production. Initially production will be sold locally, but by the second half of the 1980s it will produce around 15 million tonnes for export. Finally, Colombia has major deposits of nickel. The Cerromatoso project is likely to be under way in 1982 and to be extracting 850,000 tonnes of ore annually to be processed by the associated refinery.

Industrial development

In 1919 there were 121 manufacturing plants in Colombia (Bell, 1921), in 1927 554, and by 1930 Colombians 'considered themselves to be on the way to adulthood in the field of industrial production' (Ospina Vásquez, 1955). The principal stimulus to industrial growth was coffee production. Between 1918 and 1928 the total value of Colombian exports increased threefold: an expansion due primarily to the 'miracle' tree. The expansion of coffee generated an internal investment surplus, created foreign reserves to pay for imported machinery, stimulated domestic demand for manufactured goods, and underlined the need for better land communications. During the 1920s, Medellín was linked to both the Magdalena and Cauca rivers – the main transport arteries to the Caribbean coast. Between 1924 and 1930 the length of the railway network almost doubled to 3000 km.

Coffee had the effect of both linking Colombia more firmly to the world market and of stimulating economic growth within the country. Despite a shock in 1929 when the beginning of the world depression hit coffee exports, industrial development quickly recovered, helped by government protection: between 1933 and 1938 industrial production increased annually by 10 per cent. The Second World War slowed expansion to an annual rate of 6 per cent, but in 1945 Colombia had 7849 industrial plants with more than four employees, making a total of 135,000 workers employed in them.

In Venezuela, by contrast, the early exploitation of oil had a limited effect on industrial expansion. Unlike coffee, the organization and control of petroleum revenues did not involve large numbers of Venezuelans, both because of foreign ownership and because of an extraordinarily unenlightened dictatorship. Combined with the strength of the Venezuelan *bolívar* and the lack of governmental support for industrial growth, the country was still importing the vast bulk of its industrial consumption in 1936 (Salazar, 1976). In that year there were approximately 30,000 manufacturing workers, of whom three-quarters were in the food or textile industries (Karlsson, 1975, p. 67).

After the Second World War industrial development became a major aim of the economic élites in both countries. The standard Latin American policy of import-substituting industrialization was adopted. Heavy protection, a welcome for foreign investment and technology, and continuing export expansion gave rise to rapid industrial growth. In Venezuela industrial production increased almost fivefold between 1950 and 1969, and a further 60 per cent by 1977 (BCV, 1977). Both countries developed a wide range of industry: major iron and steel plants were established at Paz del Rio and at Ciudad Guayana; and car manufacturing, petrochemicals and engineering all blossomed to supplement the existing production of food, textiles and clothing. Much of this industry was foreign-controlled, and many of the world's major manufacturers came to be represented in one or other of the countries.

In both countries industrial development was highly concentrated in a limited number of cities. Although less spatially concentrated than most other Latin American countries, in 1945 59 per cent of Colombian manufacturing employment was located in the departments containing the three largest cities, Bogotá, Medellín and Cali; by 1974 this share had risen to 73 per cent. In Venezuela industrial concentration was still more marked; in 1953 the central states of Aragua, Carabobo and Miranda, together with the Federal District, contained 70.3 per cent of industrial value added. By 1971 this figure had risen to 73.9 per cent. While some efforts were made to deconcentrate industry in both countries in the 1960s and 1970s, most spectacularly in the Ciudad Guayana complex in Venezuela (see pp. 233–5), only limited success has been achieved.

A much more serious problem in both countries, however, has been the nature of industrial expansion. In many industries the combination of capital-intensive technology and a limited market produced over-capacity and high cost production. This has been a particularly serious problem in Venezuela with its higher labour costs, and one that has posed continual difficulties in its relations with the Andean Pact. Similarly difficult has been the employment situation. While employment increased in both nations, capital-intensive technology meant that it increased at a far slower rate than industrial production, and failed to keep pace with the rapid growth of job seekers. The obvious effect of such a situation in the context of rapid population growth has been rising levels of unemployment and under-employment. In some cities, unemployment has been little worse than in many developed countries today: the unemployment-rate in Bogotá, for example, ranged from a high of 16 per cent to a low of 7 per cent between 1963 and 1976 (Lubell and McCallum, 1978). Admittedly, unemployment is less of a problem in Bogotá than in many other Colombian cities, but it is still arguable that unemployment *per se* is not the major problem (ANIF, 1976). Indeed, Berry (1975) has argued that in a context where unemployment benefits are not available, unemployment tends to be confined to those with skills and family support who can anticipate eventually obtaining an acceptable job. The real problem is that so many people are forced to work in activities which are low-paid. The term under-employment is commonly used to describe this situation, but is inaccurate in the sense that most such people work long hours to gain a very small income. Very large numbers of Colombians and Venezuelans can be included in this category.

The inability of the manufacturing sector to provide large numbers of jobs, combined with agricultural policies which have failed to hold the rural population in the countryside, have created a situation whereby the distribution of income continues to be very unequal, and where a substantial minority of the population of both countries is extremely poor. In Venezuela in 1970 the poorest quintile of Venezuelan households received a mere 3 per cent of household income, compared to the 35.7 per cent received by the richest decile (BANOP, 1970). What is perhaps most disappointing is that there is no sign that the distribution of income is improving or that the absolute numbers of poor people are declining (Berry and Urrutia, 1975). Indeed, there are signs that the recent policies of both the Colombian and the Venezuelan governments have reduced the real incomes of poorer groups. In Colombia repressive measures towards trade unions led to a decline in real manufacturing wages between 1970 and 1975; in Venezuela, the liberation of prices of many basic foodstuffs has squeezed working-class budgets. Combined with the failure to effectively tax profits from land, property and certain business activities, the outcome has been a deterioration in the distribution of income. As we shall see, this has had major social and geographical consequences.

Population
The final major ingredient of change in twentieth-century Colombia and Venezuela has been the growth of the population. By any standard, population growth has been rapid ever since mortality-rates began to decline in the early decades of the century. *Table 5.3* shows that Colombia's population increased from a little over 4 million in 1905 to an estimated 27 million in 1980; that of Venezuela increased from 2.2 million in 1891 to around 15 million in 1980.

In Colombia mortality-rates were around 30 per thousand in the early years of the century, in Venezuela the rate in 1905 was 23 per thousand (Lopez, 1963). By 1945 the rates in both countries had fallen to a little less than 20, and by 1978 were as low as 7 in Venezuela and 8 in Colombia (World Bank, 1980). These improvements were primarily the result of eliminating major epidemic diseases such as malaria together with the slow but gradual improvement in health care, water supplies and sanitation. As a result, life expectancy at birth increased in Colombia from around 30 years in 1910 to 49 years in 1950 and 62 years in 1978.

Unlike mortality-rates, fertility-rates have actually risen during most of the century. In Venezuela they rose from 28 per thousand in 1905 to 45 in 1955 (Lopez, 1963) and in Colombia, Collver (1965) estimates that the standardized birth-rate per thousand rose from 41 in the intercensal period 1912–18 to 45 between 1951 and 1964. Since then fertility-rates seem to have declined, and it has been estimated that between 1960 and 1968 every woman gave birth on average to one less child, a decline of 14 per cent. In fact, crude birth-rates seem to have fallen between 1960 and 1978 from around 45 per thousand in both countries to 31 in Colombia and 36 in Venezuela. In Colombia this was the result of a series of private and public initiatives to slow the birth-rate; indeed, Colombia was one of the very few Latin American countries to provide official support for family planning (McGreevey, 1980). By the middle of the 1970s large numbers of women in the major cities had tried some form of birth control; half the married women interviewed in Bogotá in 1973 were practising some kind of contraception. In a 1969 survey in Bogotá it was apparent that knowledge and use had increased rapidly, 38 per cent of women had used some form of birth control in 1964 compared to 65 per cent in 1969. Rates were far higher among

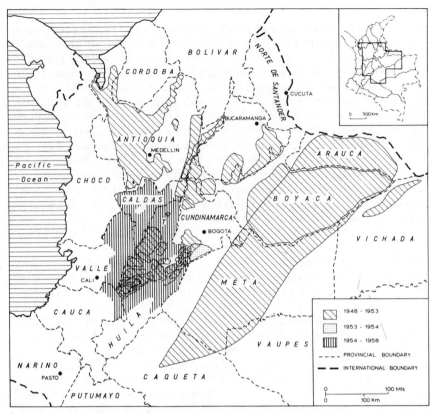

Figure 5.7　Major areas of violence in Colombia 1948–58 (after Gúzman *et al.*, 1962–4)

women with primary education or more than for those with less education: in 1969 77 per cent compared to 48 per cent (Simmons and Cardona, 1973).

Fertility-rates and mortality-rates are not of course the only factors influencing the rate of national population growth: international migration can be a vital influence. During much of the twentieth century it has had little effect on Colombia. Some skilled migrants arrived to establish businesses, but the net flow was limited. In Venezuela the buoyancy of the economy has long attracted much larger numbers of immigrants, especially from Europe. In addition, the post-1974 petroleum boom has stimulated a massive movement of people into the country, from the Caribbean, Ecuador, Peru and particularly Colombia. Most of these migrants have entered the country illegally, so that their numbers are unknown; estimates have varied from anything between 1 million and 4.5 million. In 1980 an attempt was made to register the illegal immigrants, those registering being permitted to stay. It would seem unlikely, however, that the Venezuelan authorities will either be able or sufficiently willing to stop the flood of migrants completely; the poor of Colombia are hardly likely to stop at the frontier.

If large-scale international migration is a relatively recent phenomenon, internal migration has long been one of the principal consequences of population growth, rural transformation and industrial expansion. The booms of the 1920s were the first major stimulus to urban migration, a trend that accelerated rapidly after 1940 as modernization proceeded apace. It was further stimulated in some areas of Colombia by the *Violencia* (*Figure 5.7*), a period of conflict between the two major parties which led to over 200,000 deaths and the massive exodus of rural people to urban areas in search of safety (Guzman *et al.*, 1962; Fals Borda, 1965). Migration has continued at a rapid rate and is continuing to fuel the growth of the cities. It was not until the 1960s, however, that it resulted in a decline in the rural population (*Table 5.3*) (Williams and Griffin, 1979; Cox, 1978).

The broad stimulus to migration has been the changing structure of the labour market in both countries. In Venezuela the agricultural labour force declined from 44 per cent of the total in 1950 to 21 per cent in 1971; in Colombia it declined from 54 per cent in 1951 to 36 per cent in 1973. The major expansions have been in urban activities. During the same period, the labour force in manufacturing increased in Venezuela from 11 to 18 per cent and in Colombia from 12 to 16 per cent. In services the expansion has also been fast, in Venezuela from 20 to 26 per cent and in Colombia from 16 to 20 per cent.

The obvious outcome of this shift in the labour force has been rapid city expansion. Between 1951 and 1964 all but one of Colombia's largest twenty cities were expanding at rates over 5 per cent per annum, with the three largest cities growing at annual rates of around 7 per cent. The contribution of migration is clear from the fact that in 1964 more than 70 per cent of Bogotá's population over 15 years old had been born outside the city. Although migration is still important, it is rapidly ceasing to be the major source of growth in most Colombian and Venezuelan cities. Due to the very large numbers of young people who have migrated, rates of natural increase are very high: between 1970 and 1975, 54 per cent of the growth in Bogotá's population was due to natural increase, and in 1961 49 per cent of that of Caracas (BUDS, 1974; OMPU, 1972, p. 43).

Although the growth of the major cities is now guaranteed by the structures of their populations, this has not stopped governments in either country from attempting to slow the pace of urban expansion. The inability of the authorities in most of the major cities to administer services and housing in a way that would benefit most of the poorer groups, has given the impression that the major cities are in crisis. The proliferation of self-built housing, traffic congestion, air pollution and overcrowding, whether or not such problems are directly attributable to size rather than city organization, has given rise to policies to deconcentrate employment away from the major cities.

Such efforts have been most spectacular in Venezuela. In particular, the establishment of a new industrial city in the Guayana region associated with the development of the region's iron, bauxite and hydroelectric resources has been widely publicized (Friedmann, 1966; Rodwin *et al.*, 1969). Certainly, Ciudad Guayana's growth has been spectacular, increasing in population from 29,000 in 1961 to 221,000 in 1975 (Macdonald and Macdonald, 1979), even if early projections of the city's growth anticipated that it might have had 415,000 people by the latter date. Undoubtedly, the city has attracted workers and their families,

Table 5.3 Population growth and urban development in Colombia and Venezuela during the twentieth century

(a) Colombia

Year	National population ('000)	Annual growth (%)	Urban population[1] ('000)	Rural population ('000)	Urban population[1] (%)	Population of Bogotá ('000)	Bogotá as % national population
1905	4,135	–	–	–	– (10.5)	100	2.3
1912	5,073	2.9	–	–	– (10.5)	121	2.4
1918	5,855	2.2	–	–	– (11.6)	144	2.5
1938	8,702	2.0	2,693	6,010	30.9 (16.6)	330	3.8
1951	11,548	2.2	4,469	7,080	39.6 (23.1)	648	5.6
1964	17,485	3.2	9,093	8,391	52.1 (37.5)	1,697	9.7
1973	22,500	2.7	13,719	8,781	63.1 (37.5)	2,811	12.5
1980 (est.)	27,000	2.7	18,883	9,117	69.9 (–)	4,163	15.4

(b) Venezuela

Year	National population ('000)	Annual growth (%)	Urban population[2] ('000)	Rural population[3] ('000)	Urban population[2] (%)	Population of Caracas ('000)	Caracas as % national population
1891	2,222	–	–	–	–	98	4.4
1920	2,479	0.4	–	–	–	118	4.8
1936	3,364	1.8	972	2,196	28.9	259	7.7
1941	3,851	2.8	1,207	2,334	31.3	354	9.2
1950	5,035	3.0	2,412	2,325	47.9	694	13.8
1961	7,524	4.0	7,704	2,450	62.5	1,336	17.8
1971	10,722	3.4	7,834	2,445	73.1	2,184	20.4
1980 (est.)	13,913	3.0	10,584	2,440	76.1	3,400	24.4

Sources: (a) DANE, 1977, p. 29; McGreevey, 1971, p. 110. (b) Dirección, 1974; López, 1963.

Notes: 1 Population living in Cabeceras. Figures in parentheses are the share of the national population living in what were the largest sixteen cities in 1964.
2 Population living in settlements with more than 2500 inhabitants.
3 Population living in settlements with less than 1000 inhabitants.

with four out of five migrants to Ciudad Guayana in 1975 coming from the eastern part of Venezuela (Macdonald and Macdonald, 1979). Unfortunately, like industrial expansion in the rest of the country, the creation of jobs has been more limited than had been anticipated. It had been hoped that more people would be engaged in manufacturing than in services; in fact, in 1975 three out of ten worked in industry compared to five out of ten in services. Thus, although the country's largest aluminium, steel and hydroelectric production facilities are located in the city, it has not stimulated regional development in the way that had once been hoped. In addition, agricultural development in the Guayana region has been particularly disappointing. Some Venezuelans, indeed, have been highly critical of the project: Travieso (1972), for example, has labelled it 'an error not to be repeated' because it has merely replicated the problems of the traditional cities – unemployment, *ranchos* and poverty. Whether or not his judgement is too harsh, Ciudad Guayana demonstrates clearly that even with massive injections of funds a major shift in the national settlement system is not easy to achieve. As a consequence the rates of expansion of other Venezuelan cities have not slowed, and the national distribution of population scarcely modified.

Currently, the expansion of Caracas is being restricted by encouraging industrial growth in other parts of the country, especially areas outside the Caracas–Puerto Cabello axis, and a commission is reporting on the desirability of establishing a new national capital in another part of the country. In addition, five small towns to the south and south-east of the city are being developed as satellite cities with new industrial estates and major public housing projects. Despite these efforts and incentives to encourage industrial location in more peripheral locations, it is unlikely that the social, political and economic organization of Venezuela can permit the necessary decentralization (Myers, 1978; Brewer Carias, 1975). Put simply, Caracas is the centre of national life and as such it is difficult to persuade anyone to move, however much it may be in everyone's interest. Perhaps for this reason the proliferation of regional development agencies, the establishment of a major new city, the expenditure of huge sums of money by the standards of other countries in peripheral areas, have hardly modified the country's spatial structure.

By comparison with Venezuela, Colombia's national settlement system is not specially unbalanced. Primacy is increasing, but the country's other cities are generally dynamic and attracting economic activity (Gilbert, 1975). If Bogotá's growth is rapid, it is not impeded by physical constraints or a shortage of water in the same way as is Caracas. Nevertheless, since the late 1960s the government has been making token efforts to modify the settlement pattern, in particular to slow industrial expansion growth in the cities of Bogotá, Medellín and Cali. Some attempt has been made to remedy the worst infrastructural deficiencies in the intermediate centres, and to discourage foreign manufacturers from locating in the three major cities. These efforts, however, have been very weak when matched against the continuing pressures for continued spatial concentration. Consequently, the employment and population shares of the major cities have continued to increase (Gilbert, 1975).

Inside the cities

The coexistence of opulence and poverty is one of the principal characteristics of the cities of Colombia and Venezuela. The distribution of income is the determining factor underlying most aspects of urban life. Given the affluence of as much as 40 per cent of the population in the larger cities, the roads are filled with cars, there are extensive, well-planned and well-serviced residential areas; there are good schools, hospitals and shopping centres. Given the poverty of much of the rest of the population, there are large areas of unplanned and sometimes unserviced housing, occupied by people with limited access to well-paid jobs, adequate schools and satisfactory health facilities. Without painting an unduly pessimistic view of the situation of the people living in these areas, for it is certain that although the majority are less poor than most country dwellers, levels of poverty, illness and overcrowding are still very high. In Bogotá in 1973 49 per cent of the population were earning less than $70 per month (at 1978 prices), 22 per cent of 7–9-year-olds had received no education, and 63 per cent of houses had occupancy rates of more than one person per room. In cities such as Buenaventura and Barranquilla, the situation was still worse.

Housing is, of course, one of the most obvious manifestations of the social and economic differences between classes in Colombian and Venezuelan cities. Nowhere is this more obvious than in Caracas where the proliferation of *ranchos* is obvious to even the most casual observer (Marchand, 1966). Rapid demographic growth, limited space for expansion and the lack of even rudimentary efforts at controlling land prices or the direction of urban expansion, have encouraged low-income populations to occupy all available land. The hillsides of Caracas, unstable though many of them are, are covered by the largely self-built accommodation of the poor. The same is true, even if it is less obvious, in most other cities; a majority of the population in most of these cities is housed in self-built dwellings occupying the worst land in the city. Of course, many of these dwellings have been consolidated through time in the manner described by Turner (1968) and Mangin (1967) for Lima, Peru, but many more provide an inadequate dwelling for their occupants. Such a statement is not a recommendation for destroying such housing, for that housing is a symptom not a cause of poverty. Indeed, destruction of self-help housing would merely accentuate the shortage of housing which is the immediate difficulty. The fact is that a self-help dwelling is often the only viable way in which a poor family may resolve its housing problems. It is far superior to the situation which faces a family renting a home. Of course, renting may suit some groups: families without children, single people, those dependent on a central location, but most families prefer the greater flexibility of owning their home. Ownership provides a hedge against inflation, security against eviction and freedom from paying rent. In this context it is perhaps unfortunate that so many people are required to rent homes in Colombian and Venezuelan cities. In 1971, 39.6 per cent of homes in the Federal District were rented; in 1973, 39.6 per cent of homes in Bogotá, 42.5 per cent of those in Medellín, and 41.1 per cent of those of Cali (Direccion, 1974; DANE, 1977). High levels of renting are associated both with the largest cities and with high levels of overcrowding. In Bogotá 30 per cent of dwellings have occupation densities of more than two persons per room.

Of course, more families would cease to rent if they were able to establish their own homes. Unfortunately, this is difficult in many Colombian and Venezuelan cities because of the unavailability of land. In some cities many poor people obtain land by invasion. Encouraged and helped by politicians and the authorities, they occupy public land and on occasion private land (Gilbert, 1981; Ray, 1969; CEU, 1977). In Colombian cities such as Cúcuta and Ibagué, and in Venezuelan cities such as Maracaibo, Maracay and Valencia, most poor people are living in invasion areas. But, increasingly, poor people cannot gain access to free land and are forced to purchase (Vernez, 1973; Losada and Gómez, 1976). In Bogotá and in many Colombian cities most poor people obtain land through the so-called pirate urbanization system. Entrepreneurs operating on the fringe of the law offer the poor cheap lots and credit with which to buy land in their subdivisions, but fail to provide sufficient services or wide enough roads to satisfy the standards laid down by the planning authorities. The poor are not evicted and are eventually serviced; the poor are not especially resentful of the system; but in 1978 they had to pay about £1000 for a plot in Bogotá. Needless to say, for the poor who were earning the minimum (normal) wage of £1.50 per day, this sum was difficult to save. Indeed, the major question mark against the self-help housing solution in Colombian and Venezuelan cities is not the capability of the poor to build and consolidate a home, but their access to land, materials and resources with which to do so. In an environment where the price of land and many building materials is increasing faster than wage levels, and where the poor generally occupy the worst land, this poses a major problem for the low-income groups. And, since the urban poor are increasingly a majority of both the urban and the national populations, this poses an important question about the future: will conditions for the poor deteriorate further?

Bibliography

ANIF (Asociación Nacional de Instituciones Financieras) (1976) *Empleo y desarrollo*, Bogotá.

ARCILA FARÍAS (1946) *Economía colonial de Venezuela*, Mexico.

ARMAS CHITTY, J.A. (1961) *Tucupido, formación de un pueblo llanero*, Caracas.

AVRAMOVIC, D. (ed.) (1972) *Economic Growth of Colombia: Problems and Prospects*, Baltimore.

BANCO CENTRAL DE VENEZUELA (1977) *La economía venezolana en los últim treinta y cinco años*, Caracas.

BANCO NACIONAL DE AHORRO Y PRESTAMO (BANOP) (1970) *Estudio del mercado real de vivienda en Venezuela* (MERCAVI 70), Caracas.

BELL, P.L. (1921) *Colombia: A Commercial and Industrial Handbook*, Washington, DC.

BERRY, R.A. (1975) 'Open unemployment as a social problem in urban Colombia: myth and reality', *Economic Development and Cultural Change*, 23,276–91.

BERRY, R.A. and URRUTIA, M. (1975) *Income Distribution in Colombia*, New Haven.

BEYER, R.C. (1948) 'Transportation and the coffee industry in Colombia', *Hisp. Am. Hist. Rev.*, 2, 17–30.

BOCKH, A. (1956) *El desecamiento del Lago de Valencia*, Caracas.

BREWER-CARIAS, A. (1975) *Cambio Político y Reforma del Estado de Venezuela*, Madrid.

BRITO-FIGUEROA, F. (1960) *Ensayos de historia social venezolana*, Caracas.

BRITO-FIGUEROA, F. (1961) *La estructura social y demográfica de Venezuela colonial*, Caracas.

BRITO-FIGUEROA, F. (1963) *La estructura económica de Venezuela colonial*, Caracas.

BROADBENT, S. (1968) 'A prehistoric field system in Chibcha territory', *Nawa Pacha*, 6, 135–47.

BUDS (1974) 'Bogotá urban development study, phase two', *The Structure Plan for Bogotá*, vol. 1, Bogotá.

BUSHNELL, D. (1954) *The Santander Regime in Gran Colombia*, Westport, Conn.

CARRERA-DAMAS, G. (1964) *Materiales para el estudio de la cuestión agraria en Venezuela 1800–1830, I, Estudio Preliminar*, Caracas.

CEU (Centro de Estudios Urbanos) (1977) *La intervención del estado y el problema de la vivienda: Ciudad Guayana*, CEU, Caracas.

COLLYER, O.A. (1965) *Birth Rates in Latin America. New Estimates of Historical Trends and Fluctuations*, Berkeley.

COX, P. (1978) 'Venezuela's agrarian reform at mid-1977', *Madison, University of Wisconsin*, Land Tenure Center Research Paper 71.

CRIST, R.C. (1952) *The Cauca Valley, Colombia*, Baltimore.

DANE (Departmento Administrativo Nacional de Estadística) (1977) *La población en Colombia, 1973*, Bogotá.

DIRECCION GENERAL DE ESTADISTICA (1974) *X Censo de Población y Vivienda*, Resumen General, Caracas.

EIDT, R.C. (1959) 'Aboriginal Chibcha settlement in Colombia', *Annals Assoc. Am. Geog.*, 49, 374–92.

EIDT, RC. (1968) 'Pioneer settlement in Colombia', *Geogr. Rev.*, 58, 298–300.

ESCALANTE, A. (1964) *El Negro en Colombia*, Bogotá.

FALS-BORDA, O. (1957) *El hombre y la tierra en Boyacá*, Bogotá.

FALS-BORDA, O. (1965) 'Violence and the break-up of tradition in Colombia', in VELIZ, C. (ed.) *Obstacles to Change in Latin America*, New York, pp. 188–205.

FALS-BORDA, O. (1969) *Subversion and Social Change in Colombia*, New York.

FRIEDMANN, J. (1966) *Regional Development Policy: a Case Study of Venezuela*, Cambridge, Mass.

GILBERT, A.G. (1975) 'Urban and regional development programmes in Colombia since 1951', in CORNELIUS, W.A. and TRUEBLOOD, F.M. (eds) *Urbanization and Inequality: the Political Economy of Urban and Rural Development in Latin America*, Latin American Urban Research, vol. 5, Beverly Hills, pp. 241–75.

GILBERT, A.G. (1978) 'Bogotá: politics, planning and the crisis of lost opportunities', in CORNELIUS, W.A. and KEMPER, R.V. (eds) *Metropolitan Latin America: the Challenge and the Response*, Latin America Urban Research, vol. 6, Beverly Hills, pp. 87–126.

GILBERT, A.G. (1981) 'Pirates and invaders: land acquisition in urban Colombia and Venezuela', *World Development*, 9, 657–78.

GILMORE, R.L. (1964) *Caudillism and Militarism in Venezuela 1810–1910*, Athens, Ohio.

GILMORE, R.L. and HARRISON, J.P. (1948) 'Juan Bernardo Elbers and the introduction of steam navigation on the Magdalena River', *Hisp. Am. Hist. Rev.*, 335–59.

GORMSEN, E. (1966) *Barquisimeto: eine Handelstadt in Venezuela*, Heidelberg.

GRAY, W.H. (1945) 'Steamboat transportation on the Orinoco', *Hisp. Am. Hist. Rev.*, 25, 455–69.

GUZMAN, G., FALS-BORDA, O. and UMAÑA, E. (1962–4) *La violencia en Colombia*, 2 vols, Bogotá.

HARRISON, J.P. (1952) 'The evolution of the Colombian tobacco trade to 1875', *Hisp. Am. Hist. Rev.*, 32, 163–74.

HEGEN, E.E. (1966) *Highways into the Upper Amazon Basin: Pioneer Lands in Southern Colombia, Ecuador and Northern Peru*, Gainsville.

HIRSCHMAN, A.O. (1963) *Journeys Towards Progress*, Westport, Conn.

HOUSTON, J.M. (1968) 'The foundation of colonial towns in Hispanic America', in BECKINSALE, R.P. and HOUSTON, J.M. (eds) *Urbanisation and its Problems*, Oxford, 352–90.

HUSSEY, R.D. (1934) *The Caracas Company 1728–1784*, Cambridge, Mass.

KALMANOWITZ, S. (1978) 'Desarrollo capitalista en el campo', in ARRUBLA, M., *et al.*, *Colombia Hoy*, Bogotá.

KARLSSON, W. (1975) *Manufacturing in Venezuela*, Stockholm.

LIEUWEN, E. (1954) *Petroleum in Venezuela: a History*, Berkeley.

LINDQVIST, S. (1979) *Land and Power in South America*, Harmondsworth.

LOPEZ, J.E. (1963) *La expansión demográfica de Venezuela*, Mérida.

LOPEZ, J.E. (1968) *Tendencias recientes de la población venezolana*, Mérida.

LOSADA, R. and GOMEZ, H. (1976) *La tierra en el mercado pirata de Bogotá*, Fedesarrollo, Bogotá.

LUBELL, H. and McCALLUM, D. (1978) *Bogotá, Urban Development and Employment*, International Labour Office, Geneva.

MacDONALD, J.S. and MacDONALD, L. (1979) 'Planning, implementation and social policy: an evaluation of Ciudad Guayana 1965 and 1975', *Progress in Planning*, 11, 1/2.

McGINN, N.F. and DAVIS, R.G. (1969) *Build a Mill, Build a City, Build a School*, Cambridge, Mass.

McGREEVEY, W.P. (1971) *An Economic History of Colombia*, London.

McGREEVEY, W.P. (1980) 'Population policy under the National Front', in BERRY, R.A., HELLMAN, R.G. and SOLAUN, M. (eds) *Politics of Compromise: Coalition Government in Colombia*, 413–34.

MARCHAND, B. (1966) 'Les ranchos de Caracas, contribution a l'étude des bidonvilles', *Cahiers d'Outre-Mer*, 19, 105–43.

MARCIALES, M. (ed.) (1948) *Geografía histórica y económica del norte de Santander*, vol. 1, Bogotá.

MARTINEZ, C. (1967) *Apuntes sobre el urbanismo en el Nuevo Reino de Granada*, Bogotá.

MENDOZA, D. (1947) *El Llanero: ensayo de sociología Venezolana*, Buenos Aires.

MORÓN, G. (1954) *Los orígenes históricos de Venezuela*, Madrid.

MORSE, R.M. (1962) 'Some characteristics of Latin American urban history', *Am. Hist. Rev.*, 67, 317–38.

MYERS, D.J. (1978) 'Caracas: the politics of intensifying primacy', in CORNELIUS, W.A. and KEMPER, R.V. (eds) *Metropolitan Latin America: the Challenge and the Response*, Latin American Urban Research, vol. 6, Beverly Hills, pp. 227–58.

NICHOLS, T.E. (1954) 'The rise of Barranquilla', *Hisp. Am. Hist. Rev.*, 34, 158–74.

OFICINA CENTRAL DE ESTADISTICA E INFORMATICA (1978) *Anvario Estadístico*, Caracas.

OJER, P. (1967) *La Formación del Oriente Venezolano*, vol. 1: *creación de las gobernaciones*, Caracas.

OMPU (Oficina Municipal de Planeamiento Urbano del Distrito Federal) (1972) *Plan General Urbano de Caracas, 1970–1990*, Caracas.

OSPINA-VASQUEZ, L. (1955) *Industria y protección en Colombia, 1810–1930*, Medellín.

PALACIOS, M. (1980) *Coffee in Colombia, 1850–1970*, Cambridge.

PARSONS, J.J. (1949) *Antioqueño Colonisation in Western Colombia*, Berkeley.

PARSONS, J.J. (1967) *Antioquia's Corridor to the Sea*, Berkeley.

PARSONS, J.J. and BOWEN, W.A. (1966) 'Ancient ridged fields of the San Jorge River floodplain, Colombia', *Geogr. Rev.*, 56, 317–43.

PICON SALAS (1962) *Venezuela independiente 1810–1960*, Caracas.

QUERO-MORALES, C. (1978) *Imagen-objectivo de Venezuela, Reformas fundamentales para subdesarrollo*, Caracas.

QUIJADA, J.L. (1968) 'Panaquire, pueblo de latifundio', *Boletín Histórico* (Caracas), 16, 11–37.

RASMUSSEN, W.D. (1947) 'Agricultural colonisation and immigration in Venezuela 1810–1860', *Agric. Hist.*, 21, 152–62.

RAY, T. (1969) *The Politics of the Barrios of Venezuela*, Berkeley.

REICHEL-DOLMATOFF, G. (1956) *Colombia*, London.

RICARDO, I. (1967) *Atlas de Caracas*, Caracas.

RIPPY, J.F. (1931) *The Capitalists and Colombia*, New York.

RIPPY, J.F. (1943) 'Dawn of the railway era in Colombia', *Hisp. Am. Hist. Rev.*, 23, 650–63.

RODWIN, L. *et al.* (1969) *Planning Urban Growth and Regional Development: the Experience of the Guayana Program of Venezuela*, Cambridge, Mass.

ROUSE, I. and CRUXENT, J.M. (1963) *Venezuelan Archaeology*, New Haven, Conn.

SAFFORD, F.R. (1965) 'Commerce and enterprise in Central Colombia, 1821–1870', unpublished PhD thesis, Columbia University.

SALAZAR-CARRILLO, J. (1976) *Oil in the Economic Development of Venezuela*, New York.

SAUER, C.O. (1966) *The Early Spanish Main*, Berkeley.

SCOTT, I. (1968) 'Colonial urban development in Hispanic America: the case of Santa Fé de Bogotá', *Bull. Soc. Latin Am. Studies*, 10, 20–6.

SEGNINI, I.S. de (1978) *Dinámica de la agricultura y su expresión en Venezuela*, Barcelona.

SIMMONS, A.B. and CARDONA, R. (1973) *Family Planning in Colombia*, Ottawa.

SMITH, T.L. (1967) *Colombia: Social Structure and the Process of Development*, Gainsville.

STEWARD, J.H. and FARON, L. (1959) *Native Peoples of South America*, New York.

TRAVIESO, F. (1972) *Ciudad, region y subdesarrollo*, Caracas.

TURNER, J.F.C. (1968) 'Housing priorities, settlement patterns and urban development in modernizing countries', *Journal of the American Institute of Planners*, 34, 354–63.

URRUTIA, M. (1969) *The Development of the Colombian Labour Movement*, New Haven, Conn.

VERNEZ, G. (1973) 'Bogotá's pirate settlements: an opportunity for metropolitan development', unpublished PhD thesis, Berkeley.

VILA, P. (1960–5) *Geografía de Venezuela*, 2 vols, Caracas.

VILA, P. (1966) 'Consideraciones sobre poblaciones errantes en el periodo colonial', *Revista de Historia* (Caracas), 12, 11–24.

WARRINER, D. (1969) *Land Reform in Principle and Practice*, London.

WEST, R.C. (1952) *Colonial Placer Mining in Colombia*, Baton Rouge.

WEST, R.C. (1957) *The Pacific Lowlands of Colombia*, Baton Rouge.

WILLIAMS, L.S. and GRIFFIN, E.C. (1978) 'Rural and small-town depopulation in Colombia', *Geogr. Rev.*, 68, 13–30.

WORLD BANK (1980) *World Development Report, 1980*, Washington, DC.

6 · The Guianas

D.J. Robinson and the Editors

Contrasts in the 'Land of Many Waters'

Few who look at the political or cultural patterns within South America can fail to puzzle over the anomalous group of territories known as the 'Guianas'.[1] Their origins and continued existence within a subcontinent dominated by Iberian culture invite analysis. Yet there have been few detailed studies that consider the Guianas as an entity. The elementary textbook coverage usually does little more than describe each of the political units in turn, and only Lowenthal's (1960b) study of differential population distributions, and Devèze's (1968) introductory survey pay more than lip-service to the analysis of these contiguous colonial cultural enclaves. The majority of recent publications keep strictly within the political boundaries of individual territories, thereby avoiding the problems posed by source materials in five or more languages (Kruijer, 1960; Cummings, 1969; Sanderson, 1969). Yet it is only when the Guianas are placed within the wider context of the settlement of the Guiana region that their full significance can be appreciated.

The Guianas clearly provide excellent case studies in the neglected field of colonial geography (Lowenthal, 1950). The key questions are: how did non-Iberian Europeans manage to occupy a coastal margin of South America exceedingly close to Europe? Why and when did the Dutch, French and British effect their occupation, and what significance can be attached to the different modes of settlement and development? What have been, and continue to be, the implications of these varied colonial experiences for each of the territories in terms of problems of growth, change and development during the present century?

The process of cultural differentiation, whether concerned with economy, society or features in the human landscape, has taken place within a physical setting of striking homogeneity. In general terms, the whole of the Guiana region, as its aboriginal name implies (Cummings, 1964) is surrounded and characterized by an abundance of water courses (*Figure 6.1*). The incredible Casiquiare canal closes the water circle that includes stretches of the Atlantic Ocean and the rivers Orinoco, Negro and Amazon. Topographically the core of Guiana is centred on Mount Roraima (2810 m), from which point the altitude of the land surface falls away in a series of step-like surfaces whose identity becomes increasingly indeterminate with distance from the core (McConnell, 1968). East of the depression containing the rivers Essequibo and Courantyne the land surface rises in a series of hilly ranges, which forms the Atlantic–Amazon watershed. The location of the core of high land in the north and west of the Guiana

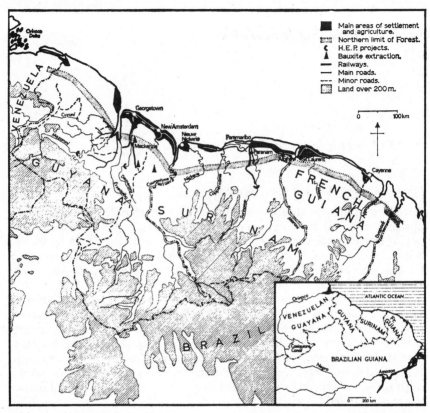

Figure 6.1   The Guianas

region has resulted in the uneven distribution of rivers and areas of broken topography. The Guianas, for example, occupy only the northern edge of the larger region, being provided with a restricted area of coastal plain in comparison with the extensive plainland periphery of the south-western edge of the Guiana region. While the Venezuelan and Brazilian territories include the long, penetrating river-valley systems, the Guianas are characterized by relatively short, north–south orientated water courses. Only in Guyana do two east–west links attain any significance: those of the Cuyuni and Rupununi. The precipitous progress of the majority of rivers towards the Atlantic coast is marked by well-developed series of waterfalls and rapids. The edge of the crystalline complex has long impeded riverine navigation. The coastal margin of Guiana sweeps in an arc between the mouths of the Orinoco and Amazon, broadening here and narrowing there in response to the orientation of the coast, which either hinders or assists the spread of Amazonian silt westwards. While French Guiana can boast of surf and sand beaches in Surinam and Guyana, as in Venezuelan Guayana, marshy vegetation anchors accretionary sediment moving westwards. The rivers disgorging from the south make their way with difficulty to the sea through lagoons and across or around coastal bars. The progressive extension of the coastal plain seawards is perhaps most noticeable in Surinam, where hundreds of ridges mark former shorelines (Eijk, 1954).

Climate and vegetation also differentiate the coastal margin from the interior of the Guianas. With the heat equator running parallel to the Atlantic coast as far east as 5°N before crossing the Atlantic towards Africa, high annual average temperatures are the rule. Diurnal temperature differences exceed those between coldest and warmest months. Except on the highest slopes of the sandstone mountain core, altitude does little to reduce the heat. Rainfall and humidity are also monotonously high. The short, drier season from August to November brings with it less cloudy skies and, therefore, higher temperatures. The 80 per cent humidity levels of the coast are made bearable only by the constant on-shore breezes.

The dominant vegetation type is dense rain-forest, which extends throughout all the Guianas. The southern margin of Guyana includes a wide area of dry savanna, which links with the Río Branco savannas of Brazil. The seasonally inundated grassy plains of southern Guyana are known by their aboriginal name, Rupununi (Eden, 1964).

The subtle natural contrasts contained within the Guiana region have been emphasized, but more particularly eclipsed, by the effect of European man. Whereas natural divisions are difficult to distinguish and merge imperceptibly one into another, the hand of man has increasingly heightened the pre-existing contrasts and, by occupying the land and exploiting its resources, has created new contrasts. The significance of contrast or differentiation within Guiana has a long history. While the prehistoric patterns are still being investigated, it is clear that since European man's arrival the coastal-interior division has been fundamental. The peripheral settlement nodes along the Orinoco, the Atlantic and the Amazon and its tributaries provided points of departure for inland exploration. Where facts were, few myths could multiply. Guiana has been, and continues to be, a land of many El Dorados (Lowenthal, 1960a; Swan, 1958).

The coast of Guiana has been the focus of human activity at least since the early seventeenth century. It was in the margin and not the interior that settlements were created and population expanded, and it was there that the primitive landscape was affected. It must be remembered, therefore, that the history of developments in the Guianas has to be compared with that of the Venezuelan and Brazilian segments if we are to appreciate the full significance of the cultural differentiation that has taken place. Indeed, the true interest in the Guianas lies in the effective role of three colonial powers impinging upon the spheres of influences of what were Spanish and Portuguese empires, as Quelle (1951) has shown. This has meant that within the coastal margin and the interior of the Guianas new cultural contrasts have emerged.

The process of cultural differentiation

Developments to 1800
Although the first exploration and attempts at European settlement of Guiana were made by the Spanish and Portuguese (Ojer, 1966, p. 202), it soon became evident during the sixteenth century that neither Spain nor Portugal was partic-

ularly concerned about the stretch of coast between the Amazon and Orinoco. A combination of circumstances, which included difficult sailing conditions, inhospitable shores, difficulties over establishing the precise limits of their legal spheres of activity and, above all, the lack of easily obtainable benefits in comparison with other regions, discouraged Spaniards and Portuguese alike from giving Guiana much attention. It was assumed that the region could be left as an unoccupied buffer zone, fulfilling a function similar to that of the Chaco wilderness in another zone of contact between Spain and Portugal in South America. It is true that fabled wealth attracted numbers of intrepid explorers to the region in the sixteenth century, and not least Sir Walter Raleigh, whose identification of the gold-bearing zone in the Yuruari plains was not appreciated until the mid-nineteenth century, but the mythical peoples and places were soon to be rationalized away in terms of Indian legend or natural seasonal phenomena. Exploration had been, by the close of the sixteenth century, essentially peripheral, utilizing the major navigable waterways of the Orinoco and Amazon. The Atlantic coast had been surveyed. But for the most part few expeditions had penetrated deep inland into the Guiana region from the main rivers. From the Atlantic coast, on the other hand, it was possible to make relatively easy journeys southwards up the Essequibo, Corentyne, Suriname and Oyapock, at least to the first major fall-line that demarcated the northern edge of the highland zone.

With Spain and Portugal occupied elsewhere, it was possible for English, Dutch and French adventurers, representatives of what were minor colonial powers in the Caribbean and Latin American context, to make the most of the possibilities in Guiana. Not unnaturally, the central portion of the Atlantic coastal margin was the prime target area; it was as far as possible from both Spanish and Portuguese enterprises, and it could be supplied and serviced if necessary by sea. From the beginning of the seventeenth century, Dutch, English and French settlers established a series of settlements at the mouths of the major rivers of the Guianas. The precise chronology and locations of the early centres of activity are extremely complicated; different colonial powers founded settlements, were dislodged, refounded them and spread out from them in intricate stages. The principal phases and characteristics may be best understood if each of the colonial ventures is considered briefly in turn, remembering that interaction between them was one of the salient characteristics.

Early Dutch interest centred on the lower Essequibo where, after being dislodged by the Spanish from Pomeroon in 1581, a small settlement had been established by 1615. Desultory colonization, however, was replaced by purposeful planned action following the incorporation of the Dutch West India Company in 1621. The key settlement was a new, fortified township called Kyk-over-al sited at the junction of the rivers Mazaruni, Cuyuní and Essequibo. During the first century of colonial activity the Dutch were as interested in trading with the Indian peoples of Guiana as in cultivating the land to produce subsistence or export crops. A wide network of trading posts was established during the remainder of the century. Trade links were established between the Atlantic coast and Indian groups living deep within what was indisputably Spanish and Portuguese territory in the middle Orinoco and Río Negro basins. Gradually, along the Essequibo and, after 1627, along the Berbice, small plantations of tobacco, cotton, coffee, cocoa and sugar were established. In 1667 a

Dutch fleet took from the English the settled zone of land at the mouth of the river Suriname, and by the Treaty of Breda of the same year Surinam, as the area was known, was ceded to the Netherlands. In exchange Britain received what is now New York State (Newman, 1964, p. 18).

During the seventeenth century a significant feature of the economic developments in the Dutch controlled areas was the introduction in increasing numbers of West African slaves. Without this supply of labour the colonies could never have developed. During the eighteenth century major modifications to the pattern of settlement and economic activities were brought about. Of singular importance was the role of a newly appointed *Commandeur* of Essequibo, Laurens Storm van Gravesande (1742–72). Gravesande was undoubtedly an able administrator; during his period of office new settlers, the majority in fact English, were attracted to free land in the Demerara region, particularly after 1746. Indeed, the process of coastal colonization, often at the expense of the up-river estates, was a fundamental feature of the eighteenth century (Swan, 1958, p. 34). After the Treaty of Utrecht (1713) the coast had become a somewhat safer place to live. There was also some evidence to suggest that the more easily settled land on higher reaches of the rivers was losing its fertility and, faced with the difficulty of clearing forest land, settlers were only too willing to benefit from company-planned drainage schemes in the coastal plain, and the resourcefulness of Dutch agricultural colonists, guided by the West India Company, was to be seen in Essequibo, Demerara, Berbice and Surinam alike (Lier, 1949).

Yet the abandonment of the river lands and the cost in cash and human labour of empoldering the fertile clay soils of the coast was only made economically feasible by the increasing significance of one of the crops – sugar. A sugar plantocracy was rapidly emerging after 1750. Disease, harsh treatment and the aftermath of numerous slave rebellions may have caused the death of hundreds of thousands of black slaves; for the plantation economy the principal preoccupation was profit. Despite Dutch political control over the coastal region during the eighteenth century, the English were always a significant element amongst private estate owners. Amidst a diked and drained polder landscape were to be found estate names that reflected the cosmopolitan nature of the colonial venture: Vryheid's Lust, Better Success, La Bonne Mère, Hampton Court, Bergen op Zoom and Gage d'Amour (Lowenthal, 1960b, p. 43). Considering the extent of British involvement in the Caribbean area during the late eighteenth century, it is not surprising that Dutch control in Guiana should have been challenged. When war broke out between England and Holland in 1781, Demerara, Berbice and Essequibo were taken by the English, but some months later were occupied by the French, who were then at war with England. The only significant feature of the 2-year period of French rule of the 'Three Rivers' was their initiation of a new town at the mouth of the Demerara, to be called Longchamps. The restoration of the colonies to the Dutch in 1783 saw their adoption of the Longchamps site as a new colonial capital – Stabroek. The troubled years that followed Dutch reoccupation in 1783 were in part the result of resentment among the planters, who were only too delighted to see the end of the West India Company in 1792, when its charter expired. Neither the savage repression of rebellions nor the improvements in public utilities could stem the tide of liberal ideology, and when, in 1796, Holland and England found themselves at war, once again the colonies were taken from the Dutch. Final control of the colonies was ceded to the British in 1814, but from 1796 British control and development was of paramount importance. By the early nineteenth century, therefore, Britain had taken possession of a sugar colony on the South American mainland which had been created by the efforts of Dutch commercial companies. Surinam, too under-developed to attract attention, slumbered on.

In what was to become French Guiana the history of economic development had been quite distinctive. Following the unsuccessful activities of seventeenth-century commercial companies supported by the French government, what restricted land occupation had taken place centred on the small township of Cayenne. To the south lay a scattering of small, isolated farming and gold-prospecting communities, out of the reach of authority and malarial mosquitoes alike. During the second half of the eighteenth century efforts were made to develop agriculture along Dutch lines but, for a variety of reasons, the attempts were abortive (Lowenthal, 1952). Whereas in their Caribbean islands the French were engaged in profitable sugar cultivation, in Cayenne, as the mainland colony was often called, the enormity of the task of empoldering the coastlands deterred all but the very brave (Papy, 1955). Cayenne was a place in which to live, not work.

Nineteenth-century changes

Significant changes in the social and economic geography of the Guianas took place during the nineteenth century. Since the colonial models had differed up to that date, so the changes that occurred affected French Guiana, Surinam and British Guiana differently. The most significant single change was the abolition of slavery, a process that was to alter fundamentally the ethnic structure of populations in all three colonies, but especially in the most economically advanced, British Guiana. At the beginning of the century, the large-scale, heavily capitalized sugar estates of the coastal plain put sugar on to the world market at a time when Jamaican production was in decline. British Guiana was part of a West Indian 'frontier' of productive agriculture (Ragatz, 1928, p. 332). Yet with costly modern milling equipment being utilized, many kilometres of canals needing continual repair; and with the perpetual need for hands in the cane fields, it was vital to the estate owners to be able to obtain cheap labour. Fox's Bill of 1807, forbidding British subjects to trade in slaves, threatened not only labour supplies but also the very existence of the sugar colony itself. When to labour scarcity were added the difficulties of selling sugar, following the abolition of Imperial preferences between 1846 and 1854, the situation became even worse (Newman, 1964, p. 21). Falling prices, rising wages and a shortage of capital among the planters meant the closure or consolidation of many estates. From a total of 380 coastal estates in 1800 the number fell sharply to 230 by 1829, to 180 in 1849 and to 64 in 1896. It was by means of purchasing cheap estates during the nineteenth century that Bookers, the present giant sugar enterprise in British Guiana, were able to enter the economic arena.

Besides reducing the labour force engaged in sugar cultivation, emancipation also involved a redistribution of the negroid element in the population of the colony. Instead of choosing to move deep into the interior, most negroes settled in the coastal zone close to Georgetown and the other large centres, on abandoned or collectively purchased estates. Without any suitable cash crop to cultivate, they remained willing to serve as part-time labourers for the estate operators. It was soon clear, however, that without a more secure and larger labour force, sugar would soon become an uneconomic industry. The search began for new labour. After small numbers of West Indians, Portuguese from Madeira, Chinese and other groups had been brought into the colony, massive introductions of indentured Indian labourers characterized the population changes during the nineteenth century. Between 1846 and 1917 some 239,000 entered British Guiana in comparison with the 35,000 entering Surinam (Lier, 1949).

Such importations of new ethnic elements into the social geography of British Guiana were bound to affect pre-existing patterns. One notable feature was the migration of the Chinese and Portuguese immigrants from the agricultural lands to take up residence and employment in the urban centres. With the negroes already in large numbers in or near the urban areas, the distribution of the later East Indians in the rural areas produced a marked differential ethnic distribution. Not only had the social pyramid been reshaped following emancipation, but the geographical location of class and status groups had also been dramatically altered.

In Surinam the nineteenth century saw relatively little in the way of fundamental changes in the economy of the colony. Since the previous 100 years had seen little agricultural development, except in the lower reaches of the Suriname, Commewijne and Cottica rivers, where defence was possible against attacks from negroes who had escaped into the interior, the disruption of agriculture following the abolition of slavery in 1863 was not as great as that felt elsewhere in the Guianas. Paramaribo certainly expanded rapidly with the migration of freed negroes from the plantations (Kool, 1956), and the indentured Hindustani labourers did add a new constituent to the social fabric of the colony (Oudschans Dentz, 1943). Surinam continued to consist of relatively small isolated clusters of agricultural communities each dependent on confined polder schemes, with a marked tendency towards separatism.

In French Guiana the only significant innovation was the establishment of the penal colony in 1852, an attempt to alleviate the shortage of labour created by the abolition of slavery in 1848. The aim was simple enough: convicts would work towards their freedom on the plantations (Gritzner, 1964). But yellow fever, malaria and the misappropriation of funds jeopardized the scheme. The notoriety of conditions in the camp at St Laurent (popularly, but misleadingly, called Devil's Island) was to tarnish the image of French Guiana for more than a century (Lowenthal, 1960a, p. 530).

During the whole of the nineteenth century little change affected the interior of any of the Guianas. Though numerous scientific expeditions penetrated deep into the upland zone, the only significant economic events were the numerous small-scale gold strikes in French and British Guiana after 1840. Of more than local importance was the gold 'boom' in the Yuruari basin in Venezuelan Guayana, which attracted large numbers of prospectors from the Guiana coastal region (Robinson, 1967, pp. 572–630). For the remainder of the century the interior regions of the Guianas were little more than pleasurable retreats for collecting plants, hunting animals and observing Amerindian cultures.

Problems of the twentieth century

As a result of their colonial status the Guianas have each continued to develop separate identities during the present century. While all three areas, even independent Guyana, still demonstrate the characteristic features of an externally orientated economy, coastal 'peripheral' population distributions in terms of their territories, metropolitan linkages in financial, administrative and other aspects of political control, each of the Guianas has continued to increase the contrasts between it and its neighbours. The precise differentiation of economic and social changes is best presented by brief views of their individual progress during the present century.

French Guiana (Guyane)

With an area one-third the size of France and an estimated population in the late 1970s of less than 50,000, French Guiana may be considered somewhat underdeveloped, though the region has development potential. However, for a variety of reasons, few attempts have been made, and few appear likely to be made, to modify the present condition of one of France's more peripheral *départements*, at least in the near future. Having acquired the status of a *département* in 1946, and having enjoyed billions of francs annually poured in by the metropolitan exchequer to provide the luxuries of public health services, free care of mothers, infants and the elderly, free education and other social services, French Guiana's residents may well view with scepticism the desire for massive development projects, whether based on agriculture or on mineral exploitation. The suggestion of massive immigration into the territory, bringing with it problems of cultural assimilation, invites the comment that to date French Guiana has avoided the conflict that has characterized other nearby political units; better, perhaps, to be happily undeveloped than suffer the inevitable pains of economic and social 'progress'. So long as France continues to subsidize its piece of Guiana, that may be a justifiable argument. The distribution of population, 50 per cent of the total living in Cayenne and its environs, and only some 10 per cent living in the interior, reflects the unchanging nature and location of activities since the early nineteenth century (Hauger, 1957).

Surinam

An autonomous member of the Kingdom of the Netherlands until 1975, when it was granted independence while retaining close trade and aid ties with the former mother country, Surinam had already experienced considerable economic development before that date. Slow but steady industrialization, based on the exploitation of bauxite and hydroelectric resources, was undertaken from the 1950s: in 1965 an aluminium smelter was opened at Paranam, some 80 km south of Paramaribo (Geijkes, Leentvar and van Donselaar, 1965; Sanderson, 1969, p. 22) while several hydroelectric schemes came on stream, notably at Brokopondo and Afobaka on the Suriname river. By the time of independence, indeed, bauxite was the cornerstone of the economy: by 1980, it accounted for 85 per cent of exports by value, 40 per cent of government tax revenues and 30 per cent of GDP, all produced by two companies, the Suriname Aluminium Co., a wholly owned subsidiary of the multi-national Alcoa Co., and Billiton Maatschappij Suriname NV, owned by Royal Dutch Shell. As primarily an export economy, therefore, Surinam is exposed to fluctuations in world trade and depression in the industrialized world; on the other hand Dutch aid is significant in current plans for the expansion and diversification of the economy from its present narrow base. Their overall aid between 1975 and 1985 is put at US$1.7 billion, part of which is for further bauxite mining and smelting ventures in West Surinam, part for farming projects, and part for extending guarantees on foreign loans. Aid is provided for joint projects approved by a Dutch-Surinamese commission.

Far more than Guyane, Surinam has witnessed the spread of economic activity along the coastal plain and southwards, though, for a country approximately five times the size of Holland, its population, estimated in the late 1970s at 370,000, is still very small. Moreover, it is far more divided ethnically than that of Guyane, and the occupational and locational differentiation of individual ethnic groups is also quite distinct from that of Guyana. Such differences reflect the particular historical process in Surinam since the abolition of slavery. Hindustani immigrants (i.e. from India, not the East Indies) arrived relatively late in the nineteenth century, and they, together with Javanese from Indonesia, found acculturation a slow and difficult process. The political parties which developed under colonial rule found their support in particular ethnic communities, which, together with the racially homogeneous village groupings, tended to promote cultural contrasts. Though racial conflict has not reached the level that it did in Guyana, racial tension has been a factor in recent political developments.

The civilian, post-independence government of Mr Henk Arron was overthrown by a military *coup* in February 1980, allegedly for inefficiency and corruption, and the army was supported by the country's largest trade union, headed by the bauxite workers. Civilian rule was soon restored, with the promise of administrative reform, economic development and a lessening of racial tension. There were difficulties, however, over the Netherlands' aid package, and sizeable emigration occurred of some 100,000 Surinamese, unhappy with the development of a strong left-wing trend in the army. Political conflict within and outside the army has persisted and, at the time of writing, it remains unclear whether Surinam will gravitate to a Cuban-style government or retain its close ties with the Netherlands and the western world in general. Whatever its future course politically, Surinam will undoubtedly seek greater national participation in bauxite, and will expect future development projects to be at least on joint-venture lines. The country's main problem will be to find markets for her potential industrial output, and to reduce her heavy dependence on food imports (40 per cent of requirements) by more efficient and productive agriculture. The territory's sparse population is matched by few natural resources, apart from large bauxite deposits and extensive forests, the exploitation of which in the production of timber is the main industry. In the past 30 years modest attempts have been made to diversify the economic base through rice cultivation schemes, livestock ranching on the southern savannas, fishing and tourism, all of which have considerable potential. But, for the most part, there is little immediate prospect of significant economic progress.

Guyana (British Guiana)

The economic history of British Guiana, as the Republic of Guyana was called before May 1966, provides a perfect target for those who wish to denigrate the activities of European capitalist entrepreneurs. As has already been seen, the colonial period produced a sugar monoculture, which was export orientated, controlled by a plantocracy, and served by an enslaved labour force. The profitability of the plantation system before the abolition of slavery and the fall in cane sugar prices, brought about by the rapid expansion of beet in the last quarter of the nineteenth century, meant that imports of all the necessary manufactured products could be afforded. There was no foreseeable motive for agricultural diversification. Given the physical conditions of the coastal plain, it was argued that only large-scale, heavily capitalized operators could make sugar pay. Gradually sugar came to dominate the entire economic productivity of the colony – that is to say, productivity measured in terms of return on capital investment, or purchasing power, rather than in terms of workers' satisfaction with their standard of living. Since, in addition, only two foreign companies controlled the entire sugar industry in Guyana until nationalized in the 1970s, it is hardly surprising that anti-colonial sentiment was widespread before independence.

Before then, however, attempts had been made to reduce the relative role of sugar in the economy, not by reducing production, but by widening the agricultural base and stimulating the growth of other industries. One of the earliest moves to diversify agriculture came in the 1940s with schemes to expand the area devoted to rice production: between 1955 and 1965 rice cultivation increased from 68,000 to 108,000 ha and, after a slow-down in growth, rose again from 1970. The Guyana Rice Board aims to raise the area planted to rice to some 200,000 ha by the mid-1980s, and to double productivity. Since Guyana does not consume the rice it grows, the staple has become an increasingly important export: already (1980) sugar and rice account for 50 per cent of foreign-exchange earnings and 19 per cent of GDP.

Other rural developments in the past 20 years have been the expansion of coconut plantations, and the improvement of stock and pasture for livestock-ranching in the Rupunini savanna region, where the spread of *pangola* grass and the eradication of foot-and-mouth disease have been important factors.

But, next to sugar and rice, bauxite has been developed as Guyana's other major product. The resources are vast, and Guyana currently produces about 20 per cent of total world output. Mining began in 1914 when George B. Mackenzie acquired bauxite-bearing land on the Demerara river, though it was 1961 before an alumina plant began operating. The bauxite-mining industry has expanded rapidly in recent years: by 1980 bauxite and alumina accounted for about 40 per cent of foreign-exchange earnings and 20 per cent of GDP. Formerly foreign-owned, the bauxite industry, like sugar, was taken into state ownership in the late 1970s. But mining, a capital-intensive industry, cannot help solve Guyana's massive unemployment problem – in 1978 50 per cent of school-leavers could not find jobs – and, as sugar and rice depend on the vagaries of climate and their value as exports on world markets, so bauxite and alumina depend on demand from the developed world. Guyana is no longer British Guiana, but its dependence economically on a small range of primary products persists.

Guyana is today a socialist state, the government controlling 80 per cent of the economy, and co-operatives and private industry 10 per cent each. The period leading up to independence in 1966 and for some years after was marked by considerable racial conflict (between those of African descent, some 37 per cent of the total, and the East Indian population, about 56 per cent), by political struggle, then strongly aligned on racial allegiance, and by personal rivalry between Dr Forbes Burnham and Dr Cheddi Jagan. It was they who led the People's Progressive Party which secured independence, but their subsequent quarrel and Burnham's ascendancy has kept Guyana's politics in a turbulent state. Economically, Burnham, who has retained power through elections which his opponents have denounced as fraudulent, and who in 1980 introduced a new, presidential constitution, has moved progressively closer to Cuba both in their interstate relations and in his model of development. He has only been able to do so, however, with the help of massive aid and loans from international institutions and private banks. The country now has a huge foreign debt, the servicing of which will put a further constraint on Guyana's future development.

Guyana has large, unexploited resources, but also large problems. Apart from the internal scene, the dispute with Venezuela over the latter's claim to the Essequibo region (over half of Guyana's territory) has been carried over from the colonial period and will not easily be resolved.

The Guianas in perspective

An almost inevitable suggestion to solve the common problems of the three political units considered above is some form of economic or political union. But geographical proximity has to be set against historical ties. It is evident that, generally speaking, the Guianas display similar features: empty, exotic interiors (Hanif and Poonai, 1968); fascinating Amerindian cultural relics (Hurault, 1963); rapidly expanding population; over-grown urban concentrations peripherally located with respect to their potential hinterlands. Yet many of these similarities disappear if a stronger lens is placed before the eye. Differences in attitudes and ideals may be discerned, and contrasts in landscapes, dress, language, diet and religious belief can be identified. The old Amerindian regional term no longer serves a useful purpose: the 'land of many waters' has given way to a land of many peoples.

Notes

[1] It should be noted that here the term 'Guianas' is used to describe collectively the three political units of Guyana (formerly British Guiana), French Guiana and Dutch Guiana (Surinam). Complexities arise when one meets French and Spanish terms. In French the 'Guianas' are Les Guyanes, and French Guiana is La Guyane Française (occasionally Cayenne); in Spanish 'La Guayana' refers to the Venezuelan portion of the Guiana shield, Las Guayanas to the 'Guianas'. The most recent addition to the regional terminology is Guayana Essequiba – namely, that portion of Guyana presently claimed by Venezuela.

Bibliography

AUTY, R.M. (1968) 'The demerara sugar industry, 1930–1965: a geographical study of rural change', unpublished MA thesis, University of Toronto.

AUTY, R.M. (1980) 'Transforming mineral enclaves: Caribbean bauxite in the 1970s', *Tijds. Econ. Soc. Geografie*, 71, 169–79.

CUMMINGS, L.P. (1964) 'The name Guiana: its origins and meaning', *Journal of British Guiana Museum*, 38, 51–3.

CUMMINGS, L.P. (1969) *Geography of Guyana*, London.

DEVÈZE, M. (1968) *Les Guyanes*, Paris.

DEW, E. (1979) *The Difficult Flowering of Surinam: Ethnicity and Politics in a Plural Society*, The Hague, Boston and London.

EDEN, M.J. (1964) *The Savanna Ecosystem – North Rupununi, British Guiana*, Montreal.

EIJK, J.J. VAN (1954) *De landschappen van noord Suriname*, Paramaribo.

GEIJKES, D.C., LEENTVAR, P. and VAN DONSELAAR, J. (1965) *Biological Brokopondo Research Project, Surinam. Progress Reports*, Utrecht.

GRITZNER, C.F. (1964) 'French Guiana penal colony: its role in colonial development', *Journal of Geography*, 63, 314–19.

HANIF, M. and POONAI, N.O. (1968) *Wildlife and Conservation in Guyana*, Gainsville.

HAUGER, J. (1957) 'La population de la Guyane française', *Annales de Géog.*, 66, 509–18.

HAWATH, J.J. (1956) 'The economic–geographical structure of Surinam', *Tijds. Econ. Soc. Geografie*, 47, 165–6.

HURAULT, J. (1963) *Les Indiens de Guyane française*, The Hague.

KOOL, R. (1956) 'Paramaribo: het economische leven van een stad in tropische land', *Tijds. Econ. Soc. Geografie*, 47, 276–88.

KRUIJER, G.J. (1960) *Suriname: en zijn buurlanden*, Meppel.

LIER, R.A.J. VAN (1949) *Samenleving in een grensgebied: een sociaal-historische studie van de maatschappij in Suriname*, The Hague.

LOWENTHAL, D. (1950) 'An historical geography of the Guianas', unpublished MA thesis, University of California.

LOWENTHAL, D. (1952) 'Colonial experiments in French Guiana', 1760–1800', *Hisp. Am. Hist. Rev.*, 32, 22–43.

LOWENTHAL, D. (1960a) 'French Guiana: myths and realities', *Transactions of the New York Academy of Sciences*, 523–40.

LOWENTHAL, D. (1960b) 'Population contrasts in the Guianas', *Geogr. Rev.*, 50, 41–58.

McCONNELL, R.B. (1968) 'Planation surfaces in Guyana', *Geog. J.*, 134, 506–20.

MANLEY, R.H. (1979) *Guyana Emergent*, Boston.

NEWMAN, P. (1964) *British Guiana: Problems of Cohesion in an Immigrant Society*, London.

OJER, P. (1966) *La formación del Oriente Venezolano*, Caracas.

OUDSCHANS DENTZ, F. (1943) *De Kolonisatie van Guyana*, The Hague.

PAPY, L. (1955) 'La Guyane française', *Cahiers d'Outre Mer*, 8, 209–32.

QUELLE, O. (1951) 'Die Bevölkerungsentwicklung von Europäisch-Guyana: eine antropogeographische Untersuchung', *Die Erde*, 3, 366–78.

RAGATZ, L.J. (1928) *The Fall of the Planter Class in the British Caribbean, 1763–1833*, London.

ROBINSON, D.J. (1967) 'Geographical change in Venezuelan Guayana, 1600–1880', unpublished Ph.D. thesis, University of London.

SANDERSON, I.T. (1969) *Surinam*, New York.

SINGH, J.L.R. (1967) 'A geographical study of rice in Guyana', unpublished MSc essay, University of London.

SMITH, R.T. (1962) *British Guiana*, London.

STRACHAN, A.J. (1980) 'Water control in Guyana', *Geography*, 65, 297–304.

SWAN, M. (1958) *The Marches of El Dorado*, London.

VERKADE-CARTIER VAN DISSEL, E.F. (1937) *De mogelijkheid van land bouwkolonisatie voor blanken in Suriname*, Amsterdam.

7 · The Central Andes

Clifford T. Smith

In common with other Latin American countries, Ecuador, Bolivia and Peru share an overwhelming concern with the problems of poverty and under-development. The central Andes as a whole is the poorest major region of South America, at least in terms of average income per head. National averages are low, but they conceal great regional differences within each of the countries concerned. As in other Latin American countries, the sharpest gradients in living standards are those between certain sectors of the major cities, but nowhere else, perhaps, is there so brutal a contrast between one region and another. In Ecuador the commercially orientated and expanding agriculture of the coast can be set against the poverty and subsistence farming of the sierra and the emptiness of the *oriente*; in Peru the contrast between the coast and the sierra in terms of urbanization, living standards and productivity is even more emphatic; and even in Bolivia the grinding poverty of the *altiplano* has to be measured against the modest prosperity of the *valles* and the opportunities of the *oriente*.

Harsh and abrupt contrasts in regional income and economic development are, in part, related to the existence of large Indian populations in the inter-montane basins of the Andes. The problem of integrating the Indian into the economic, social and political life of the nation has been a recurrent theme in the history of all three Andean countries, and in none of them can it yet be said that these problems have been resolved. The presence of so large and compact a block of Indian population is a distinctive characteristic of the central Andean countries. As in other parts of Latin America agrarian reform is at present a dominant issue, but the problems posed by the maldistribution of ownership, anti-social land tenures and the need to raise agricultural productivity are rendered more acute by the harsh difficulties of the Andean environment and the Indian problem.

Finally, all three of the central Andean countries contain within their boundaries some share of the undeveloped interior lowlands of South America and, for Peru and Bolivia at least, the colonization and settlement of new land in the east has become an important issue since 1950, all the more so since the opening up of new land seems to present opportunities for relieving local overcrowding in the sierra while diverting attention from pressures aimed at the breaking up of the *latifundia*. But within each country the role of the eastern lowlands is quite different: Ecuador has a similar but more accessible environment in its coastal zone; for Bolivia, with no Pacific coast, the east is the only part of the national territory that can produce tropical foods and raw materials; and in Peru the role of the *selva* is more complex and more controversial.

Environmental contrasts

The coast

The coastal zones of Peru and Ecuador are the most important economic regions of their respective countries, by reason of their accessibility as much as by their inherent resources. Yet coastal Peru is one of the most arid regions in the world. Lima, for example, has an average annual rainfall of only 41 mm, much of it falling as light drizzle (*garúa*) during the winter months when low stratus cloud and fog normally envelop much of the coastal region. Aridity is the product of circumstances similar to those found in other tropical west-coast countries, such as south-west Africa, southern California or Western Australia, and is closely associated with the existence of cold water off-shore and location with respect to tropical high-pressure cells. Off Peru two cold currents are distinguished: the outer or oceanic current, varying in temperature from 14° to 18°C from northern Chile to northern Peru; and the inner or Peruvian coastal current, which is colder, moves north or north-west at a faster rate, and is a product of the upwelling of deep, cold, bottom-water to the surface. Subsiding air from the south Pacific high-pressure cell, already cool and stable, is further cooled as it passes across cold coastal waters, producing a deep inversion layer of cool and humid air that laps up against the barrier of the Andes. Temperatures in the coastal region are anomalously low for the latitude (Lima at 12°S has a mean annual temperature of 19°C compared with 25°C at Bahia, 13°S); relative humidities are high (averaging 87 per cent at Lima) and, during the southern winter, overcast skies are common in the coastal region, though there are important regional variations, winter cloudiness decreasing away from the coast itself and northwards towards the Equator. During the southern summer insolation is sufficient to disperse clouds in-shore, and weather is warmer, though still humid. But even during the summer convectional development over warm land is inhibited by the permanence of the temperature inversion, thus preventing heavy rainfall in all but those most exceptional years when cold upwellings off-shore are weak, warmer water off-shore permits the development of unstable air, convectional rainfall occurs and disaster comes to the coastal region, adjusted as it is to arid conditions.

One other aspect of the coastal ecology has been of fundamental importance to Peru in historical and modern times. Cold, upwelling bottom-water is rich in mineral nutrients and supports a dense population of plankton, which in turn supports a fish population of enormous quantity and variety. Under the off-shore cloud blanket, diatoms thrive better nearer the surface than under sunlight so that there is an abundant food supply for pelagic species of fish, which in turn support abundant bird life: cormorants, guanay, pelicans, gannets and many other species follow the shoals of fish, but the most important from the point of view of Peru's development have been the species that nest in great communities reaching densities of 130,000/ha on the twenty or more off-shore islands and the isolated peninsulas of the coast. Vast thicknesses of bird droppings, rich in nitrogen (14 to 17 per cent) and phosphates (*c.* 11 per cent), have accumulated to form workable deposits of guano, highly valued from early times as a fertilizer. Just as the fishing industry has been a major activity and source of external revenue in recent years, so the guano deposits were the resource that drew Peru into the mainstream of international commerce in the nineteenth century.

The possibilities of the Peruvian coast for agriculture clearly depend on irrigation and therefore on available water supply in relation to the extent of suitable soils and gently sloping terrain. The supply of land suitable for irrigation varies in each of three morphological regions. In north-western Peru relief is low and tabular with much faulting; much of the area is covered by sands of recent origin; and fairly level land suitable for irrigation exists if water can be brought to it. In the central coast as far south as Ica the sierra penetrates to the coast in many places. Valleys are separated by the lower spurs of the Andes, which often end in a magnificent cliffed coast. Narrow, steep-sided Andean valleys debouch in a series of alluvial aprons containing thick infill brought down during fluvial/glacial episodes in the high Andes. Modern streams are slightly incised into this infill, and in some areas alluvial aprons coalesce to form more-or-less continuous, gently sloping land near the coast. But in general potentially irrigable land is limited. Surplus water from one valley can sometimes be transported by canal to a neighbouring area of deficiency, as, for example, the water from the Pativilca river is carried over to the Fortaleza valley. In southern Peru the character of the coastal region changes. A coastal range of igneous rocks is exposed in the Ica region and continues southwards into Chile. The trough between the coastal massif and the main Andean range is filled with sediments and volcanic rocks. Streams break through the coastal massif in narrow gorges, but upstream have incised themselves deeply into the softer rocks, carving flat-floored and irrigable valleys. In the south, then, irrigable land is scarce, and in contrast to the northern and central regions it is also, for the most part, remote from the coast itself and was fairly inaccessible until the building of the Pan-American Highway in 1940.

Water supply is the other major limiting factor in coastal agriculture and is governed by the regime of Pacific-flowing drainage deriving from the high Andes. In central and northern Peru the high sierra above 3000 m is moderately well watered, but rainfall decreases southwards, and the western flanks of the Andean ranges are increasingly arid. Arequipa, for example, receives only 115 mm of rain annually. Precipitation is also seasonal, with a wet season from October to March and a marked dry reason in June, July and August. River regimes faithfully reflect this pattern, and streams fed solely by rainfall frequently dry up completely during the southern winter. In general it is only the streams nourished by snow-melt, glacier ice or natural lakes that have a substantial winter flow. Of these the Rimac and the Santa rivers are the most important, the latter being the only river to break through the western cordillera to drain an important interior basin.

In the neighbourhood of the boundary between Peru and Ecuador the coastal zone changes with striking abruptness from the aridity of northern Peru to the hot, wet, equatorial climate characteristic of much of coastal Ecuador. The cold Peruvian current turns westwards and the coastal waters off Ecuador are relatively warm. The south-western part of Ecuador receives a seasonal rainfall from January to April of about 200 mm, and average temperatures are higher than in coastal Peru. But towards the Andes and the north rainfall increases to 1700 mm a year and the rainy season extends from January to June. Temperatures are generally higher, and at Guayaquil, for example, mean temperatures range from 24°C and 27°C between the coldest and warmest month.

The Andes

The scale, majesty and complexity of the Andean ranges, overwhelming when seen from the air, are no more than dimly suggested by the contours of an atlas map. Some 120 km wide in Ecuador, the Andean region spreads out southwards to a zone some 800 km across in Bolivia; its highest peaks, rising to over 6000 m, have fired the imagination of travellers and have attracted mountaineers since the time of Whymper's exploits in the volcanic peaks of Ecuador in the nineteenth century.

The Andean countries are built around the axis of the great Cretaceous batholith of acidic igneous rocks, which forms a large part of the western cordillera of Ecuador, Peru and Bolivia, though it is frequently obscured in southern Peru, western Bolivia and in Ecuador by overlying sedimentary and volcanic rocks. To the east of the Cretaceous batholith, sedimentary rocks ranging in age from Palaeozoic upwards have been folded, faulted and meta-morphosed, and culminate in the high ranges that form the eastern cordillera of Ecuador and the central and eastern cordilleras of Peru and Bolivia. In general the more recent rocks form the most easterly of the folded and faulted ranges that subside into the sediments of the Amazon basin, themselves largely composed of the debris produced by the weathering and erosion of the uplifted Andes.

Structural troughs and basins of tectonic origin reach their maximum extent in the central plateau of Bolivia, are continued into southern Peru in the basin of Lake Titicaca, and are represented further north by the central trough of Ecuador. Similar but smaller and less continuous troughs form zones at less than 3500 m in other parts of Peru, and these are of the greatest importance for settle-ment and agriculture. Most of these tectonic basins, and especially the *altiplano* of Bolivia and southern Peru, are partially filled with relatively undisturbed sediments, lavas and volcanic ash of fairly recent origin; some have formerly supported lakes, which are now represented by areas of lacustrine sediments. The *altiplano* of southern Peru and Bolivia is still, indeed, a region of internal drainage to the basin of Lake Titicaca, Lake Poopo and the saline flats of south-western Bolivia. Varied in size, accessibility and the quality of their soils, the tectonic basins are the agricultural oases of the Andean regions in all but the most arid and saline plains of the south-west, for there are few other zones, except for the steeply sloping sides and narrow floors of deeply incised river gorges, that fall below the altitudinal limits for productive arable farming.

The landscapes of the Andes consist only in part of the high, glaciated massifs rising to over 5000 m or the tectonic basins and troughs. There are, too, spec-tacular volcanic features such as the impressive cones that flank the central trough of Ecuador, including the famous peaks of Cotopaxi and Chimborazo, or the volcanoes of southern Peru and western Bolivia, and volcanic landscapes also extend over much wider areas in the form of lava-capped plateaux and thick deposits of highly erodible volcanic ash and tuff.

Over much of central and southern Peru and in parts of Bolivia there are, how-ever, large areas of rolling relief and occasionally monotonous high plains at approximately 4200–4400 m. In Bolivia and Peru this has been named the *puna* surface after the predominant vegetation of high-altitude tundra, and is probably an erosion surface of sub-aerial origin formed near sea-level before the major uplift of the Andes to their present position. In southern Peru it is flexed downwards towards the coast, declining to *c.* 1000 m; in northern Peru it can be traced at an altitude of about 3600 m, but is much more strongly dissected than further south. But whatever the origin of the *puna* surface, it is certainly one of the great misfortunes of Peru and Bolivia that its altitude is too high for cultivation and too high even for the grazing of sheep on a large scale. The barren, tundra-like pastures of the *puna* are useful only for the grazing of llama and alpaca.

Terrace features and erosion surfaces of more limited extent have been traced at lower levels in Bolivia and Peru, perhaps marking phases in the irregular uplift of the Andes, and these are of local importance in providing moderately level surfaces for arable cultivation, e.g. in the basins of eastern Bolivia. The other major features of Andean landscapes are the river gorges dug deeply into the Andean surfaces, particularly in the eastern and central parts of the area where precipitation and therefore headward erosion from the Amazon basin have been much greater than in the more arid west and south-west. They sometimes offer narrow strips of potentially cultivable land, but more often they present difficult barriers to communication, especially from east to west across the region, for the major drainage trends are parallel with the major strike of Andean faulting and folding. Pacific-flowing rivers rarely penetrate beyond the crest-line of the western cordillera, partly because of the continuity and relative resistance to erosion of the predominantly igneous rocks, but also because of the relative aridity of the western cordillera. The watershed between Pacific and Atlantic-flowing streams lies as near as 150 km to the Pacific coast in some places, and except in northern Peru presents a continuously high barrier to communications.

Yet in a number of ways communications have been easier and less costly between the coast and the high Andes than between the Andes and the eastern plains: the absence or sparsity of vegetation on the Pacific slopes has facilitated easier movement by men on foot, by llama herds or by mule traffic, whereas dense vegetation to the east has always presented a difficult barrier to movement. The short Pacific-flowing streams cut a direct passage from highland to coast, but in the lower eastern Andes the intricate dissection of softer rocks under hot and humid conditions makes for steep slopes and circuitous routes of access, quite apart from higher costs of clearing, construction and maintenance.

It is as difficult to generalize about the climates and vegetation of the Andes as it is about the landforms and terrain. Differences over small distances often overshadow the general tendencies. In Ecuador there is a drier season between June and September and a maximum from November to May. There is a general tendency for precipitation to decrease southwards: Quito has a rainfall of 1000 mm, Cuenca 925 mm and Loja 768 mm. This tendency is continued in Peru, where wet and dry seasons are more strongly accentuated, and the wet season lasts from October to March. Isohyets are roughly parallel with the coast in central and northern Peru, swinging inland in southern Peru and across Bolivia in a generally north-west to south-east direction. South-west Bolivia is truly arid country. But even in central and northern Peru deep Andean valleys are often surprisingly arid, and precipitation is highly variable from year to year in most of the highland basins to the south of Ecuador. Drought is one of the major agricultural hazards. But in the eastern zone of the Andes there is very often a sharp transition from a landscape of semi-arid aspect with worn pastures and xerophytic vegetation to luxurious jungle. In Peru this is the *ceja de la montaña* (the brow of the mountain), a zone of 'cloud-forest', in which precipitation increases sharply and a dense growth of small trees, epiphytes, ferns and grasses may appear up to 3300 m, giving way eastwards to true tropical forest at lower altitudes. In Bolivia it is represented by the sharp transition to the *yungas*, as for example to the north-east of La Paz.

In terms of farming and land use, the effect of altitude on temperature regime is of major importance throughout the region. Mean annual temperatures show the effect of altitude directly and mean annual range of temperature tends simply to reflect latitude, with very low ranges of 1° or 2°C near the equator, increasing southwards to 4.4°C at La Paz. But what is much more important is the diurnal range of temperature. At Huancayo in central Peru, at 3300 m, the diurnal range may be as much as 13°C in the wet season, rising to 19°C in the dry season. Night frosts are regular from May to August and may occur in any month except at the height of the wet season in January to March. Further south in Peru and on the Bolivian *altiplano* the frequency of night frosts is even higher. High diurnal ranges limit the period during which temperatures are above the minimum required for plant growth. The growing season needed by plants, measured in terms of days, is longer than at lower altitudes, and the growing season available is limited by the occurrence of night frosts. The short growing season needed by barley, for example, is a major reason why it is suitable for cultivation at high altitudes. Barley, potatoes and indigenous root crops and grains such as *cañahua* and *quinoa* will thrive up to 4000 m and potatoes even above this; the effective limit for wheat is usually about 3500 m and for maize 300 m lower. In addition to drought, frost at unexpected times is a major hazard to farming, particularly in November and December during the growing season; hail is the third major hazard.

In the Peruvian Andes grasslands of varied composition are the major form of vegetation, and woodland is rare except for the survival of *quinual* in sheltered places up to 4500 m and the occurrence of planted eucalyptus, usually below 3600 m. At high altitudes there is a transition to tundra conditions with mosses, lichens and grasses of poor nutritive value, among which tussocky grasses such as the *ichu* are dominant. In the drier regions of the south the low, tough, xerophytic *tola* shrub is widespread. In Ecuador the high level grasslands are more humid and support a richer flora, including many cacti, and are described usually as the *páramos.*.

The eastern regions
Although there is sometimes a sharp and definitive transition from the Andean front to the plains of the Amazon basin, the outer ripples of the Andean storm have left traces in the form of low ranges well to the east of the main Andean front. The plains themselves are by no means uniformly level. In Peru a distinction is made between the *selva alta* and the *selva baja*. The *selva alta* (400–1000 m) has a substantial range of relief and there are broad, open valleys with good agricultural possibilities on relatively immature soils. Below 400 m the *selva baja* consists for the most part of three terrace-like regions: a marginal

zone of well-dissected terrain; the high terraces, in general some 60 m above the lower terraces, slightly dissected, and frequently capped by leached, mature and agriculturally poor soils; and thirdly, the low terraces (*restingas*), fairly level and undissected, but with considerable micro-relief reflecting the course of former meander scars, and in part liable to flood (Pulgar Vidal, n.d.).

Climate and vegetation show significant variations through the region. In eastern Ecuador rainfall is high and well distributed through the year; further south totals are greatest near the eastern flanks of the Andes, reaching as much as 4000 mm, but there is a rainfall maximum from November to May and a minimum from June to October. The tendency to a wet-and-dry regime is accentuated southwards into south-eastern Bolivia, where total amounts are less, and there is a transition towards the extreme seasonal variations of the Gran Chaco. Forest is the natural vegetation of most of the area, with a broad variation from scrub and thorny woodland south of Santa Cruz to semi-deciduous forest and palm savanna to the true rain-forests of north-east Bolivia, eastern Peru and Ecuador. But even within the rain-forest areas there are often fairly large spreads of savanna grasslands and even thorn scrub, which are only now becoming fully known as a consequence of systematic aerial photography of these empty regions.

Historical antecedents

The physical environments of the central Andean countries are highly contrasted and in many ways difficult for human occupation, apparently destined to consist of very many nuclei of population widely separated by distance and by barriers of high relief, desert wastes or thick forests. Yet political and cultural unity has been repeatedly imposed over large parts of the region, not least in pre-Columbian times during the Inca empire and also during the earlier Tiahuanaco period (*c.* AD 900–1100) and even before this, if only in a cultural sense, during the Chavin period (*c.* 1000–500 BC) The continuing strength and vigour of the Indian population owes a great deal not only to the isolation of scattered communities, but also to the legacy of high Indian cultures and the sheer weight of their numbers in comparison with those of their Spanish conquerors. More than anywhere else in Latin America the social and economic problems of integrating the Indian population into national life have proved difficult and intractable. Some have thought to restore the dignity of the Indian by an appeal to the great traditions of the Indian past, and have considered the Incaic tradition as an important element of distinctive nationality; *indigenismo* is a common theme in the literature of the central Andean countries. But the dominant theme has, of course, been the hispanicization of Indian groups – by the 'formal' processes of conquest, conversion to Christianity, and education, or by 'informal' processes such as racial mixture, conscription, migration to the cities and the penetration of commercialization to remoter regions.

The Indian heritage

When the Spaniards arrived in Peru they found an empire that extended from northern Ecuador through the central Andes into north-west Argentina and central Chile. Coastal and Andean cultures alike had been welded into a highly centralized and despotic empire administered from Cuzco, the capital city and cradle of the Incas. The administration may have been less uniform than was once thought and regional variations considerable, but a remarkable unity had been created in less than a hundred years. An excellent system of communications by road was as important to the Incas as it had been to the Romans in western Europe as an instrument for administrative and military efficiency. Roads were carefully engineered, often with surfaces of stone and rubble, sometimes tunnelled into precipitous slopes, or paved and stepped to facilitate the rapid movement on foot of messengers and armies, since there was no knowledge of the wheel. Suspension bridges crossed rivers and deep gorges, and pontoon bridges were supported by bundles of reeds over broader, shallower rivers such as the Desaguadero at the foot of Lake Titicaca. On the coast Inca roads replaced or supplemented previous systems, and often consisted of walled highways in irrigated areas and of cleared tracks marked with stones in the open desert.

Other mechanisms had helped to create a unity of the Inca empire, notably the system of massive forced-labour migration by which young men of 17 to 30 were liable for military and labour service in any part of the empire; whole populations were resettled or exchanged to ensure the subjugation of recently conquered or potentially rebellious territory. If possible such movements were organized between areas of similar climates and agricultural possibilities: peoples from Ecuador were moved into the Cajamarca region, and there was some settlement in the Quito area by people from the Titicaca basin; loyal subjects from central Peru were moved to southern Bolivia. The means by which a fairly homogeneous empire was created and by which Quechua was spread to regions as far apart as north-western Argentina and Ecuador were often ruthless, but they were effective enough to create among the Spaniards the impression of a monolithic structure, and to make it possible for Spanish rule to replace that of the Incas with remarkable rapidity.

The Inca empire was, however, but the culminating phase of a long cultural development in the central Andes in which intensive use had been made of potential resources on the coast and in the sierra. It is an area of major world importance as a hearth of domesticated plants and animals, though some cultivated plants were undoubtedly introduced from Central America, and there is still room for doubt as to the precise location of domestication of others. Squash and beans, with chili peppers, cotton and various indigenous fruits, were the basis of the earliest agriculture in the coastal region. The introduction of maize, probably from Central America *c.* 1400 BC, ushered in a great expansion of cultivation and was also associated with the cultivation of other new crops such as peanuts, avocado pears and, later, manioc or *yuca*. In the highland regions *quinoa* and *cañahua*, grain crops still of considerable importance in the region, were first cultivated between about 300 BC and AD 200. The potato must have opened a new range of opportunities for cultivation throughout the high Andes, just as it was to do later in western Europe in the seventeenth and eighteenth centuries. Another wave of domestication took place during the period of cultural and demographic expansion approximately dated from the first century AD to 600–900. Sweet potato, *papaya*, the pineapple and other crops were added to the agricultural repertoire in coastal and warmer regions, and root crops such as *mashua*, *oca* and *olluco* appeared in the highlands. Finally, the domestication of the llama and the alpaca was probably an accomplished fact by 1000 BC, probably originating in the Lake Titicaca region where there is the greatest variety of wild and domesticated types. Grazing of llama and alpaca gave value to high, natural pastures, which were useless in other parts of pre-Columbian Latin America.

The rural landscapes of the central Andes still bear the imprint of the painstaking effort of pre-Columbian Indian peoples to make the maximum use of resources in a difficult land. Water control was carried to a high pitch of efficiency, particularly in the desert coast. Archaeological studies of coastal Peru have traced the elaboration of water-control systems and their close association with developing socio-economic patterns of settlement and regional control. Early settlement was scattered, sparse and uncontrolled, depending on shallow water-tables for the cultivation of beans and squash, but controlled irrigation made its appearance with the cultivation of maize, initially on a small scale and near river courses or the highland margins. Water control on a much larger scale came later, involving the improvement of water distribution to cultivation plots, the building of stone-faced terraces, and large-scale canal-building, which involved the territorial organization of whole valleys as single units. The elaboration of irrigation systems continued to about AD 900 when their maximum extent may have been reached in some of the coastal valleys, and when the cultivated area of individual valleys may have reached limits not yet surpassed, or exceeded only since the mid-twentieth century. In the Chicama valley the pre-Columbian canal of La Cumbre, some 120 km long, is still used to irrigate part of the valley.

Less is known of the extension of the cultivated area in the sierra in regions where rainfall is adequate for farming. Massive and elaborately constructed terraces are found in close association with some of the most important Inca sites, faced with worked stone, equipped with staircases and stone canals for the distribution of water, and with level 'treads' even on steep slopes. Such terraces are to be found at Pisac or Machu Picchu, for example, and they seem to be associated with Inca garrisons and fortresses. But terraces of a more informal and irregular character are widely spread in many other areas around the margins of Lake Titicaca on terrain of moderate to steep slopes; in the Bolivian *altiplano*, and in many of the upper sections of the coastal valleys of Peru. In the sierra no less than on the coast, the existence of terracing points to a carefully controlled attempt to maximize the area under cultivation, to conserve water and soil, and to secure successful harvests at the highest limits of the altitudinal range for particular crops.

Similar preoccupations are also to be seen in the practice of ridged cultivation on flat, badly drained land in the neighbourhood of Lake Titicaca, where flat-topped ridges were artificially raised above flood-level for cultivation. In all, some 82,000 ha of land organized in various patterns of ridged cultivation can still be traced in the area, again pointing to a substantial pressure of population on the arable resources of the region in pre-Inca times (Smith, Denevan and Hamilton, 1968). Elsewhere in the central Andean region similar features occur near Guayaquil and in the Mojos region of north-east Bolivia, which were both areas outside or marginal to the Inca empire, possessing dense population and well-developed cultures before the arrival of the Spaniards. Both are regions liable to seasonal flooding (Denevan, 1966).

The organization of agriculture and of rural society was much more varied in the Inca empire than is suggested by the simplifications of some of the post-Conquest witnesses. The conventional view is that land was divided among the community, the Inca and the Sun, but local chiefs (*curacas*) sometimes held land on their own account, and many chiefs retained land even after their conquest by the Incas. Peasant communities were organized into *ayllus*, which had a territorial connotation and within which the extended patrilineal family was the most important kinship group. It is, however, difficult to say how far land was collectively operated. The periodic redistribution of land within the *ayllu*, often noted by chroniclers, may not have been widespread. Individual ownership of at least the usufruct was normal near Lake Titicaca in the 1560s, annual redistribution was not practised, and peasant holdings were inherited by surviving children, and only redivided among the *ayllu* if there were no surviving issue. Grazing of llama and alpaca was also organized in *ayllus*, but in addition to communally owned flocks there were others individually owned by chiefs and peasant households. But all grazed freely on land held for common use. Finally, the *ayllus* were also units within which systems of mutual labour aid were operated, and although this is a custom that has survived to the present, it is constantly threatened by increasing commercialization and growing economic differentiation within peasant groups.

In an indirect way the *ayllu* still has a role to play. Modern *comunidades* are numerous in all of the central Andean countries. In Bolivia there are about 4150, in Ecuador *c.* 1200, and in Peru some 1600. They have a corporate legal entity and most have common grazing lands, the defence of which against encroachment is often a major *raison d'être* of the *comunidad*. Some, but by no means all, have common property, including arable land, and a few, such as Muquiyauyo in central Peru, have made considerable economic progress. Some authors have seen the *comunidad* as a lineal descendant of the *ayllu*, and also as a hope for the development of co-operation in farming and thus as a potential institution making for economic growth. Both of these views are debatable, for many *comunidades* can trace their existence no further than the beginning of the twentieth century, but the *idea* of a lineal descent is important in so far as it encourages a spirit of local cohesion and collective aspiration.

The indigenous heritage thus persists in the rural landscape and in rural tradition, but one of the most important achievements of the central Andean cultures, the development of an urban society, has left few comparable traces except at Cuzco and in the ruins of ancient cities. Urban evolution has been traced from the grouping of rural settlements in a fairly clear relationship to ceremonial and pyramidal mounds to the more complex proto-urban structures of Tiahuanaco times, which were much more than gathering places for defence or religious celebrations, since they had acquired administrative functions and were centres for the collection of tribute in kind, and probably in labour, from surrounding rural areas.

In the coastal regions the great phase of the city-builders which followed early expansion occurred about 250 years before the Inca conquest, when the coastal empires were ruled by urban élites who created for themselves elaborately planned cities, generally rectangular in basic conception with symmetrically arranged streets and canals. These, too, were often segmented into large, walled compartments containing gardens, pyramids, reservoirs, storage bins, palaces and houses. Chanchan, the largest of the coastal cities and capital of the Chimú empire of the northern coast covered an area of some 15 km². But it was, of course, Cuzco, heart of the Inca empire, that was the greatest city of them all, perhaps containing a population of 50,000 at the time of the Spanish Conquest, and built of stone with that meticulous attention to craftsmanship and finish that can still impress the modern visitor. Temples, palaces and fortresses were the main foci around which were grouped craftsmen, warriors and priests, as well as the massive drafts of forced labour brought in from the provinces to serve in construction, military service, domestic tasks and the like.

When the Spaniards arrived, therefore, they were confronted by a numerous population, highly skilled in the agricultural arts as well as in the production and working of precious metals, and, above all, highly urbanized and socially stratified, ruled and exploited by an urban élite. And it was the urban élite, the religious and administrative structure, that was the main target of the Spanish Conquest; the indigenous urban tradition was replaced by the Mediterranean urban culture of the Spaniards.

Conquest and the colonial regime

The conquest of Peru by Pizarro and his men in 1532 ushered in a new phase in the developing geography of the central Andes. The Spaniards approached the Andean environment with a different perception of its resources, new attitudes to the exploitation and control of labour and new techniques in farming, industry and trade; above all, they brought contacts with a wider world and new, devastating diseases. A new regional balance emerged in which neither coastal Peru, as in the time of the city-builders, nor the Cuzco region, as in the time of the Incas, but the capital city of Lima and the mineral deposits of the Viceroyalty of Peru and especially Potosí were the major foci around which development took place.

It may well be that diseases of European origin had preceded the Spaniards in the Inca empire and that smallpox, arriving between 1524 and 1526, had already been responsible for great losses of population. High mortality among the Indians continued at least until the 1560s, most probably as a result of influenza, measles and smallpox, as well as of the disturbances due to conquest and civil war. In 1571, when 311,257 Indians paid tribute from an area corresponding roughly to the coast and sierra of Ecuador, Peru and Bolivia, the total population was perhaps 1.5 million. Guesses at the pre-Conquest population range from some 6 million to between 30 million and 35 million, but an estimate of the order of 9 million to 12 million is probably nearer the truth (Smith, 1970), a figure not again reached until the present century. Depopulation was probably most drastic in the coastal region. In the sierra numbers declined less catastrophically. The coastal regions, formerly the site of powerful empires and flourishing cities, had less than 10 per cent of the Indian population by 1571, and chroniclers remarked again and again of the shortage of Indians, the ruined buildings and the abandoned fields.

As elsewhere in Latin America, conquest was followed immediately by the foundation of towns. They were symbols of imperium, centres of administration, foci for the work of the church, centres from which *encomenderos* commanded the labour of Indians under their control, and centres to which flowed the produce of the countryside and the labour of its inhabitants. As regional centres they had a reciprocal relationship of sorts with the countryside, though the flow of goods and services was heavily unbalanced in favour of the Spaniards in the towns, with little tangible flow in the reverse direction. From the beginning the towns were the instrument by which the *conquistadores* and their successors were able to establish themselves as citizens (*vecinos*) and thus to lay claim not only to urban property for houses and gardens, but also to substantial agricultural holdings and common rights of grazing in the land that was allotted to the newly founded township. Given that good arable land was so scarce in the central Andes, a careful choice of town site may frequently have allowed the annexation of a significant proportion of available arable land in many districts.

In the central Andes poor initial choice of town sites was often easily corrected in the sixteenth century when buildings were still makeshift and little investment had yet gone into the embellishment of churches and public buildings. In some cases the initial site was unhealthy, in others movement was from cold, high-altitude sites down to warmer climates and more abundant agricultural land. Huánuco el Viejo, for example, was founded in 1539 near an Inca city, but was soon abandoned because of the civil wars, the cold climate and the lack of wood. It was refounded in 1543 in the attractive and fertile valley of the Huallaga several thousand metres below. Huamanga (Ayacucho), Ica and Arequipa were other major settlements that were shifted for various reasons. The capital of the Viceroyalty was initially to have been at Jauja at *c*. 3000 m in the rich and densely peopled Mantaro basin, but accessibility to the sea was regarded as more important, and in 1534 the capital was shifted to Lima on the river Rimac in the midst of fertile irrigated land equipped with a supply of Indian labour nearby – a decision of the utmost importance for the subsequent development of the central Peruvian Andes.

Few important colonial towns in Peru were founded directly on Indian centres of importance (Cuzco, Cajamarca and Quito were the major ones), though the proximity of Indian labour supplies was essential, and many towns were founded near to former Indian cities: Trujillo near Chanchan, Lima near Cajamarquilla, Ica near Tambo Colorado. By the end of the sixteenth century an urban pattern had been established, which has remained remarkably constant. Very few towns of modern importance were founded after the end of the sixteenth century, and, although all of them were tiny by modern standards, there has been relatively little change in the *ranking* of important places, except for the changing fortunes of mining centres and the growth of La Paz, Guayaquil, Iquitos and Huancayo.

Agrarian life was fundamentally changed by the Conquest, most radically in the coastal region, much more slowly in the sierra. New crops, domesticated animals and agricultural techniques were introduced and assimilated to Indian systems of farming. There was some specialization of production for the market. And there were basic changes in the structure of landholding and the patterns of rural settlement.

In the coastal zone of Peru a new agriculture was created. Land was available for the creation of Spanish estates, partly through the disposal of municipal land in the neighbourhood of Spanish town foundations, and partly because the extent of Indian depopulation must have meant that much irrigable land had gone out of cultivation. The coastal valleys were accessible to markets for agricultural produce not only in Lima and Guayaquil, but also further afield in other parts of the west coast of Latin America. Spanish farming enterprise was attracted, and by the 1620s there were at least a few Spaniards with farms in almost all of the coastal valleys, though very few of them possessed *encomiendas*. Indian labour was brought down from the sierra, but negro slaves played an important role on coastal plantations, serving to change significantly the racial composition of the region.

Maize was quickly adopted as a staple grain by the Spaniards, but wheat was grown, and, although the high humidities of the coastal climate must have made wheat liable to wilt and rust, flour and biscuit were exported to Lima and Guayaquil. Alfalfa was widely introduced as a fodder crop, nourishing the mule trains used for coastal transport from Paita to Lima and for transport into the sierra. But the major specialities of the coast were sugar and vines, both introduced into Peru before 1550. Sugar did well in the upper parts of coastal valleys where water supply was good and insolation higher than in lower zones, but by the end of the sixteenth century the northern coast near Trujillo and in the Chicama valley had acquired the reputation for sugar production that it still enjoys. The regions around Ica and Pisco were already well known for their vineyards and for their wines, which are still a speciality of the area. Wine was exported along the west coast and to the sierra mining regions as far afield as Potosí.

Change was, perhaps, less radical in the coastal regions of Ecuador, but cacao plantations were set up by the residents of Guayaquil for trade along the west coast, though they were less carefully tended than those of Central America, and the major industry was the extraction of timber for export to Lima and for the local shipyards, for this was the last point southwards at which timber was available until central Chile was reached. Henequen was cultivated for the manufacture of cordage, and the area was already known for bananas and avocado pears.

In the sierra Spanish settlement was much more limited to the towns, except for some rural settlement in the Mantaro valley and in the Callejón de Huaylas. Most agriculture outside the towns remained the responsibility of the Indians, who served their Spanish masters either by the provision of service labour in *encomienda* or as *mita* (the institution of forced labour taken over from Inca practice), or by the payment of tribute in kind or in cash. Under such circumstances, agricultural change and the introduction of European crops and stock were highly selective. European crops had to find their niche, for example, in the careful altitudinal zoning of traditional Indian farming. Thus, barley could fit into the high altitude crop range as a supplement to potatoes, *cañahua* and *quinoa*, chiefly because of its short growing season, and it was slowly assimilated into Indian farming up to altitudes of 4000 m. Wheat was a bread grain in demand by Spanish settlers and fitted well into the system between 3000 m and 3500 m, where it yielded at least as well as maize. By the early seventeenth century wheat had spread fairly widely in the inter-Andean basins of Ecuador, in northern and central Peru, and in the *valle* region of Bolivia near Cochabamba and Sucre. But in general the Indians remained faithful to their ancient practices and, even where Spanish demand for European crops induced change through higher prices or through the exaction of tribute, there often developed a dual agriculture by which Indian crops were grown for subsistence and European crops for exchange.

Sheep were widely and rapidly accepted by the Indians, and even in the 1560s were increasing in numbers in areas such as southern Peru, where there was little Spanish settlement. Sheep could graze at altitudes almost as high as the llama and were easy to rear and more productive of wool. Their unit cost was low, and they supplied raw material not only for domestic cloth production along traditional lines, but also for the woollen mills of northern Peru or Pasto and Riobamba in Ecuador. The rearing of cattle and horses remained much more the prerogative of Spanish-controlled estates. The plough, too, was only slowly accepted by Indian farmers, and its use is even now by no means universal. It was regarded very much as a concomitant of hacienda agriculture and of mestizo rather than Indian farming. The plough was unsuitable for the cultivation of very steep slopes, and its use also implied investment in stock and an attendant revolution in the rhythm of traditional farming, which Indian communities were unable or unwilling to undertake.

The evolution of the hacienda was one of the most important features of the colonial regime to have an enduring effect on the life of the central Andes. In early colonial times it was control over labour rather than the possession of land that mattered and this was achieved by the system of *encomienda*, by which Spaniards were granted the right to labour services in various forms from the Indians of specified areas or villages. In the central Andes most *encomiendas* were relatively small, and there were no less than 538 by 1570. Initially a right to labour, the *encomienda* soon became a right to tribute and then to a money payment, so that there was, in principle, no transition from *encomienda* to land-ownership as such. The annexation of land to newly founded townships formed the nuclei of some Spanish estates, and the crown also had the right to dispose as it wished of the lands formerly belonging to the Inca, the lands of the Sun and waste land or abandoned lands. The decline of population must have been accompanied by abandonment of cultivated land, and the operation of the *mita*, the forced labour system by which a proportion of Indians were liable to be drafted from their native villages for labour in the mines, in construction work, in factories and in the towns, often resulted in further abandonment of cultivated land by Indians who never returned to their native villages. And it may well be that much land was granted, or simply occupied, which was properly the unenclosed grazing land used, or formerly used, by Indian communities. And some may have been fallow for long periods under the Andean system of shifting cultivation. At all events, the hacienda was firmly established as the dominant form of landownership in many parts of the central Andes by the end of the colonial regime, though important stages in the further concentration of holdings were yet to take place in the nineteenth century.

It was chiefly as a storehouse of precious metals that Spain and the rest of the world regarded the Viceroyalty of Peru. Once the riches of the Inca empire had been looted, the central Andes were ransacked for gold and silver ores. Widely scattered mining centres attracted Spanish and, later, Portuguese miners and supported urban settlements, local agricultural specialization and route networks to serve as supply lines. All were based, of course, on Indian labour. Quito had its sources of precious metals, and provincial centres such as Cuenca and Zaruma exploited gold and silver. Rich sources of alluvial gold stimulated settlement in north-eastern Peru near Moyobamba and Chachapoyas; in central and southern Peru the Callejón de Huaylas, Castrovirreyna and Carabaya to the north of Lake Titicaca were centres of mining for silver or gold. In Bolivia Oruro was founded in 1607 shortly after the discovery of silver there, and a few years later had more than a thousand Spaniards.

But it was Potosí that stood out like a giant among the producers of precious metal. Silver ores were discovered in 1545 in the ridge of Potosí at an altitude of over 5000 m. By 1570 it had a population of about 120,000 and was by far the largest settlement in the whole of the Americas, though its Spanish population was not much greater than 4000 at the height of its prosperity. The silver ores were rich, but at such an altitude that the environment presented difficulties for mining, which were only overcome by incurring heavy costs in life, labour and capital. Lack of wood meant that grass was used as fuel and the area was devastated for many kilometres around. The short rainy season meant that a complex series of reservoirs was needed to supply water-power for crushing mills. *Mita* labour was drawn from as far afield as Cuzco. From 1553 a mercury amalgam was used in Mexico to extract silver more efficiently, and 10 years later cinnabar ores were discovered in central Peru at Huancavelica. The mercury ores of Huancavelica rapidly assumed such strategic importance in relation to Potosí that the mine was taken over by the crown. Potosí and Huancavelica became 'the poles on which this kingdom is supported'.

The impact of Potosí on the economic geography of the central Andes and, indeed, on the whole of Spanish America, was considerable. Lima was the greatest single beneficiary as the centre of administration, because the official routes for the transport of silver and for the supply of imported European goods passed through Lima and then by sea to the isthmus of Panama. The route from Potosí to the coast by way of Arequipa or Arica stimulated the production of foodstuffs and of fodder for the mule trains. Wines and sugar were imported from the Peruvian coast. Cattle trails linked Potosí with Chile and north-western Argentina; and the illegal 'backdoor route' through Buenos Aires, of increasing importance through the seventeenth century, foreshadowed the reorientation of the area towards the Atlantic. But within the central Andes it was Potosí (and Oruro), through its demands for foodstuffs and manufactures, which stimulated the growth of agriculture, population and trade in the basins of central Bolivia at altitudes low enough to produce wheat, fruits, meat, sugar and the like. Cochabamba and Tarija grew, but it was Chuquisaca (Sucre) that reaped the greatest benefit, since it was the most important basin near to the wealth of Potosí. The colonial architecture of Sucre still hints at the former wealth it enjoyed.

Within a hundred years of the Conquest major changes had been wrought in the geography of the central Andes. The sierra was still largely Indian with a small, largely urban minority of Spaniards and mestizos; the coast had experienced a major structural change as a result of the catastrophic decline of its Indian population and the growth of Spanish, negro and mestizo groups. A hierarchy of new towns had been created, dominating the countryside and linked by networks of trade and communications which had little in common with the former patterns of the Inca empire. The impact of colonial society on land-use systems and agrarian structures was by no means fully worked out, but the foundations had been laid. Not until the second half of the nineteenth century were new changes of equivalent magnitude to be brought about, and only then as a result of the impact of modern industrial demands from western Europe and North America for raw materials and foodstuffs, and it is to the implications of this later movement that we must now turn.

Modern development

After the achievement of independence in the early nineteenth century, each of the three central Andean states that emerged was confronted with similar problems. Each was drawn into the international commercial world after a period of revolution, war and political upheaval following declarations of independence. Export economies were superimposed on an ex-colonial society that was singularly lacking in its capacity for entrepreneurship, capital formation or financial or industrial expertise, particularly since the movement for independence had removed the initiative and experience provided by the *peninsulares* from Spain itself. Success in the production of foodstuffs, minerals and raw materials for export made it possible to finance investment in railways, roads, urban construction and port facilities, but in general terms material progress resulting from the expansion of export economies was restricted to limited geographical sectors of each country. Precise geographical limits can rarely be assigned to the 'modernized' and 'traditional' sectors of dual economies resulting from the growth of metropolitan regions and the expansion of export sectors, nor can the 'spread effects' associated with them be worked out in the detail that one would like. But the concept is a useful one, particularly since regional inequalities in the distribution of income have been all the more acute because of the relatively sharp social and geographical cleavage between predominantly white and mixed or Indian groups.

Industrialization, stimulated by events during and immediately after the Second World War, and since then by deliberate policy, has made some progress, especially in Peru, but the prospects for industrial growth, notably in Bolivia and Ecuador, have been severely limited by the size of the *effective* internal market, much lower than total populations seem to imply because of the persistence of poverty and of large sectors of predominantly subsistence farming. On the whole, industrialization has also served to accentuate regional imbalance, primarily because of the concentration of industry in the capital cities or large towns.

In common with other Latin American countries the central Andean republics face the consequences of a rapid growth of population, which has inevitably meant an increasing pressure on the land, particularly among small farmers, and the fragmentation of land into *minifundia*. There has been an increase in rural under-employment and a search for alternative sources of income, often involving seasonal employment elsewhere or permanent migration to regions of greater economic opportunity. Lima, La Paz, Quito and Guayaquil are the great magnets for population. Urban populations have swollen beyond the capacity of new industry to employ the thousands who are attracted by the amenities of the city and the chance rather than the promise of employment. Distress and poverty are to be seen in the makeshift settlements on the fringes of the cities in the *barriadas* (euphemistically the *pueblos jovenes*) of Lima, in the regularly flooded slums of Guayaquil or in the upper quarters of La Paz, but it may be argued that the urban migration is, nevertheless, a positive force making for change both in the towns and in the countryside.

The growth of numbers and the spread of commercialized farming have also served to sharpen the contrast between the large estates on the one hand and the fragmentation of smallholdings at the opposite end of the agrarian spectrum. Land reform is widely felt to be a possible solution to the social inequities in the distribution of land, but only in Bolivia, and more recently in Peru, have events forced rapid progress in this direction. Agrarian reform is seen not only in political and social terms as a means of raising the status of under-privileged and dependent peasantry, but also in economic terms. To many in Latin America, agrarian reform implies the raising of agricultural productivity at least as much as a thorough-going land redistribution. And it is therefore argued that if the agricultural productivity of peasant farms can be substantially raised, imports of foodstuffs will be reduced and an internal market will be created for cheap industrial goods that can be produced by a domestic manufacturing industry. But agrarian reform is rarely seen in geographical terms in relation to the spatial and regional implications of land reform and agrarian improvement.

Finally, one other issue of importance to all three countries is the question of colonization. All three have large, undeveloped territories in the eastern regions which are potentially sources of tropical crops ranging from rice and manioc to rubber and coffee, and which may have great potential for cattle-raising. Planned colonization is expensive, and heavy investment is needed in transport in order to open up new land and provide access to markets. For Peru and Bolivia the east has greater importance than it has for Ecuador, where the accessible coastal region has similar environmental potential. But for all three countries the question of whether or not scarce capital should be invested in eastern settlement poses a number of fundamental problems. Is there a political necessity to settle the national territory effectively? Where can markets be found for tropical crops once internal domestic markets have been satisfied – and, if so, for what crops? Is the return on capital and effort in eastern settlement likely to be greater than the return on similar expenditure incurred in programmes of agricultural intensification in older settled regions? And, finally, is the existence of available land for peasant settlement in the east an adequate alternative to and substitute for a programme of land reform in older settled areas?

Peru
For 20 years after Peru achieved its independence from Spain in 1824, the country was ravaged by war and its economy stagnated. Mining had declined, and Lima had lost much of the brilliance that its population of *peninsulares* (about a third of the population in 1776) had formerly helped to give it. Agriculture served mainly domestic needs and local trade. But from 1840 Peru was jerked into the mainstream of international commerce. Its coastal guano resources suddenly gained value at a time when the age of high farming in England and the beginnings of agricultural chemistry in Germany were creating a demand for fertilizers that could easily be satisfied by the simple quarrying and

loading of guano from the off-shore islands. Guano production increased from 7000 tonnes in 1830 to about 400,000 tonnes in 1860, when guano was the largest single export from any Latin American country. Directly and indirectly, revenue from guano exports or loans secured on export revenue served ultimately to finance public works. *Criollo* entrepreneurship was stimulated and new immigrants, particularly to Lima and other coastal towns, brought a new range of occupations and new attitudes to trade and commerce.

Railways were of undoubted importance in opening up the sierra and the coast to new commercial currents. The Southern Railway was opened in 1870 from Mollendo to Arequipa and to Puno in 1874, thus linking the *altiplano* of northern Bolivia to the coast by way of steam navigation on Lake Titicaca, and making possible the further extension to Cuzco. In central Peru the dramatic and ingenious engineering of Henry Meiggs created the line from Lima to Oroya over a pass at 4780 m, later extended to Huancayo and northwards to Cerro de Pasco. The optimistic hopes that accompanied the planning and execution of the railways into the Andes were never fully realized, but the railways brought new life to old sierra mines, such as Cerro de Pasco, and made possible the exploitation of lead, zinc, copper and other non-ferrous metals in addition to the traditional mainstays of silver and gold. A new commercial interest began to infuse haciendas in the sierra when the profitability of merino and alpaca wool was realized. Indeed, one of the major consequences of the impact of a commercial economy in the sierra, in conjunction with the legal encouragement given to the individualization of land tenure, was to give a monetary value to extensive natural grazings, which had formerly been regarded as waste or common land. Large areas of natural pasture were incorporated into haciendas, and a new impetus was given to the purchase of Indian lands and the disintegration of traditional *comunidades*.

Unwarranted optimism about the possibilities of the forested east led to desultory attempts to encourage European immigration and to set up agricultural colonies in the *montaña*, e.g. the German settlement at Pozuzo, but very little of permanent importance was achieved in the face of difficulties presented by isolation, ignorance, disease and official neglect. In the east high profits from the collection of wild rubber and cinchona produced a wave of ruthless exploitation both of labour and resources. But this brief boom in activity ended with the successful introduction of cultivated rubber and cinchona (as a source of quinine) in the East Indies, and it was, in any case, part of an economy focused on the navigable rivers and the boom town of Iquitos, which had more connection with Manaos and Brazil than it had with the coastal and Andean economy of Peru.

The major impact of new commercial enterprise was in the coastal region. Sugar had been a staple from early colonial times, but production increased rapidly in the late nineteenth century with the introduction of steam-power, largely by foreign immigrants. The industry now became strongly localized on the northern coast, retreating from the upper coastal valleys elsewhere. At first the ownership of sugar *centrals* was independent of sugar plantations, but with the increasing concentration of landownership in the late nineteenth century both cultivation and processing fell into the hands of a few large enterprises, some of which continued to be foreign-owned. In the Chicama valley, for example, sixty-five haciendas had been amalgamated to seven by 1918 and to four by 1950. Cotton cultivation began on a larger scale at the time of the American Civil War, when the cotton famine in Europe initiated a search for alternative sources. Peru was fortunate not only because of the accessibility of its coastal valleys for export, but also because of its possession of highly valued long-staple varieties: Piura cotton in the north, and the very fine high-grade Tangüis type. Output expanded with the Lancashire boom of 1900–10, and even before 1914 a small cotton textile industry was established in Lima.

By the end of the century the guano resources, grossly over-exploited, had ceased to have importance in Peru's foreign trade, and the boom in nitrates from Tarapaca was of no further significance to Peru after the annexation of the province to Chile at the end of the War of the Pacific (1879–83). By 1914 Peru's external economy had become firmly based on a variety of exports both of agricultural and mineral origin. The varied composition of Peruvian exports certainly helped to reduce its vulnerability to price fluctuations in individual commodities, though it has never shielded the country from cyclical movements of international trade. Sugar and cotton from the irrigated coastal valleys have been of enduring importance until the mid-1970s; coffee from the *montaña* has increased in recent years, but the wool of the sierra has lost ground. Minerals have been of fundamental significance, particularly copper, together with silver, lead and zinc. Oil exports were of importance until the mid-1960s since when Peru ceased, until very recently, to be self-sufficient. Iron-ore exports became important after 1953, but the most important recent trend has been the rise and fall of fish products. Exports of fishmeal, growing dramatically from modest levels in the 1950s to be a major export commodity until their collapse in 1972, reflect the exploitation of the remarkable ecology of the coastal waters, like the guano boom of a century before. Over-fishing and some change of the ecology of coastal waters have brought the boom to an end, though it remains a major industry.

The external sector remains fundamental to the economy of the country. Several points may be stressed, however. First, exports continue to be almost entirely of basic raw materials and foodstuffs, and 'non-traditional' exports of manufacturing industry are still relatively small, though tourism makes an increasing contribution to the external sector. There has been steady advance towards the more complete processing of minerals before export, notably in copper, zinc and iron ore. Secondly, the expansion of exports has slowed perceptibly in recent years. Between 1950 and 1962 the real value of exports increased almost threefold, but apart from the apparently short-lived boom in fishmeal exports between the late 1960s and 1972, the real value of exports has grown very little since the early 1960s.

From 1950 to 1976 the economy as a whole maintained a high and fairly steady average rate of growth of over 5 per cent per year, sustained in general by buoyant export performance. Since 1976 the Peruvian economy has been sharply hit by the impact of higher oil prices, international recession and the decline of some formerly important export commodities (*Figure 7.1*). GDP declined in 1977 and 1978, and growth has only recently been resumed. Component elements of GDP may be seen from *Figure 7.1* to have changed substantially, pointing in general to a relative decline in agriculture and a continued advance of industrialization.

In 1950 agriculture contributed 22.6 per cent of GDP and manufacturing industry only 13.6 per cent. Mining made up a further 4.5 per cent. By 1962 the relevant proportions were 19.3 per cent for agriculture, 17.0 per cent in manufacturing industry and 6.0 per cent in mining. But by 1979 agriculture had declined to 12.6 per cent, mining had grown to 10.1 per cent, and manufacturing industry had by then increased to 24.5 per cent. The most depressing feature has been the poor performance of the agricultural sector. Even when agricultural output was increasing at the rate of over 3 per cent per year between 1950 and 1962, production for the domestic market was stagnating and high performance was mainly in the export sector. Since then agricultural output has grown very slowly indeed, and has in fact stagnated since 1972. During the 1970s agricultural production has fallen short of the growth of population. Imports of foodstuffs, a matter of concern even in the 1920s, have continued to rise to the point at which they represented 20 per cent of the value of commodity imports.

Nevertheless, in spite of setbacks in the late 1960s and in the late 1970s GDP has risen faster than population, though painfully slowly. *Per capita* GDP has risen by rather less than 50 per cent since 1959, and even over the period of maximum overall growth from 1963 to 1976 the average annual rate *per capita* was only 1.8 per cent. As in so many other countries in the developing world population is growing so rapidly that the economy must show a rapid expansion in order to make progress in *per capita* terms. And in Peru population is growing with a rapidity characteristic of much of tropical Latin America. In the years between the censuses of 1940 and 1961 the population rose from 6.2 million to 9.9 million at an average rate of growth of 2.5 per cent per annum. The 1972 census recorded just over 13.5 million people and a higher rate of growth at 2.9 per cent per annum, and estimates for 1979 put the total at 17.5 million. Birth-rates are high at 41.0 per thousand in 1970–5, but have fallen slightly since the 1960s, chiefly in the cities. Death-rates have fallen to a level of 11.9 per thousand, and are comparable with the death-rates of more developed countries. It is a relatively young and expanding population with some 44 per cent aged 15 or less in 1972, implying a relatively low proportion of 'economically active' people, a high degree of potential mobility in geographical terms, and also a potentially high burden to the community in terms of education, social welfare and health services.

Major institutional changes underlie the economic trends noted above, however, and have had fundamental repercussions of great relevance for Peru's social and economic geography. Until the 1960s the Peruvian economy had been relatively uncontrolled, responding to export opportunities for its agricultural and mineral products, welcoming the investment of foreign capital, but vulnerable to the international winds of economic change and with a disappointing record of industrial growth and social inequality. In 1968, however, at a time of financial and political crisis, a military regime under the leadership of General Velasco took over the reins of government and instituted a series of measures designed to accelerate the 'modernization' of the economy and to reduce its dependence on foreign capital. Interpretations of the military government have tended to be conflicting, but its undisputed effect was certainly to increase vastly the role of the state in economic and social affairs. Heavy industry, the railways, the major banks and mining companies, the fishing industry and fishmeal

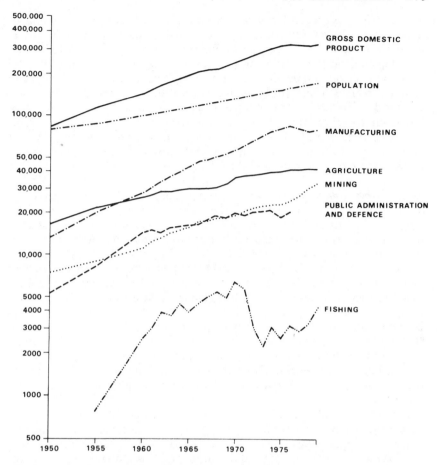

Figure 7.1 Peru: economic growth, 1950–79. GDP, by sectors, 1950, 1955 and 1960–79 (million soles at 1970 prices); population, 1950–79

production were all nationalized. Various forms of co-operative institutions were initiated or encouraged, and a far-reaching programme of agrarian reform was put into effect. From 1975 a shift in the balance of power within the military government occurred, and some of the reform programmes were halted or reversed. Agrarian reform has ground to a halt, many of the co-operative ventures in industry have been abandoned, and the fishing industry, though not the manufacture of fishmeal, has been returned to private ownership. Representative government was re-established in 1980, and more 'orthodox' policies have since been followed. But many of the changes instituted between 1968 and 1975 are likely to be irreversible, especially in the agrarian sphere, and there is no doubt that the participation of the state in the economic life of Peru will continue at a much higher level than before 1968, with corresponding implications for regional development.

The characteristics of regional development in Peru could be outlined in various ways. The basic regional contrasts in terms of environmental character, land-use systems and even cultural characteristics are those between the coastal zone, the sierra and the eastern region. The Peruvian Institute of National Planning has adopted a fourfold division of the country, recognizing the individuality of the Eastern region, but dividing the remainder of Peru into North, Central and Southern regions, which cut latitudinally across the coast and the sierra. Each kind of division has its merits for particular purposes, but it here seems more important to stress, first of all, the contrast between metropolitan Peru, consisting of the conurbation of Lima and Callao, or Greater Lima, and the rest of the country, within which the most useful division that can be made is that between the coast, the sierra and the *selva*.

Greater Lima – metropolitan Peru In terms of its size and wealth, and its range of social and economic activities, Greater Lima is so clearly differentiated from other urban centres in Peru that the contrast between metropolitan Peru and the rest of the country is worth emphasis. Its growth continues at an astonishing rate: in 1940 the provinces of Lima and Callao had 645,172 people; by 1961 numbers had almost tripled to 1,845,910; by 1972 the total was 3.3 million and current estimates speak of 5 million. The primacy of Lima in the urban structure of Peru is nothing new, of course, and may be traced from the early colonial period, but the proportion of the total population living in the metropolitan area has increased substantially, rising slowly at first from 3.7 per cent in 1876 to 4 per cent in 1908 and then fairly rapidly in the inter-war years to 10 per cent by 1940, 19 per cent in 1961, and no less than 24 per cent in 1972. Lima/Callao, or Greater Lima, is over ten times larger than the second largest city in Peru, and the fifth largest city in Latin America after Mexico City, Buenos Aires, São Paulo and Rio de Janeiro.

In terms of wealth, income and welfare, the metropolitan region has a clear leadership over the rest of the country, far greater even than the size of its population would suggest. Wages are higher than the national average, at least for those who have jobs. In 1972 it contained two-thirds of the nation's doctors and almost half of the hospital beds; it had two-thirds of all Peruvian households fully equipped with bathroom and toilet facilities and some three-quarters of all households possessing any of the following: TV set, refrigerator, radio, sewing machine. Its leadership in industry, commerce and finance is undisputed, with over two-thirds of those employed in finance, and about a half of all those employed in Peru in manufacturing industry, construction, commerce and transport. In 1968 it generated 65 per cent of the gross value of industrial production in the country, and according to Levano (1969, pp. 176–7) Lima/Callao produced no less than 82 per cent of the consumer goods made in Peru for the national market and 92 per cent of the output of durable consumer goods and capital goods. Furthermore, its predominance in manufacturing appears to have increased during the 1960s. Out of the total national increase of 74,254 employed in manufacturing industry between 1961 and 1972, no less than 61,475 or 83 per cent were in Lima/Callao.

As a result of the metropolitan concentration of wealth, Greater Lima is much more important as a market for industrial goods than population figures alone would suggest. It is not surprising that new industries oriented to the substitution of imports should be located at the largest single market and near the major importing port. Much of recent industrial expansion has been of this character, but the production of consumer goods and the final stages of manufacture and packaging of imported semi-finished manufactures are increasingly supplemented by more basic industries producing capital goods and consumer durables. In addition the region has developed industries associated with traditional export activities, and some industrial plans rest on the prospects of export to other countries of the Andean Pact. The list of industries is impressive, though many are on a relatively small scale: wide-ranging engineering industries, including shipbuilding, marine and electrical engineering, the assembly of motor vehicles, leading with government help and direction towards the local manufacture of components, metal-refining (notably zinc at Cajamarquilla, near Lima), cement, chemicals, including fertilizers, some oil-refining, man-made fibres and plastics, pharmaceuticals, textiles, food-processing, fishmeal production and fish-canning, furniture, footwear and clothing. National private capital has played a substantial role in this process of industrialization, but in recent years both state and direct foreign investment have increased rapidly, especially in capital-intensive industries. Indeed, it is important to make a distinction between a 'modern' industrial sector which is relatively capital-intensive, and worked by a unionized labour force, and on the other hand a very large number of small-scale artisan workshops and individual outworkers who are to a varying extent integrated into the operations of large-scale industry and commerce, e.g. in footwear, clothing and the manufacture of motor-car parts (Fitzgerald, 1976).

Industrial growth has given new shape to the urban pattern. Small workshops and domestic crafts are scattered throughout the built-up area, but larger-scale industries are thrusting out towards and even beyond the margin of the city. Much of the older industrial growth was in Callao, along the roads linking Callao with Lima, and in transpontine Lima across the Río Rimac to the north, and there was also early growth of industry along the Central Railway and the Central Highway towards the east. These continue to be preferred locations, but recent growth has pushed far out to the north and south along the Pan-American Highway running parallel to the coast.

There are many other reasons, besides the existence of a large and expanding market, why industry should be located in Lima rather than elsewhere in Peru. Some of these are bound up with the idea of cumulative causation suggested by Myrdal and others (Myrdal, 1965; Hirschman, 1966). There are external economies to be gained from the presence of a fairly wide industrial base and the presence of associated industries, banking expertise, marketing structures, a pool of labour with varied skills and an adequate professional and managerial class. International middle-class standards of housing, shopping facilities, professional services and recreation make it attractive (for many the *only* attractive location) for entrepreneurs, executives and businessmen. But it is also evident that proximity and personal access to the seat of government and the highly centralized bureaucracy is also important, particularly with the increasing role of the state in the economy.

Growth has put continuing pressure on the urban infrastructure and particularly on the provision of adequate supplies of water and power. Irrigation, domestic and industrial water supplies are derived from the basin of the River Rimac on which Lima is sited, and from neighbouring valleys, the Lurín and the Chillón, or from wells that tap the water-table in the alluvial apron on which Lima is situated. Even in 1956 water supplies were inadequate for the demands made on them by the city and the irrigable area nearby, and there was adequate irrigation water for only 16,000 of the 24,000 ha then in cultivation. The natural drainage basin of the Rimac has been artificially extended by the Marcapomacocha scheme, whereby lake waters at 4400 m above sea-level, formerly draining into the Mantaro valley to the north-east of Lima, have been diverted through a tunnel some 10 km long to the valley of the Santa Eulalia, a tributary of the Rimac. Future demands require additional supplies by the diversion of Mantaro water, a systematic attempt to make greater use of the high-season flow of the Rimac itself, and the possibility of desalination plants has been mooted. Even now, domestic piped water is not available in many of the *barriadas* or *pueblos jovenes* of recent growth, and water is supplied to them by trucks at a cost of some forty times the price of piped water. The expansion of power supplies presents further problems, which are closely linked to water supplies in that the bulk of Lima's electrical power is of hydroelectric origin, including the power derived from the Marcapomacocha scheme and the Mantaro Valley project. Here, too, the predominance of the metropolitan region stands out, for 75 per cent of Peru's public supply of electrical power was, in 1968, devoted to the supply of Greater Lima.

Pressure on urban services has been occasioned by the explosion of the population, and migration has played a dominant role in this expansion. In 1972 41 per cent of the population of Greater Lima had been born in other parts of the country. The characteristics of the migrants and the reasons why they move are the subject of continuing investigation, in part to confirm or refute the view that they are typically of illiterate peasant origin, constituting a rural or semi-rural enclave in the city, difficult to assimilate, and threatening the very fabric of urban life by their numbers and because of the difficulties of finding adequate housing or employment. A greater proportion of the migrants to Lima do indeed come from the sierra than from the coast, reversing a former pattern, before 1940, of coastal predominance. But a fairly high proportion of migrants come from provincial towns rather than directly from the countryside. Migrants tend to be young, and the pioneers from a particular area tend to be more ambitious and receptive of change than their stay-at-home neighbours. In common with migrants to other Latin American capitals, migrants to Lima tend to be among the better educated and of relatively high social status in their region of origin. Unemployment rates for migrants in Lima tend to be lower than those for non-migrants, and those who come from the larger towns to Lima seem to have the greatest chance of success in their search for jobs. In short, the evidence does not suggest that migrants are at a particular disadvantage, nor are they illiterate, nor does it suggest that migrants are unusually resistant to assimilation to an urban way of life, though they do keep strong associations with their home towns or provinces.

The physical expansion of the urban area to accommodate the swollen population of Greater Lima has been dramatic. Shanty towns, squatter settlements or *barriadas*, officially re-labelled as *pueblos jovenes* in recent years, have grown up, sometimes literally overnight, on the fringes of the old built-up area. They are not all new, the first being traced to 1924, but they are mostly of recent origin (142 out of 182 such settlements were started between 1955 and 1966), and they have been variously estimated to contain between 20 and 40 per cent of Lima's population. Such squatter settlements are, of course, a common feature in the developing world, but they have spread very widely in Lima, not only because of explosive population growth, but also because the arid environment of the coast makes their proliferation on uncultivated desert lands a relatively simple matter. Bare and arid hillslopes, *quebradas* and alluvial plains, devoid of vegetation and agriculturally useless, have been occupied without let or hindrance. Poor and temporary constructions of reed matting, similar to those commonly found in rural settlements along the coast, are later replaced by solid brick and adobe houses as the settlement matures. Uncompleted buildings are a normal feature as owners wait to save enough to install window frames, a second storey or some other improvement. Older *barriadas* take on the appearance of mature working-class residential areas. But in general, living conditions are certainly bad, housing is poor, and piped water, sewerage and electric power are often not available, though the military regime made efforts to improve their status in the early 1970s, and conditions are at least as good as they are in many parts of the sierra or in other parts of the coast.

Nevertheless, these are the sections of the urban population, sometimes labelled as the marginal sector, whose prospects for employment are most precarious even when the economy is booming, and many are in a desperate situation at times of recession. Under-employment is widespread, and although very difficult to calculate, the estimate has been made that 19 per cent of those employed in commerce, 25 per cent of those employed in industry and 49 per cent of those in the service sector are under-employed. The growth of population has certainly outstripped the capacity for employment provided by manufacturing industry. In spite of substantial growth, manufacturing industry is employing a steadily decreasing proportion of the labour force (down from 23 to 20 per cent in Greater Lima between 1961 and 1972). Over the same period no less than 70 per cent of the increase in the labour force went into commerce, transport and the general and amorphous census category known as 'services', and only 15 per cent into manufacturing industry.

Metropolitan dominance is a Peruvian fact of life, but the rest of Peru, however much it may be relegated by Limeños or by planners to a satellite status, is rich in its variety of contrasts. The coast, the sierra and the *selva*, so obviously distinctive in ecological terms, are also divisions traditionally recognized in Peruvian regional consciousness; each has its own cultural tradition expressed in literature and music; and each has a popular image, not always accurate or complimentary, for the inhabitants of the other two. There are, of course, important differences within each of these zones which are in part related to environmental and historical circumstances, and in part to the strength or weakness of Lima's pervasive influence. But one general characteristic of location in the distribution of population is common to almost all of Peru: nuclei of population and economic activity are small and discontinuous. Locally dense populations exist, as for example round Lake Titicaca, in the Huancayo basin or in irrigated coastal ·valleys, but they are separated by large areas of virtually unpopulated country; the oases of the coast are separated by barren stretches of desert; in the sierra, zones of settlement are separated by deep gorges or high barren *puna*; and in the *selva*, settlements and cultivated land are but islands in a sea of forest. To overcome the problem of distance must always be a major preoccupation of policy, therefore, quite apart from the inherent cost of building and maintaining good communications.

Striking regional differences are revealed by various social and economic criteria, and they are by no means coincident with regional divisions based on ecological criteria. The distribution of population shows two major trends: the growth of urbanization and the relative shift of population from the highlands down to the coast, and to a lesser extent towards the unpopulated eastern regions. The relative shift of population from the sierra is a trend which has been going on for centuries. After the demographic catastrophe of the sixteenth century it was likely that the coast contained no more than 10 to 15 per cent of the total population, and although the balance may have shifted slightly in favour of the coast by the early nineteenth century, and then more swiftly with participation in international commerce after 1840, the coast still had only 24 per cent of the total at the time of the 1876 census, and the sierra some 71 per cent, with 5 per cent in the eastern regions. By 1940 the proportions had changed a little further, with 30 per cent on the coast compared with 60 per cent in the sierra (*Table 7.1*).

Table 7.1 Peru: distribution of population, 1876–1972

	Total population	Coast	%	Sierra	%	Selva	%
1876	2,659,000	629,000	24	1,903,000	71	127,000	5
1940	6,208,000	1,848,000	30	3,959,000	64	401,000	6
1961	9,907,000	4,039,000	41	5,017,000	51	851,000	8
1972	13,538,000	6,550,000	48	5,651,000	42	1,337,000	10

Since 1940 the swing to the coast has gathered pace, and by 1972 it registered a larger population than the sierra. The *selva*, too, now has a much more significant proportion of the total population.

Everywhere in Peru the rate of urbanization has increased: from 4.4 per cent per annum between 1940 and 1961 to 6.4 per cent in the 1960s. Lima's primacy increases, but a rank-size diagram reveals the rapid growth of a group of regional centres which had more than 100,000 people in 1972. They are mainly the long-established regional capitals: Arequipa is the largest of them, followed by Chiclayo, Trujillo and Piura on the northern coast, Cuzco and Huancayo in the sierra and Iquitos on the Amazon, but they also include the mushroom town of Chimbote, a creation of the boom in fishing in the 1960s and of the iron-and-steel industry, and which grew from being a village in 1940 to having a population of 159,000 in 1972, though it has since fallen on hard times. Smaller towns have increased both in number and size, but show no marked differentiations as a group (see *Figure 7.7*).

The distribution of rural population has begun to show signs of important changes. Rural population was still growing in the 1940s and 1950s at a rate of

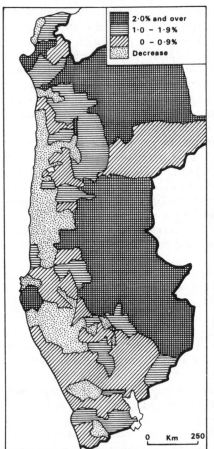 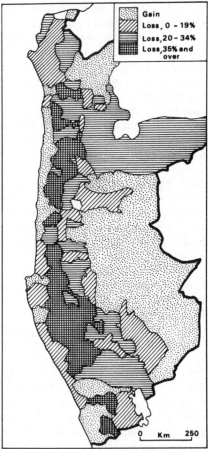

Figure 7.2 Peru: average annual rate of change of rural population, 1961–72

Figure 7.3 Peru: net migration balance, 1972. The map is based on the residence of population born in other provinces. Percentages indicate the net gain or loss by migration as a percentage of the population of each province

1.2 per cent annually, and with the greatest increases of over 2.0 per cent in the *selva* and in some of the sugar and cotton-growing areas of the coastal region. Growth was already sluggish in most of the sierra provinces, especially in the south, but a turning point seems to have been reached in the 1960s. Overall rural increase was only 0.5 per cent per year, and although rates of growth remained high in the *selva*, and in a few areas of expanding irrigation on the coast, especially in the north and near Ica in the south, many rural areas have registered a fall in population. Rural decline is clearest in the sugar regions of the northern coast and in the sierra hinterland of Lima, but also in some of the big and sparsely populated provinces of the southern sierra (*Figure 7.2*).

The factor most closely associated with relative changes in the pattern of urban and rural population is, of course, migration, and *Figure 7.3* is an attempt to show the spatial incidence of migration. It shows the population born in each province as a function of its population in 1972, and therefore relates to a total resultant of migration over a generation. The population of each province comprises its native-born and immigrants from elsewhere, so that the percentage figures shown indicate total *net* loss or gain by migration. It is evident that the population is highly mobile: in 28 provinces net out-migration was 40 per cent or more of the population in 1972. The areas gaining by migration are of two kinds: urban centres and some of the eastern regions in the *montaña* and *selva*. Areas losing population most heavily correspond, as one would expect, with weakly urbanized provinces suffering a decline of rural population between 1961 and 1972, and they include sierra provinces on the Pacific slopes of the Andes which are relatively easily accessible to the coast and to Lima. It is interesting that poor and remote provinces of the inner sierra, to the east and south of Cuzco, for example, and to some extent in northern Peru, have not been so strongly affected by currents of out-migration, and their rural populations have continued to increase at a modest rate.

The direction of migration is dominated by movement to Lima. *Figures 7.4* and *7.5* represent the net balance of people moving from their department of birth. In 1972 only 2 million out of a total population of nearly 3.5 million in the Department of Lima (59 per cent) were born there. A further 10 per cent had migrated to it from other coastal provinces, but of the remaining 31 per cent the overwhelming majority had come from the sierra. The net movement into the Department of Lima and the Province of Callao is shown, by departments for 1961 and 1972, in *Figure 7.4*. During the 1960s, both the rate of migration to Lima and the field from which it draws its migrants have increased. More migrants to Lima by 1972 had come from further afield: from Cuzco and Puno in the south and from the remote departments of the northern sierra; and a smaller proportion from departments in the immediate neighbourhood of Lima or from departments with alternative poles of attraction: the growing cities of Arequipa, Chiclayo and Trujillo and new mining areas in the south. *Figure 7.5*, showing net inter-departmental flows other than to Lima, is complementary to *Figure 7.4* and confirms that rapidly growing urban centres have attracted net population flows from a regional hinterland: Chiclayo, Trujillo, Arequipa and Ica in the coastal region. But it also shows a significant movement within the sierra to the central region dominated by Huancayo and the mining areas, drawing migrants from the poor departments of the south-central sierra. Compared with these inter-regional flows the movement from the sierra to the relatively empty zones of the *montaña* and the *selva* is relatively small, though it should be mentioned that in some important areas of settlement in the *montaña*, movement is predominantly within departments and therefore not indicated on the map.

The movement of population from the Peruvian sierra is a response to its poverty, in turn a result of rural pressure on the land, low productivity and the lack of alternative opportunities. Poverty and deprivation are difficult to map, but the Peruvian census includes tables relating to the possession of material goods which are a useful surrogate measure of the distribution of income in the absence of more direct evidence. The data relate to the possession by households

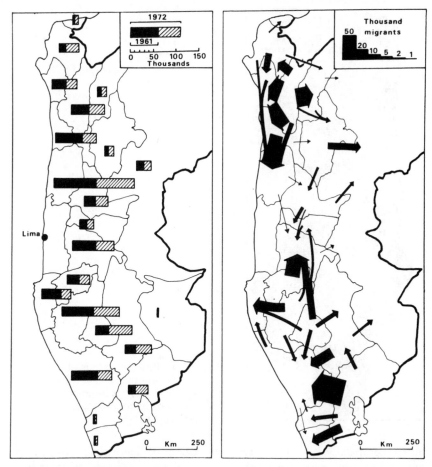

Figure 7.4 Peru: net migration to the Department of Lima and the Province of Callao, 1961 and 1972. The map shows the population living in Lima/Callao who were born in other departments, less the population living in other departments who were born in Lima/Callao

Figure 7.5 Peru: net migration between departments, excluding Lima/Callao, 1972 (The map is constructed on a similar basis to *Figure 7.4*)

of durable consumer goods: radios, sewing machines, refrigerators and TV sets. *Figure 7.6* shows, by provinces, the proportion of households with *any* of these goods – radios and sewing machines are, of course, the most common possessions. It displays, most obviously, the relative prosperity of the coast, the poverty of the sierra, and the intermediate level of the eastern regions (where the figures may well be suspect as a result of the omission of remote communities). But within the sierra there is a very important major distinction to be made between the central sierra and the rest. The central sierra includes mining and agricultural areas in the Mantaro valley which show a level of material goods

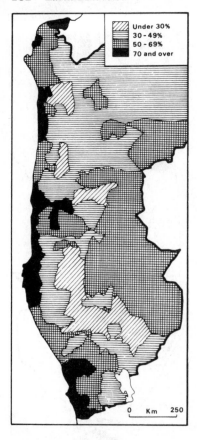

Figure 7.6 Peru: proportion of households possessing one or more consumer durables (radio, sewing machine, refrigerator, TV) according to the census of 1972

almost comparable to coastal areas. Two regions of deprivation stand out, in which 70 per cent or more of households did not even possess a transistor radio in 1972: the southern sierra including parts of Cuzco Department; and a second, less continuous zone in the inner regions of the northern sierra in the drainage basin of the Marañon. Remoteness and isolation are the common characteristics of both of these areas. Comparison with *Figure 7.2* suggests that they correspond roughly with the areas which have not lost population quite so heavily by migration as the Pacific slope provinces of the sierra. Although increasing, migration has failed to provide a relief from poverty and an escape from subsistence farming.

The coastal region is the richest and most productive area of Peru. In wealth, income, the degree of industrialization, urbanization, literacy and welfare it is the leading geographical sector of the country. In agricultural terms it contributes the greater part of Peruvian exports of agricultural origin, its yield in value per hectare is twice that of the sierra, and nearly five times greater in terms of the value of agricultural production per economically active person in agriculture. Agricultural modernization has proceeded much further than elsewhere, and not only on the large estates.

Except in the extreme north, near Tumbes, irrigation is a necessity for crop production in the coastal region. Since the late nineteenth century the area under irrigation has doubled; between 1906 and 1965, 120,293 ha have been newly irrigated and water supply has been improved on a further 178,000 ha. Much was done by traditional methods to increase the area under sugar and cotton by the construction of simple take-off canals from uncontrolled rivers, then later by the sinking of tube wells, the number of which increased very rapidly after 1945. In the Ica valley there were only 49 in 1938, 80 in 1944 and 476 by 1956. Where underground water is sufficient at suitable depths, tube-well irrigation has had the advantage of liberating farmers from irksome rules and traditional usages, but it has favoured entrepreneurs with capital and in some areas water-tables have been lowered significantly as a result of over-exploitation (Smith, 1960). Since 1950 attention has turned to more ambitious projects involving the building of dams to regularize flow, the construction of longer canals to distribute water to areas of deficiency, or the diversion of Atlantic-flowing drainage across the watershed to the Pacific slope. Some of these schemes (e.g. at Marcapomacocha above the Rimac valley, or the Choclococha scheme above the Ica valley) have involved the diversion of high-level lakes to new outlets across the relatively level country of the *puna* surface at over 4000 m. Relatively short tunnels and diversion channels have been needed, but construction work is often costly because of the inaccessibility of the sites. Some of the more recent projects to be undertaken, such as that at Olmos in the north, involve much longer and more expensive tunnelling operations. The cost of irrigation of new land in the coastal area is therefore high, and new projects tackling more difficult engineering problems are even more expensive.

There is local specialization in a number of the coastal valleys, some of it already established in colonial times, such as the concentration on viticulture in the Ica valley, but much of it quite recent. Rice cultivation is strongly localized on the clay–loam soils of the Lambayeque valley and neighbouring areas in northern Peru; market-gardening, fruit and vegetables are concentrated in the Lima region as one would expect, and dairy and poultry farming have made rapid progress in recent years in response to urban markets. Further south, the irrigated area around Arequipa is devoted to intensive production of vegetables, potatoes and dairy production on the basis of irrigated alfalfa. But it has been sugar and cotton that have dominated coastal agriculture since the late nineteenth century. Sugar occupies only 48,000 ha on the coast, but by value represents 13 per cent of coastal output, and yields are about three times the world average. The industry is efficient, using modern techniques and high-yielding varieties of cane. Insolation and water supply for all-the-year-round irrigation have been major factors locating sugar on the coast to the north of Lima. North of Lima the cloud cover characteristic of the desert coast in the Peruvian 'winter' breaks up, and insolation is high throughout almost the whole year. Most of the rivers flow throughout the year so that irrigation, moderately high temperatures and sunshine can support continuously vigorous cane growth with a high sucrose content. Controlled irrigation makes possible a rotation of harvests such that a continuous supply of cane can keep the mills in full operation throughout the year except for a few weeks for cleaning and maintenance. Linkages with other industries have been strong since *bagasse* is the raw material for the manufacture of paper, and a small chemical industry has been created to supply raw materials for sugar-refining. Until recently sugar was produced by a handful of large estates, a few of them foreign-owned, on which sugar cultivation was integrated with the refining mills so as to control the supply and cultivation of cane. Slavery and Chinese immigrant labour preceded the use of labour contracted by agents from the northern sierra (by a system known as '*enganche*' – the hook) from the late nineteenth century. Sugar in Peru became identified as a highly capital-intensive, export-oriented form of production with a professional management structure and a permanent labour force, becoming more and more unionized in the twentieth century (except for seasonal and temporary labour). The sugar estates were the first target for agrarian reform in 1969 and have been converted, not without a difficult transition, to large-scale co-operatives.

Cotton is the other great traditional staple of Peruvian coastal agriculture, located mainly to the south of Lima and in the far north in the Department of Piura. Unlike sugar it is a crop with seasonal water requirements which are relatively modest, and 'summer' temperatures are everywhere high enough in the coastal region. Cotton cultivation could expand, therefore, into areas irrigated from rivers with a highly seasonal flow. It is less profitable now than it was 20 or 30 years ago, however, and the area under cultivation has dropped in recent years from about a third to less than a fifth of cultivated land in the coastal area, and output in 1979 is only a half of what it was in the early 1960s. The relations of production generated by cotton cultivation have differed substantially from those of sugar production. Large-scale landownership was slightly less dominant than in sugar, and although some production was on large-scale units directly operated by the owners, the major difference was the predominance in cotton cultivation of share-cropping tenancies, by which farm operators of modest means grew cotton on holdings of a few hectares and paid the landowner with a share of the yield. Techniques of cultivation are less demanding than in the case of sugar, requiring less highly organized management, and share-cropping was also a means whereby large landowners could expand production rapidly with a nominal investment of capital. Like sugar, however, cotton cultivation was integrated with the sierra through its demand for labour, but mainly by way of highly seasonal demand for the cotton-picking harvest. But in cotton cultivation, too, agrarian reform has replaced the large landowners and share-cropping systems by co-operative structures since 1969, and in many areas cotton has been replaced by other crops directed to internal consumption.

The growth of the fishing industry from insignificant proportions in 1950 to become the world's largest until 1970 in terms of the tonnage of fish landed, has made a considerable contribution to the national income and to the export sector. Output declined after a record catch in 1964 of over 9 million tonnes, and fears of over-fishing led to the establishment of a close season during June, July and August. Levels recovered for a few years, with record catches in 1968, 1970 and 1971 of over 10 million tonnes, but in 1972, with incursions of warm oceanic water from the north (El Niño), and as a result of over-fishing in previous years, the yield dropped catastrophically, and has not since recovered its former levels, hovering between 2 million and 5 million tonnes. Measures are regularly taken to foster recovery by selective bans on fishing and the declaration of close seasons, though fishing has often been allowed to continue in southern waters, less affected by the warm current of El Niño. At the peak of the industry some 2400 fishing vessels were employed, operating from no less than forty-seven ports, many of which are tiny settlements landing no more than a few hundred tonnes. With decline, however, the fishing fleet has been steadily reduced to a level of about 500 ships. The major fishing ports are Chimbote, Tambo de Mora, Pisco, Callao, Supe and Ilo. Over 90 per cent of the catch is the small, sardine-like anchoveta used for the manufacture of fishmeal for fertilizer and as an animal feeding stuff with high protein content. Expansion of fishing has thus been closely associated with the building of fishmeal-processing plants at the major ports, though the number of plants has fallen in recent years from 100 to 42. Both the fishing industry and fishmeal manufacture have created important backward linkages to other activities, quite apart from their intrinsic importance in creating employment, providing exports, and contributing to national income (Roemer, 1970). A modest shipbuilding and repair industry has sprung up, chiefly in Callao. Firms specializing in the construction and manufacture of fishmeal plants were set up, and in other ways the manufacture of nets, pumps, centrifuges, rope and carpentry have been stimulated. In a regional sense, fishing has helped to spread new activity to a number of places such as Casma, Ilo, Supe and above all, Chimbote, a boom town which became an important focus for immigration from the sierra. But the decline in fishing and fishmeal production for export has created unemployment and contributed to the balance-of-payments crisis. It is unlikely that former levels of activity will recur in the foreseeable future, though conservation measures are being actively pursued. It is difficult to see how far the nationalization of the fishing industry and of fishmeal production in 1973 was a response to the crisis in fishing or to other considerations of policy, but ownership of the fishing fleet was returned to small firms in 1976 after a period of contraction and rationalization in accordance with the shift in government policy. Future prospects rest on the success of conservation and on the possibility of expanding the production of white fish for human consumption. Some progress in this direction is being made, with a 50 per cent increase in the catch of food-fish from 1973 to 1977, to 450,000 tonnes, and an increase in the production of canned fish from 28,000 tonnes in 1973 to 48,000 tonnes in 1977.

Mining, too, has made its contribution to the economic growth of the coastal region, though the relevant mining areas have been mainly located on the Pacific slope of the Andes rather than on the coast itself. The open-cast exploitation of low-grade copper ores at Toquepala in the south made a major contribution to mineral exports from the early 1960s, and involved the building of a new railway to the coast in the south. Other deposits of low-grade copper ores were opened up at Cuajone in 1976 and at Cerro Verde, near Arequipa, the following year. The copper refinery at Ilo is being expanded to cater for increased output. The mining of high-grade iron ores at Nazca in southern Peru came into operation in 1953, and other iron-ore deposits at Acari nearby are also exported from the port of San Juan, created to handle iron-ore exports, mainly to Japan, South Korea, and the USA, but also to the Peruvian steel industry at Chimbote. Northern Peru's mineral production from the coastal zone has a long and complex history from the late nineteenth century when oil was discovered at Talara, and exploited with British and American capital, though later nationalized.

Production from mainland wells has been declining in recent years, but off-shore drilling has been successful and currently produces a substantial share of Peru's expanding oil production. Settlements like Talara associated with oil production have been relatively isolated enclaves in the desert, but recent developments in northern Peru envisage the creation of an industrial complex at Bayovar, a new oil terminal and cargo port, revolving mainly about chemical and petrochemical industries. Phosphate deposits in the Sechura desert and oil from eastern Peru by way of the Northern Andean Pipeline (completed in 1976), of which Bayovar is the terminal, provide the basic raw materials. This and the expansion of the irrigated area in the Chira-Piura project may lay the foundation for a revival of growth in the north.

Greater Lima has the lion's share of Peruvian industry, but the rest of the coastal region, particularly if Arequipa is included, has most of the remainder. Much of coastal industry outside Lima/Callao has been concerned with the processing of agricultural or mineral products, and the industries derived from them either by backward or forward linkages: sugar-refining, fishmeal production, copper-refining, cotton-processing and dairying, but a small chemical industry (related to the sugar industry), textiles, cement, light engineering and woollen industries in Arequipa were also established before the recent drive towards industrialization with state enterprise and intervention, and a movement towards regional development. Arequipa and Trujillo are seen as major regional industrial centres. Heavy engineering, electronic industries, and the production of man-made fibres are being grafted on to Arequipa's existing industrial tradition, which has been largely concerned with woollens, dairying and cement production, and uses hydroelectric power (HEP) derived from the River Chili. Trujillo is an old regional capital in the heart of sugar-producing country which has been selected for the decentralization of the motor industry away from Lima, and is concerned with vehicle assembly, machine tools, diesel engines, tyre manufacture, and the production of pulp and paper nearby at Santiago de Cao (on the basis of *bagasse* produced by the sugar-refining process). In addition, one may note the creation of a new greenfield industrial complex at Bayovar, the extension of the chemical industry at Paramonga to include plastics, and the development of some heavy engineering and tinplate manufacture at Chimbote as a complement to its iron-and-steel industry, established in the 1950s and using HEP from the River Santa, iron ore from Nazca and imported coking coal.

Coastal Peru is the 'leading geographical sector' of the country in economic terms, and has been so from at least the mid-nineteenth century. It is also distinctive in social and cultural terms. Even in colonial times the sparsity of Indian population and the use of negro slaves in coastal plantations meant that the area rapidly acquired a different racial character from the sierra. In the nineteenth century the immigration of Chinese, either as traders or shopkeepers or as coolie labour for the exploitation of the guano quarries and the building of the railways, added a new strain to the racial character of the coastal zone. There was a revival of immigration from Europe, but it has always been the Indian immigrants from the sierra who have predominated in numerical terms to accentuate the generally mestizo character of the area. There are still occasional enclaves of traditionally mestizo culture, relatively static in economy, mobility and society, and such groups have been described at Moche and Santiago de Cao and can be

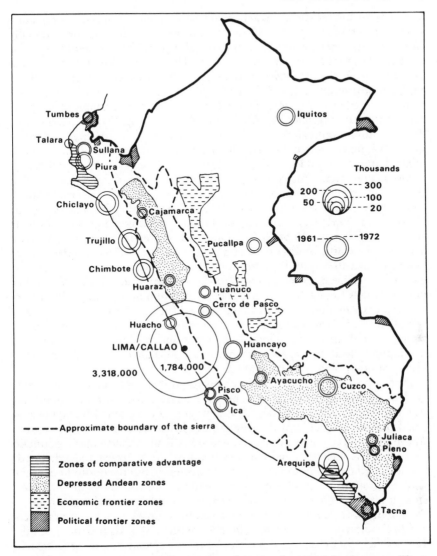

Figure 7.7 Peru: urban population, 1961 and 1972 and regional development. The populations of all towns of over 30,000 are shown, together with the designation of zones for regional development (1977) adopted by the Peruvian Institute of Planning

traced in the Ica valley, but the norms of coastal Peru are more nearly those of an urbanized society. Levels of literacy are higher than in the sierra, and the labour force is more articulate and more highly organized. But the relative prosperity and higher income levels of the coast are constantly undermined by the immigration of cheap, unskilled labour from the reservoir of poverty and population in the sierra.

The poverty of the sierra is evident from *Figure 7.7*. In terms of urbanization, possession of durable consumer goods, provision of services and education, non-agricultural employment, agricultural productivity and literacy, the sierra falls far below the average standards of the coastal region. The backwardness of the sierra has been interpreted in various ways. The debate is complex and stems from fundamentally ideological standpoints in the social sciences, though there has been little examination of the geographical implications of differing view-points. One group of arguments represents the sierra as an example of a dependent peripheral or satellite region subordinated to metropolitan interests within Peru, in turn subordinated to the finance capitalism of the advanced industrial world. Within the sierra, emphasis has been given to concepts of marginality, urban domination and internal colonialism, and the ways in which rural modes of production are articulated to capitalistic forms. Some of these views involve geographical corollaries to doctrines basically concerned with class exploitation and the expropriation of surplus values. Others, in the light of Myrdal's (1965) or Hirschman's (1966) views, also argue, without accepting radical views in their entirety, that the growth of Greater Lima and the coastal zone has involved a drainage of wealth, profits, skills and manpower by 'backwash' effects more powerful than the 'spread' or 'trickling-down' effects of modernization and growth created by the development of market opportunities. Finally, three other interpretations cannot be entirely ignored: a deterministic geographical view resting an explanation of poverty in terms of a difficult or hostile environment; a neo-Malthusian view focusing on the pressure of population on limited resources; and an outdated ethnic determinism, attributing the failure of development to Indian apathy and reluctance to change in the face of a long history of oppression and exploitation.

There is no doubt that the sierra presents very difficult conditions for efficient, modern, farming methods. Reasonably flat land is scarce at altitudes suitable for crop farming; in many areas rainfall is deficient, or more seriously, highly variable from year to year, and the extension of irrigation is much to be desired. Above 3500 m early or late frosts and hail are difficult hazards; many mountain soils are severely eroded, and pastures frequently over-grazed and exhausted. Cultivable land is scarce and patchy in its occurrence. But the possibilities of the sierra environment should not be too readily dismissed. There are important areas in which progress has been achieved, particularly in central Peru where the accessibility of the Lima market or the proximity of markets in the mining areas and in Huancayo have served to stimulate commercial peasant farming. In the Tarma valley, for example, a careful and intensive system of farming is oriented to the production of flowers and vegetables for the Lima market. The Mantaro valley is also a zone of relatively intensive farming, where incomes and material standards of living are significantly higher than in other parts of the sierra. There are efficient pastoral estates in central Peru and in the south, and small-scale cattle-fattening enterprises around Lake Titicaca are efficient at a low level of technology. Possibilities exist for the intensification of farming in the sierra of northern Peru on reasonably good, well-watered land. Much of the sierra is necessarily committed to pastoral farming and productivity is generally low. Yet there is considerable potential for them improvement of pastures by controlled grazing or re-seeding, the cultivation of fodder crops and the extension of irrigation, the introduction of hardy beef stock to improve poor *criollo* cattle, and the improvement of sheep, alpaca and llama. The use of ecological niches at different altitudes by Andean communities displays a complex and skilled relationship with a 'difficult' environment, and given better communications and market organization it has always seemed possible for the sierra to produce temperate commodities in tropical latitudes at a distance of only a few hundred kilometres from its markets. Nevertheless, agricultural progress in general has been slow and hesitant. The 1972 agricultural census revealed that technical assistance reached only 3 per cent of farmers, rural credit only 4 per cent. Only 16 per cent use fertilizers on their lands (including animal manures), and 20 per cent used purchased seed. Yields of the major crops (potatoes, maize, wheat, barley and indigenous grains and root crops) are miserably low.

Attitudes towards the possibility of change and reluctance to accept innovation among the Indian peasantry have frequently been debated. The cultivation of attitudes more receptive of change was one of the aims of the Vicos experiment conducted by Cornell University, and proved moderately successful, though under special circumstances (Dobyns, Doughty and Lasswell, 1971). Anthropologists and sociologists have shown that Indian communities are responsive to opportunities for change under favourable circumstances, e.g. in Huaylas (Doughty, 1968), Muquiyauyo in the Mantaro valley and in the Chancay valley (Matos Mar, 1969). The factors triggering change are very varied: the building of a new road, education, experience away from the village as a result of temporary migration or military service, communal mobilization against a common enemy (usually an encroaching landowner), or information and help from expatriates of the community in Lima. The dissemination of news and ideas by radio and the organization of peasant unions are all factors making for change and the acceptance of change, though not without disturbance of the community organization and an increasing social differentiation of the community; and change is by no means always for the better. Nevertheless, agricultural innovation and production for the market instead of for subsistence involve risk, and for small farmers at the margin of subsistence they involve risk to the very means of survival, not merely a monetary loss. Market production exposes the farmer to whole new areas of risk besides those natural hazards to which he is accustomed: price fluctuations, negotiations with middlemen, the vulnerability of indebtedness (for many peasants the unacceptable face of rural credit facilities vaunted by the modernizers). For those with land inadequate to support a family and with poor or unreliable access to markets, the rational and most advantageous strategy has often proved to be migration rather than agricultural improvement. The land remains a source of subsistence, but family effort is devoted to a search for income from the export of labour rather than from agricultural production for the market. People, in short, are more mobile than the agricultural commodities they can produce. For peasant farmers isolated from the market, then, with too little land and negligible savings, the retention of archaic forms of land use, and subsistence agriculture combined with the export of their surplus resource – labour – is a rational, optimizing strategy.

In geographical terms, the critical factor is that of accessibility to markets – for agricultural commodities and for labour. Where market opportunities exist by reason of accessibility to good roads, railways, mines, or the urban markets of the coastal region, the peasant farmers of the sierra have responded, modifying traditional methods and specializing for the market in ways which have frequently threatened and even substantially changed the social framework of so-called traditional communities. The geography of agricultural modernization within the sierra, even among farmers with only 1–5 ha, shows a clear relationship with accessibility to major markets and with the degree of urbanization. And this pattern, in turn, shows some degree of relationship to levels of welfare and education, the loss of population by migration, and the intercensal rate of growth of rural population.

There is a high degree of statistical correlation among all these factors at the provincial level. The regional pattern is one in which the central sierra stands out as the most highly developed zone, with values for urbanization, literacy, agricultural modernization, and material possessions and the size of the permanent wage-paid labour force in agriculture most closely approaching those of the coastal regions. Secondly, the sierra provinces of the Pacific slope, roughly corresponding to the Pacific side of the Andean watershed, have the character of a transition zone between coastal growth and prosperity and the poverty of the inner sierra. Accessibility to coastal labour and commodity markets is fairly good, and this is reflected in high out-migration, decline in rural population between 1961 and 1972, and intermediate levels of literacy, material welfare and agricultural productivity. The Puno region benefits from its close contacts by rail and road to Arequipa, the second largest city of Peru. It has a long history of commercial development, particularly of sheep and alpaca wool, and it has a high density of population and flat, cultivable land around Lake Titicaca. Values for development are about average for the sierra as a whole. Finally, there are two areas which are low on all scales of development and officially regarded as the depressed areas of the Andes in Peru (see *Figure 7.7*) – the inner sierra of Northern Peru, grouped around the basin of the Marañon, and the south-central sierra, comprising most of the Departments of Huancavelica, Ayacucho, Apurimac and western Cuzco. Urban growth is weak, communications are difficult and levels of agricultural development, literacy and national welfare are the lowest in Peru.

The growth of population beyond the capacity of the sierra environment to support peasant farming at a tolerable level of living is a fact of importance. It has undoubtedly led to the fragmentation of peasant holdings, the cultivation of sub-marginal land, and a great deal of soil erosion or exhaustion through the cultivation of excessively steep slopes or the overgrazing of limited pastures. In 1961 there were *c*. 600,000 agricultural units of less than 5 ha in the sierra region; by 1972 this figure had increased to 884,209 agricultural units, and nearly half of these were less than 1 ha. Growth of population has created, too, the urgent pressures to supplement farm incomes by other means, either by employment on haciendas, in local towns or in craft industries, or by migration to the coast, to Lima or to the *selva*. Within the sierra, however, there is relatively little non-agricultural employment and few prospects for any immediate increase. Industrial possibilities are limited, and although there are modest enterprises in Huancayo, Cuzco, Puno and Juliaca, chiefly concerned with food, drink, textiles and construction, they rest for the most part on the special needs and tastes of the sierra Indians and are thus dependent in the last resort on agricultural productivity in the region itself. Mining is locally important, particularly in the central sierra along the Central Railway to Cerro de Pasco, and also in many scattered sites throughout the region, but employment offered by mining is relatively little, and the impact of mining on regional agricultural economies has not been great (except perhaps in the central zone). Small-scale craft industries exist in many parts of the sierra, and whole villages often concentrate on the production of a range of goods for which they are well fitted by resource or traditional skills. But possibilities of expansion are clearly limited, again, by peasant income levels and by the growth of the potential tourist market. But many would argue that the pressure of population is not the heart of the problem in the sierra. Migration has been high and the intercensal growth of population in the sierra has been quite modest. Many have seen the fragmentation of holdings and the cultivation of sub-marginal land as not only a product of the lack of land and resources, but primarily as the result of the monopoly of land by a privileged few, effectively preventing the expansion of peasant farming.

Indeed, it is the question of land tenure which has been seen as the major obstacle to fundamental agricultural and social change in the sierra and as symptomatic of the injustice and inefficiency of the social structure of the region. The polarization of ownership between a small number of large estates with most of the land, and many small farmers with access to a few hectares or less has been commonly labelled as the *latifundia/minifundia* complex. The *latifundia* of the sierra have been characterized by the exploitation of a dependent peasantry, the extensive use of resources, lack of capital investment and a low level of technology, and by mechanisms which drain rural earnings to absentee landowners in the cities. Traditional tenures involved the granting of a small plot to peasant farmers in return for which they owed their landlord labour services on the land he worked directly, in his household and in the cartage of his goods. On pastoral estates, *colonos* were allowed to graze their animals on the hacienda in return for services in caring for the hacienda flock; indebtedness was a mechanism retaining the dependent peasantry on the estates. If wages were paid they were minimal, so that the landowner secured a production from his land with the minimum of cash outlay. Landownership itself was valued as much for power and prestige and as a hedge against inflation as much as for the income it generated. With some notable exceptions, landowners made little effort to maximize income or raise productivity, partly because revenue from estates could be more profitably invested elsewhere – in commerce, urban land, in industry or even abroad. The migration of capital from the large estates has been the counterpart of the migration of the peasantry. In many areas, therefore, extensive forms of exploitation prevail on the large estates, while peasant farmers in neighbouring districts coax a subsistence crop from a hectare or two of poor and over-exploited land.

In the Peruvian sierra the reaction of peasant communities, first to the appropriation of their communal land by the expansion of the haciendas in the late nineteenth and early twentieth centuries, and later to the harshness and injustice of their condition, was increasingly hostile. Rural rebellion has had a long history in Peru going back to the colonial period. In the early 1920s the Puno region was a focus of rebellion as a result of the appropriation of land by the expanding haciendas, and in the early 1960s there were widespread outbreaks of land invasions to occupy under-used portions of haciendas claimed as former communal lands. Land invasions and the organization of peasant unions are among

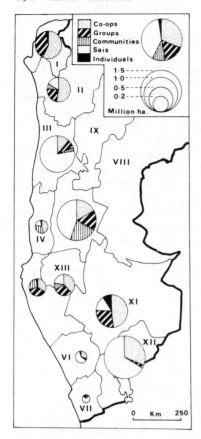

Figure 7.8   Peru: agrarian reform at July 1976. Circles show the area adjudicated in each Agrarian Reform Zone and the proportions allocated to co-operatives, groups, *comunidades*, the 'agricultural societies of social interest' (SAIS) and to individuals

the factors leading to increasing pressures for land reform, though by no means the only ones.

Tentative efforts at land reform in 1962 and 1964 were followed by the much more radical programme of the military regime from 1969. Between then and 1976, when the expropriation of large estates ground to a halt, some 8 million ha were expropriated for the benefit of 280,000 farm families (*Figure 7.8*).

A major feature of the Peruvian agrarian reform was its emphasis on co-operative structures. Relatively little land was adjudicated to individuals (2 per cent), but rather more to peasant groups (16 per cent) seen as the precursors of co-operative structures, and about 10 per cent was adjudicated to peasant communities (*comunidades*). Many of the former estates were reorganized into co-operatives (31 per cent) and their membership was constituted by former tenants and labourers on the estates. In some areas, particularly in the sierra, an attempt was made to draw in neighbouring communities, which were to participate with former estate workers and tenants in larger units labelled as Agricultural Societies of Social Interest (SAIS). Both co-operatives and SAIS were to share the profits among the membership and to accumulate capital for investment as well as to repay the government for land, stock and installations. Agricultural production was to be intensified and techniques modernized under the aegis of professional officials and trained agronomists. The reality in some cases, however, falls short of the ideal.

The application of reform has not been uniform across the country, and the areas most affected by land reform were precisely those in which the general process of modernization in the sierra had proceeded furthest: the central sierra, the Puno region and the Pacific slope. The majority of smallholders have not been affected by the reform. They still confront the old problem of land hunger and fragmentation of their land, and sometimes resent the new-found relative wealth of former labourers and tenants, now the beneficiaries of the reform sector. There was, in any case, never enough land to go round and to provide the sierra peasantry with adequate holdings: one calculation concluded that there was only enough land in large estates to provide 9 per cent of sierra farmers with adequate holdings. There are frictions between the professional cadres of management in the co-operatives and the former dependent peasantry in the reformed sector. Many of the former tenants on the estates have been reluctant to abandon to the co-operatives their personal rights to land or to the pasturing of stock which they formerly had in labour-service or share-cropping tenures. Furthermore, it is still a matter of some doubt as to whether the new structures will go very far to resolve one of the major deficiencies of the old system: the tendency for large units to be worked extensively in the face of land hunger among the peasantry at large.

The co-operative structures have not been wholly successful. They have been most easily accepted where former estates were worked by a substantial permanent wage-paid labour force, notably on the coastal sugar estates and in a few of the sierra estates. But even here there have been many problems: co-operative members are jealous of their newly-won privileges and make exploitative use of casual labour. In the sierra many co-operatives exist only in name, and are effectively divided into peasant holdings. In general, and in spite of land reform, the *minifundia* remains; and in a sense, the *latifundia* still persist, though under new management and with a closer identification with the interests of the state. Land reform in Peru has left many problems still to be solved, not least that of increasing the volume and efficiency of food production, but nevertheless, a major and probably irreversible threshold has been crossed. The long-term impact of Peru's land reform is likely to be considerable.

To promote the settlement and colonization of the empty, forested regions of eastern Peru has seemed at times a possible solution to some of the country's pressing problems. From the 1940s road construction into the *montaña* has been followed by, and sometimes anticipated by an extension of settlement, especially in the central zone (along the road from Cerro de Pasco to Tingo Maria and thence to Pucallpa, and in the zone of Satipo, Oxapampa and Chanchamayo) and to a smaller extent in the area north of Cuzco in La Convención.

During the Belaúnde regime of 1963–8 settlement of the *selva* was a major target of government policy. It was argued that eastern colonization would increase national production, create new sources of external revenue, substitute domestic sources for imported foodstuffs and raw materials, relieve over-population in the sierra, and help to divert the current of migrant population away from Lima and the coast. Detailed resource studies, an ambitious road-building programme and a number of colonization projects were undertaken. The Grand Design of

Belaúnde's *Carretera Marginal* (the Marginal Highway) was to link the eastern regions of Andean countries from Venezuela to Paraguay, and more specifically to link up existing settlements in eastern Peru by a longitudinal highway.

The economic function of such a route has repeatedly been questioned and the heavy investment involved could scarcely be justified, though individual links of the *Carretera Marginal* have value in so far as they provide road access to isolated settlements from the coastal region by way of existing trans-Andean links. Construction work on the *Carretera Marginal* was virtually abandoned during the military regime, which concentrated its attention on the implementation of agrarian reform in the *selva* and elsewhere, though some colonies, including a few military colonies near the Brazilian border, were established. But Belaúnde's second term of office has seen the revival of interest in the settlement and development of the eastern regions, new promotion of road-building schemes, and the revival of colonization projects (see *Figure 7.7*), particularly in the central *montaña* region.

Of the major crops produced in the *selva*, coffee has proved its worth on international markets and exports have increased significantly in the last decade. But international agreements, poor prices and high transport costs limit possibilities of much further growth. Tropical fruits such as pineapple, citrus, passion fruit and papaya have limited possibilities for supporting an export trade as fruit juices or in canned form, but are unlikely to be highly important. Bananas are an important *selva* crop, but they face coastal competition and cannot compete in international markets with, for example, Ecuador. Rice is a good prospect for the home market, especially from the northern area around Bagua where it is already well established, and beef may have good prospects as a future industry if domestic prices are maintained at a higher level than at present or if imports of foreign beef are restricted. In the *montaña*, coca has been a traditional crop for centuries, supplying the modest needs of the sierra Indians for the dried leaves as a stimulant, but in recent years it has made illicit fortunes with the expansion of the drug traffic, though not on quite the scale of Bolivia. Many other crops are grown (e.g. rubber, palm-oil, cocoa), some on no more than an experimental scale, but in general the difficulties and expense of transport to the coast, or by navigable water to Iquitos, greatly restrict competitiveness. Some products must compete with coastal production more favourably located: sugar, cotton, rice and fruits, for example. Beef faces competition from the sierra as well as from the coast. The nearest, most obvious and accessible market is, of course, the sierra itself, but as long as the sierra market is limited by the poverty of its inhabitants, there is little hope in this direction, except for the small-scale marketing of coffee, citrus, fruits and the like.

Indeed, the major achievements in the settlement of the eastern region have been accomplished as a result of the piecemeal, individual movement of highland peasant farmers, carving new farms for themselves from the forests. It is this process which is given the unlovely label of 'spontaneous colonization' to distinguish it from the officially sponsored, and often very expensive organized colonization schemes. In the movement from the *altiplano* to Tambopata, for example, highland Indians have colonized new lands, maintaining social and commercial links with their villages of origin, but they adapt surprisingly well to totally different environmental conditions, switching from barley or potatoes as a staple crop to maize or manioc, and producing a small surplus for sale. They have, in general, 'squatted' without title to the lands they occupy, and practise forms of shifting cultivation, clearing a piece of land for a few years' crops before abandoning the plot and clearing afresh.

Attitudes towards the colonization of the eastern region have shifted violently in the recent past. In the 1950s, large land concessions were still being made to foreign companies (e.g. at Tournavista, near Pucallpa) and European immigrants or Peruvian nationals were encouraged to establish plantations for the cultivation of coffee, tea, citrus fruits, etc. Spontaneous settlement was, on the whole, ignored. In the 1960s emphasis was given to road-building and the establishment of organized colonies, seen as the nuclei from which settlement would spread; spontaneous settlement was to be encouraged and provided with educational and health facilities. During the military regime the full vigour of the agrarian reform laws was not applied to the eastern regions, but efforts were, however, made to provide small and medium farmers with title and access to credit facilities. But in recent years the trend has been towards the rehabilitation of the large estate as the means by which economic exploitation is to be encouraged (Scazzocchio, 1980). In Madre de Dios large-scale concessions are linked with programmes of deforestation and the conversion of land to pastures for beef production along Bolivian or Brazilian lines. There are new contracts for the extraction of timber, and fears are expressed not only for the ecological consequences of deforestation, but also for the future of the scattered indigenous groups of Indians whose communal lands are threatened.

Bolivia

Of all South American countries it is Bolivia that faces the most intractable geographical and economic problems, not only because of its poverty and its excessive dependence on exports of tin and associated minerals, but also because of its position as a land-locked state and its lack of internal cohesion. In spite of their regional contrasts, Peru and Ecuador have identifiable 'leading geographical sectors' which focus on the nodal regions created round Lima, Guayaquil and even Quito. But neither the export economy nor metropolitan dominance have succeeded in creating for Bolivia a strongly established nodal region that might serve to counteract the centrifugal tendencies of its peripheral regions. And until very recently it has also demonstrated in extreme form that dichotomy between traditional subsistence economy and export-orientated activities which may be taken as a hallmark of the 'dual economy'.

Its poverty within South America is evident from those indices that are often taken as a general measure of economic and social development. Of the ten countries of South America (excluding the Guianas), Bolivia has the distinction of being at the bottom of the league table of development, with a *per capita* GDP of $363 in 1978 (1970 prices), an urban population (in settlements of more than 5000) of only 38 per cent and a weakly developed manufacturing sector. An illiteracy rate of 62 per cent and an average of 2500 people per physician, are symptomatic of inadequate provision for health and education. Land-locked and with a population of only 4.6 million in 1976, yet faced with the problem of integrating a large and sparsely populated territory of extremely diverse and difficult terrain, Bolivia is clearly one of the weakest states of Latin America.

Political and economic weakness have been evident from the time when Bolivia achieved its independence in 1825, carved from the Viceroyalty of La Plata and consisting essentially of what had formerly been the *audiencia* of Charcas. At its creation, Bolivia contained approximately 2.3 million km², stretching from the Pacific to a rather indefinite northern boundary in the Amazon basin and across the Gran Chaco to an equally indefinite boundary in the south-east. Now, with an area of some 1.1 million km², it covers less than half of its former extent as a result of the loss of territory to all its neighbours. Brazil and Peru appropriated unexplored and sparsely populated rain-forest, swamp and savanna in the north and north-east during the nineteenth century when the search for cinchona and wild rubber gave sudden but temporary economic value to these remote forested areas. To Paraguay Bolivia lost much of its claim to the thorn and scrubland wastes of the Chaco as a result of its humiliating and crushing defeat in the disastrous War of the Chaco as recently as 1935. But the most important loss, to which Bolivia has never been entirely reconciled, was the loss of its window on the Pacific to Chile as a consequence of the War of the Pacific (1879–83). Bolivia thus became the land-locked state it now is, and the creation or the maintenance of external links by the building of railways and roads or by the improvement of river navigation has continued to preoccupy Bolivian policy.

No less important than the vulnerability of a land-locked state is the internal lack of cohesion within the existing boundaries. Regionalism is a force that has, at times, threatened the unity of the state, particularly in the eastern regions: 'a citizen of La Paz or Santa Cruz is apt to speak of himself and think of himself as a *Paceño* or a *Cruceño* rather than a *Boliviano*' (Edelmann, 1967). In part this lack of cohesion is associated with the harsh realities of terrain and climate and the possibilities of settlement.

Four major regions of settlement are usually identified in Bolivia (*Figure 7.9*): the *altiplano* in the west of the country, the low-lying *oriente* in the east and northern regions, and two intermediate zones between them, the *valle* region and the *yungas*. Each is associated with highly contrasting conditions of population, settlement and agriculture.

The *altiplano*, enclosed between the western cordillera, which forms the boundary with Chile, and the eastern cordillera, which contains the major zone of mineralization, is itself divided by climatic boundaries, for rainfall diminishes rapidly towards the south and south-west, which is truly arid country of salt-pans (*salares*), internal drainage and a sparse and scattered population. It is only in the north-east of the region, therefore, that the *altiplano* can support a dense rural population, especially near and to the south of Lake Titicaca, core of the Aymará-speaking Indian population of Bolivia.

Beyond the eastern cordillera the Andean massif is deeply and intricately dissected, a barrier to communications and barren of population except in a few favoured basins at approximately 2000 to 2700 m above sea-level, supporting an adequate range of food crops and a dense rural population. Three of these *valle* regions have played an important role in Bolivian development since colonial times, when the area around Sucre (Chuquisaca) was the nearest area of supply for the city and mines of Potosí and the nearest zone of comfortable temperate climate for Spanish residents. Tarija to the south was never so intensively

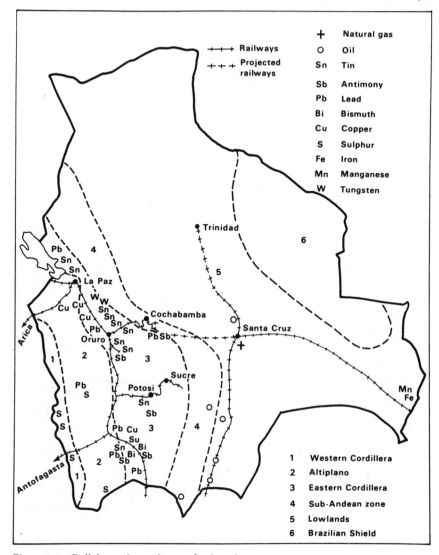

Figure 7.9 Bolivia: major regions and minerals

developed, but the Cochabamba basin, largest of the three, and accessible from the Oruro mines and even from La Paz, has always been a major zone of Indian and mestizo settlement.

The *yungas* is roughly equivalent to the *montaña* in Peru and identifies the highly dissected, humid, cloud-forest or tropical-forest zones at altitudes of less than 2500 m on the eastern slopes of the Andes. Like the *montaña* in Peru it has been the scene of recent clearing and colonization for the production of tropical and sub-tropical crops. It is best developed to the north of La Paz and to the north and north-east of Cochabamba, especially in the region of the Chaparé river.

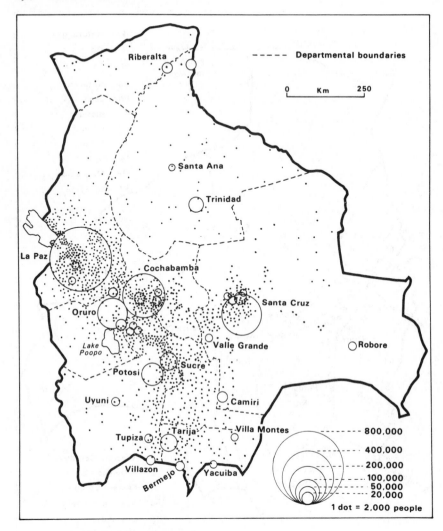

Figure 7.10   Bolivia: distribution of population, 1976

Taken together, the *altiplano*, *valle* and *yungas* regions contain about 70 per cent of the total population of Bolivia.

The *oriente* consists of the low-lying zones of eastern and northern Bolivia, in which climate and vegetation range from Amazonian rain-forest and massive areas of swamp and savanna grasslands to the highly seasonal rainfall and harsh scrub and thorn forests of the Chaco along the border with Paraguay and Brazil. This eastern and northern region contains some 30 per cent of the population, much of which is concentrated in the expanding zone of settlement in the south-east around Santa Cruz, the department of that name, containing 15 per cent of the total population. The north and north-east are still virtually empty, except for a few riverine settlements of no great importance. Indeed, except for the urban agglomeration of La Paz (650,000 in 1976) and the dense rural population near Lake Titicaca, the existing nuclei of population are relatively small, widely scattered and difficult or expensive to link together by reason of terrain as well as distance.

The differentiation in agricultural economy and distribution of population in the *altiplano*, the *valle* region and the eastern lowlands is accentuated by the composition of the population. Estimates of the 1950 census put the Indian population at 50 per cent of the total and the *cholos* of mixed blood at 35 per cent, with 15 per cent white or near-white. The *altiplano* is overwhelmingly Indian, except La Paz, with an important distinction between the Aymará country near Lake Titicaca and to the south, and Quechua-speaking populations elsewhere in the sierra, and in the *valle* region, where mestizos are more numerous, especially in the towns. Whites and mestizos, with a greater proportion of Guaraní blood, are more frequent in the eastern regions. The Aymará in particular retain a sense of community, and prevailing attitudes of whites and mestizos to the Indians, or of the Indians to the mestizos, have at times encouraged the regionalism inherent in Bolivian geography.

Yet the regional balance of population has been changing quite dramatically in the recent past. The centre of gravity of Bolivian population, and of its social and economic interests, has been shifting steadily eastwards (*Figure 7.10*). Considerable effort has been put into the integration of the country as a whole by the improvement of its internal communications, and by the economic expansion of the *oriente*. The effects are apparent in the changing population structure of the country as a whole, and are characterized in part by the familiar process of increasing urbanization, but also by an eastwards movement from the *altiplano* to the *oriente*. The basic elements become apparent from *Table 7.2*. La Paz retains its status as a fast-growing capital city, but the predominantly Indian *altiplano*, with its long-standing preoccupation with mining and high-altitude subsistence farming, is losing ground rapidly to the new interests of the *oriente* which increased its share of the national population from 20 per cent in 1950 to 29 per cent by 1976. Rapid urban expansion in Santa Cruz has meant that it has replaced Cochabamba as the second largest city of Bolivia. For various reasons it has been the 'boom town supreme' of Bolivia in the last 20 years. Indeed, a new dynamic economic axis linking La Paz, Cochabamba and Santa Cruz in an east-west direction has now largely replaced the former export-oriented axis linking La Paz with the mining areas of the eastern cordillera by way of Oruro to the south, and thence to Antofagasta in Chile.

Table 7.2 Bolivia: distribution of population, 1950 and 1976

Region	1950	%	1976	%
La Paz (province)	290,731	11	695,566	15
Altiplano	1,124,120	42	1,519,780	33
Valle	735,988	27	1,081,702	23
Oriente	553,326	20	1,350,768	29
Total	2,704,165	100	4,647,816	100

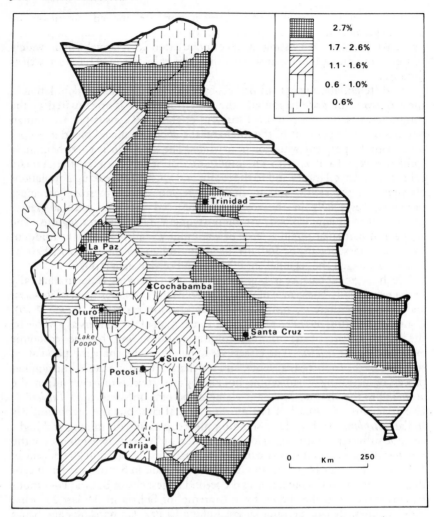

▓▓▓	2.7%
≡≡≡	1.7 - 2.6%
╱╱╱	1.1 - 1.6%
▦▦▦	0.6 - 1.0%
┊┊┊	0.6%

Figure 7.11   Bolivia: average annual rate of growth of population, 1950–76, by provinces

The map of population distribution suggests the existence of both of these axes, but should be seen in conjunction with *Figure 7.11*, showing the rates of change of population from 1950–76. The average rate of population growth in Bolivia has been relatively low for Latin America, at 2.1 per cent per year. But large areas of the *altiplano* have increased very slowly at rates of growth of less than 1 per cent per year, including both the sparsely populated, arid region of the *salares* in the south-west, and also the densely populated Aymará zone of the *altiplano* in the neighbourhood of Lake Titicaca. Only in and around the cities of the *altiplano* has there been fairly rapid growth – in and near La Paz, Oruro and the mining areas. On the other hand, the provinces of the *oriente* have generally registered growth rates of over 2 per cent/year and Santa Cruz itself has been growing at the rate of 7 per cent/year, its population increasing from about 50,000 in 1950 to 257,000 by 1976.

Centrifugal tendencies of the eastern lowlands and the *valle* region were also encouraged in the past by the extent to which the pattern of communications has tended until recently to link nuclei of settlement more strongly to the outside world than to each other. Railway-building policy was primarily oriented from the beginning to the creation of external links by which Bolivia's mineral wealth could be tapped. Completion of the Southern Peruvian Railway to Puno in 1874, and the introduction of steam navigation on Lake Titicaca provided the first link to La Paz, though the railway link from Guaqui to La Paz was only finished in 1903. The Antofagasta–Bolivian Railway to Oruro was built by 1892, and reached La Paz in 1911; the Arica line was completed in 1913; and the railway across the *altiplano* from Uyuni to Villazon and thus to Argentina was finished only in 1925. The western railways to the Pacific ports, built primarily to handle exports of Bolivian tin and the import trade for La Paz and the mining centres of the *altiplano*, still remain the most important in terms of traffic carried, but they are, in general, in poor condition and badly maintained.

An eastern railway system has only begun to take shape in the recent past with the completion, as late as 1955, of a line from Santa Cruz eastwards to Corumba where it links with the Brazilian system, and a line, completed in the early 1960s, from Santa Cruz southwards to Yacuiba and onwards to Argentina. Argentine capital is also involved in a new construction project, the first two stages of which have been completed, from Santa Cruz northwards to the Mamoré river at Trinidad. All of these relatively new lines stress not only the recent emphasis which has been placed on the development of oil and agricultural resources in the *oriente*, but also the strategic importance for Bolivia, as a land-locked state, of its external links. These concerns are also echoed in extensive road-building programmes in the eastern region, linking the main urban centres and focused especially on Santa Cruz and Cochabamba, and the building of pipelines to export natural gas to Argentina and to Brazil.

While the external links of Bolivia's road and rail network are adequate, the same cannot truly be said of its internal linkage. The eastern cordillera presents difficulties of terrain and relief between the *altiplano* and the eastern region, and it was only in 1953 that a most important modern road linked Cochabamba with Santa Cruz, completing the connection from La Paz to the east; but there is still no rail connection between Cochabamba and Santa Cruz, a link which would complete a transcontinental railway connection from Santos in Brazil to Arica in Chile. New roads from Cochabamba and La Paz have opened up new fields for settlement in the *yungas* regions around Caranavi and in the Chaparé river area, but much of northern and eastern Bolivia is still dependent on mule traffic or river trade on the 12,000 miles of navigable waterways, chiefly on the Beni, Mamoré, Guaporé and Chaparé rivers. In the far south-east a new river port, Puerto Busch, is being built to facilitate navigation on the River Paraguay. But although overland links, especially in the east and north of Bolivia, are very poor, air transport is fairly well developed, and many large and remote ranches have airstrips for the export of beef to urban and international markets.

To a large extent the weakness of internal integration until very recently may be associated with the failure of the traditional export economy to generate spread effects in the economy as a whole. Indeed, until the 1950s Bolivia could be regarded as a classic example of a dual economy in which a traditional and largely subsistence economy had not been greatly modified by the super-imposition of a modernized sector associated with mineral production. 'Whatever infrastructure was present in Bolivia prior to the 1952 revolution was almost entirely produced, either directly or indirectly by the mining industry' (Zondag, 1966, p. 21). Labour for the mines was certainly drawn from subsistence farmers in the *altiplano*, but low wages and therefore low purchasing power induced few secondary changes in the farming of surrounding areas and most of the food requirements for the mining population were in fact satisfied by foreign imports.

> One might say that two civilizations gradually came into existence in Bolivia. One was highly mechanized and highly capitalized, employing large numbers of foreign technicians who tended to build their own world with modern comforts, while the other civilization was based completely on the old way of life, which had not changed for hundreds of years. Mining camps constituted practically the only link between the two. (Zondag, 1966, p. 21)

It is not surprising, therefore, that the main thrust of Bolivian policies, especially since the early 1950s, has been towards the integration of a national economy, economic diversification and the expansion of the modernized sector in industry and agriculture.

Yet mining is still the most important basis of the Bolivian economy. Mineral products accounted for 91 per cent of Bolivia's official exports in 1978 and 8.5 per cent of the GDP. Oil and natural gas provided 17.7 per cent of all exports, and tungsten, silver, zinc and antimony together provided almost exactly the same proportion, but tin exports alone made up 56 per cent of the total. Bolivia was the second-largest world producer of tin after Malaysia, though its position is threatened by Thailand and Indonesia, and it is also a major world producer of antimony, tungsten and bismuth. Recent years have also seen the rise of eastern Bolivia as a producer of oil and natural gas, and there are huge deposits of high-quality iron ore in the far south-east of the country at Mutún which still remain largely untapped.

Most of the wealth in non-ferrous metal ores lies in a highly mineralized zone in the eastern cordillera stretching from the Peruvian border east of Lake Titicaca to the frontier with Argentina in the south. Tin ores are here associated with other minerals, including silver, tungsten and bismuth; silver, lead and zinc ores occur in close association. There are important deposits of antimony; copper ores underlie the *altiplano* south-west of Oruro, and sulphur deposits in the border region with Chile are of lesser significance. About a half of Bolivia's non-ferrous mineral production comes from mines within 100 km of Oruro, but there are other important groups of mines near La Paz, Uncia and Potosí.

Tin mining began seriously about 1895, rapidly replacing silver as a major product with the building of railways and the increase of tin prices following the growth of its use in the canned-food industry. The growth of production and exports from 10,000 tonnes in 1900 to 40,000 tonnes in 1930 was closely bound up with the fortunes of the tin empires of the Patiño, Hochschild and Aramayo families. The Patiño empire, for example, became an international enterprise involving the finance of tin-smelting and processing in Britain and the USA, but in Bolivia itself the Patiño empire alone provided 50 per cent of the public finance of the country and over 80 per cent of Bolivia's foreign currency (Klein, 1965).

In one of Bolivia's more important revolutions, however, the MNR (National Revolutionary Movement), strongly backed by the miners' unions, came to power in 1952 and nationalized the greater part of the mining industry. The family tin empires were expropriated to form COMIBOL (Bolivian Mining Corporation), which is still the dominating element in the mining industry, especially in the production of tin. Three branches of the mining industry are now recognized in Bolivia and are distinguished by different laws and tax structures: large-scale mining, controlled by COMIBOL, employing about 25,000 people and responsible for about 60 per cent of Bolivia's mineral output in 1980; medium-scale enterprises, employing about 6000–8000 people, privately owned and accounting for 25 per cent of output; and the small-scale and co-operative sector, characterized in general by low technology and primitive methods (though not exclusively so) and employing about 30,000 people. The medium-scale private sector is largely Bolivian-owned, though there are some foreign-owned and joint enterprises, and foreign capital has been attracted to some of the more efficient enterprises which have, for example, pioneered Malaysian techniques of tin-dredging in Bolivia, and have achieved unusually high recovery rates for tin and tungsten. It produces most of Bolivia's antimony and nearly half its tungsten. The medium-scale sector appears to be the most dynamic of all three categories, though the price of success is sometimes feared to be nationalization and incorporation into the large-scale sector. The small-scale sector includes countless tiny mining operations conducted with little technical expertise or equipment, though some of the co-operatives are better founded, and it is, in general, the sector most exposed to price fluctuation for its product.

COMIBOL, the large-scale sector, inherited a difficult situation from the period of nationalization. The mines themselves had been run down before nationalization, with few reserves and little new investment; there was a flight of technical expertise, largely foreign, which has only recently been replaced; over-manning has been a constant problem, and labour relations with powerful unions have been difficult. Furthermore, the nationalized mining sector has been heavily taxed by governments in various ways, reducing the revenues that might otherwise have gone to capital investment and more intensive exploration. Moreover, chronic political instability and wide fluctuations of policy towards the labour unions and investment have not helped to create a stable environment. Nevertheless, it has had some success in maintaining production and has managed to keep down costs at the minehead (Fox, 1978).

Formerly, all of Bolivia's major minerals were exported as ore concentrates, but in the last 15 years there has been a concerted effort to establish a domestic smelting and refining industry. The completion of the first tin smelter near Oruro in 1971 was regarded as a major step forward, and progress has been made towards the establishment of refineries and smelters for bismuth, lead and silver (near Potosí) and antimony (near Oruro). A growing proportion of Bolivia's minerals is being exported as metals rather than in the form of concentrates.

For a variety of reasons Bolivia has long been a high-cost producer of tin, with consequent vulnerability to changes in demand and price on world markets. Tin ores have declined in quality from a metal content of 16–18 per cent before 1914 to about 1 per cent at the time of nationalization, and to as little as 0.5 per cent in some important mines in recent years. Tailings from old workings often yield higher-grade ores than many of the primary ore veins and are being reworked. Veins are thin and tortuous, so that mining is difficult and involves the removal of much waste material. The recovery rate is relatively low because much metal is lost as dust during production of ore concentrates. Furthermore, transport hauls across the Andes to Pacific ports are long and costly. But in addition, taxation, royalties and the provision of education and welfare facilities by COMIBOL, desirable though they are, all tend to raise costs.

Since the 1920s exploration for oil in south-east Bolivia has been attended by great hopes for the future development of the region, and for Bolivia as a whole. Camiri was the earliest field to produce, from 1927, and still remains one of the most important areas of production, which now comes from a variety of fields in the departments of Santa Cruz, Tarija and Chuquisaca. Output grew rapidly in the late 1950s and reached a peak in the mid-1970s, since when production has fallen as a result of declining output and the failure to explore and develop new sources. Current estimates of reserves are relatively low at no more than 10 years' consumption at current levels, and the size and productivity of new finds have fallen short of expectations. From the late 1960s oil exports seemed to lead a badly needed diversification of Bolivia's external trade away from the preponderance of non-ferrous metals, but declining production and very rapidly rising internal consumption in the late 1970s suggest that Bolivia is likely to be a net importer of oil in the early 1980s. Nevertheless, the oil boom has given a considerable impetus to the development of the east and south-east. A network of pipelines links Camiri and Santa Cruz with Sucre, Cochabamba, Oruro, La Paz and Arica in Chile. Oil refineries are located at Santa Cruz and Cochabamba, and there are smaller ones elsewhere, with implications for the development of petrochemical industries at Cochabamba and Santa Cruz. Oil revenues have provided departments with funds to improve urban infrastructure and communications and to invest in industrial development. Yet it is hard to resist the conclusion that the oil industry in Bolivia has suffered from political uncertainties and the vicissitudes of policy. Oil companies were nationalized in 1937; in 1956 Bolivia was again opened up to foreign companies, but in 1969 the major foreign producer was expropriated and foreign participation only became possible again from 1972. Of fifteen foreign companies attracted by these new arrangements, only two remain in 1980.

Compared with the vicissitudes of the oil industry, natural gas has been highly successful. Vast reserves exist in south-east Bolivia near Santa Cruz and in the departments of Tarija and Chuquisaca, and deposits have also been found south-west of Oruro. A pipeline to Argentina has been completed from Santa Cruz and carries exports which are significant to the Bolivian economy as a whole. A further pipeline eastwards to the Brazilian border is being planned, and when completed will permit the export of gas to São Paulo.

There is no shortage of potential sources of energy in Bolivia. Oil and especially natural gas are more than sufficient for internal needs, and there is a vast potential for hydroelectric power. Hydroelectric power, generated by state-controlled enterprises and by private capital, serves the major cities of La Paz, Cochabamba and Oruro; Santa Cruz is served by thermal electric power using local natural gas. But apart from small private generator plants, the provision of electricity is restricted mainly to the towns and the mines. Bolivia's *per capita* consumption of electricity is, indeed, lower than that of any other South American country with the exception of Paraguay, and a large proportion of the industry's output (40–50 per cent in the early 1970s) is directly related to consumption by the mining industry, and only 15 to 20 per cent is consumed by other industry.

Manufacturing industry is limited in part by the low purchasing power of the vast bulk of Bolivian population. Only 9 per cent of the economically active population, according to the 1976 census, were then engaged in manufacturing industry, and a further 6 per cent in construction, and manufacturing industry contributed only 14 per cent of GDP. A very large proportion of those engaged in manufacturing are, in fact, small-scale artisans employing up to two or three workers. Large-scale industry is still very limited in spite of the efforts of the National Institute of Investments (INI) to promote industrial development. In general, industries are of three kinds: those directly dependent on the mining and oil sector: smelting and oil refining, a small petrochemical industry and industries supplying simple equipment for the mining industry; consumer goods industries concerned with textiles, food, drink and tobacco; and a small, but carefully nurtured sector concerned with durable consumer goods and engineering. Cochabamba has a metal-working industry producing refrigerators and the like, and it is also the locale for motor-assembly plants being set up by Ford and Renault. But plans for the establishment of heavy industry in Bolivia, on the basis of Mutún iron ores and natural gas in the south-east of the country, are still hardly at the drawing-board stage. La Paz, Cochabamba and Santa Cruz are the major industrial centres. La Paz, as capital, has the greatest concentration of purchasing power to attract consumer industries and the advantage of proximity to government. Santa Cruz is growing more rapidly and has an industrial potential deriving from the proximity of raw materials – oil and natural gas, but also food production (sugar-refining is already well established). But Cochabamba has potential for future development. It is located at an intermediate position between the capital and Santa Cruz, and it lies in a productive, densely populated basin with a temperate climate, but with easy access, too, to sources of tropical foodstuffs from the *yungas* region of the Chaparé river. It is a centre of textile production, dairying, and other food-processing industries and metal-working, and has satisfactory electricity supplies and substantial HEP potential in the process of being realized.

As in Peru and Ecuador, growth of the internal economy must rest on the improvement of agricultural productivity and output. Agriculture still occupies 45 per cent of the economically active population, and although agricultural production has increased steadily in the 1960s and 1970s, most of the increase of about 2 per cent per year can be attributed to an extension in the area cultivated rather than to increases in yields. Rural incomes tend to be low, and Bolivia has, even for Latin America, an unusually high disparity between the level of rural and urban incomes.

There are fundamental contrasts in Bolivian rural development which reflect in part the physical environment and in part the social and historical context of different areas. Climate severely restricts the range of crops that can be grown on the *altiplano* at altitudes of over 3800 m. The growing season is short, and high diurnal temperature ranges involve high risks of damage from frost and hail. Rainfall is highly seasonal and also unreliable from one year to another. Conditions around Lake Titicaca are better than elsewhere because of the ameliorating effects of the lake itself – slightly higher rainfall and slightly fewer risks from frost and hail, but climate becomes more arid and temperatures more extreme towards the south-west of Bolivia, so that crops require irrigation and natural pastures become more impoverished. In the *valles*, temperate conditions and higher rainfall permit the cultivation of wheat, maize and beans, temperate fruits and horticultural crops, as well as offering better pastures for dairying. Both of these areas may be regarded as the old-settled regions of Bolivia, and it is in these areas that land reform has had a profound effect since 1953. The concentration of landholding and the persistence of the traditional hacienda economy and of servile tenures were formerly at least as oppressive in the Bolivian *altiplano* as anywhere in Latin America. There was the same contrast between the large estates and the inadequate holdings of the Indians in *minifundia* and in *comunidades indigenas* as in Peru and Ecuador. According to the census of 1950, 90 per cent of the private agrarian property was held by 4.5 per cent of the total number of landholders, yet 70 per cent of the farm units together comprised only 0.41 per cent of the area exploited. In the *altiplano* the familiar pattern of the central Andes was normal whereby *colonos* were allowed a plot of land in return for their labour and the labour of their families on the hacienda lands. Share-cropping systems existed by which peasants often provided seed, manure, transport and labour, and the owner provided only the land in return for a half-share of the crop. In the *valle* region both forms of tenure existed, together with cash renting, and the fragmentation of holdings occasionally reached extremes, as in the Cochabamba valley.

Land reform was undertaken as a part of the revolutionary policy of the MNR in 1953, though it is difficult to establish how far the agrarian reform law was initiated from above or was simply a result of pressure from the indigenous population and a recognition of the fact that land invasions were taking place already on a large scale, especially in the Cochabamba region and near La Paz (Heath, Erasmus and Buechler, 1969). Expropriation of the haciendas took place rapidly, often achieved by hastily formed peasant unions and inspired by middle-class urban dwellers, miners or peasant leaders, but legal recognition and the issuing of titles were often long delayed. Legal title to land, quite apart from assuring the owner of security, is important as a qualification for receiving bank credit, but it was only after 1960 that the issue of titles reached a substantial annual rate. By 1978 it was claimed that 74 per cent of the rural population held title, and that 22.8 million ha had been adjudicated. Many farmers in the *altiplano* and the *valle* regions have benefited from the reform. Areas redistributed per family were sometimes small, too small indeed to provide an adequate family farm; for example, over 8000 families in the Cochabamba area received less than 1 ha. But it is also clear that in densely populated areas there was not enough land to provide adequate family farms. The effects of reform have been very considerable, both in social and psychological terms and in terms of agricultural production. Early reports suggesting that land reform was followed by a fall in agricultural output were certainly based on unreliable statistics and took no account of variability of output due to weather conditions. But over a longer period there has been some expansion in agricultural production and a great responsiveness of peasant producers to market opportunities: near La Paz and Oruro peasant farmers have turned towards more intensive market-gardening, for example. The area under cultivation in the *altiplano* and the *valle* regions has increased, but yields remain low and traditional methods of cultivation persist. The quality of stock seems to have fallen following land reform, and there is evidence of over-grazing and exhaustion of soils from various parts of the *altiplano*. Credit and extension services have been lacking and substantial agricultural improvements require substantial investment in, for example, irrigation, the improvement of stock breeds, controlled and rotated grazing and the cultivation of fodder crops and improved grasses. There is a greater use of fertilizers and pesticides in some areas, for example in the Upper Cochabamba valley, where the land under cultivation increased rapidly after the land reform. Within the severe limits imposed by climatic risks, distance from urban markets and the lack of capital, advice and credit, peasant producers have shown adaptability to market opportunities, though production for subsistence frequently continues to be an over-riding necessity, and rural poverty is widespread.

Marketing structures were profoundly altered. Under the former system the *hacendados* often received their rents in kind at the door of their town houses for disposal to buyers. They thus performed a middleman function which some developed further when their estates were lost, buying produce from their former dependents instead of merely receiving it as rent. But the peasants themselves have also organized their own new outlets for the sale of surpluses through markets and fairs which have sprung up in new locations more conveniently placed in the *altiplano* at closer intervals than the old centres (Preston, 1969). New towns have begun to emerge in the *altiplano*, and these reflect the existence of greater surpluses and greater purchasing power in the hands of the peasants as well as the social changes following land reform. Peasant trade unions have often taken the lead in the establishment of new towns of this kind, which are also seen to have a social as well as a marketing function as foci for the building of schools. Significantly, most of the new towns are located within a 100-km radius of La Paz, where marketing opportunities are better than elsewhere in the *altiplano* (von Marschall, 1970).

Other changes have been of equal if not of greater importance. The traditional leadership of peasant communities has been challenged through the existence of peasant unions, which have often persisted as a result of the continuing need for negotiations with authority because of the long legal processes whereby titles are finally secured. And the participation of active and younger men makes for greater receptivity to other innovations. The possibilities of social and geographical mobility have increased. Those *hacendados* who did not fall to the ranks of the peasantry, working the rump of a former estate, have of necessity had to find a new outlet for enterprise in trade and commerce or occasionally in industry and handicraft. Material welfare of the *campesinos* seems to have improved, and observers have remarked on the increase of *calamina* roofs, bicycles, transistor radios and the use of western style of dress, though these signs of material progress may not necessarily be attributable, of course, to agrarian reform alone. The pejorative or at best negative term '*indio*' has been replaced by '*campesino*', a change symbolic of the rise in status of the peasantry, and of a recognition of personal dignity and worth.

> One of the most striking features of Bolivia's land reform and social revolution is the degree to which *campesinos* have begun to assume the roles of citizens and to participate in social systems that were not only closed but virtually unknown to them a decade ago. (Heath *et al.*, 1969, p. 387)

But it is also clear that change has been slowest in remote areas where subsistence agriculture still predominates and an archaic social hierarchy persists, for example in Sucre and Tarija (Heyduk, 1974). Much remains to be done, particularly in education and in raising agricultural productivity, improving methods of production and providing rural credit, but the changes in social structures, attitudes and values associated with land reform may have created a fertile soil in which such innovations may take root and spread, if and when the market opportunities exist.

Great hopes are entertained for the settlement of new lands in the east, more important to Bolivia than to either of her northern neighbours as the only zone with a substantial potential for lowland tropical crops. As in the more northerly countries, colonization of the empty lands is seen as at least a partial solution to several problems: as a means of redistributing the overcrowded rural population of the northern *altiplano*; as a means of reducing the imbalance of population density between the *altiplano* and the rest of the country; and as a means of increasing domestic food supplies, thus reducing imports and releasing foreign-exchange reserves for other, more important imports of capital goods. But in Bolivia the settlement and colonization of the east also serves to strengthen national unity, creating an integrated state in which the link between Santa Cruz, Cochabamba and La Paz would form a strong and interdependent economic axis. Santa Cruz is the major nucleus around which new settlement has been encouraged in recent years, though successful settlement has also been achieved in two important areas of the *yungas*: to the north-east of La Paz in the Alto Bení region and in the Chaparé basin to the north-east of Cochabamba.

In the Chaparé spontaneous settlement by migrant squatters, chiefly of Quechua-speaking Indians, received relatively little attention by the state until recently (Edelmann, 1967). Before 1952 squatters or *tolerados* settled and cleared land on the large estates, taking a few crops before being evicted by the landowner. Since the revolution of 1952 and the land reform that assured migrants into the *yungas* of their right to land, the stream of settlers has grown, viable settlements have been created (notably near Todos Santos) and road networks and schools have been built with the minimum of assistance from government. Locally organized peasant unions have played an important role in community development and often organize the allocation of land to new settlers. Government help has increased in recent years, but the Chaparé settlements have proved much cheaper than more ambitiously organized projects of colonization.

Similar spontaneous settlement north-east of La Paz took place after the building of the road to Caranavi in 1958. In 4 years about 3000 families moved into the area, quite apart from the planned settlement of the Alto Bení. The Alto Bení project near Caranavi, at 450–1000 m above sea-level and 240 km by road from La Paz, was financed by a USAID programme and the Bolivian Development Corporation from its inception in 1959 (CBF, 1965). Preliminary assessments of soil quality, agricultural possibilities, health and educational needs were made; roads were built; selected colonists were provided with 10–12 ha of land (1 ha of which is previously cleared and planted in rice, maize, bananas and *yuca*); and they were given temporary housing, tools, clothing and seed. Schools and clinics have been built, and allocations of land made in conjunction with a planned settlement pattern that reflects the preference of the Aymara Indians for a fairly nucleated structure. Rice, maize and *yuca* are the main subsistence crops, bananas provide a quick yielding cash crop, and coffee of good quality has found a market in La Paz with limited possibilities of export; coca, cacao and citrus are also exported from the region.

In both of these areas of the *yungas*, and indeed in the Santa Cruz region as well, shifting cultivation is widely practised for subsistence crops in conjunction with tree or shrub crops such as coffee or bananas. In general, as in Peru, high-land Indians have adapted fairly well, though at a low level of technology, to the totally different environmental conditions of the *yungas* and the *oriente*. There is, however, considerable mobility among these latter-day pioneer colonists. Although it has been said that the failure rate of colonists in the Alto Bení project was low (Edelmann, 1967), this is not the universal experience. Many colonists in the planned schemes have moved on, and sometimes back to the *altiplano*, for while agricultural incomes tend to be higher in the *yungas* there is less opportunity for additional cash income from temporary urban labour.

The major area of settlement, however, has been in the region of Santa Cruz, particularly since the completion of the road to Cochabamba in 1953. In the 1950s foreign colonies of Mennonites, Italians and Japanese were encouraged to settle near Santa Cruz (Crossley, 1961), and in general have achieved a reasonable degree of success, particularly in the production of timber, rice, sugar and cotton. But however successful such foreign colonies were in raising agricultural production and in acting as demonstration models for Bolivian farmers, it is with the latter that the future must rest. A number of other planned colonies were established by the Bolivian Development Corporation, but special mention should be made of the role of the army in development. A military colony was set up in 1955 for soldier-colonists, largely from the *altiplano*, but modern plans have become more ambitious. It is relevant that the army has been in power in Bolivia since 1964 except for very brief periods in 1969 and in 1979–80, and that military thinking places a high value on the settlement of frontier regions near the Brazilian and Paraguayan borders. Eleven military projects were under way in 1977 in the Department of Santa Cruz. The barracks are the nuclei around which the army has been responsible for the building of schools, provision of water supply, the building of roads, churches, markets and landing strips. Military enterprise has penetrated deeply into the economic life of Santa Cruz, however, and strategic interest in establishing frontier agricultural colonies has shifted towards the establishment of large-scale agro-industrial enterprises, with massive concessions of land at Bermejo in Tarija department near the Argentine frontier, north of Puerto Suarez on the Brazilian frontier and at Abapo-Izozog where, with the help of external finance, irrigation schemes are planned for the large-scale production of wheat, soya beans and cotton (Rivière d'Arc and Lavaud, 1979).

In the Santa Cruz region, as elsewhere in Bolivia, however, spontaneous colonization has played an important role, but as land has been taken up by large landowners or medium-scale capitalist farmers, near Santa Cruz and along the increasingly elaborate network of roads, spontaneous peasant settlement has pushed out to a fringe area to the north in which rice is one of the few crops which will withstand high costs of transport to the mills in Santa Cruz. But rice is a crop which yields low returns, unlike the more profitable production of cotton and sugar, which are located near to the sugar-mills and cotton-processing plants near Santa Cruz itself.

Cattle ranching is one of the important elements of land use in the Santa Cruz region, but it is also much more widespread in the sparsely settled regions to the north, in Bení Department and to the south in the Departments of Tarija and Chuquisaca, and has increased rapidly since the 1960s with high beef prices and the possibility of export markets in Chile, Peru and Brazil. There is a wide range in size and efficiency from the small and rudimentary ranch to highly organized and highly capitalized large-scale enterprises, run from La Paz, with herds of over 15,000 head of cattle and equipped with up-to-date installations, airstrips and private aircraft to ship beef to export markets. Economies of scale, particularly in the ability to provide full cargoes for air transport, and the monopoly of air transport make smaller ranches vulnerable to large and efficiently run units, and there is an increasing tendency towards concentration of landownership in big ranches. But it is sometimes a ruthless business, with exploitation of wage-paid labour, conflicts over boundaries and legal rights, and the dispossession of tribal Indians. 'Many tribes have been forced to move from regions they have occupied for generations because cattle ranchers turn their cattle into planted lands, burn their houses, and prohibit access to water holes and hunting areas' (Clark, 1974).

No discussion of settlement and colonization in the *yungas* and *oriente*, or of the Bolivian economy, for that matter, would be complete without reference to the importance of illegal trade. Even in the *altiplano*, the smuggling of contraband cloth, consumer goods and foodstuffs into Bolivia has been a source of income for communities on the border with Peru in the Lake Titicaca region (Preston, 1969), and in south-west Bolivia on the border with Chile (Rivière, 1979). But by far the most lucrative illegal trade in recent years has been the expansion of the illicit drug trade. The coca shrub has long been cultivated in the eastern regions of the Central Andes as a source of the leaves which provided a stimulant to the highland Indian. But it is also a source of refined cocaine, and its cultivation has expanded rapidly in recent years in Colombia, the Peruvian *montaña*, and in the Bolivian *yungas* and the *oriente* as a result of the growth of international criminal connections. It has been estimated that Bolivia produces something of the order of 35,000 tonnes of coca leaves a year (1980) of which three-quarters is used for illicit trafficking. In the Chaparé region coffee and citrus plantations have been cut down to make way for coca. And observers suggest that Bolivia's 'earnings' from the cocaine traffic may be not less than US$600 million – a figure approximately equivalent to the *total* value of legal exports in 1978 — and other estimates put the figure at more than double this amount.

Ecuador

Before proceeding to a discussion of the regional contrasts within Ecuador and to a discussion of recent trends, accelerated since 1972 by Ecuador's appearance as a major exporter of oil, it seems necessary to comment briefly on some of the underlying characteristics of the country: its territorial framework and the implications thereof; the predominance of social and economic inequalities; and the increasing role of the state.

The territorial framework of Ecuador has its roots in the status accorded to Quito as a royal *audiencia* from 1563, at first within the Viceroyalty of Peru, and later as a *presidencia* within the orbit of New Granada. This northerly orientation was confirmed in the years immediately after the achievement of independence from Spain in 1822, when the area was included within the Federation of Gran Colombia with Colombia and Venezuela. But since 1830 Ecuador has remained an independent state, jealously preserving its separate identity against the territorial inroads made by its neighbours. At its independence from Gran Colombia, Ecuador had an area of some 706,000 km^2 and its boundaries extended eastwards along the navigable Amazon to a point well below Iquitos. But the eastern territories have been eroded: by Brazil in 1904; by the cession of a zone to the south of the River Putumayo to Colombia in 1916; and as a result of the annexation by Peru, after a short and disastrous war in 1941–2, of much of the territory then remaining to Ecuador in the eastern lowlands of the Amazon basin. Ecuador has never been completely reconciled to Peru's acquisition of its eastern territories, however, and Ecuadorian maps, for example, have never recognized the adjusted boundaries. From time to time the issue still sours political relationships with Peru. Ecuador controls the Galapagos Islands, and fairly recently has successfully claimed a 320-km limit for its territorial waters, and like Peru, lays claim to the marine resources within these boundaries.

Ecuador, therefore, emerged as one of the smallest of the countries which were the legacy of the Spanish colonial empire in South America, both in terms of population and area, and this fact of size has had important implications in the political and economic geography of the country. It has meant, for example, a reliance on external trade to an even greater extent than in the case of Peru, and with this has gone an inevitable dependence on foreign capital, and imported skills and technology. Industrialization is limited by the small size of the potential domestic market, which is in any case closely circumscribed by the poverty of the great majority of the population. Planners and industrialists are therefore inclined to see potential benefits from economic integration, first in the Latin American Free Trade Area (LAFTA) and later in the Andean Pact, since the creation of a common market or a free-trade area would, it is thought, provide a wider market for Ecuador's industries, especially if protective devices could provide Ecuador with a privileged status as one of the smaller countries involved.

Secondly, Ecuador shares with Peru, Bolivia and many other Latin American countries, a social structure which is highly stratified. The 'traditional' landowning oligarchy still retains more effective power than it does in some other

Latin American countries, but the power structure has been changing, partly by the diversification of landowning groups into other activities and away from land as the direct primary source of power and income; but mainly by the participation of other groups: merchants, industrialists, professional men and a growing bureaucracy. Income and power are strongly concentrated; much of the peasantry, the landless labour, the unskilled labourers in the cities and many of those involved in the urban tertiary sector live at the margin of subsistence. The top most prosperous 20 per cent of the population received nearly three-quarters of the national income in 1971, but the bottom 60 per cent of the population received only one-eighth of the total income, equivalent to a *per capita* income of about $50 at a time when the average *per capita* income was about $250 (Griffin, 1976). Average *per capita* income has been rising fairly rapidly with the oil boom, but even in 1978 it was no more than $627. Inequalities in the social distribution of income and wealth clearly have important implications for regional inequalities, bearing in mind the concentration of middle and upper-income groups in the cities and the preference of the well-to-do for urban life, but it is difficult to give precise details (Bromley, 1977).

As in Peru and Bolivia, and for similar reasons, there is an ethnic dimension to social inequalities. The rural population in the sierra is predominantly Indian in dress, customs and language. The peasantry and landless labour tend to be Indian; the landowners, traders and merchants mestizo, but a few Indian groups, notably the Otavalo, have retained a sense of identity and status, deriving from their widespread activity in domestic artisan production and itinerant trading, particularly in cloth. Nevertheless, the Ecuadorian Indians of the sierra have, on the whole, been less successful than their counterparts in Peru and Bolivia in retaining a collective sense of self-identity. Migration, both to the coastal zone and above all, to the cities, has tended to blur the clarity of regional and social distinctions a little, though it remains true, in general, that to be Indian is to be poor. In the eastern regions there are still isolated Indian groups retaining traditional forms of culture and land use, but as elsewhere, in the Amazon basin they are threatened by the advance of the trader, the oil-man, the Andean settler and the hand of the state. In the coastal zone the ethnic mix is more complex and less homogeneous. Slavery has left its legacy in a mixed population as it has in northern Peru, but immigration of the Indian from the sierra, and a colonizing white and mestizo population with the expansion of export-crop production have produced a coastal society which is not only ethnically complex, but is also much more individualistic, mobile, and less firmly attached to the land by ties of community and custom than that of the sierra.

Finally, one other element conditioning the economic and social geography of Ecuador is the increasing role of the state. This is a universal tendency in the modern world, of course, but in Ecuador, the participation of the state, directly or indirectly, in industry and agriculture as well as in more traditional spheres of activity such as road-building, the provision of public utilities and education, has been given new strength by the income accruing to the state as a result of the revenues generated by the rapid increase of oil exports from 1972. The government of Ecuador has been dominated in the recent past either by populist presidents or by military rule, both tending to articulate the interests of the landowning, commercial, industrial and professional élite. There has been relatively little pressure, at least in comparison with Peru or Bolivia, for radical social change. The increasing power of the state has, correspondingly, been directed primarily towards technical modernization, both in agriculture and industry, the encouragement of capital investment from foreign as well as domestic sources, and a consequent expansion of a powerful state bureaucracy.

The historic core of the Ecuadorian state is the sierra. Quito was the stronghold and capital of the northern margin of the Inca empire, and the Andean region was much more intensively settled and cultivated in pre-Columbian times, and more coherently organized, than the coastal region. The prehistoric cultures of the coastal region were fairly advanced, particularly in the Guayas basin, but they lacked the brilliance and organization of the coastal Peruvian cultures. The Spanish colonial settlement, attracted by an abundant Indian labour force and by the mild and temperate climate of the inter-Andean basins, reinforced the primacy of the sierra and maintained the continuity of Quito as the capital of the region. Guayaquil remained a relatively isolated outpost, important though it was as a port of call *en route* for Peru and as a maritime outlet for the highlands, as well as a shipbuilding centre on a coast where there were virtually no accessible timber resources between it and central Chile. Between the eastern and western cordilleras of the Andean system, the string of intermontane basins provided land and labour for the support of urban centres which were the foci of Spanish settlement: north of Quito at Ibarra; to the south of it at Latacunga, Ambato, Riobamba, Cuenca and Loja. They were, and to some extent still are, local administrative and ecclesiastical centres, controlling a subordinate Indian population, but with important commercial functions, and with increasingly significant artisanal and industrial functions (Bromley, 1979). The relative ease of movement between the intermontane basins from north to south perhaps made it possible, at least in colonial times, for the integration of the sierra region around its capital, Quito, in a way which would never have been possible in the more complex and larger-scale relief structures of Peru. And in modern times the retention of capital functions in Quito, despite the economic growth of the coastal region and particularly of Guayaquil, has meant the retention of an important economic growth pole in the sierra, and has perhaps helped to reduce the regional inequality between coast and sierra which is so strikingly evident in Peru.

Apart from some short-lived activity in gold-mining near Quito in the colonial period, the economy of the sierra has rested directly or indirectly on its agricultural potential. In colonial times, Quito exported hides and leather, and even some foodstuffs, to the coast, but it was the production of woollen cloths which brought external income to many of the towns in the sierra. Quito, Cuenca, Latacunga and Otovalo were major foci of textile production and trade, based on local wool production. Competition from imported English cloth from the time of the Industrial Revolution destroyed much of this textile tradition, though textile factories survived in Latacunga in the mid-nineteenth century (Bromley, 1979), and the industry has revived in modern times in Quito, Ambato and elsewhere. From the middle of the nineteenth century Ecuador was renowned for its so-called Panama hats, made from the straw of the *toquilla* palm. Originally concentrated in the coastal region, manufacture was encouraged in the 1860s in the southern sierra, and especially at Cuenca which is still the major centre, but of an industry which has declined with changes in fashion.

The farming of the sierra is still much more completely devoted than that of the coast to subsistence, the supply of local urban markets, or to the supply of the coastal region with commodities such as potatoes, onions, wheat, flour and meat. Marketing structures are complex and frequently inefficient (Bromley, 1973), and sierra agriculture has failed to expand at a rate fast enough even to supply the needs of the sierra region itself. Maize and potatoes are the major arable crops of the sierra, together with wheat and barley and indigenous crops. The high *páramos* above 3000 m are grazed by sheep and cattle, though much has yet to be done in the improvement of breeds and of pastures. The inter-montane basins are areas of dense rural population and excessive pressure on arable land, especially where soils based on weakly consolidated volcanic ash have been badly eroded. There are some efficiently run and intensively used large estates, for example between Cayumbe and Latacunga, well-organized dairy farming near Quito, and intensive, irrigated fodder production in some of the drier intermontane basins, but one of the characteristic features of sierra land use in Ecuador, as in Peru, is that the most intensive use of land for arable crops takes place on the smallest farms. The dominant scene is one of extensively worked, under-exploited *latifundia*, and of small-scale farming on inadequate holdings by an increasing, and increasingly discontented, population of *minifundistas*. Agrarian reform may have reduced the onerous conditions of labour tenancy under which many of the sierra Indians formerly worked, but it has done very little to solve the problem of the *minifundia*.

No less than 57 per cent of the land is held by only 1.5 per cent of farm units, while 75 per cent of farm units occupy only 10 per cent of the land area. Onerous labour tenancies, by which dependent peasant farmers owed labour services to a landowner in return for the right to cultivate a small plot of land, were the object of attack by agrarian reform laws from 1964. Yet by 1970 only 21,000 labour service tenants out of 176,000 had been freed from their labour obligations and confirmed in their possession of land. The new agrarian reform law, brought into force in 1974, legislated for the abolition of these neo-feudal tenancies, but the effect of abolishing labour-service tenancies has been to increase the social differentiation in rural areas and to increase the pressure on land. The crisis of the *minifundia* is expressed in the fact that 1,850,000 peasant farmers must exist on only 1,240,000 ha. Farms of less than 1 ha increased from 93,000 to 206,000 between 1954 and 1968. Fragmentation of small farms is increasing, and although some landowners have found it profitable to sell land to smallholders in order to invest the proceeds in the urban economy, there has been no serious attempt by the agrarian reform to redistribute land on a more equitable basis. Whatever the rhetoric, the aim of the agrarian reform laws of 1964 and 1973 has been to increase agricultural productivity, particularly on the large estates, but not to attempt any serious redistribution of land.

The coastal region of Ecuador has been the major growth area of the country in the last 100 years. Until the late nineteenth century it was sparsely inhabited outside Guayaquil and along the routes of communication between it and Quito, but the hot, humid region of the coast possessed both the climate and the accessibility to participate in the growing world trade in tropical crops and raw materials which was expanding rapidly from about 1870 with growing demand from western Europe and North America. The active participation of Ecuador in world trade was delayed until the 1870s, but when it occurred, it was almost entirely based on the export of tropical crops from the coastal region. Various products have enjoyed a period of boom conditions, but the economy has been all the more vulnerable to fluctuations in world prices until the recent oil boom because of an excessive dependence on only one or two products at any one period. By the end of the nineteenth century a distinction was emerging very clearly between the commercially oriented coastal region and the traditional agrarian structure of the sierra, between the ancient colonial capital of Quito and the thriving commerce and business activity of Guayaquil and between the old-settled heartland of Ecuador and the frontiers of new colonization and settlement in the coastal region. It was a regional distinction which found political expression in the conflict between traditional, Catholic and conservative groups associated with Quito and the sierra, and the progressive and commercially minded liberal groups associated with Guayaquil, who achieved power in 1896 and dominated the government of Ecuador until 1925.

Cacao was the first of the boom crops, dominating exports from the end of the nineteenth century until 1924. Exports increased from 5540 tonnes in 1838–40 to 11,194 tonnes in 1871–80 and reached a maximum level of *c.* 40,000 tonnes a year between 1910 and 1924 when Ecuador was the largest exporter of cacao. Most of the plantations were located in the neighbourhood of Guayaquil in two main zones: to the north, above the regularly flooded plain of the River Guayas (the *cacao de arriba*), and on the eastern margins of the Gulf of Guayaquil (the *cacao de abajo*). All but a few thousand tonnes were channelled through the port of Guayaquil, the boom town of the period. In 1910, with a population of 60,000, it had already surpassed Quito in size, and by 1920 the population had topped 100,000. Almost the entire economy of Ecuador was built on the 'golden grain' of cacao before 1914. Export revenues derived from it helped to build the railway to Quito, not finally completed until 1908. But the cacao boom, which had supplied three-quarters of Ecuador's exports, ended abruptly in the 1920s, hard hit by low prices and by the ravages of witchbroom disease after 1922. By 1924 production was down to a third of former levels and remained low until the end of the Second World War. Since then, post-war shortages, high prices and government encouragement have brought cacao back into favour, and in recent years production has recovered to some 76,000 tonnes in 1976 from about 214,000 ha of plantations. Cacao now contributes about 13 per cent by value of total agricultural exports.

The fall of cacao in the 1920s brought about some changes in landholding structures as large plantations were occasionally broken up, providing an opportunity for some small and medium landholders and for some foreign buyers. But it also instigated a search for alternative crops, and in the lean years of the 1930s it was coffee and rice that were favoured. Many small producers turned to coffee, the production of which has expanded considerably in recent years and which contributed some 20 per cent of agricultural exports by value in 1975, and for which Ecuador is allocated a quota by the International Coffee Association. Rice, cultivated mainly in the Guayas lowlands and further north in the neighbourhood of Puerto Viejo, first produced an export surplus in 1928 and occasionally still does so, though the internal market for rice has expanded considerably. The area under rice cultivation has increased rapidly from some 53,000 ha in 1954 to 279,000 ha in 1968, and there has been a striking trend towards the raising of productivity, especially on the large estates, with the expansion of technically advanced methods and new strains of rice.

But since the Second World War it was, above all, bananas which became the mainstay of Ecuador's external economy until the advent of oil exports in late 1971. Ecuador had been able to make progress in international banana markets after the war at a time when Central American plantations were threatened by disease, damaged by hurricanes and hampered by labour troubles. Exports rose from a mere 9000 tonnes in 1925 to over 600,000 tonnes in the early 1960s and have been over 1 million tonnes in recent years, rising to 1.4 million tonnes in the mid-1970s. Formerly responsible for 50 to 60 per cent of Ecuador's exports, they have now been displaced by oil as the major export, but they still make up over 40 per cent of the value of agricultural exports.

Although Ecuador is still the world's largest exporter of bananas, the possibilities of continued expansion have repeatedly been questioned and attempts made to cut back the area cultivated in bananas. Central American and Caribbean producers have always been more favourably placed in relation to markets in the eastern USA and Europe, not only because of distance and the expense of refrigerated shipping, but also because shipment to Atlantic ports must face charges on passage through the Panama Canal, and also, perhaps, because the international banana trade is largely in the hands of multi-national companies who also control Central American production. In the 1960s Ecuador's leadership was threatened by the expansion of Central American production using the new high-yielding and disease-resistant Cavendish variety of banana, and with a switch from the traditional method of exporting bananas in stems to careful packing in cartons or boxes. Ecuador was losing ground in US markets and in western Europe and sought alternative outlets in Japan, eastern Europe and other South American countries, to which there was a substantial increase in exports (Preston, 1965a).

Central American production is still dominated by the United Fruit Company and the Standard Fruit Company, and large plantations are normal. Ecuador's production is not, for the most part, in the hands of foreign-owned plantations. The average size of the producing unit (47 ha) is substantially larger than the average farm in Ecuador, but it represents an average of producers ranging widely from substantial plantations, which make a more than proportional contribution to exports, and a large number of medium and smallholders who are particularly important in newer zones of production, but who make greater use of the banana as fodder for animals and as a staple foodstuff.

In view of the uncertain outlook for bananas and the intensely competitive international market, encouragement has been given in a variety of ways, mainly through the Dirección Nacional del Banano, established by government. Panama-leaf disease remains a problem, but Sigatoka, the other major threat, has been fairly successfully controlled since 1950 by spraying from the air. Most Ecuadorian plantations cultivated a variety of banana known as Gros Michel, but a campaign is under way to convert to the more disease-resistant and higher-yielding Cavendish variety, and with some success, for over half of output was of this variety in 1971. The quality of export products has improved with changes in transport, grading and handling, and the export of bananas in boxes rather than in stems. But much fruit is wasted, and only about a third of production is exported. Some is used for domestic consumption, but about 45 per cent is lost or rejected. There is a constant search for more efficient usage of that part of the crop that is not up to the exacting standard required for export. Plants have been established for the production of dehydrated banana, banana flour and banana wine, but nearly half the wastage, or about 20 per cent of total production, is fed to livestock, the rearing of which has consequently become a significant coastal occupation (Preston, 1965a).

About 60 per cent of banana production is in the Guayas lowlands near Guayaqui, and the northern zone is less important now than it was in the 1940s. Expansion of the area under bananas has followed the building of new roads – from Quito to Santo Domingo from 1947, along the road to Quevedo, completed in 1961, and near the roads from Santo Domingo to Esmeraldas and Chone. Indeed, cultivators have frequently anticipated the building of new roads in their haste to take advantage of new opportunities, and it is appropriate to conceive of a banana-frontier colonization in the coastal hinterland. With irrigation, plantations have expanded, too, in the coastal zone to the south of the Guayas estuary in an area from which export is cheap by way of Puerto Bolívar, and in which the incidence of disease is reduced by the existence of lower humidities during the dry season.

The coastal region as a whole contributes almost all of Ecuador's agricultural exports and makes a substantial contribution, particularly in rice, fruits and some vegetables, to Ecuador's own food requirements. It has been the scene of successive agricultural frontiers based on specific boom crops, commercially oriented from their inception. The elaboration of the road network linking coastal towns with each other and the sierra has opened up new possibilities for settlement on virgin land, of which the most notable examples are to be found in the neighbourhood of Santo Domingo, a flourishing bustling frontier town since the 1960s.

Agrarian structures reflect these conditions. Large estates exist, but are not as overwhelmingly predominant as in the sierra. Frontier settlement on the coast was a product of perceived economic opportunity, not a desperate search for subsistence farming, and it attracted a substantial component of settlers with access to some financial resources and able to acquire medium-sized holdings. The banana frontier, for example, helped to erode the predominance of the large estate, but not so much by peasant settlement as by the investment of an urban-based middle class (Redclift, 1978, p. 50). Land tenancies were evolved which were directed towards the production of export crops, and these frequently took on the character of sharecropping tenancies, most commonly in the production of cacao and rice (Redclift, 1978, p. 53). Since the aim of the landowner was to secure a marketable product for sale with the minimum of direct expenditure in investment or wages, sharecropping forms of tenure had an immediate advantage, though hardly conducive to the improvement of agricultural productivity and the adoption of more advanced techniques. It is this aspect of sharecropping tenancies, rather than a desire for social justice, which has guided the application of agrarian reform in the coastal region. From 1970 share-cropping tenures in rice production were abolished and replaced by co-operative structures in which the state plays a significant role. The reform was a response to the crisis of rice production triggered off by droughts in 1967 and 1968, and was inspired by the perceived need to introduce technical innovation to increase productivity to levels comparable to those of the large estates. There is thus a contrast between the prevalence, at least until recent times, between labour-service tenures in the sierra and sharecropping tenures in the coastal region. Finally, on large and medium holdings alike, there is a greater proportion than in the sierra of wage-paid agricultural labour (representing a greater penetration of capitalism in agriculture) either on a permanent basis or to satisfy seasonal demands. Migration, especially from the sierra, has thus been deeply involved in coastal development, providing seasonal and permanent labour, sharecroppers, squatters and permanent, independent settlers. But in both areas the process of agrarian reform has, in recent years, modified the pattern of tenures.

The *oriente* constitutes the third of Ecuador's mainland regions. In terms of population and settlement the eastern region is still undeveloped. According to the census of 1974 it possessed only 2.6 per cent of the national population, in spite of a rate of growth which has been much more rapid than that of the coast or the sierra; its population of 168,000 in 1974 was more than double that recorded in the census of 1962 (75,000) and well over three times that of 1950 (46,500). Indian groups still make up an important fraction of the population. The small, isolated and relatively primitive groups of lowland Indians, declining in numbers and increasingly grouped in nucleated settlements round religious mission settlements, were estimated at some 16,000 in 1959. The lowland Quechua Indians represent a second group settled on the lower slopes of the eastern cordillera and, although increasing in numbers (*c.* 27,000 in 1959), are as dependent on the large estates as their highland counterparts (Bromley, 1972). Spontaneous agricultural settlement by migrants from the highlands has long taken place on a small scale, creating isolated communities mainly devoted to subsistence farming with small surpluses of fruit, (especially *naranjilla*, a distinctive Ecuadorian fruit), alcohol, or semi-refined sugar (*panela*) for sale in the highland markets. Many link roads across the broad, high and forbidding eastern cordillera have started by local initiatives, but none of these has been completed, and road building has depended on nationally supported programmes. Three penetration roads had been completed as short links to the upper *oriente* by 1970 from Loja and Cuenca in the south and from Ambato in the centre – together with some sections of the Carretera Marginal, but there were still only 431 km of road by that date in the whole of the *oriente* (Bromley, 1972). Directed programmes of agricultural colonization were planned in the 1960s, but very little has been achieved, and assistance to spontaneous settlers has been minimal.

The events which have changed the prospects of the *oriente*, and, more importantly, of the Ecuadorian economy as a whole, have been the discovery and exploitation of substantial oil reserves near Lake Agrio in north-eastern Ecuador. Oil strikes were made in 1967 by a Texaco–Gulf consortium, and oil production began to assume a significant volume by the end of 1971, when exports began. Exploitation has involved the construction of pipelines from Lake Agrio to other parts of the field, and the trans-Andean pipeline from Lake Agrio to Quito, and thence to an off-shore marine terminal and oil refineries near the port of

Esmeraldas. It has also involved a substantial programme of road-building in the north-east linking the oilfields with storage facilities at Lake Agrio and with Quito, thus opening up the region to the possibility of more intensive agricultural colonization. Yet it seems unlikely that the site of oil production in the *oriente* will constitute a growth pole for the generation of a productive and prosperous agricultural colonization in the region. Construction work has familiarized a fluctuating immigrant labour force with the area and some have remained. Temporary construction camps along the roadside have formed a nucleus for small groups of permanent settlers. But the permanent employment provided by oil operations in the area itself is, however, quite small and much of it highly skilled and highly paid labour more or less insulated from the local economy. The new roads are axes of settlement along which colonists have claimed their rights to smallholdings of 50 ha, nominally organized into co-operatives as the law requires, but many of them are part-time holdings of men with land or seasonal occupations elsewhere in the sierra. Technical assistance, credit or other help from government has been singularly lacking, and problems of clearing the forest, the maintenance of cultivated land, and difficulties of marketing or shortage of labour have frequently meant that only a few hectares of land out of a 50-ha holding are actually cultivated. But parallel with this kind of sporadic small-scale settlement, large areas of land have been acquired, often as speculative holdings, by senior officials of contracting companies, military officers and others (Bromley, 1972). Rice, bananas, maize and yuca are basically the subsistence crops of the region, but the basic problem faced in the expansion of settlement is the lack of marketable products: tea has met with some success, and the prospects for beef cattle are reasonably good, but not without radical improvement in methods of production. There are possibilities for tropical fruits, and even, perhaps, for import substitution of crops like rubber, jute, sisal and oil palm. But results are likely to be slow, and the long-term agricultural importance of the *oriente* to the economy of Ecuador must surely be viewed in the light of the fact that abundant land is still available in the coastal region which is endowed with a fundamentally similar natural environment as far as most tropical crops are concerned, and is much better placed in relation to external and internal markets.

Regional contrasts are reflected in patterns of population distribution and growth, and in patterns of income and welfare. The population of Ecuador as a whole is growing more rapidly than that of most Latin American countries at a rate of 3.3 per cent per year. Between the censuses of 1962 and 1974 population increased from 4,476,000 to 6,501,000, but the population of the coastal region, and of the *oriente*, has grown much faster than the average. Between 1962 and 1974 the population of the coastal region increased by 49 per cent from 2,127,000 to 3,169,000; that of the sierra provinces by only 38 per cent from 2,271,000 to 2,140,000; and that of the *oriente* more than doubled from a mere 75,000 in 1962 to 168,000 in 1974. For the first time the population of the coastal zone has surpassed that of the sierra, following a trend similar to that established in Peru over a longer period. But in all Ecuador it is the towns, and above all Guayaquil and Quito, that have grown most rapidly: Guayaquil from 259,000 in 1950 to 511,000 in 1962 and to 861,000 in 1974; Quito from 210,000 in 1950 to 355,000 in 1962 and to 565,000 in 1974. The distribution of urban population is bi-primate in the sense that no town remotely approaches these two, and in 1974 the nearest, Cuenca, had only 105,000 people. On a scale rather smaller than in some other Latin American countries, Ecuador is experiencing a marked metropolitan migration of population, and the squalid squatter settlements of Guayaquil bear witness to the scale of movement. Indeed, reliance on seasonal, temporary and permanent migration from the peasant communities of the sierra to the towns, and to the coastal region as a whole, have become so integrated with rural life that for many small-scale farmers their exiguous plots have become less significant as a source of income, and more an insurance against bad times and a source of subsistence foodstuffs, with little stimulus to improve agricultural practices (Preston, 1978).

Metropolitan dominance, though shared in Ecuador between Quito and Guayaquil, is as marked as anywhere in Latin America. Together, the provinces in which they are located (Pichincha and Guayas) have over three-quarters of all industrial establishments and of the industrial employment in the country. Wages in permanent employment are higher than elsewhere; the proportion of households with piped water or electricity is higher than in the rest of the country, and so too is the degree of literacy. In 1973 tax revenue per head was 2.7 times higher than the national average in the province of Pichincha, and twice as high in Guayas (Bromley, 1977, p. 35). For some purposes, Quito and Guayaquil serve fairly distinctive spheres of influence – in terms, for example, of migration fields, newspaper circulation, bus services and road-traffic flows. Guayaquil's service hinterland thus includes much of the coastal region, the southern highlands and the southern *oriente*, though as a port it serves almost the whole of Ecuador, of course, in conjunction with Esmeraldas to the north. Quito serves northern Ecuador and the central sierra. Together they contain much of Ecuador's industry, though in 1965 Quito had a rather larger labour force than Guayaquil, but Guayaquil had a greater value of industrial output. But Quito is, above all, the seat of government and of the national bureaucracy, while Guayaquil still predominates in commerce and banking. Metropolitan dominance reflects, of course, the concentration of income in the hands of a small proportion of the population and the preference of these groups for metropolitan styles of living. Bromley (1977, p. 34) suggests that 'this pattern is becoming more marked through time, increasing the disparities between the two major cities and the remainder of the country'. Yet it may also be suggested that in comparison with Peru, Ecuador has perhaps been relatively fortunate in the historical and geographical circumstances which led to the retention of the seat of government in the sierra. Quito and Guayaquil share functions which in Peru are concentrated in the single urban agglomeration of Lima–Callao. It is perhaps because of the continuing vigour of Quito that the Ecuadorian sierra is not wholly relegated to a peripheral status.

In 1974 Ecuador joined the select band of OPEC countries as the second-largest oil producer in South America after Venezuela, though one of OPEC's smaller producers. Exports of oil from the Lake Agrio field began on a large scale in 1972, and the value of exports increased rapidly in the following years, from $60 million in 1972 to $615 million in 1974, falling slightly in the following year. In comparison with major world producers Ecuador's contribution has been relatively small, but enough to produce fundamental changes in the economy of a small country. In the early 1970s the external economy was radically changed; almost overnight the traditional and vulnerable dependence on exports of tropical crops, chiefly bananas, coffee and cocoa, was replaced by oil as the dominant export. Total value of exports rose almost fivefold between 1971 and 1975, when crude oil made up 57 per cent of the total. The effect on the growth of GDP was to create a sudden surge to a new level. In 1972 the rate of growth of GDP per head had been 2.9 per cent, but in 1973 and 1974 the rate of growth increased to 15.0 per cent and 10.6 per cent, subsequently levelling off in the late 1970s.

The regional impact of oil in the *oriente* has been outlined above, and the direct effect of oil production on employment has been relatively small, though a large but temporary demand was created for construction work in connection with the building of roads and pipelines, and later the construction of an oil refinery near Esmeraldas with a capacity of 50,000 barrels a day. The major effects by far have been the expansion of Ecuador's capacity to import and of its international credit-worthiness, together with the enormously expanded capacity of the state to intervene in the economy through the development of a public sector fed by tax revenues on oil exports. Tax revenues from oil exports have been channelled by government in various directions, including military expenditure and other non-productive activities, but great efforts have been made to stimulate the 'modernization' of the economy by substantial direct investment in physical infrastructure, industrialization, agrarian reform, and the improvement of agricultural productivity. Increasing foreign investment, stimulated by both the credit-worthiness of Ecuador's new found wealth and by the potential profitability of an 'oil-rich' country, has gone on parallel with the increasing role of the state.

Recent plans have involved heavy investment in electrical energy with the aim of supplying electricity to 45 per cent of the population in a few years. A massive hydroelectric plant is being built at Pisayambo, but this is merely the largest of a number of projects. The improvement of communications has long been a priority in development policy. The road network has been extended from 5350 km in 1935 to 10,750 km in 1958 and to some 18,000 km in 1977. The basic pattern of communications is simple and an adequate basis for elaboration: a longitudinal road through the Andean sierra (part of the Pan-American Highway) from Colombia to Peru, a more or less longitudinal road in the coastal region from Esmeraldas to the Peruvian border, together with a radial spread from Guayaquil, and no less than eight transversal links from the intermontane basins to the coastal zone (including the railway from Quito and Otavalo to San Lorenzo, completed in 1957). Three spur roads, including the road from Quito to the oil-fields, provide access to the *oriente* and the navigable tributaries of the Amazon.

Industrial expansion has been given considerable encouragement by the state since the mid-1950s. Until recently Ecuador had very few industries, its traditional agricultural exports having led to relatively little processing industry. Coffee and bananas need little preparation before export; rice-milling made some progress, of course, and it is only recently that steps have been taken to develop the manufacturing stages of cocoa production. A modest textile industry has existed since colonial times and still provides much of the national demand, chiefly in Quito and Ambato. The manufacture of Panama hats was once the main concern of the coastal provinces, and then later of the southern provinces of the sierra, but they have suffered from the changes of fashion. But from 1955 tax incentives, high allowances for depreciation, credit facilities and import regulations favourable to the import of raw materials, capital goods and semi-finished manufactures encouraged industrial growth, and national bodies were making feasibility studies of industrial possibilities. But since 1972 and the oil boom both foreign and public investment have flowed into industrial development on a much larger scale. The state itself participates in large-scale, capital-intensive undertakings such as steel, shipbuilding and petrochemicals. The fishing industry is growing with government help, and the exports of prawns, flatfish, tinned tuna fish and sardines have grown in value from $14.6 million in 1971 to $54 million in 1976. The manufacture of food products and consumer goods accounts for about two-thirds of the total of manufacturing activity, with progress in sugar-refining, flour-milling, brewing, soft drinks, cocoa manufacture, meat and fish-canning, together with a sustained interest in textiles and the establishment of a pharmaceutical industry. But there is frequently too great a reliance for comfort on imported raw materials, capital goods and foreign investment. Average incomes have been rising and internal demand is greater, but it seems that inequalities are increasing, and there is still the enduring problem of a relatively small domestic market. And it remains true that some three-quarters of manufacturing industry is located in Quito and Guayaquil. Many would argue that the oil boom has increased the dependency of the Ecuadorian economy on international capital and finance. Oil has, directly or indirectly, relieved the constraints previously set by the shortage of capital for investment, and the distribution of income from oil has enlarged the internal market, but in its attempt to foster industrialization and 'modernization' the state must also look to the possibilities of the external markets which may be generated by the success of the Andean Pact, and to the enlargement of the internal market in the rural and agricultural sectors both by increasing agricultural productivity and by raising levels of income among the rural poor.

Bibliography

BLANKSTEIN, C.S. and ZUVEKAS, C. (1973) 'Agrarian reform in Ecuador', *Economic Development and Cultural Change*, 22, 73–94.

BROMLEY, R.D.F. (1979) 'The functions and development of colonial towns: urban change in the central highlands of Ecuador, 1698–1940', *Transactions, Inst. Brit. Geog*, 1, 4, 30–43.

BROMLEY, R.J. (1972) 'Agricultural colonization in the Upper Amazon basin: Ecuador', *Tijds. Econ. Soc. Geografie*, 63, 278–94.

BROMLEY, R.J. (1973) 'El intercambio de productos agricolas entre la costa y la sierra ecuatoriana', *Revista Geográfica* (Rio), 78, 15–33.

BROMLEY, R.J. (1974) 'The organization of Quito's urban markets', *Transactions, Inst. Brit. Geog.*, 62, 45–69.

BROMLEY, R.J. (1977) *Development Planning in Ecuador*, London.

CBF (Corporación Boliviana de Fomento) (1965) *Reseña histórica del Proyecto Alto Beni*, La Paz.

CHUNGSUK CHA (1969) *El rol de la selva en el desarrollo agrícola del Perú*, Lima.

CLARK, R.J. (1974) 'Landholding structures and land conflicts in Bolivia's lowland cattle regions', *Inter-American Economic Affairs*, 28, 15–38.

CLEAVES, P.S. and SCURRAH, M.J. (1981) *Agriculture, Bureaucracy and Military Government in Peru*, Ithaca, NY.

CROSSLEY, J.C. (1961) 'Santa Cruz at the cross-roads', *Tijds. Econ. Soc. Geografie*, 52, 7–21.

DENEVAN, W.M. (1966) 'The aboriginal cultural geography of the *Llanos de Mojos* in Bolivia', *Ibero-Americana*, monograph 48, University of California, Berkeley.

DIETZ, H.A. (1980) *Poverty and Problem-Solving under Military rule: the Urban Poor in Lima, Peru*, Austin, Texas.

DOBYNS, H.F., DOUGHTY, P.L. and LASSWELL, H.D. (eds) (1971) *Peasants, Power and Applied Social Change: Vicos as a Model*, Beverly Hills.

DOLLFUS, O. (1965) *Le Pérou*, Peris.

DOUGHTY, P.L. (1968) *Huaylas*, Ithaca, NY.

EDELMANN, A.T., (1967) 'Colonization in Bolivia: progress and prospects', *Inter-American Economic Affairs*, 20, 4, 39–54.

EIDT, R.C. (1962) 'Pioneer settlement in Eastern Peru', *Annals Assoc. Am. Geog.*, 52, 255–78.

FIFER, J.V. (1967) 'Bolivia's pioneer fringe', *Geogr. Rev.*, 57, 1, 1–23.

FIFER, J.V. (1972) *Bolivia: Land, Location and Politics*, Cambridge.

FITZGERALD, E.V.K. (1976) *The State and Economic Development: Peru since 1968*, Cambridge.

FLETCHER, G.R. (1975) 'Santa Cruz: a study of economic growth', *Inter-American Economic Affairs*, 29, 23–43.

FOX, D.J. (1967) *The Bolivian Tin-mining Industry*, International Tin Council, London.

FOX, D.J. (1978) 'The Mining industry in Bolivia', *BOLSA Review*, November, 12, 594–606.

GRIFFIN, K. (1976) *Land Concentration and Rural Poverty*, London.

HEATH, D.B., ERASMUS, C.J. and BUECHLER, H.C. (1969) *Land Reform and Social Revolution in Bolivia*, New York.

HEATH, D.B. (1973) 'New patrons for old: changing patron–client relationships in the Bolivian *yungas*', *Ethnology*, 12, 75–99.

HEYDUK, D. (1974) 'The hacienda system and agrarian reform in highland Bolivia, a re-evaluation', *Ethnology*, 13, 1–11.

HIRSCHMAN, A.O. (1966) *The Strategy of Economic Development*, New Haven, Connecticut.

LEVANO, C. and ROMERO, E. (1969) *Regionalismo y Centralismo*, Lima.

LLOYD, P. (1980) *The 'Young towns' of Lima*, Cambridge.

LONG, N. and ROBERTS, B.R. (eds) (1978) *Peasant Cooperation and Capitalist Expansion in Central Peru*, Austin, Texas.

McEWEN, W.J. (ed.) (1969) *Changing Rural Bolivia*, Research Institute for the Study of Man.

MARSCHALL, K.B. von (1970) 'La formación de nuevos pueblos en Bolivia', *Estudios Andinos*, 1, 23–37.

MATOS MAR, J. (ed.) (1969) *Dominación y cambios en el Perú rural*, Lima.

MILLER, R., SMITH C.T. and FISHER J.R. (eds) (1976) *Social and Economic Change in Modern Peru*, Centre for Latin-American Studies, University of Liverpool.

MYRDAL, G. (1965) *Economic Theory and Underdeveloped Regions*, London.

ORLOVE, B.S. (1977) *Alpacas, Sheep and Men*, New York.

PRESTON, D.A. (1965a) 'Negro, mestizo and Indian in an Andean environment', *Geog. J.*, 313, 2, 220–34.

PRESTON, D.A. (1965b) 'Changes in the economic geography of banana production in Ecuador', *Transactions, Inst. Brit. Geog.*, 37, 77–90.

PRESTON, D.A. (1969) 'The revolutionary landscape of highland Bolivia', *Geog. J.*, 135, 1, 1–16.

PRESTON, D.A. and TAVERAS, G.A. (1977) 'Emigration, land tenure and agricultural change – areas of Indian population in highland Ecuador', *Working Paper* 180, Department of Geography, University of Leeds.

PRESTON, D.A. (1978) *Farmers and Towns: Rural-urban Relations in Highland Bolivia*, Norwich.

PULGAR VIDAL, J. (n.d.) *Geografía del Perú*, Lima.

REDCLIFT, M.R. (1978) *Agrarian Reform and Peasant Organization on the Ecuadorian Coast*, London.

RIVIÈRE, G. (1979) 'Evolution des formes d'échange entre *altiplano* et vallées', *Cahiers des Amériques Latines*, 20, 145–56.

RIVIÈRE d'ARC, H. and LAVAUD, J.P. (1979) 'Les fonctions civiles de l'armée bolivienne', *Cahiers des Amériques Latines*, 20, 159–80.

ROEMER, M. (1970) *Fishing for Growth*, Cambridge, Mass.

SALGADO, G., (1970) *Ecuador y la integración económica de America Latina*, Buenos Aires.

SCAZZOCCHIO, F.B. (ed.) (1980) *Land, People and Planning in Contemporary Amazonia*, Cambridge.

SMITH, C.T. (1960) 'Aspects of agriculture and settlement in Peru', *Geog. J.*, 126, 397–412.

SMITH, C.T. (1968) 'Problems of regional development in Peru', *Geography*, 53, 3, 260–87.

SMITH, C.T. (1970) 'Depopulation of the Central Andes in the 16th century', *Current Anthropology*, 11, 1–12.

SMITH, C.T. (1974) 'The promise of Eldorado', *Geographical Magazine*, 10, 46, 545–9.

SMITH, C.T., DENEVAN, W.M. and HAMILTON, P. (1968) 'Ancient ridged fields in the region of L. Titicaca', *Geog. J.*, 134, 3, 354–67.

THORP, R. and BERTRAM, G. (1978) *Peru, 1890-1977: Growth and Policy in an Open Economy*, London.

WENNERGREN, E.B. and WHITAKER, M.D. (1975) *The Status of Bolivian Agriculture*, New York.

WOOD, H.A. (1972) 'Spontaneous agricultural colonization in Ecuador', *Annals Ass. Am. Geog.*, 62, 599–617.

ZONDAG, C.G. (1966) *The Bolivian Economy, 1952-65*, New York.

8 · Brazil

J.H. Galloway

Brazil has emerged as a country of great diversity and regional contrast, and some outlines of the diversity are easily recognized. The distribution of population divides Brazil into settled and virtually unsettled areas. Two-thirds of the country still have a population of less than 10 per km². For the most part Brazilians have hugged the coast, and this coastal distribution of population has been one of the constants of the historical geography of the country (see *Figure 8.15*). A second broad division is that between traditional and modern Brazil. This division began to emerge during the nineteenth century, and today the line separating these two Brazils is not easily drawn. Traditional Brazil is rural and agricultural; the population is largely of Indian and African origin; people of European origin are but a small infusion, and usually at the top levels of society. Society is strongly hierarchical, there is a high degree of illiteracy and the standard of living is low. Modern Brazil is, by comparison, more urbanized, more industrialized, its agriculture more productive. A large part of the population is of European origin, and is generally more literate and more prosperous. Interior Brazil is traditional, with an economy based on pastoralism, hunting, collecting and shifting cultivation, enlivened only by an occasional flash of sudden and ephemeral wealth. The coastlands, at least as far south as Rio, are also part of traditional Brazil with their plantation agriculture and long colonial past. The boundary between the traditional and the modern is blurred. Paved roads now cut across Amazonia, and well-financed ranchers take charge of the land along the Araraquaia in northern Goiás. Belo Horizonte is not part of traditional Brazil, nor is Brasília, but most of rural Minas Gerais, the old gold towns and the state of Goiás are. The Paraiba valley, inland from Rio, has suffered from the exploitative agriculture associated with traditional Brazil, but is being drawn into part of modern Brazil, mainly through the establishment of industry. The most modern parts of Brazil are the industrial south-east and the south.

Regional disparities in standards of living and rates of population growth are very strongly marked (see *Table 8.5*). The extremes in 1970 were the north-east and the south-east with *per capita* incomes of 40 per cent and 151 per cent of the national average respectively. An obvious concern of the Brazilian government is to try to prevent these regional disparities in income from increasing. To reduce the gap between rich and poor will require massive investment and extraordinary economic growth in the most deprived regions, but it is probably unrealistic to expect a marked change in their relative economic positions. A region's economy must grow more rapidly than the national average if it is to increase its share of

Figure 8.1   The states and territories of Brazil

the national income. In the years since the 1940s, a period of rapid economic growth for Brazil as a whole, there was in fact little change in the share each region received. Understandably, discrepancies in regional income have influenced internal migration in Brazil. People move to those regions that offer greater opportunity, notably the agricultural frontier in the central west and the industrializing south.

The unfolding of the Brazilian experience is taking place on an enormous stage. Brazil covers an area of 8.5 million km², which amounts to almost one-half of South America and makes it the fifth largest country in the world. Much of Brazil is composed of two extensive plateaux, the Guiana highlands and the Brazilian highlands (*Figure 8.2*). These highlands are composed of pre-Cambrian rocks, in places overlain by a mantle of sedimentaries, particularly sandstones and limestone. Along the east coast there is a steep escarpment fronting the Atlantic called the Serra do Mar. In the north-east of the country this escarpment is known as the Borborema escarpment and the plateau behind it as the Borborema plateau. On these highlands, where the crystalline bedrock is exposed, the relief is gently undulating; more resistant crystalline rocks form ranges of hills or *serras* while the remnants of the sedimentary mantle stand out as tabular uplands. In the southern states, the land rises in a series of escarpments, separating plateaux or *planaltos*. The Amazon basin is the most extensive area of lowland in Brazil. Only a small part of the Paraná–Paraguay lowlands comes within Brazil, nothing more than a narrow fringe along the south-western

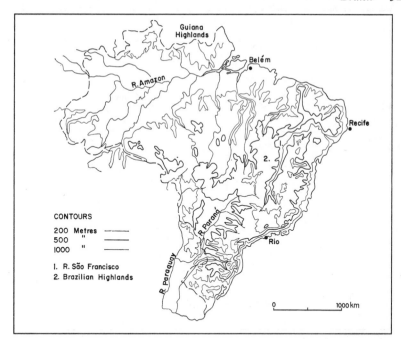

Figure 8.2 Brazil: relief

frontier and the seasonally flooded region known as the Pantanal in Mato Grosso. Elsewhere in Brazil lowland is restricted to the narrow plain between the Serra do Mar and the sea, and to a few river valleys.

Brazil is in large part a tropical country with a range of tropical climates (*Figure 8.3*). All of Amazonia has a humid tropical climate, but there are variations in the amount and seasonal intensity of rainfall from one part of the basin to another. Humid tropical climate is found along the Atlantic coast to about the latitude of Rio de Janeiro, where seasonality becomes more marked and the annual range of temperature greater. South of Rio the climate grades rapidly through subtropical to temperate. The climate of the Brazilian highlands is moderated by altitude and over much of the highlands there is a marked drier season of the year. An anomalous climate in Brazil is that in the interior of the north-east, reaching from northern Minas Gerais to the coast of Ceara and Rio Grande do Norte. Here, where a more humid climate might have been expected, the climate is semi-arid. Not only is rainfall sparse, but it is extremely variable from year to year. This region is often referred to in Brazil as the *polygono das sêcas*, the 'polygon of drought'. Only the southern states of Brazil have a temperate climate. In northern Paraná frosts occur and further south they become more frequent, which causes a distinct change in agriculture.

The vegetation of Brazil is shown in generalized outline in *Figure 8.4*. The natural vegetation of large regions of Brazil is forest. As is to be expected from the size of the country and range of climate, the type of forest varies greatly, from the rain-forest of the Amazon basin to the Araucanian or Paraná pine forests of

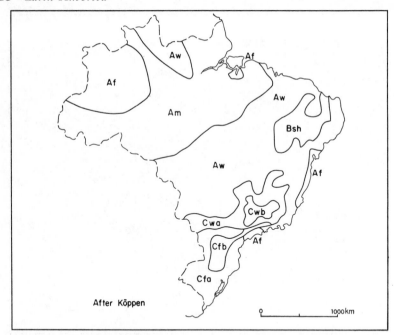

Figure 8.3 Brazil: climate
Af, Am   Wet, with no cool season and no really dry season
Aw   Wet, with no cool season and distinct dry seasons
Bsh   Hot, semi-arid
Cwa   Wet, with mild, dry winters and hot, rainy summers
Cwb   Wet, with mild, dry winters and cool, rainy summers
Cfa   Wet, with mild winters, hot summers and no dry season
Cfb   Wet, with mild winters, cool summers and no dry season

the temperate south. The vegetation of the 'drought polygon' is xerophytic forest, a dry scrub woodland referred to often as *caatinga*. Over much of the Brazilian highlands and in parts of the Amazon basin and of the south, the forest cover is not continuous; various gradations of vegetation cover are found between forest and open grassland. The origins of these savannas or grasslands are still disputed. At least three explanations have been offered attributing the grasslands to climate, to edaphic and geomorphological conditions, and to the activities of man. Whatever the origin, man has certainly been responsible for extending the area of grassland during recent centuries through cultivation and grazing cattle.

The fact that the country is largely tropical has underlain much writing and thinking about Brazil. The tropical climate has permitted the argument of climatic determinism to explain the lack of development and, even if the crude application of this doctrine is seldom now made, aspects of the tropical environment appear to have been handicaps to economic progress. The climate does favour the spread of disease. Tropical soils are often fragile and their fertility easily exhausted. Western man has not mastered these soils as he has the soils of temperate regions. The physical geography has in other ways hindered the development of the country. The Serra do Mar is a barrier to movement inland,

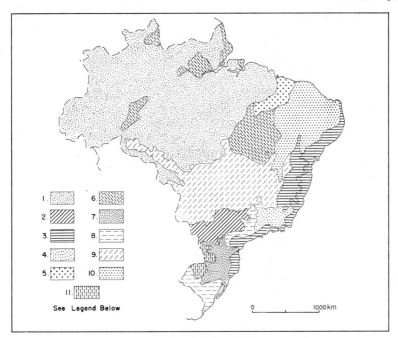

Figure 8.4 Brazil: vegetation (after *The Physical Geographic Atlas of the World*, Moscow, Academy of Sciences, 1964, pp. 170–1)

1 Constantly humid and variably humid, evergreen, equatorial Amazon-forest

2 Brazilian variably humid, evergreen, tropical forest

3 East Brazilian variably humid, evergreen, tropical forest

4 South Brazilian moderately humid, evergreen, tropical mountain-forest

5 Brazilian palm forest

6 Brazilian tropical, tall-grass savanna (Campos Limpos)

7 Araucanian forest

8 Brazilian tropical, treeless savanna

9 Brazilian savanna with xerophytic trees and shrubs (Campos Cerrados)

10 East Brazilian spiny shrub and cactus, tropical woodland (caatinga)

11 Humid, evergreen, subtropical forest

the rivers of the Brazilian highlands drain to the Paraná–Paraguay system and do not provide direct routes towards the Atlantic stream of commerce; the usefulness of the São Francisco river as a route is severely impaired by the falls at Paulo Afonso, which block navigation from the sea. The fact that Brazil has no west coast, a goal to draw people into and across the interior, may be of some psychological significance, and also the sheer size of the country may be a drawback to development.

The colonial period 1500–1800: the land use of a tropical colony

Cabral, in 1500, was the first European known to have visited Brazil; he named

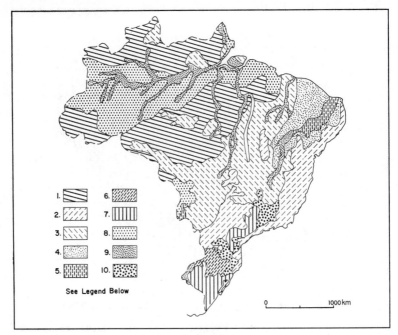

Figure 8.5   Brazil: soils (after *The Physical Geographic Atlas of the World*, Moscow, Academy of Sciences, 1964)
  1 Red-yellow lateritic (mainly ferrallitic) soils of constantly humid tropical forest
  2 Red lateritic (ferrallitic and ferritic) soils of seasonally humid tropical forest
  3 Red lateritic (ferrallitic and ferritic) soils of tall-grass savanna
  4 Brownish-red laterized (ferrallitized and ferritized) soils of xerophytic tropical forest
  5 Red-brown soils of dry savanna and reddish-brown soils of desert-like savanna
  6 Yellow soils and red soils of humid, subtropical forest
  7 Reddish-black soils of subtropical prairie
  8 Tropical bog soils
  9 Alluvial soils
  10 Yellow and red mountain soils

this new land Vera Cruz – land of the True Cross – and claimed it for Portugal. After cruising briefly along the coast he continued his voyage to India. Cabral was followed to Brazil by Dutch, English and French as well as by other Portuguese; all came to explore and to log the tropical forest for valuable timbers. The name of Vera Cruz was forgotten and the land became known from the most sought-after of these woods – Brazil wood. The active Portuguese settlement of Brazil began in the 1530s, the aim in part being to reinforce Portugal's claim to the country by effective settlement and to forestall claims by rival European powers.

Portugal was hardly well equipped to undertake such a gigantic task as the settlement of Brazil. Portugal's population was small and, despite the eastern trading empire, it was relatively poor. The colonization of Brazil proceeded slowly. The crown tried to divest itself of the burden of colonization by dividing the coast between the Amazon and southern Brazil into *donatarias*, each containing a limited stretch of coast and reaching back into the interior, and awarding them to wealthy nobles and adventurers. The resulting subdivision of Brazil is shown in *Figure 8.6*. It was the responsibility of the grantee or *donatario* to organize, finance and adminster his domain, and he was permitted to recoup his expenses through taxes, duties and the distribution of land, though some sources of revenue were reserved for the crown. The scheme failed. In several of the *donatarias* settlements were not even established and some settlements were destroyed by Indians. Only three – São Vincente, Bahia and Pernambuco – survived as agricultural colonies, in part through luck, in part because the *donatarios* were able men and commanded substantial financial resources. In 1549 the crown assumed responsibility for administering Brazil. All that survives of the *donatarios* in modern Brazil are the place names and, possibly, some influence on state boundaries, especially in the north-east. On the plateau of southern Brazil, however, at São Paulo, there arose a settlement that, in the early years of colonization of Brazil, was really beyond government control. São Paulo was a community of Indians and Portuguese, which became the home base of the *bandeirantes*,[1] men who roamed in large groups over the interior of Brazil, exploring, and enslaving Indians.

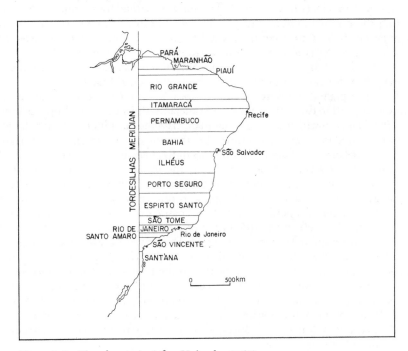

Figure 8.6 The *donatarias* (after Holanda, 1960)

The sugar coast

When the Portuguese began to colonize Brazil, they had long been familiar with sugar cultivation. Sugar had been introduced to the Mediterranean lands possibly as early as the eighth century by the Arabs, and it had become an integral part of medieval Mediterranean agriculture. In the west it was grown in Sicily, southern Spain and Portugal itself, in the Algarve and even as far north as Coimbra. Sugar, too, was associated with the earliest phase of Portuguese expansion overseas; it was grown on Madeira and by the mid-fifteenth century was being exported from there to Europe. In Madeira the cultivation of sugar was still on European lines. It has to be irrigated and much of it was grown on small estates or farms. In the second sugar colony established by the Portuguese, São Tomé, off the west coast of Africa, the humid tropical climate made irrigation unnecessary, and the crop was grown entirely on large estates worked by African slaves. It was the type of sugar plantation evolved in São Tomé that the Portuguese carried with them to Brazil. The plantation of Brazil represented, therefore, not an experiment in tropical land use, but the transfer across the Atlantic of an already tested and successful institution of the Portuguese commercial empire.

The narrow plain that fronts the Atlantic along the east coast of Brazil has a climate suitable for sugar cultivation from Natal to Florianopolis, but differing circumstances determined the location of sugar colonies. The Indians of what is now Espírito Santo were warlike and able to prevent colonization until after the end of the colonial period. To the south, distance from Europe was a major handicap. São Vincente and Rio de Janeiro were secondary producers of sugar during the colonial period, and proximity to Europe, as well as sparse Indian population, helped to make the coastal plain in the north the most favoured location. From Pernambuco and São Salvador sugar cultivation spread north and south, until by 1800 sugar was grown from Rio Grande do Norte to southern Bahia. Here, for two centuries, was the heart of colonial Brazil.

The sugar plantations were called *engenhos*, a word that literally means mill, but was applied to the entire complex of caneland, the mill for crushing the cane and the factory in which the sugar was manufactured. Landholdings were large, to be measured in square kilometres rather than in hectares, and had their origin in grants of land or *sesmarias* awarded to individuals by the crown or colonial governors as rewards for services rendered, and in the hope that the new landowners would encourage settlement. The plantation itself occupied only a part of these vast estates. Plantations, therefore, were often widely separated and even at the end of the colonial period much of the coastal plain was still in forest. Three factors limited the effective exploitation of a *sesmaria*: accessibility, the size of the labour force and the capacity of the mill. The prevailing level of technological knowledge of the planters, together with the choice of power for the mills – wind, water or animal – severely limited the size of mills. Moreover, since transporting cane was costly, it was more economical to build a new mill some distance from the old, rather than extend the plantation around the old mill. Thus, new plantations were created, which frequently kept the name of the parent plantation, but added a distinguishing suffix or prefix such as *novo* (new), *cima* (above) or *baixo* (below).

By building a new plantation a wealthy landowner could increase his acreage of cane; through the lease of land, landowners without ready capital could pass to others the burden and expense of clearing land and founding new plantations. Two forms of lease came into common use. In one, land was leased on a share-cropping basis, the tenant undertaking to grow cane as his main crop, to send it to be crushed in his landlord's mill and to surrender the molasses and one half of the sugar as rent. In the second form of lease the tenant had the use of the land for 8 to 12 years, rent free, but in return he had to clear the land and build a plantation complete with mansion, mill and slave quarters, which at the end of the lease became the property of the landlord. Both types of tenants were known as *lavradores* – literally, cultivators – and obviously they were people of some means with movable goods such as slaves and oxen. *Lavradores* commonly owned between six and ten slaves, and the importance of this class is suggested by the fact that in Pernambuco at the end of the colonial period there were two or three of them attached to each estate. There was another poorer group of tenants known as *moradores*. Usually they were settled at remote points on an estate to keep an eye on the landlord's property and were permitted to build a cabin and cultivate a few provision crops. *Moradores* were retainers in a society that measured a man's importance by the breadth of his land and by the number of people he kept about him. Towards the end of the colonial period and in the nineteenth century the nature of this tenancy changed. As slaves became more difficult to obtain and more expensive, *moradores* were increasingly drawn into the running of the plantation, being required to work in the fields and mills a given number of days a week.

But throughout the colonial period, slaves worked the plantations. In the early years of settlement attempts were made to coerce the Indians into the labour force. However, they could not be obtained in large enough numbers: the Indian population was small, could easily retreat into the interior and there was no indigenous market, as there was in Africa, for slaves and dealers from whom slaves could be bought. Expeditions had to be mounted to capture the Indians. Slaves were more easily obtained from Africa than in Brazil, though the impressment of Indians continued until the eighteenth century. Slaves were brought from the west coast of Africa between Guinea and Angola, and even from Mozambique. The slave trade and the close ties between Brazil and Africa continued until the mid-nineteenth century.

The size of the slave labour force varied greatly from plantation to plantation, but on the larger plantations there were gangs of about 150. Whatever the size of the plantation and whether or not it was operated by a *senhor de engenho* (owner of a plantation) or *lavrador*, the pattern of land use was the same. The cleared land was divided into canefields, pasture and provision grounds. The cane was grown from cuttings planted in trenches or in rows of holes dug across the fields. Shoots appeared within 12 to 14 days and the cane was harvested a year or more later, depending on the type of soil and the weather. Following a harvest the roots of the cane were left in the ground to produce further crops, a custom known as ratooning. Over the years, ratoon crops gave progressively poorer yields, the better soils producing five to six profitable crops, the poorer soils three at the most. Only in fields on the most fertile alluvial soils in river valleys were the old roots dug up and new canes planted; elsewhere the fields were abandoned to rough pasture and the forest. No deliberate attempt was made to preserve soil fertility. The manure of plantation livestock was not used; and cattle pens, such as those moved across the fields of West Indian plantations, were unknown. The only fertilizer the soil received was ash from the burning of the debris of clearing and of harvesting. Land was plentiful, and so seldom could a landowner, even with the help of his tenants, cultivate all his land that there was no incentive to preserve the fertility of the soil. It was easier and cheaper to clear new fields and let the old revert to forest as soon as ratoon crops no longer gave a worthwhile return. New clearings and patches of second growth forest soon became characteristic features of the landscape of the sugar coast.

On each plantation there were pasturelands and fields of provision crops. The staple provision crops were maize, manioc and beans, all borrowed from the Indians, and yams, which were the most significant African contribution. The pasture was required for the plantation livestock, but no attempt was made on the plantation to breed livestock, which were raised in the interior of the country and sold to planters at fairs that developed along the margins of the sugar-growing region. In some districts of the coast a second cash crop was grown. Tobacco was cultivated in Bahia for export either to Europe or to Africa, where it was exchanged for slaves; while towards the end of the eighteenth century some plantations on the drier margins of the coastal plain began to grow cotton in response to the demands created by England's Industrial Revolution. Both tobacco and cotton were also grown by smallholders as well as by planters, and cotton in particular, because it gave better yields in a dry climate, led to the extension of cultivation westwards on to the Borborema plateau.

The productivity of the agriculture in the north-east of Brazil during the colonial period was almost certainly low. Perhaps because of the apparent abundance of land, a wasteful and exploitative approach to resources developed early. Soils were not fertilized, shifting cultivation was adopted and timber was used for fuel (for boiling the cane juice) rather than burning *bagasse* (the crushed cane-stalks). The Portuguese also abandoned the plough for the digging stick used by the Indian and African labourers. Throughout the colonial period there was no important change in agricultural technology: cultivators unthinkingly followed the methods of their forefathers and a society was moulded from which the idea of progress was absent. During the seventeenth century knowledge of the cultivation and manufacture of sugar had flowed from Brazil to the West Indies, but, by the end of the eighteenth century, the plantations of the West Indies were far more productive.

This was an economy that even in the early years attracted few immigrants. Sugar colonies are not lands of opportunity except for those few individuals with influence and capital to acquire land, build mills and obtain labour. The migration from Portugal to the sugar-growing regions of Brazil was a small one in which men were predominant, and they mingled easily with the Indians and Africans. The blending of races from three continents in the north-east began the cherished Brazilian tradition of racial tolerance. The population was almost entirely rural. The only towns of any size were the administrative centres and sugar-exporting ports of Salvador and Recife. The sugar industry of the north-east of Brazil enjoyed a period of prosperity during the sixteenth and early seventeenth centuries, when the Portuguese still maintained a near monopoly of supplying sugar to Europe. This was followed by a prolonged decline and lapse into poverty. The Dutch invasion of the north-east, beginning in the 1620s, led to thirty years of guerrilla warfare, during which extensive damage was done to the sugar plantations. While in Brazil the Dutch learnt the techniques of sugar production, and carried this knowledge to the West Indies. From the mid-seventeenth century onwards, the plantations of the north-east found it difficult to compete with the West Indian plantations, which were closer to Europe. The first 'boom' in the development of Brazil was over.

Gold

The discovery of gold in the last years of the seventeenth century began the second 'boom' in the history of Brazil, leading to a major gold rush and to the opening up of a large part of the country's interior. Since the beginning of colonial times the hope had existed that precious metals and gems would be discovered, but for two centuries the exploring and slaving expeditions of the *bandeirantes* had wandered over the rich mineral zones of the Brazilian shield without coming across or recognizing gold or gems in paying quantities. This failure can be attributed partly to bad luck and partly to ignorance, for most *bandeirantes* had little or no idea of what precious metals looked like in their natural unsmelted state, nor did they know of the most likely places to find them. The exact date of the first discovery of rich deposits of gold is uncertain, but by the early 1690s gold in large amounts was known to exist in the valleys of the Serra do Espinhaço in the present state of Minas Gerais (literally the General Mines). This was the first of many discoveries of gold and diamond fields during the eighteenth century in Goiás and Bahia, in Minas Gerais and in the Mato Grosso. The richest strikes of all were those around the Serra do Espinhaço and on the northern margins of the Pantanal in the Mato Grosso. Gold mining became the mainstay of the Brazilian economy in the eighteenth century, and these two regions in particular were the centres of activity.

Most of the gold obtained in Minas and elsewhere in Brazil during this century was alluvial or placer gold recovered in the simplest manner. Wealthy miners employed slaves to pan for them, and there were instances of entrepreneurs damming and diverting streams to permit a thorough search of the stream bed. Actual gold mines – shafts following veins of gold into the rock – were few and small. The mining technology of the Portuguese was extremely limited, and there were no mines in colonial Brazil that could be compared with those of Potosí or Zacatecas in Spanish America.

The discovery of gold made Brazil a mining colony in addition to being a sugar colony and turned it into a land of opportunity for the first time. To become a planter required capital; to become a miner required very little capital and little or no skill: a poor man could hope that with luck he might become rich. This chance of a short cut to wealth brought a rush of people to the gold workings – from São Paulo and the north-east as well as from Portugal. The densely populated provinces of Minho and Douro in northern Portugal, the Azores and Madeira provided most of the Portuguese emigrants. So large did this exodus of agricultural workers become that the government of Portugal became alarmed and endeavoured to restrict it. Possibly as many as 800,000 Portuguese came to Brazil in the 100 years following the first discovery of gold. Not all went to Minas; many made their way to mining camps further in the interior while others stayed in the coastal towns.

The first people to the mines were *Paulistas* – people from São Paulo – who were gradually outnumbered by the arriving Portuguese. Male emigrants from Portugal by far outnumbered the female, and this imbalance was made even worse by the practice common among the well-to-do Portuguese of putting daughters into convents, either in Salvador or in the home country. An outcry against this practice resulted, in 1732, in legislation to prevent women from leaving Brazil without official government sanction. This legislation was relaxed a year later to permit wives to leave with their husbands without first having to obtain official permission! Nevertheless, there were few Portuguese women in Minas Gerais and the miners, therefore, found wives and mistresses among the Indians and Africans, with the result that the mining camps were just as much racial 'melting pots' as were the plantations of the north-east.

By the mid-eighteenth century, the population of Minas Gerais was over 300,000, widely distributed through the valleys of the south-central part of the state, particularly in ephemeral mining camps. Camps near exceptionally rich deposits of gold, on important routes, or selected as administrative centres, evolved into permanent settlements and even into sizeable towns. The towns had a distinctive appearance; they grew up haphazardly, in deep valleys and along steep hillsides; the streets were narrow, cobbled and often stepped; the houses of the rich stood flush with the street and were large and solidly built, two or three storeys high, with heavy wooden balconies, and many were decorated with the traditional Lusitanian tiles. The faithful and the fortunate endowed many churches, ornately decorated with gold leaf and sculptured soapstone. Here, in the wilds of Minas in the eighteenth century was produced some of the best baroque architecture in Brazil. Such were the towns of São João del Rey, Congonhas do Campo, Sabara, Mariana and, above all, Vila Rica do Ouro Prêto – Rich Town of Black Gold – the capital of Minas Gerais, which was the largest town in Brazil at the height of the gold rush with a population of about 60,000.

The fact that these towns, and indeed the entire gold-rush area, lay in the interior of the country and in a formerly sparsely populated part of it created special difficulties: the provision of the population with food, the export of gold and import of consumer goods, and the enormous demands for beasts of burden to transport the goods raised new problems. Travel to and from the mines was never easy and always slow. There were several possible routes to the coast – all were long and all had their drawbacks (*Figure 8.7*). The São Francisco valley provided a route between the Minas and the *sertão*[2] of the north-east. So many cattle in fact were driven along this valley to the Minas that a shortage of cattle was experienced on the sugar plantations. However, as a route to the coast the São Francisco had the great handicap of the Paulo Afonso Falls, which blocked access to the sea. Travellers left the river well above the falls and reached Salvador by a long trek across hot and difficult terrain. The route from the mines to the coast via São Paulo was circuitous, but drove roads crossed the Captaincy of São Paulo to the plains of southern Brazil, another source of cattle and of mules for Minas. The shortest route to the coast was down the valley of the Rio Doce, but it was made unusable by the warlike Indians of Espírito Santo. The comparatively short route leading to Rio de Janeiro meant crossing two mountain ranges – the Serra da Mantiqueira and the Serra do Mar – but it was this last route that was selected

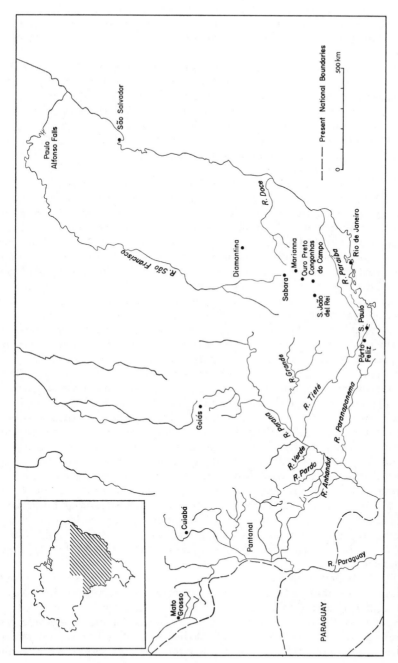

Figure 8.7 Brazil: the gold country

as the official route between the coast and the gold mines, and along it the government attempted to direct the gold and imports from Europe to facilitate the collecting of crown revenues. With Minas as its hinterland, the small port of Rio de Janeiro grew rapidly in size and importance and in 1763 replaced Salvador as the capital of Brazil.

The demands of the mines for beasts of burden and beef encouraged the growth of ranching in the backlands of much of eastern Brazil, but other food-stuffs were expensive to transport and had to be produced locally. Food in the mining areas was always expensive and never plentiful. The Indian crops of maize and manioc were grown in the traditional slash-and-burn manner. Agriculture and the need for timber and charcoal led to a concerted assault on the forests, with the result that during the course of the gold rush the hills of central Minas were stripped bare.

The economy of Minas Gerais during the eighteenth century was an exploitative one: the mining did not generate lasting industrial or agricultural development. Gold production reached its peak probably during the 1750s with an annual output valued at £2.5 million. Thereafter there was a decline as the richer washings were worked out, and no further bonanzas were found. By 1780 the annual production of gold had dropped to less than £1 million in value. Less gold meant that fewer people could be supported, and during the second half of the century there was a slackening in trade and a decline in population. A drift of people to the coast or back to São Paulo set in, though some miners, buoyed up by the hope of finding new El Dorados, ventured further into the interior. Towns were depopulated and became villages: Ouro Preto by 1800 had only 8000 to 10,000 inhabitants. Those people who did remain in Minas eked out a living by prospecting, by washing and rewashing the gravel in the streams of the Serra do Espinhaço for a small return, by subsistence agriculture and by ranching. The forests had gone, the hills were bare, the beautiful baroque towns stagnated, marking time, unchanged by new activities, and in the minds of the people there arose a nostalgia for the past – the source of a distinctive regional consciousness.

The course of the gold rush to Mato Grosso and Goiás has many similarities to that of Minas Gerais. There was a rapid assembling of a racially heterogeneous population, the more successful mining camps grew into towns such as Cuiabá and Mato Grosso, forest was felled for timber for fuel and to make way for the cultivation of food crops, and cattle ranching began. With the exhaustion of the placer gold there was a similar story of decline, of stagnating towns and dwindling population, of the decay of commercial agriculture into subsistence shifting cultivation and largely subsistence pastoralism. After the gold rush a sparse population survived in a land scarred by traces of hasty and exploitative settlement.

Like Minas Gerais, the mines of Goiás and Mato Grosso were discovered by the *bandeirantes* – those of Goiás in the late seventeenth century and those of Mato Grosso in 1718 and in following years. The Goiás discoveries were eclipsed by those of Minas Gerais and attracted comparatively small numbers of people; far more went to Mato Grosso, for news of gold here came after the novelty of Minas Gerais had worn off and, even more important, after some of the richest washings in Minas had been worked out. Cuiabá, the main gold camp in Mato Grosso, grew so rapidly that by the mid-1720s it had a population of 7000, of whom 2600 were slaves, and it had been awarded the legal status of a town (*vila*). The gold rush to Mato Grosso led to a marked decline during the 1720s in the population of São Paulo.

Transport to these distant mines was even more of a problem than to those of Minas Gerais (see *Figure 8.7*). The route from São Paulo to Goiás was long and tedious, leading northwards through the present towns of Campinas and Uberaba, a trip of many weeks. From Goiás one could continue westward to Cuiabá, but the usual route to Cuiabá was via a network of rivers. Starting at Pôrto Feliz, near the town of São Paulo, the route descended the Tiête to the Paraná, down the Paraná to its confluence with the Pardo, up the Pardo to its headwaters, then a portage to the basin of the Paraguay and so up the Paraguay and its tributaries to the gold camps. The trip was made by canoe, usually dugouts hollowed from a single trunk, and these were stronger and more durable than the bark canoes sometimes used by the *bandeirantes* on their explorations. The river route passed through lands held by warlike Indians, whom for long the Portuguese were unable to subdue. For the sake of defence, therefore, the canoes set out in convoys for Cuiabá, with as many as several dozen canoes to a convoy. The convoy that accompanied the first governor of Mato Grosso to Cuiabá in 1726 comprised 108 canoes and 3000 people. There was only one convoy a year in each direction, and this usually left Pôrto Feliz sometime between March and mid-June, taking 5 to 7 months to reach Cuiabá. The return journey took only two months: the loads were lighter – gold instead of imported goods – and there were fewer passengers. These annual journeys were known as 'monsoons' (*monções*), an apt borrowing of a term to suggest clearly a seasonal event and an annual arrival of the utmost importance. 'Monsoons' continued with little change until the first decades of the nineteenth century, when they became smaller and increasingly irregular. The last convoy for Cuiabá set out about 1838, the year of a severe epidemic of typhoid along the banks of the Tiête, which carried off many boatmen and river pilots. But by this date, also, the alluvial gold of Mato Grosso had been largely washed out.

The gold rushes of the eighteenth century were short lived, but they had a lasting impact on the development of Brazil. The discovery of gold, in addition to its repercussions on the finances of Brazil and Europe, attracted immigrants, led to the growth of Rio de Janeiro and extended Portuguese influence and settlement into the interior of the country. The gold camps of the Serra do Espinhaço, Goiás and Cuiabá formed the nuclei of the present states of Minas Gerais, Goiás and Mato Grosso. These camps were springboards for further explorations, and by the 1730s expeditions from Cuiabá had reached into the Amazon basin. The search for gold and the wanderings of the *bandeirantes* permitted Portugal to stake a firm claim to the vast area that is now the interior of Brazil.

Cattle

In the first phase of its development Brazil was a sugar colony; in its second phase a gold mining colony; both sugar and gold mining created a demand for beasts of burden and the population required food – beef – so that on the margins of the sugar and gold country and, indeed, everywhere the Portuguese and the *mameluco* wandered they raised cattle. For the sparsely populated frontier lands of Brazil ranching was an almost ideal form of land use. It required only a small labour force, it was tolerant of a wide range of climatic conditions and variations in quality of pasture, the animals could be driven to market and the hides provided a valuable export. Finally, ranching was a stand-by form of livelihood that the population could turn to when other means of earning a living failed. Ranching incorporated both Brazilian and European traits. Ranching was of Iberian origin, and the cowboy, the lasso, branding and the round-up were brought across the Atlantic. Abundance of land in Brazil made for vast ranches; the cowboys were of Indian or of mixed Indian and Portuguese descent, and the cowboys of southern Brazil borrowed the Indian bolas instead of using the lasso. In two regions of Brazil during colonial times ranching became the predominant economic activity and lent powerful support to the economy of the country: the *sertão* or backlands of the north-east and the plains to the south of São Paulo, between the Paraná and Uruguay rivers and the sea, a region which became known variously as São Pedro do Rio Grande do Sul, the Banda Oriental (the east bank of the Uruguay) or the Vacarias do Mar (the Cattle Ranges of The Sea).

In the north-east ranching began as an offshoot of the sugar economy of the coast to satisfy the need of the plantations for oxen. At the very beginning of the colonial period cattle were raised on the coastal plain, but with the extension of sugar cultivation ranching retreated inland to the Borborema plateau and eventually to the dry plains of the interior. With few predators and little competition for the pasture the number of cattle increased rapidly. By the early eighteenth century an estimated 800,000 head roamed the *caatinga* to the north of the São Francisco, while in Bahia and the valley of the São Francisco a further 500,000 were reported. These cattle were descended from Portuguese stock, but in the harsh environment of the *sertão* they evolved into relatively small animals, hardy, and almost feral.

As with the sugar plantations, the ranches were established on large land grants or *sesmarias*. The headquarters of a ranch or *fazenda* were located where possible by a river, and scattered over the land grant at sites where water was available were subsidiary centres known as *ranchos* or *currais*. At each *rancho* there was a herd of between 200 and 2000 head with a cowboy or *vaqueiro* in charge. It was his responsibility to tend to the herd and protect it from wild animals and even from Indian attack. On some *fazendas* there were twenty or more *ranchos*. *Vaqueiros* were not usually paid during the first few years of their employment, but then received one of every five calves born in the herds under their charge. A *vaqueiro* could then accumulate a herd of his own and, given the abundance of land on the ill-defined and unpoliced Luzo-Brazilian frontier, could aspire to become a landholder and rancher in his own right.

Drove roads led from the cattle lands to fairs along the margins of the sugar-growing region. These fairs marked the boundary between the predominantly pastoral districts of the north-east and the predominantly crop-growing districts. With the extension of the cultivated area the fairs were forced further into the interior. The drove road from the *sertão* of Paraíba and Ceara once penetrated to Iguaraçu, only a few kilometres from Recife, but the fair was removed first to Goiana, and eventually, in the nineteenth century, to Campina Grande, in Paraíba. During the colonial period cattle from the interior of Pernambuco were sold at Vitória de Santo Antão at the foot of the Borborema escarpment where this escarpment juts eastwards towards Recife, while in Bahia, Feira de Santana became the main cattle fair. Many of these fairs led to the growth of small towns, and thus have had a lasting effect on the settlement pattern.

Cattle ranching in the north-east provided a means of settling and realizing some wealth from a harsh environment. From an economic point of view, ranching was essentially a support of the sugar industry, but it did also produce a valuable by-product in hides. At the end of the colonial period, when the sugar industry was in relative decline, hides were the third most important export of Recife, after cotton and sugar, and counted for 10 per cent of the value of all exports.

By contrast, cattle ranching in the Vacarias do Mar did not develop as an appendage of another agricultural economy. This vast region between São Paulo and the River Plate remained throughout the colonial period a sparsely populated borderland between the Spanish and Portuguese empires, with neither gold nor precious stones nor a cash crop to attract population and official attention. During the sixteenth and early seventeenth centuries the Jesuits founded missions on the borders of this region in southern Paraguay and in what is now the Misiones Province of Argentina, missions that were pillaged from time to time by armed bands from São Paulo searching for slaves. The herds of the Vacarias do Mar descended from cattle that had escaped from these missions. With nearly empty fertile humid lands to roam over, the cattle rapidly increased in number to the extent that by the end of the seventeenth century there were estimated to be 4 million. Cattle were hunted by a sparse population of mixed Spanish, Portuguese and Indian origin; the hides were exported, and the carcasses left to rot upon the ground.

The eighteenth century brought a significant change to the cattle economy of the Vacarias do Mar, occasioned in part at least by the gold rush in Minas Gerais, which created a market, thereby giving a greater value to land and livestock. Instead of merely hunting feral cattle, the people of the region now laid claim to ownership of lands and herds. The ranch or *fazenda*, the round-up, branding, the cowboy – here known as the *gaucho* – in fact the paraphernalia of Iberian cattle culture long familiar in the *sertão* of north-east Brazil, now appeared in the south. Hides continued to be exported in large numbers, but cattle were also driven north. The breeding of mules on a large scale began, the mules being employed in transporting goods between the mines and the coast.

An appreciation of the growing importance and potential wealth of the territory led Portugal to reinforce its claim of possession. A fortified settlement, the Colonia do Sacramento, was founded on the river Plate across from Buenos Aires, while a government-sponsored immigration from the Azores resulted in the establishment of several villages along the coast of Santa Catarina. This attempt at the integration of the Vacarias do Mar into Brazil was only partly successful: dispute of possession continued into the nineteenth century with, finally, Brazil retaining the larger share of the region, and the southern fringe becoming the independent state of Uruguay. At the end of the colonial period, the Vacarias do Mar was still very sparsely populated; there were no towns or ports. Possibly the greatest change in the landscape to have taken place was the extension in the area of grassland.

The end of the colonial period provides a convenient point at which to attempt an assessment of what had been accomplished in Brazil by the Portuguese over three centuries. The work of exploring the country was over in the sense that the lie of the land and the course of the rivers were known in major outline. This had been accomplished in large measure by the *bandeirantes*. But in effect the Portuguese colony occupied only about half of the territory of what is now Brazil, for over the Amazon basin and other remote regions of the interior Portugal's control was nominal even though its claim was recognized. In the far south its claim to possession was still in dispute. There was a multi-racial population of approximately 2.5 million, exclusive of Indians still unassimilated into Luzo-Brazilian society. This society was already distinctive, formed out of Indian, African and Portuguese contributions, yet with a sufficient sense of identity to maintain Brazil as a unit following independence. It was, however, far from being a technological society, and in its agriculture it was conservative, backward and wasteful. Already by 1800 the landscape of Brazil was scarred by traces of the destructive use of resources.

Economic activity and population were very unevenly distributed. In only two regions of the colony, the forest zone of the north-east and the goldfields of Minas, had there been any approach to intensive exploitation, and it was in these two regions that large numbers of people had been assembled (*Figure 8.8*). The north-east contained about 40 per cent of the population of Brazil, nearly all on the coast, and 28 per cent of the population (about 400,000 people) still remained in Minas Gerais. In the cattle-ranching regions the population was very

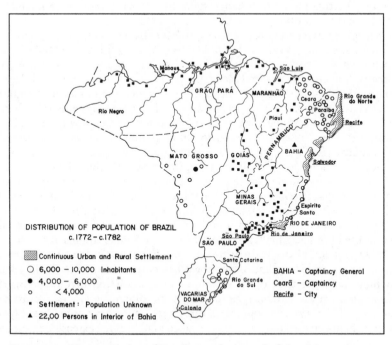

Figure 8.8   The population of Brazil towards the end of the eighteenth century (Alden, 1963)

sparsely distributed, and in the far north of the country the Luzo-Brazilian population numbered only a few thousand. In Maranhão a small number of planters and traders lived in and around São Luiz; in the Amazon there were some missions and garrison posts, such as Belém and Manaus (both of comparatively recent foundation – 1609 and 1721 respectively). The population of Brazil in 1800 was overwhelmingly rural. The largest towns were on the coast, ports and administrative centres: Rio de Janeiro with an estimated population of 100,000, São Salvador with about 50,000 and Recife with about 25,000. All three of these towns were growing, Rio the most rapidly. The gold-mining towns were in decline, some by now all but deserted. Elsewhere there were a number of small centres which had the legal status of a *vila* or *cidade* because of some official function performed there, but these were little more than a collection of a few huts, a church or two, perhaps a garrison, with little or no commercial function being carried on. The town of São Paulo, as distinct from the captaincy of the same name, even though it was 260 years old, was in reality still only a village with few inhabitants.

The outlook for Brazil at the end of the colonial period was not good. Its sugar economy was depressed, the richest deposits of gold had been washed out and though cattle ranching provided a livelihood, it did not provide a very good one or much revenue for Portugal. The government was keenly interested in discovering new cash crops and reviving trade in the old. The nineteenth century was to bring more speculative 'booms', a new flow of immigrants and a departure, in parts of Brazil, from the pattern of development established during the colonial period.

The nineteenth century: the emergence of two Brazils

Brazil passed from colony to independence gradually and comparatively peacefully during the early years of the nineteenth century. In November 1807 the Prince Regent of Portugal, Dom João, with his family, government and retainers fled Lisbon, as Napoleon's armies neared it, for Rio de Janeiro, which became in 1808 the temporary capital of the Portuguese empire. One of the first acts of the exiled government was to throw open the ports of Brazil[3] to the shipping of friendly nations, and other long-standing, restrictive mercantilist measures were relaxed. In 1815 João, now king, raised Brazil to formal and legal equality with Portugal. João appears to have liked Brazil; certainly he was loath to leave it and did not do so until 1821. His eldest son, Dom Pedro, remained behind, and it was under him that Brazil was proclaimed an independent empire in 1822, a fact not recognized by Portugal until 1825. Brazil remained a monarchy until 1889 and then became a republic. Perhaps the greatest achievement of imperial Brazil was that it managed to weld Portuguese America into a single nation. Opposition to independence by Portuguese garrisons was speedily overcome, and the regional rebellions of the first half of the century, which usually had a secessionist tinge to them, all failed. In 1828 the long-standing dispute over possession of the lands to the north and east of the River Plate was finally settled, with the creation of the independent buffer republic of Uruguay.

Brazil was now in charge of its own destiny, and this political change, therefore, did have some effect on the historical geography of the country. Perhaps this can be seen most clearly in immigration policy. Attempts were made to attract immigrants not only from Portugal but from all Europe and to settle them in the south. Portugal had prevented the establishment of industries in Brazil; now industries could be founded. The nineteenth century saw a quickening of the economic life of the country: railways were built, small ports grew into large commercial cities, the population increased from about 2.5 million in the late 1700s to over 17 million in 1900. But the approach to land use, ingrained during the colonial period, also persisted as economic booms were followed by slumps, and resources were wastefully used. In fact, what can be seen emerging during this century are two Brazils, one conservative, technologically backward and with a low standard of living, the other progressive, technologically more advanced and reaching towards a higher standard of living. The contrast between these two Brazils was to produce political and economic problems in the country.

The north-east: a deepening crisis

The north-east has come to form part of traditional Brazil. The social and cultural characteristics of the *Nordestinos*, which made them so unresponsive to the need for change and helped to make their agriculture so wasteful and unproductive, persisted throughout the nineteenth century. Few attempts were made to revive the economy of the region, and it is probable that the standard of living, never high for the mass of the population, suffered a decline. The relative decline, therefore, of the economy of the north-east *vis-à-vis* the rest of Brazil, already noticeable before the end of the colonial period, continued; the political power of the region underwent eclipse, and by the twentieth century the north-east had sunk to the level of a depressed region. There has been a tendency until recently to regard these years between the end of prosperity in the eighteenth century and the present era of development plans as a period in which little or nothing of importance happened in the north-east. It is true that sugar, cotton and livestock continued to be the dominant constituents of the regional economy, but the agriculture of the region did, in fact, undergo some modifications which brought with them changes in population distribution and settlement patterns.

Developments in the sugar industry governed the changes that took place in the coastal region. Even in 1800 this part of the north-east was still comparatively sparsely populated and much land remained in forest, making possible a large expansion in the acreage of sugar. Despite the poor economic climate, the number of plantations continued to increase, by both the traditional method of subdividing large estates through inheritance and the carving out of new plantations along the agricultural frontier. About mid-century, plantations were being built along the foot of the Borborema escarpment, a fact which suggests that by then all land in the coastal region had passed into private ownership. In Pernambuco alone the number of plantations increased from about 300 in the late eighteenth century to about 1000 in 1830. The wasteful methods of the sugar industry together with the low level of productivity at the beginning of the century left plenty of scope for improvement. The first advance was the introduction in 1810 of the Bourbon variety of cane to replace the Creole variety that had been cultivated since the beginning of colonial times. The Bourbon cane provided higher yields per hectare while its juice contained a higher sucrose content. Indeed, the advantages of Bourbon cane were so readily apparent that within a decade of 1810 it had replaced the Creole. The use of ploughs became more widespread; by mid-century they were to be found on the better-managed plantations. Manuring, too, became more common, producing higher yields and probably allowing fields to be cultivated for a greater length of time than before. The capacity and efficiency of some mills were improved by the replacement of the old-style vertical rollers by heavy horizontal presses. A few planters even installed steam engines, but their cost, the great demands they made on fuel, and the difficulty of finding mechanics to make repairs, persuaded most planters to stick to animal or water power. The collective result of these improvements, and of the extension in the acreage of sugar, was a large increase in exports. During the first half of the century, exports of sugar from Recife rose from 10,000 to 50,000 tonnes annually.

Clearly, there was still a market for Brazilian sugar abroad. The experience of the industry during the first half of the nineteenth century does suggest a revival from a nadir reached some time at the end of the colonial period. However, it appears probable, judging by complaints planters made of their financial position, that a very low level of profit was accepted. Part of the expansion can perhaps be attributed not to economic forces but to the innate conservatism of a society that could bring itself to accept no other form of livelihood, even though capital might have produced higher returns if invested in enterprises other than sugar, outside the region. Indeed, the founding of some plantations required not so much liquid capital as labour. Though, admittedly, slaves cost money, the existence of the *morador* class meant that work was to be obtained in return for the use of land, which planters had in plenty. This continuing demand for Brazilian sugar may also be attributed in part to the difficulties competitors were experiencing, difficulties that the north-east was to encounter only in the second half of the century.

In the nineteenth century producers of cane sugar not only had to contend with competition from beet for the first time, but also had to meet two further basic problems that were to transform the industry: a labour crisis brought on by the end of slavery and a massive reorganization of sugar production occasioned by advances in the technology of milling and manufacturing sugar, which rendered the old style plantations obsolete. The labour problem was the first encountered. The successful slave revolt in St Domingue, now Haiti, in the late eighteenth century brought about the destruction of what was the most productive sugar industry in the Caribbean, while abolition in the British West Indian islands led to a movement of ex-slaves away from the plantations and so to a decline in production, particularly in Jamaica. In Brazil slavery continued until 1888, although during the 60 or so years preceding abolition slaves came to form a smaller and smaller proportion of the labour force in the sugar industry. Slaves were expensive, particularly with British attempts to suppress the Atlantic slave trade, and planters had turned to the *moradores* for labour, who outnumbered slaves on most plantations well before abolition. Abolition in the forest zone of the north-east was merely the final act of a long period of transition, and it was followed by a movement of ex-slaves leaving one plantation to become *moradores* on another, or going to the towns or to the interior to try to better their lot. The decline in sugar production was slight and temporary, but abolition did have an important effect on the settlement pattern. Slaves lived in quarters adjacent to the plantation houses, *moradores* at scattered points on the estates. The decline of slavery and its eventual abolition led to a dispersal of the population from the plantation nuclei. The plantations ceased to be important elements in the pattern of population distribution; with the reorganization of the sugar industry, they also ceased to be the basic units of production.

By the mid-nineteenth century technological advances had culminated in the building of factories for the processing of sugar, and these could mill far more cane per day and extract more juice from the cane than the old mills. Moreover, new refining techniques permitted the factories to improve on the *muscovado* and *clayed* grades of sugar that the old mills produced. But the factories were expensive to build, and to operate at capacity they required far more cane than the usual plantation grew. They were, therefore, normally built by companies rather than by individuals and, to ensure supply of cane, the companies bought up the plantations or contracted with planters for their crops. In Brazil the first factories (referred to as *usinas*) were built in the last quarter of the century. Difficulties of financing in the north-east slowed the introduction of the system, and well over half a century elapsed before the bulk of the sugar of the north-east was milled and manufactured in *usinas*.

Where it was adopted the new system brought changes in agricultural practices, in the class structure and in the settlement pattern. Large companies could afford new equipment and could conduct research more easily than individual planters. By 1920 tractors and ploughs were far more common in the districts where the central factory system held sway than where it did not. The purchase of land by *usinas* and agreements to buy cane from planters meant that more and more of the traditional mills went out of use. Many planters withdrew from the countryside to Recife or Salvador or even to Rio. Their crumbling plantation houses and abandoned mills are nostalgic reminders in the landscape of today of the passing of a system of sugar production and of a way of life that by the end of the nineteenth century was already more than 300 years old. The planters who did remain to grow cane for the *usinas* became known as *fornecedores* (suppliers of cane) and their former position at the peak of the class pyramid was assumed by the owners of the *usinas*, the *usineiros*. *Moradores* were not immediately affected, though hired labour did become more common. But the *lavradores* as a class did suffer. No longer did they have a role in establishing new plantations, and the *usinas* preferred, where possible, to cultivate their own cane rather than to rely on share-cropping arrangements. Around the *usinas* there grew up new nuclei of population in small company towns and villages with regularly laid out streets, churches and company stores, and inhabited by a wage-earning proletariat.

During this period of transition the amount of sugar grown and manufactured in the north-east continued to increase and markets continued to be found. In the middle years of the nineteenth century, competition from other producers of cane sugar as well as from beet weakened the north-east's hold on its traditional markets in Europe and North America, but the rapid growth of population in southern Brazil and Argentina was creating an alternative market. This change in the pattern of exports was accelerated by the introduction of the central factory system. The qualities of sugar it produced were not well received in the traditional markets, but were readily accepted in the new markets to the south. Ironically, the modernization of the sugar industry of the north-east did not lead to the retention of its traditional markets, but contributed to their loss.

This large increase in the amount of sugar produced and the reorganization of the industry did not bring a return of prosperity to the forest zone. Compared with many other sugar-producing regions, sugar production in the forest zone was still costly and inefficient. Even in the domestic Brazilian market competition from sugar plantations in the states of Rio de Janeiro and São Paulo was beginning. No alternative form of land use to sugar was in the offing, yet the monoculture of sugar was more widely practised than ever before. During the nineteenth century the commitment of the forest zone to sugar had been strengthened, making change even more difficult. From a social point of view the abolition of slavery was probably the only favourable development. Landownership and wealth had become even further concentrated, the ranks of the landless peasantry had greatly swelled, food shortages associated with the monoculture of a cash crop had not been eased, and it appears probable that, despite the rise in sugar production, the increase in population meant a decline in *per capita* income. Of the forest zone's problems at the end of the colonial period, only that of slavery had been solved: the others had increased.

In the *sertão* of the north-east the nineteenth century was a period of very rapid change. There was an intensification of settlement, a great extension of the cultivated area and a depletion of some of the resources, but also there were first attempts to tackle the basic problem of the *sertão*: drought. The *sertão* was still sparsely settled at the opening of the century, but the population grew rapidly both by natural increase and by the arrival of runaway slaves and freemen. The *sertão's* share of the total population of the north-east during the nineteenth century rose from about 10 per cent to about 50 per cent. One of the *sertão's* products, cotton, came for a time to rival sugar as an export, and became permanently established as an important cash crop. It was during this century that the *sertão* came to occupy a more significant place in the economy and society of the north-east than it had before.

The attraction of the *sertão* lay in its open spaces; cultivable land was still available and was more easily come by than near the coast. The agricultural frontier in the north-east, which at the end of the eighteenth century had lapped along the foot of the escarpment of the Borborema plateau, was pushed westwards across the plateau, until by the end of the century it had reached the margins of the driest districts of the north-east. Further inland the lack of rainfall was to limit cultivation to the dry beds of seasonally flowing streams, to valleys of the Cariri in southern Ceara watered by artesian wells, and to ranges of hills that caught orographic rainfall. The relatively humid portion of the *sertão* acquired its own local name – the *Agreste*. Manioc, beans and maize, the typical food crops of the north-east, were grown in the *Agreste*, some to be sent to the forest zone, and cotton became the main cash crop. The strong demand for Brazilian cotton, which began in the second half of the eighteenth century, continued through the early years of the nineteenth. In the *Agreste* some cotton plantations were established, but much of the cotton appears to have been grown by smallholders, employing only family labour. In the mid-years of the century, however, the *Agreste* suffered from the competition of the southern United States, and it is probable that the acreage of cotton declined. The *Agreste* produced poor-quality cotton, and the primitive method of ginning often broke the fibres. The American Civil War led to a revival of cotton cultivation, and during the last decades of the century the place of cotton in the economy of the *Agreste* was firmly established as demand became more stable with the creation of a domestic textile industry.

This extension of cultivation inland to an area in which cattle-raising had previously been the predominant form of land use led to the familiar conflict between rancher and farmer. Ranching on the *Agreste* continued to be an integral part of the local economy, but the landscape became criss-crossed with fences to separate animals from crops. The spread of cultivation also led to the removal of the cattle fairs from the borders of the forest zone to the *Agreste* and even further inland to the margins of the dry *sertão*. Ranching itself underwent a change. Formerly cattle had been the most numerous livestock, with some raising of horses and mules, but by the early 1900s sheep and goats had come to outnumber cattle, in the backlands of Pernambuco at least. Sheep and goats can make do with poorer feed than cattle, and their appearance, therefore, in such large numbers during the century is indicative of a deterioration in the quality of the grazing. Over-grazing may have resulted from the cycle of wet and dry years, the herds increasing in size in the wet years and grazing the land bare in the dry. Soil impoverishment and soil erosion caused by shifting agriculture and deforestation may have produced in some districts a vegetation cover in which only sheep and goats could find fodder.

The growth of population in the interior of the north-east meant that the droughts which from time to time afflicted the region now caused great suffering and loss of life. The drought of 1877–9 took a terrible toll. Of the one million inhabitants of Ceara it is believed that one-half died. Many *sertanejos* fled to the coast, though, once there, thousands died from exhaustion and disease. Some refugees emigrated to the Amazon. Yet after the droughts many of those who had fled returned to their homes. In fact, the threat of drought has not been a deterrent to the settlement of the *sertão*. The disaster of 1877–9 did force the government of Brazil to start thinking of ways of mitigating the effects of drought, but little was accomplished. Some dams to create reservoirs were proposed, though it was not until 1906 that the first one was completed, and only in 1909 was a government agency formed to prepare for drought. Drought and what to do about it has been since 1877 a persistent theme in the politics of the north-east.

A further aspect of change in the north-east during this century was the growth of towns. This was not only a result of more trade and more people, but also of the beginnings of a rural–urban migration, which has greatly accelerated in the present century. By 1900 Salvador and Recife both had a little over 100,000 inhabitants; these cities continued to be as they were in the colonial period, the principal ports, and during the last decades of the nineteenth century they were linked to their immediate hinterlands by railways. These were typical 'colonial' lines in that the primary purpose was to facilitate the export of a cash crop by linking areas of production with the ports. Fortaleza, Natal, Maceio, Aracaju, only minor garrisons or villages at the beginning of the nineteenth century, became sizeable towns, provincial capitals and ports. Away from the coast, however, even as late as 1900, there were no large towns, only many small ones, which were dependent on a cattle fair or a railhead or on the trade generated by the local agriculture, or which were, in the forest zone, attached to a *usina*. These towns consisted of little more than a sparse network of unpaved streets, a square and church.

By the end of the nineteenth century few of the problems the north-east had inherited from the colonial period had been solved and new ones, such as the need to deal with droughts in the interior, had been added. The rapidly increasing population, then reaching 6 million (one-third of the country's total), made the task of raising the standard of living all the more difficult. In 1900 the north-east was already, compared with other parts of the country, 'depressed' and backward.

Coffee: the development of São Paulo

As the production of sugar and gold had dominated the economy of Brazil during the colonial period, so coffee became the mainstay of the economy in the nineteenth century. Coffee was brought to the Americas at least as early as the beginning of the eighteenth century. It became an important export crop in the French colony of St Domingue, but elsewhere was grown only in small quantities: in Jamaica and Central America and in Brazil on the hillsides around Rio de Janeiro. Throughout the eighteenth century there was never a strong demand for coffee, and it was only during the nineteenth century that it became a popular drink in Europe and North America. In the United States alone, between the years 1821 and 1844, the *per capita* consumption of Brazilian coffee rose from 28.35 g to 2.27 kg. This suddenly acquired but lasting taste for coffee led in Brazil to the spread of coffee cultivation across the south-eastern part of the country from the state of Rio de Janeiro to Paraná, to a large inflow of settlers, many of them from Europe, and to an accumulation of wealth which was used to build the city of São Paulo as well as to finance the beginnings of the industrial development of Brazil.

From the early nineteenth century to about the 1880s the Paraíba valley, across the Serra do Mar from Rio de Janeiro, was the principal region of coffee production in Brazil. The beginnings of Luzo-Brazilian settlement here really only date from the previous century and the decline of the gold rush, when a backwash of *mineiros* abandoned the exhausted diggings and moved into the valley to raise cattle. Few of these settlers had a good legal claim to the land they used, and boundaries to properties were at best only vaguely known. There was plenty of land and, so long as the agriculture was predominantly subsistent rather than commercial, little value was attached to land. With the onset of the coffee 'boom' and with the realization that the Paraíba valley, so close to the port of Rio, was suited to coffee cultivation, a scramble for land took place. The wealthy and the powerful soon claimed large holdings, dispossessing squatters and smallholders alike to make room for plantations. These first appeared on the south side of the valley, in the state of Rio de Janeiro, and gradually coffee cultivation was extended up the valley towards São Paulo and into the hills on the north side, in Minas Gerais.

At first the quality of the coffee produced was poor, for the planters had no previous experience of coffee cultivation and had to learn the business more or less through trial and error. Only gradually were improvements made, but by mid-century the pioneer period was over. There was by then a dense rural population and an active business and social life in the prosperous small towns of the valley, while planters could afford to build themselves elegant mansions and play prominent roles in the political affairs of the country. This prosperity, however, was not to endure. There were two serious flaws in the economy: the exploitative agricultural methods employed and, as the plantations were worked by slaves, the labour crisis that the impending abolition of slavery would bring. At the time slavery was the issue most talked about and its abolition generally considered the most serious threat, but the real, though ignored, problem and the one that brought lasting damage was soil erosion. The forests were gradually cut down to make way for coffee, the trees being planted in rows aligned downhill. After 3 years the first harvest was collected, but it took 6 years for the trees to mature. During these early years food crops were grown between the rows of trees. As the groves of trees began to age and their yields decreased, further stretches of forest were cleared and new groves planted. Whenever the hillsides were left without an adequate cover of vegetation the topsoil was washed away, and the practice of planting the coffee trees in rows downhill helped channel the waters of the subtropical downpours into rivulets, which rapidly deepened into gullies. These eroded lands were invaded by weeds, coarse grass and ants, and so made uncultivable. When no more forest remained to be cleared in the valley and no new coffee plantings could be made, coffee production began to decline. Sir Richard Burton, who travelled across the valley in 1862, commented on the extent of erosion (Burton, 1869, I, p. 42); by 1888, the year of the abolition of slavery, many plantations were well on the way to being ruined by the mismanagement of land. Rather than spend their profits on belated attempts to repair and conserve, the planters founded new plantations, opening a frontier of coffee cultivation further west, and left behind them a ravished valley, with a depressed and declining population, which turned to making a living from that traditional Brazilian stand-by in times of economic collapse – cattle-ranching.

The new frontier opened up during the second half of the nineteenth century lay to the north and west of the city of São Paulo, as shown in *Figure 8.9*. In this region of varied geological structure, relief, soil and vegetation, planters learnt to be selective in siting their coffee groves. It was soon realized that coffee grew better on the wooded hillsides than on the open savanna, and that the so-called *terra-roxa* soils gave the best yields. These soils have developed on flows of diabase and are reddish-purple in colour, porous, deep and rich in humus. The abundance of *terra-roxa* in the neighbourhood of Ribeirão Preto and São Claro made these towns centres of coffee production. For a time Campinas, at the focus of routes leading from these two districts of the frontier to the coast, gained population more rapidly than the city of São Paulo itself. In the last two decades of the century, another region of *terra-roxa* soils around Botucatu was planted to coffee. On these new plantations the methods of preparing and grading coffee beans greatly improved, but the exploitative methods of cultivation that had characterized the plantations of the Paraiba valley were continued, with the same results of soil erosion and soil exhaustion. The productive life of a plantation appears to have been of the order of 30 to 50 years. The coffee frontier, therefore, continued to move, and during the twentieth century pushed northwards into Minas Gerais and Goiás, as well as westwards along the *terra-roxa* soils into the state of Paraná.

The expansion of this frontier was only made possible by the building of rail-

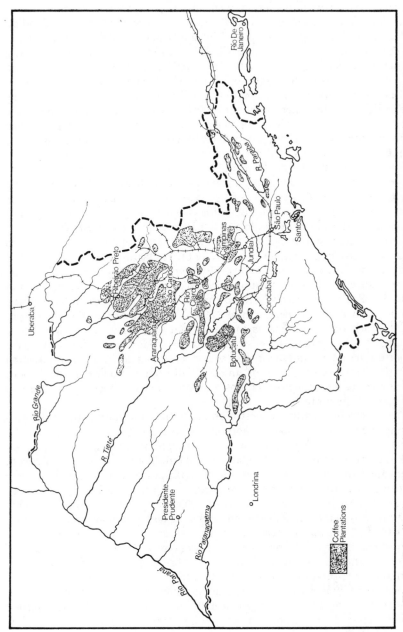

Fig. 8.9 Distribution of coffee in São Paulo at the beginning of the twentieth century (after Monbeig, 1952)

ways, for distances from the coast were now becoming too great to make the use of mule trains practical. Indeed, for a time the lack of adequate transportation appears to have slowed the development of the frontier around Campinas. The first railway line, opened in 1867, led from Santos through São Paulo to Jundiaí, and was financed with British capital. The rapid extension of the rail network was for the most part financed by the *Paulista* planters themselves. As soon as a district was producing a fair amount of coffee, agitation began for a rail link. Though the companies had ambitious plans to build railways into the far interior of the country – some of which were eventually realized – the role of the railway in São Paulo in the nineteenth century was to follow and support the coffee frontier, rarely to precede it and open new lands. The railways converged on São Paulo city and its port of Santos, which in the last years of the century was greatly extended and improved, and became the world's largest exporter of coffee.

A major contrast between the plantations on the new coffee frontier and those of the Paraíba valley was the employment of free rather than slave labour. The new frontier, after all, was 'opened up' when the transatlantic slave trade was over. The regions into which the coffee frontier was moving were sparsely populated and could not provide all the hands that were needed. The planters looked to Europe for immigrants, forming their own immigration companies to subsidize passages or persuading the state of São Paulo to pay subsidies. The trickle of immigrants arriving in the mid-years of the century turned into a flood towards the end, and continued into the twentieth century. Between 1887 and 1900 São Paulo state received more than a million immigrants, with nearly 140,000 arriving in the peak year of 1895. The conditions on the plantations under which these immigrants were expected to work were far from good. Tenancy, poor wages and lack of schools and medical facilities combined to make opportunities for advancement seem so remote that many immigrants retreated from the plantations to the towns, and especially to the city of São Paulo. Labour was also recruited for the plantations from other parts of Brazil, particularly in the states to the north.

The nineteenth century witnessed an astonishing transformation in the geography of São Paulo. A poor, remote and backward state in 1800, it was by the end of the century the most prosperous and progressive in all Brazil. Its population increased from about 150,000 to 2.25 million; it had a rail network unequalled in Latin America outside the pampas of Argentina; there were numerous sizeable, prosperous towns, and a capital, São Paulo, which had increased in size from 30,000 in 1870 to 239,000 in 1900. The profits from coffee were invested in industrial development; by the end of the nineteenth century, the city of São Paulo had become the financial and industrial centre of Brazil. There were, nevertheless, problems. The economy of the state and, indeed, of Brazil as a whole had come to rely very heavily on the export of coffee, which was soon to be over-produced and difficult to market. Coffee also led to the phenomenon of the 'hollow frontier', a frontier in which the population is denser and economic activity greater along or near the frontier than in the older settled regions behind it. The frontier moves not by pressure of population behind it, but by the exhaustion of resources along it, as well as by the promise of greater rewards to be earned in the new lands beyond it. How often and how closely the actual conditions in São Paulo approached this description of the ideal 'hollow frontier' is difficult to say. The Paraíba valley may provide the best example. Certainly, in the old plantation tracts of São Paulo there was a period of readjustment and a running down of economic activity. In the Paraíba valley, if not also in other districts, there was a decline in population. People turned to cattle-ranching and to the cultivation of sugar, cotton and food crops. Smallholdings, owned and operated by immigrants, became much more numerous.

The south

For the three southern states of Paraná, Santa Catarina and Rio Grande do Sul, the nineteenth century was a period of transition from traditional to modern Brazil. In 1800 these southern states were remote and sparsely inhabited, with an economy based on cattle-ranching; by the end of the century, they had acquired a substantial population, much of it through immigration from Europe. There was an active agricultural frontier, and one of the most productive agricultural economies in Brazil was in the making. Compared with the nineteenth-century migration to North America or even to São Paulo, the numbers of immigrants reaching southern Brazil were comparatively small. This immigration was sponsored in the first instance by the government of Brazil and later by the state governments[4] as well as by private colonization companies. Immigrants in the first half of the century were mostly German, and in the 1870s and 1880s Italian, while in the last years of the century the national origins were much more diverse. These immigrants settled in the forested lands, avoiding the grasslands, for the forest soils were considered, generally correctly, to be the more fertile.

The government of Brazil in 1824 founded the first of its European colonies at São Leopoldo in Rio Grande, to strengthen Brazil's hold on these southern reaches of its territory. São Leopoldo lies in the fertile, forested land between the Jacui river and the escarpment of the *planalto*. Two more German colonies were founded in 1829, at Rio Negro on the borders of Paraná and Santa Catarina, and at São Pedro de Alcantara, near Florianopolis. These two colonies, much more isolated than São Leopoldo, remained small, while by 1830 São Leopoldo had 5000 colonists and became the nucleus of an expanding region of German settlement. Progress was hindered by the civil war in Rio Grande of 1835–45, but in the late 1840s settlement spread westwards through the forestlands on the north side of the Jacui valley. The colony of Santa Cruz was founded in 1849. German immigration to Rio Grande was checked in 1859 by the so-called Heydt rescript. By this order the Prussian government, shocked by the treatment accorded Germans in Brazil, especially on the coffee plantations of São Paulo, stopped further Prussian emigration to Brazil, and this example was followed by other German states.[5]

The nucleus of Italian settlement in Rio Grande was to the north of São Leopoldo, on the *planalto*. Such colonies as Caxias, Garibaldi and Veranópolis founded in the 1870s and 1880s quickly attracted a population of over 20,000. During the later years of the century colonizers of German, Italian and Luzo-Brazilian descent were moving into the valleys of the upper Jacui and Ijui rivers and their tributaries. A German colony, Santo Angelo, was founded in the Ijui valley as early as 1855. This frontier was supported by the building of a railway from Porto Alegre, which reached Passo Fundo in 1895 and Erechim in 1910. By the beginning of the twentieth century the only large stretch of forest in Rio

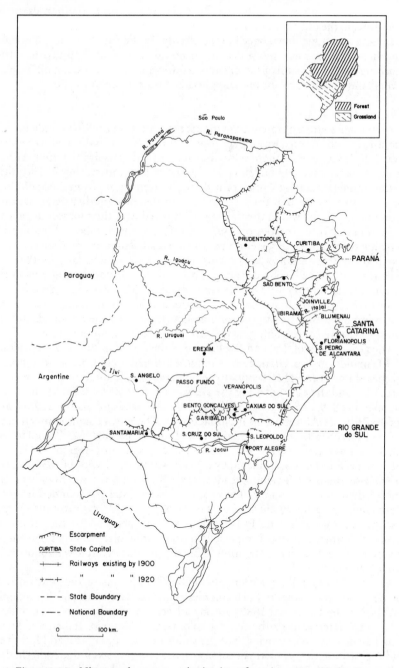

Figure 8.10 Nineteenth-century colonization of southern Brazil (after L. Waibel, *Capitulos de Geografia Tropical e do Brasil*, Rio de Janeiro, Serviço Gráfico do Instituto de Geografia e Estatística, 1958, facing p. 208)

Grande still unpenetrated by colonists lay along the Uruguay river in the northern part of the state.

In Santa Catarina a second nucleus of German colonization, privately sponsored, was established in a region of small valleys and dissected escarpment near the coast to the north of Florianopolis. In 1849 the Kolonisationsverein von Hamburg bought land from the Prince of Joinville, a connection of the Brazilian royal family, and founded the Dona Francisca colony with headquarters at the town of Joinville. The colony prospered, its population increased and settlement began to spread inland. By 1910 a railway linked Joinville with São Bento. In the Itajaí valley Dr Herman Blumenau in 1850 founded the town which bears his name. This colony, too, was successful and by 1882 contained a population of 16,000, of which 70 per cent were German. From Blumenau settlers began to move inland. In 1897 an affiliate of the Hamburg Company acquired land on the upper Itajaí where it founded Hamônia (now Ibirama), linked to Blumenau by rail in 1909. From Ibirama settlement spread into the valleys of the tributaries of the Itajaí.

European immigration to Paraná began late. The state had only had a separate existence since 1853, and the active promotion of colonization did not begin until the 1860s. Small groups of Germans, Italians and Poles were settled on the lower *planalto*, a region of grassland and forest, many in the vicinity of Curitiba, the capital. Towards the end of the century, Ukrainians began to arrive in large numbers, founding a large colony in 1896 at Prudentopolis on the second *planalto*, across which a railway was then being built from São Paulo.

By the end of the nineteenth century the only forested districts beyond the agricultural frontier in southern Brazil were in western Paraná and Santa Catarina, as well as the Uruguay river country of Rio Grande. The clearing and cultivation of these districts was to be the work of the twentieth century, when a second frontier was also opened in an attempt to cultivate the grasslands.

In all three states colonists cultivated small properties. In Rio Grande, during the course of the century, there was a reduction in the size of property awarded to the arriving colonist. The original settlers at São Leopoldo had been given about 77 ha (160,000 *braços* square); after 1851, colonists were awarded 48 ha and, in 1890, 25 ha became the standard grant in both government and private colonization schemes in that state. The colonists raised large families and inevitably, through inheritance, the properties rapidly became subdivided. Indeed, excessive subdivision of land resulted in the creation of *minifundia* – properties too small to provide the owner with an adequate living – and was to become a major problem in the twentieth century. Sons who could not hope to inherit an adequate parcel of land in the home colony moved to the frontier, thus helping to provide the manpower for the settlement of new regions. In contrast to the frontier in São Paulo, the frontier in these states was not 'hollow'.

The standard of agriculture in these regions of nineteenth-century settlement was at first disappointingly low. In the strange environment and in the midst of the work of clearing the land, the colonists were unable to use European agricultural practices; they copied instead the techniques and crops of the Luzo-Brazilian, growing the traditional crops of maize, manioc and beans under a system of shifting cultivation. This subsistence agriculture supported only a very meagre standard of living. Along the far frontiers of settlement and in isolated areas, where colonists were few in number, this acculturation to a Luzo-Brazilian way of life was hardest to resist and most complete, and it persisted longest. Wherever there was access to a market, improvements in agriculture could be made and colonists began to cultivate potatoes, rye, upland rice and even wheat, to keep a cow or two and to raise hogs. With livestock on the farms it became possible to manure the soil, and evolve from rotation of land to rotation of crops. A further stage in the improvement of agriculture, but one reached by only a small percentage of colonists before 1900, was well-developed mixed farming, based on crop rotation and stock-raising, supplying city markets as well as local creameries and cheese factories. On some farms tobacco was a cash crop, and the Italians of Rio Grande began to specialize in viticulture. The keys to success for this European colonization appear to have been fertile forest soils, access to a market and settlement in large, compact groups. Success was achieved most markedly in three areas: around Curitiba, in the older German and Italian districts of Rio Grande, and in the vicinities of Blumenau and Joinville. Here there existed, by the end of the nineteenth century, small towns and villages built in a distinctive European as opposed to Luzo-Brazilian style, and a population speaking German, Polish or Italian in preference to Portuguese, living from the cultivation of small properties, and enjoying a standard of living far above that to be found anywhere else in rural Brazil.

While the colonizing of the forestlands of southern Brazil was taking place, there was little change on the grasslands. They were still sparsely populated, and were the domain of large estates, of traditional cattle-raising, and of the gaucho. To the south, in Uruguay and on the pampas of Argentina, there had been a revolution in livestock-ranching during the second half of the nineteenth century (see Chapter 9), but in Rio Grande, for reasons not yet clear, this revolution had scarcely begun by 1900. Despite the introduction of some pedigree livestock from Uruguay in the 1880s, the Brazilian register of pedigree livestock was not opened until 1906, and by 1915 only 301 animals had been registered in Rio Grande.

The transformation of the southern states of Brazil from a sparsely populated and scarcely exploited outpost of a colonial empire into the 'bread basket' of modern Brazil had begun by 1900: the agricultural potential of the forestlands was being realized, and a start had been made on turning the lower Jacui valley and the land around the Lagoa dos Patos into irrigated rice fields. Although the agricultural revolution on the grasslands had still to begin, by 1900 traditional Brazil was yielding to modern in these southern states.

The Amazon

For much of the interior of Brazil the nineteenth century brought few changes. Across the hills and plains from Minas to Mato Grosso, a scant population drowsed through a life of cattle ranching, shifting cultivation and half-hearted searching for gold and diamonds. The old towns continued to decay from their eighteenth-century baroque magnificence. In Minas there were only rare and isolated sparks of new activity, such as the smelting of iron ore, the beginnings of a cotton textile industry at Juiz da Fora and, at the very end of the century, the building of the new capital Belo Horizonte. In one large region of the Brazilian hinterland, the valley of the Amazon, there was a dramatic, even frenzied, burst of activity.

The Amazon was the last large region of Brazil to be drawn into the national economy. This came late in the second half of the nineteenth century and provides yet another example of the cycle of 'boom and bust', which has characterized the development of Brazil. Until this time little serious attempt had been made to exploit the region. Some garrisons had been established at a few strategic locations, and here and there along the main stream and its tributaries missionaries had attempted to assemble the local Indians into villages. There was some export of tropical woods and of what were sometimes called 'backland drugs' such as cinnamon, cloves and sarsaparilla. Around Belém small quantities of sugar, rice and, in the eighteenth century, coffee were grown. There was some cattle ranching, especially on Marajó Island. The commerce of the Amazon had been discouraged by the ban on foreign shipping, which was not lifted until 1867. The population was small, consisting of some Christianized Indians, a few slaves, some missionaries, soldiers and traders, and, deep in the forest, Indian societies still survived. In fact, the major result of European contact with the Amazon had so far been to disrupt and reduce the indigenous population through disease and slavery rather than to bring prosperity and progress.

The 'boom' in the Amazon was based on rubber. *Hevea brasiliensis*, the species of tree that is by far the major source of natural rubber, is native to the Amazon basin and is widely distributed there. Until the discovery of the process known as vulcanization in 1839, rubber was of limited value, for it was difficult to use, melting easily when warm and becoming brittle in cold weather. Vulcanization corrected these deficiences, thereby greatly increasing rubber's usefulness, and this improved product found an enormous, expanding market in an industrializing world. Exports of rubber from the Amazon increased from a few thousand kilogrammes a year before vulcanization to 42,000 tonnes in 1912.

The demand for labour created by this 'boom' could not be met by the local population, but it attracted the poverty-stricken peasantry of the north-east, labouring under the desperate conditions of life in the *sertão*. Accurate statistics of this migration are hard to obtain, but it seems probable that between 1870 and 1910 200,000 people left the north-east for the Amazon. Few of these *Nordestinos* 'made good'. Rubber-collecting was arduous, dangerous work, the land was already in private ownership, and the landowners and exporters controlled the trade, reaping the profits. The rubber-collector or *seringueiro* was advanced money for food and equipment and assigned a stretch of forest. He tapped the rubber trees and, at his base camp on some creek or tributary, moulded the latex into large balls, which were then collected and taken to a port of export. The latex the *seringueiro* collected was somehow never enough to pay off his debt to his employer.

Belém and Manaus became the centres of the rubber trade. Both grew rapidly. In 1865 Manaus had a population of 5000; by the end of the century its population was 50,000 and that of Belém about 80,000. Nearly a third of the export trade of Brazil, by value, was made up of rubber passing through these two ports. They were cosmopolitan cities, with large numbers of English, French, Germans, Syrians and Lebanese involved in the rubber trade. The enormous profits were spent on luxuries and ostentatious buildings, private mansions and government palaces, theatres and elaborate public squares. Little thought was given to the future. No money was invested in improving the rubber industry or in other resources in an attempt to diversify the economy. Even agriculture was neglected, so that food had to be imported from other parts of Brazil or even from abroad at great expense.

The end of the 'boom' came rapidly, in the early years of the twentieth century. Seeds of *Hevea brasiliensis* had been taken to Kew Gardens, London, in 1876 and there successfully planted. These plantings provided the stock to establish rubber plantations in south-east Asia during the 1890s. Plantation rubber was found to be cheaper to produce and of a higher and more uniform quality than the wild rubber of the Amazon. On plantations it was relatively easy to introduce quality control, the trees could be cared for and made to yield a greater quantity of rubber per year than trees in the Amazon, and labour costs could be reduced, one man being able to tap far more trees per day on a plantation in Malaysia than was possible in the forests of the Amazon. As the acreage of plantation rubber increased in south-east Asia, Amazonian rubber lost its hold on the world market. In 1905 less than 1 per cent of the world's rubber came from plantations; by 1910, this share had risen to 10 per cent and to 93 per cent by 1922. In the Amazon trade slid almost to a halt. In Manaus and Belém the theatres closed, palaces began to crumble and grass grew in the streets. Once more in the history of Brazil, a population assembled by a 'boom' turned to ranching and cultivation when the 'boom' was over. Only a trickle of exports now came from the Amazon: nuts, hardwoods and a little rubber.

Brazil in the twentieth century

Physical geography still presents Brazil with problems of size and distance, and the varied environments offer different opportunities. The past has left a strong tradition of primitive, wasteful cultivation and the need for an enormous amount of repair work. Political decisions have been made with far-reaching consequences about the use of mineral resources, encouragement of industry and exploitation of frontier areas. Brazil, too, has more than shared in world population growth: the annual average rate of population increase between 1900 and 1950 was about 2 per cent, but between 1970 and 1980 it was 2.8 per cent, rates above the world average, with the result that the population of Brazil has increased from some 17 million at the beginning of the century to about 125 million at present. There is a movement of people from countryside to towns to the extent that some regions are now more urban than rural. The human geography of Brazil is very much in a state of transition.

Agriculture
Even though industry has been gaining a prominent place in the Brazilian economy, agriculture still remains of fundamental importance. In the late 1970s agricultural products accounted for 65–70 per cent of the value of Brazilian exports, and Brazil holds a very high rank in the table of exporters of agricultural produce, in 1977 second only to the USA. Agriculture is still the largest employer, occupying in 1970 about 32 per cent of the Brazilian work force. As is usually the the case in developing countries, agriculture forms a very conservative section of the economy. Agricultural techniques remain backward and yields are low. In much of the country little attention is given to manuring, seed selection

or rotation of crops; shifting cultivation is the rule, the hoe and machete the usual tools. It is in agriculture that the Indian element in the cultural amalgam of Brazil is most pronounced. Agriculture really forms part of traditional Brazil, except in some districts of São Paulo and the southern states where the most productive and the most mechanized agriculture in the country is to be found. According to the 1970 census, in São Paulo there was one tractor for each 313 ha, but only one for each 4623 ha in Pernambuco, and for each 6267 ha in Paraíba. During the present century there has been a sharp increase in the quantity of agricultural goods produced. This increase has come more from an extension in the cultivated area – from 6.6 million ha in 1920 to over 30 million in the 1970s – than from improving yields per hectare.

Much of the farmland is either in large estates or in very small holdings. In 1975 52 per cent of the farms in Brazil were less than 10 ha in size and comprised only 2.8 per cent of the agricultural land, while farms of over 1000 ha, only 0.9 per cent of the total number of farms, accounted for 42.8 per cent of the farmland. The smallholdings are usually too small to provide an adequate living and the large estates are only partially exploited. If rural Brazil is to become part of modern Brazil there will have to be a far-reaching agricultural reform in the very broadest sense of the phrase. There must be a rationalization of the land-tenure pattern to consolidate smallholdings into viable economic units, and to detach unused land from the large estates to make it available to people who will use it. Education of the rural worker, diffusion of better agricultural techniques, improved transportation and storage and packaging facilities for agricultural products must all be part of the reform.

Nowhere in Brazil is the need for reform more desperate than in the north-east, where generally more than 60 per cent of the gainfully employed work in agriculture – a figure which rises to over 80 per cent in some districts. The levels of productivity are low, and the standard of living for the vast majority of the population is abysmal. Part of the low productivity can be blamed on the climate, for in much of the region the scant and unreliable rainfall severely limits the agricultural possibilities. But the manner in which the land has been settled and exploited is also to blame. Deforestation and over-grazing have diminished the resource base, man has relied on wasteful and backward agricultural techniques, and patterns of land tenure and the types of agriculture have severely contributed to the present crisis.

Land tenure in the north-east is characterized by large landholdings or *latifundia*, some dating from colonial *sesmarias*, and by smallholdings or *minifundia* with origins in squatters' plots. In Pernambuco, according to the 1970 census, properties measuring 100 ha and larger, which amounted to only 3.5 per cent of all agricultural properties, covered 60 per cent of the agricultural area. The same census revealed the marked extremes in land subdivision in the north-east: the 111 *latifundia* of more than 10,000 ha each covered in all about 2 million ha while over 1 million *minifundia* shared 3.5 million ha. The *latifundia* are not intensively exploited; only a small proportion of their land is cultivated, a proportion which declines from the humid coast towards the interior where in the *sertão* much land is uncultivable, but there are also large areas of poor-quality pasture and neglected scrubland. In medieval Portugal the *sesmaria* was designed to ensure cultivation of land; in Brazil it has led to a system of land tenure that has the opposite effect. There has evolved a tradition of prestige in landownership in which status derives from the extent of land owned as well as from the income gained, and this tradition helps to maintain large, partially used estates. Furthermore, the tendency in Brazil to favour squatters' rights, permitting a person who has cultivated a plot for some time to claim ownership or at least immunity from eviction, discourages renting of land and particularly the granting of long leases. Smallholdings or *minifundia* also limit agricultural production. Agricultural techniques on the *minifundia* are more backward than those on the *latifundia* and yields are lower. *Minifundia* at best only marginally enter the commercial economy, and many are subsistence holdings. The pattern of land tenure is thus a major obstacle to increased agricultural production.

Agriculture is geared to the cultivation of commercial cash crops. The production of food crops on large estates is a sideline; they are usually grown by the smaller farmers and the *minifundistas*. Hence, food is in short supply and, given the local wage-level, expensive. The north-east is still an agricultural region in which the people are ill-fed. In the coastal region, sugar is still the major source of income, but by world standards production is inefficient. Yields are low, the industry unmechanized and labour-intensive.[6] The economy of the interior of the north-east revolves around cotton, sisal and livestock. The cotton supplies the Brazilian textile industry. Sisal was introduced from Mexico and began to be an important crop during the Second World War. It is well suited to the climate, finds a good market and has in some areas displaced cotton. The livestock industry is only gradually being improved. Zebu cattle have been interbred with the original stock, and the feed has been improved since the 1930s with the planting of palma, a South African forage crop, which, mixed with cotton-seed cake, is used during the dry season. None of these commercial forms of agriculture provides a high income or permits the paying of good wages.

Since the end of the last century attempts have been made to mitigate the problems of the north-east. In 1900 the government of Brazil established the National Department of Works Against Droughts, responsible for stockpiling food against severe shortages during droughts and for building small reservoirs in the *sertão*. But it has functioned only fitfully, and has made no attempt to improve the quality of cultivation or tackle the issue of land tenure. All too frequently the reservoirs, though built at public expense, have benefited only the already wealthy landlords who own the surrounding property. In 1948 the government founded the São Francisco Valley Commission and the São Francisco Hydroelectric Company to develop the resources of the São Francisco valley. Hydroelectric power is now generated on the São Francisco and an extensive programme of rural electrification has been nearly completed, one of the few concrete achievements to the Commission's credit. The most promising attempt yet to tackle the problems of the north-east began in 1959 with the founding of The Superintendency for the Development of the North-East (SUDENE), which was charged with the planning of all aspects of the regional development of the north-east. The main features of the agricultural section of the regional plan drawn up by SUDENE are the rationalization of land tenure, and an agricultural extension programme aimed at raising the educational level of cultivators as well as introducing new varieties of crops and improved techniques. SUDENE set out to increase the yield of sugar per hectare, so that land could be released from sugar cultivation without causing a decline in the total sugar crop, and this land then used for food crops. By the mid-1970s cane cultivation had become concentrated in coastal Pernambuco and in Alagoas; Bahia, for so long a major producer of sugar, now contains rather less than 1 per cent of the area planted to sugar cane in Brazil and no longer makes a significant contribution to the industry. Despite this rationalization of the north-east's sugar industry, the region accounts for a declining share of Brazilian sugar production: 33 per cent in 1979 compared to 53 per cent in 1945.

SUDENE has also been administering a programme to attract industry to the north-east. However, reformed agriculture and industrial development have not provided sufficient employment for the increasing population; migration to São Paulo and other points in central and southern Brazil has been going on for many years and can be expected to continue. People from rural areas are also flooding into the cities of the north-east. A small number of people have been encouraged to settle in the still sparsely populated state of Maranhão, and others have made their way into Amazonia. SUDENE has achieved some success, but the amount of work to be done and the social, political and economic problems to be overcome in raising the standard of living of the north-easterners are still formidable.

The problems bequeathed by past land use in São Paulo are of a different order and scale than those in the north-east. To simplify, it is possible to see the agricultural development of São Paulo in the twentieth century as revolving around two themes: stabilizing the coffee frontier and solving the problem of the excess production of coffee, and secondly, the creation of new forms of land use in the old plantation tracts.

During the early years of this century the coffee frontier continued to advance south-westwards along the course of the tributaries of the Paraná, keeping to the *terra-roxa* soils (*Figures 8.9* and *8.11*). In the 1920s and 1930s coffee cultivation crossed from São Paulo into the state of Paraná, and this was the last southward advance of the coffee frontier. In northern Paraná climate varies with altitude from subtropical to temperate, winters are cooler than in São Paulo, radiation frosts are common, and occasional cold spells do occur. There is in northern Paraná risk of damage to coffee trees from frost, but further to the south the incidence of frost is so frequent that coffee cannot be cultivated with any realistic hope of success. The development of this region has been guided by a colonization company, originally English-financed, known as Paraná Plantations Ltd. The company was organized in 1925 and had by 1927 acquired over 10,000 km² of land. Two subsidiary companies were established: one, the Companhia de Terras Norte do Paraná, to subdivide and administer the lands; the second, the Companhia Ferroviária São Paulo-Paraná, to build a railway to link the region with São Paulo and the port of Santos. The English management of this settlement scheme ended with the Second World War. In 1939 the Brazilian government acquired the railway, and in 1944 a group of *Paulista* capitalists bought the land company, renaming it the Companhia de Melhoramentos do Norte do Paraná.

The region the company controlled consists of a deeply dissected plateau. The soils are mostly *terra-roxa*, but there are areas of sand. Roads were laid out along the interfluves, converging on selected points where towns or villages were planned. The grid of road and service centre was such that no part of the

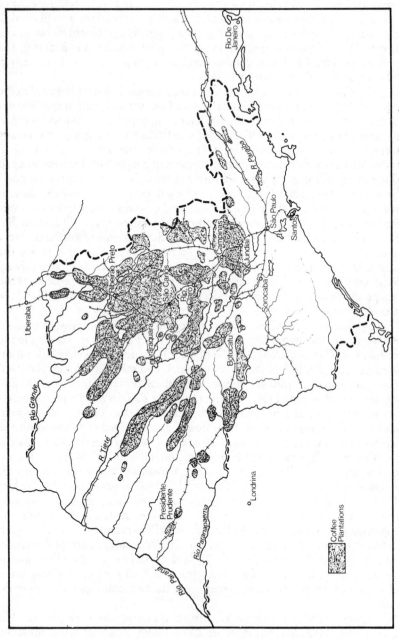

Fig. 8.11 Distribution of coffee in São Paulo, later 1920s (after Monbeig, 1952)

company land was more than 15 km from a centre. The land was divided into long and narrow allotments, laid out parallel to each other between roads on the interfluves and rivers in the valley bottoms. This subdivision of land gave each settler access to transport, to water and to a range of soils, vegetation types and microclimates. It was a pattern of land subdivision subsequently copied on government colonization schemes elsewhere in southern Brazil. The company also took the precaution when originally acquiring its land to buy out all claimants to a given property, even if this meant, as it often did, paying for land several times over, to ensure that the company's settlers would have clear titles to their plots and so be spared the legal wrangles over ownership and threats of dispossession that have marred many settlement schemes in Brazil. Land was sold to colonists, and following the change in the ownership of the company in 1944, large holdings have been allowed for the cultivation of coffee.

It had been the original intention of the English planners to cultivate cotton, but coffee soon became the predominant crop. Cotton, however, quick to mature, served the useful purpose of providing farmers with a cash income while the coffee trees were growing. Coffee was planted on the valley sides, particularly on the slopes facing the north. More care has been taken with these groves than with those in the old plantation tracts. The soils have been manured, the trees planted along the contour rather than aligned downhill, and attention given to new varieties, with the result that yields are high and less damage has been done to the soil. Other crops are grown according to the local climate: oranges and cotton in the warmest districts, wheat, maize and other mid-latitude crops in the cooler districts. Land is also used for grazing cattle. This diversified agriculture, with its variety of cash crops, the security of tenure possessed by the farmers and good transport has been the key to the successful planning of northern Paraná. There is here now a dense and prosperous rural population and several rapidly growing towns such as Londrina. The coffee used to be exported via São Paulo and Santos, but is now sent from the Paraná port of Paranagua. The revenues from coffee have been invested in promoting the economic development of Paraná, helping to make it one of the richest states of Brazil, and transforming the capital, Curitiba, from a dull, provincial town into a bustling, modern city.

Coffee has also spread into Mato Grosso do Sul, into Goiás, and there has been some revival of coffee cultivation in the state of São Paulo. Coffee is also grown in Minas Gerais and Espírito Santo and in several other states.

During the 1960s, in terms of value of production, coffee was the most important crop. Brazil in the 1960s was growing nearly half the world's coffee crop, and coffee accounted for nearly 50 per cent of Brazil's export earnings; but the government has been trying to reduce the country's dependence on this crop and to diversify agriculture in predominantly coffee-growing districts. Substantial financial inducements were offered to reduce the number of trees, and during the 1960s the area planted to coffee decreased greatly. In the late 1970s Brazil was still the world's leading grower of coffee, but it now contributes only 25 per cent of the value of Brazilian exports.

In São Paulo diversification of agriculture began with the passing of the coffee frontier. There was a search for alternative crops, and cotton, citrus fruits, manioc, maize and rice have all become important. Sugar is now a major crop. Fields of cane cover an impressive area in the state and reach across the

Paranapanema River into Paraná where, at Porecatu, north of Londrina, one of the largest sugar mills in the world has begun to function. São Paulo's share of Brazil's sugar production increased from 21 per cent in 1945 to 45 per cent in 1979. The quality of pasture has been improved, new breeds of cattle introduced and now dairy farming and the raising of beef cattle are major activities. Towns, such as Presidente Prudente, have grown rapidly serving these agricultural areas in western São Paulo. The worn-out soils of the old plantation tracts have gradually been made productive with the aid of fertilizers, and some of the steep slopes have been terraced. There has also been a change in landownership. When the coffee plantations became run down, land values declined and immigrants or their descendants were able to buy land and so became farmers in their own right. Japan has been a second source of farmers. The Japanese have been emigrating to Brazil in comparatively small numbers during this century and most have gone to São Paulo, where they have set themselves up as market-gardeners. The agriculture of São Paulo is now more advanced and productive than that of any other state in Brazil. The huge urban population of south-central Brazil provides an accessible market for the produce.

In southern Paraná, Santa Catarina and Rio Grande do Sul the course of agricultural development has continued along the lines laid down during the last years of the nineteenth century. The long-established European colonies are now prosperous regions of mixed farming, with an important production of wine and tobacco in Rio Grande do Sul. The distinctive cultural landscape survives, and the ancestral languages are still spoken. The birth-rate in these colonies has been high. One result of this has been the subdivision of landholdings to produce *minifundia*; a second has been a movement of population to frontier regions. Today, in the half of Rio Grande north of the Jacui river, 65 per cent of all land-holdings measure less than 100 ha. This is an excessive subdivision of land, which retards the development of really efficient agriculture and contributes to soil exhaustion and soil erosion. In the more recently settled districts, where rotation of land is still practised, some holdings are so small that land must be re-cleared and cultivated before the soil has been able to regain its fertility. The settlement frontier is now along the upper Uruguay river in northern Rio Grande, in western Santa Catarina and south-western Paraná. On the frontier the national origins of settlers are still discernible from language and landscape, but the amalgamation with Luzo-Brazilian culture has proceeded much further than in the old colonies. Settlers on the frontier work in lumbering, in collecting the leaves of *Ilex paraguayensis* for yerba maté or Paraguayan tea, which is a popular drink in southern Brazil as well as in Paraguay and Argentina, and in cultivating small properties. Kidney beans, soya beans, manioc and maize are the main crops, and this frontier region has also become noted for pig-rearing. Brazil is second only to the United States as a producer of soya beans, and during the last decade they have become a very significant item in the export list. An attempt has been made to cultivate wheat on the forest–grassland borders. For several years following 1945 the harvests were promising, but yields have since often been disappointing. A succession of bad summers is partly to blame, but the incidence of disease in the wheat has been high, a reflection of the generally poor methods of cultivation. The southern states account for nearly all the wheat grown in Brazil, but this amounts to only a small proportion of Brazil's annual consumption.

In the Jacui valley, immediately to the south of the German colonies, people of Luzo-Brazilian descent were starting to cultivate rice by the beginning of the nineteenth century. Rice cultivation now extends through the river valleys of central Rio Grande and also around the coastal Lagoa dos Patos. The rice fields are irrigated, but cultivation is not very intensive, for the land is periodically left fallow to be used as pasture for cattle. Rice is an increasingly important food in Brazil and rice from Rio Grande is sent to the urban populations of the south-centre of the country.

The grasslands of the south, and especially those of the *campanha* along the borders of Uruguay, are still the domain of the large estate and of livestock-ranching. The quality of the livestock has been greatly improved, especially since the 1930s, through importing pedigree European livestock for cross-breeding. There is still room for further improvement in the stock itself, in the pasture and in methods of packaging and processing the meat. Much of the meat is now frozen and exported, but some carcasses are still turned into *charque* (dried meat).

The two-thirds of Brazil that include the Mato Grosso, Goiás, the interior of Minas and the Amazon basin contribute as yet only a small portion of the total agricultural production of Brazil. This vast interior of Brazil can be regarded as one immense frontier region: not a frontier where Luzo-Brazilian settlement is taking place for the first time, for these territories have been roamed over by cattle-ranchers, shifting-cultivators and others for two centuries or more, but a frontier of improvement. Frontier work today consists of improving transport, transforming subsistence shifting-cultivation into stable commercial agriculture, and raising the quality of ranching. In fact, the aim is to exploit these lands more effectively, and to mesh them firmly into the Brazilian economy, realizing thereby a long-standing Brazilian ambition of a *marcha para oeste* – a march to the west.

Improvements in transport began in the first decades of the century with the building of railways westwards from São Paulo across the southern Mato Grosso to Corumbá and Ponta Pora and northwards into Goiás. These lines opened up a small part of the interior to the markets of south-central Brazil. Railway towns such as Campo Grande and Anápolis became important regional commercial centres, while the areas served by the lines became the foci of economic activity in the states of Mato Grosso and Goiás. The state of Goiás went so far as to transfer its capital from the remote, gold-mining town of Goiás to a new capital, Goiânia, built during the 1930s near the railway. In recent years the building of Brasília and the linking of this capital through a network of roads to the other major cities of the country has given a further boost to the agricultural development of the interior. Brasília itself is a market for food crops, albeit a limited one, and the roads permit cheap transport of goods to the coastal centres of population. The Belém–Brasília highway, completed in 1964, has made the eastern margins of the Amazon basin relatively accessible. This was the first of a number of north-south highways linking the Amazon with southern Brazil. The east-west Transamazonian Highway was begun in 1970 and completed in 1975 (*Figure 8.12*).

The opening up of the Amazon has been a major planning initiative on the part of the Brazilian government. The motives have been mixed: desire to

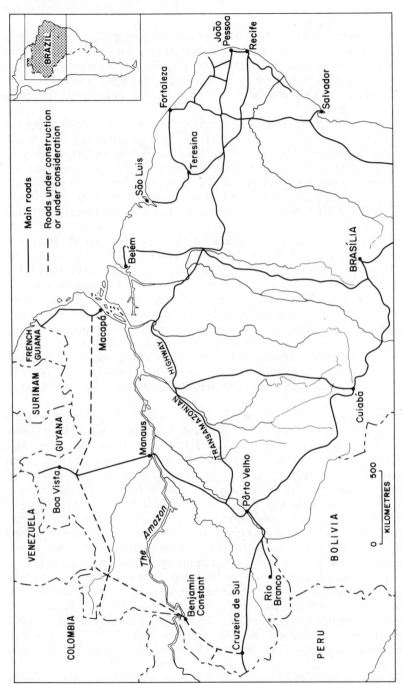

Figure 8.12 Northern Brazil: main highways

demonstrate effective occupation of Brazilian territory, a wish to integrate the Amazon into the national economy, to give access to mineral wealth, to provide land for agricultural expansion and to diffuse the population problem of the north-east through settlement schemes along the Transamazonian Highway. The result has been one of the most controversial attempts at regional development in the modern world. Brazilian planners have been criticized on economic grounds: possibly, indeed, the capital and personnel for agricultural development might have brought better returns had they been invested in improving agriculture in the accessible coastal regions with large urban populations, than in spreading or trying to improve agriculture in distant reaches of the country. The planners have also been criticized for a callous disregard of the interests of the tribal Indians and for a short-sighted indifference to the environmental effects of their activities in Amazonia. On the other hand the political arguments against ignoring distant regions of a country are compelling. Given a decision to exploit the resources of Amazonia for the benefit of the Brazilian state and its increasing population of over 100 millions, change in Amazonia was inevitable. Questions, nevertheless, remain as to whether the wisest choices have been made and whether sufficient care has been given to the problems of the environment. Speed, enthusiasm, even opportunism, have taken a toll, but it will be some years yet before the costs and benefits of this venture can be tallied, and the measure cannot simply be a financial one.

Already there are disappointments. Prompted by the drought in the north-east in 1970 the government decided to settle some 100,000 families along the Transamazon where it passes through Pará south of Belém. Families were to be allocated 100-ha plots and a hierarchy of settlements built along the Highway to provide different levels of service for the new rural population. Yet problems of storage, transportation, and the fact that the settlers can ill afford fertilizers, herbicides, insecticides and machinery have hindered the development of commercial agriculture. Yields of crops have been low. Indeed, the settlers might fare better and do less damage to the environment by adopting some aboriginal agriculture practices. Only 5000 to 6000 families had been settled by the government along the Transamazon by 1977, and less than half of these had come from the north-east. Unplanned, unsupervised squatter settlement continues apace where there is access to land. This type of piecemeal clearing of land and the application of traditional agricultural techniques will not improve Brazil's food supply nor solve the problems of the north-east. By contrast, around the southern margins of Amazonia, in Goiás and Mato Grosso, cattle-ranching has begun to flourish on large, well-financed estates. The beef is sent to the cities of the south and even abroad.

This movement of peasants and profiteers, miners and road-builders into regions hitherto very difficult of access has brought to the fore once more the conflict of interest between the tribal Indian and the Brazilians. This contest for land and resources – for space in which to live a particular way of life – has been unequal since the beginning of colonial times, but probably never more so than today. Disease, the greed of an acquisitive society and economic forces are ranged against the Indian, while his only defence of distance has gradually been whittled away. The Brazilian frontier has been, still is, a brutal place for the weak and defenceless, and there is no denying the tragedy that has overtaken many tribes in recent years. The government of Brazil has replaced the discredited Indian Protection Service with the National Indian Foundation and given it a mandate to provide medical care for the Indians, to administer Indian parks both actual, such as Xingu National Park, and proposed, and in general to look after the interests of the Indians. Such a mandate is easily given, but deciding what the Indians' best interests are, and then achieving them, are very difficult assignments. The human and economic problems of Amazonia will be a major test of the quality of the Brazilian government during the foreseeable future.

Industry: the resource base
Brazil is rich in mineral resources, but the mining industry is poorly developed and mining accounts for only 2 per cent of the GNP. As yet, moreover, the country has been imperfectly explored, and in all probability more discoveries of natural wealth will be made. The shield is the source of a wide range of minerals and precious stones. *Garimpeiros* (prospectors) are still to be found in the backlands, panning for gold and diamonds. Gold is mined today, as at the Morro Velho mine at São João del Rey, but the main mineral wealth of the shield now is high-grade iron ore. The iron-ore reserves of Brazil are among the richest in the world; ore bodies are known in Piauí, Ceara, southern Mato Grosso, the Serra dos Carajás in southern Pará, and in the so-called Iron Quadrangle, south and east of Belo Horizonte in Minas Gerais. It is the ores of the Quadrangle that are being exploited, for they have a high iron content, are easily mined and are near to the centres of industry. These ores have been mined on a small scale since the beginning of the nineteenth century, but large-scale mining dates only from the Second World War, with the growth of a domestic iron-and-steel industry and the decision to encourage the export of ore, a reversal of the traditional Brazilian policy of preserving resources for future domestic needs. The export of ore from Minas is largely controlled by the Companhia Vale do Rio Doce, a government-owned company. A railway connects the Iron Quadrangle, via the valley of the Rio Doce, with the coast of Espírito Santo. Exports of ore via this route began in 1942, 1.5 million metric tons were exported in 1951 and 11.5 million in 1967. The new port of Tuburão, near Vitória, the capital of Espírito Santo, was opened in 1966, and exports reached over 40 million tonnes by 1979. A further expansion in ore exports, over the short term at least, may be difficult to achieve. There is competition from other suppliers which the slow-down in the world economy during the second half of the 1970s and early 1980s has made intensive. Already changing world demand has led to a delay in a joint Brazilian–United States–Japanese venture to export the Carajás ores.

The demands of the iron-and-steel industry, domestic and foreign, have led to a great increase in the mining of manganese, and the value of manganese exported doubled in the early 1970s. The largest mines are those of the Serra do Navio in the territory of Amapá to the north of the Amazon. The Bethlehem Steel Company of Baltimore has an important interest in these mines. The United States Steel Company, through a subsidiary, has begun to mine the manganese ores at Mount Urucum near Corumbá, and manganese is also mined in Minas Gerais. Lead is produced in Bahia, phosphorus at Olinda, near Recife, and, along the arid coastland of Rio Grande do Norte and Ceara, most of Brazil's salt is made by evaporation from the sea. Brazil also produces small quantities of a wide variety of minerals such as zinc, copper, tungsten, nickel and bauxite. Minas Gerais remains the most important mining state, not only because of the ore bodies there, but because of its better communications and proximity to industrial centres. Distance and expense of transport as well as lack of demand mean that the exploitation of many Brazilian ore bodies is still uneconomic.

The forests of Brazil have been and still are a valuable resource. Wood is the traditional fuel of Brazil. It was used in the sugar mills and to power the railways. Even today, wood is still the fuel of the countryside, and many of the blast furnaces of Minas Gerais depend on charcoal. The demand for wood for a fuel has led to extensive deforestation. Tropical woods were a staple export in colonial times, and today Araucanian pine is possibly the most important commercial tree. It has given rise to a wood-working industry in the southern states, manufacturing such commodities as boxes, furniture and plywood. In Rio Grande do Sul and Santa Catarina there is a new but expanding pulp and paper industry.

Coal has played a very modest part in the industrial development of Brazil. Known coal reserves are small, the quality of the coal is poor and unsuitable for coking, and there is little likelihood of new coalfields being discovered. Coal is mined in each of the three southern states, and by 1979 total annual production was of the order of 14 million tonnes, 80 per cent of which came from Santa Catarina. The coal is used to generate electricity, but now the iron-and-steel industry is the major industrial consumer. Brazil imports coal, particularly from the United States and Germany.

The oil industry in Brazil is controlled by a government corporation, Petroleo Brasileiro SA, known as Petrobras. Petrobras has built and operates refineries, owns filling stations and manufactures petrochemicals and fertilizers. Until very recently it also held the monopoly of exploration for oil, but in the 1970s Petrobras invited foreign companies to bid on 'risk concessions' to explore and develop oilfields off the coast. The search for oil in Brazil has been disappointing. Exploitation of an oilfield in the *recôncavo* of Bahia, near the city of Salvador, began in 1939. Small quantities are now pumped at Carmópolis, in Sergipe, and in Alagoas, but traces only have been found in the Tocatins valley and near Manaos in Amazonia, in Maranhão as well as in the Paraiba valley. During the 1970s oil was found in commercial quantities some 60 km off the coast of Campos, Rio de Janeiro, but the development of these fields, which may not be as rich as was initially thought, is proving to be expensive and difficult. Oil consumption in Brazil has been rising more rapidly than oil production: in 1968 domestic oil accounted for about 36 per cent of Brazil's requirements, in 1978 for only 18 per cent, that is for some 160,000 barrels a day (b/d) out of a total consumption of 1,100,000 b/d. By 1985, as the oil from the Campos Basin comes on shore, domestic production should reach between 350,000 and 500,000 b/d, but Brazil could then be paying as much as US $15 billion on oil imports. This situation helps explain Brazil's keen interest in sugar-cane alcohol to supplement oil in fuel for motor vehicles.

With coal and oil in such short supply, Brazil has come to rely on hydroelectric power. The hydroelectric potential of Brazil is among the highest of any country in the world, and, as yet, only a small portion of it has been developed. Use of hydroelectric power began in Brazil at the very end of the nineteenth century to light the cities of São Paulo and Rio, to power their tramways and to supply the nascent industries. In 1901 a plant was built on the Tieté river, 32 km downstream from São Paulo; by 1912 the capacity of this plant had been increased sixteen times, and work began in 1913 on realizing the hydroelectric potential of the Sorocaba river. Power plants were also built in the mountains around Rio, and in the 1920s a major project was undertaken on the Serra do Mar, near Santos. The westerly flowing Rio Grande river was dammed, diverted and dropped 700 m over the serra to generating stations at Cubatão. There has, however, often been a gap between the increasing demands for electricity and the building of new facilities.

Before and after the Second World War power failures were a familiar feature of Brazilian life. The production and distribution of hydroelectricity is coming increasingly under government control, and for the past several years the government has been putting into effect an extensive programme for building new hydroelectric stations. This programme has been aided by the continuing improvement in the methods of transmitting power, permitting the harnessing of rivers at greater and greater distances from the consuming centres. Several large projects have already been completed. On the São Francisco river, power generated at Paulo Afonso, where the river drops over the lip of the Brazilian plateau, is used for new industry in the north-east and for rural electrification; the Tres Marias plant provides power for Minas Gerais. Altogether Brazil expects to build thirty-four new plants by 1988. The projects include the development of the power potential of the Parnaíba river on the border of Maranhão and Piauí, of the Tocantins river in Amazonia, and of the Paraná basin. The largest hydroelectric station in the world is under construction at Itaipú on the Paraná. It should begin to generate electricity in 1983, but will take a further 5 years to complete. The exploitation of the hydroelectric potential of the Paraná is requiring careful negotiations between Brazil, Argentina and Paraguay. Given these resources, Brazil has so far made only a modest commitment to nuclear power by building a plant at Angra dos Reis, on the coast west of Rio de Janeiro.

A final major resource to be considered is manpower. There is no shortage of labour in Brazil – quite the reverse – but the general level of education and technological competence is low. In the early days of Brazil's industrial development technicians and even skilled labourers were brought in from abroad; now in the established industrial centres a skilled labour force has been built up. However, the lack of training and even the wrong frame of mind for industrial employment handicaps migrants from the countryside in the search for jobs.

There are some weaknesses in Brazil's industrial resource base. The failure to find oil in large quantities has been the keenest disappointment. The shortage and low quality of coal, a major drawback in terms of nineteenth-century industrialization, is less and less serious as world industry turns to other fuels.

Industrial development
The mercantilist policies pursued by Portugal successfully stifled any attempt at manufacturing in Brazil during the colonial period. In the early nineteenth century iron-working, based on local ores and using charcoal for smelting, was begun on a limited scale in Minas Gerais, and a cotton-textile industry was established later. During the twentieth century Brazil has provided an example of industrialization aimed at import substitution. Gradually Brazil has been chang-

ing from a country exporting agricultural products and importing manufactured goods to one capable of manufacturing many of its own industrial requirements. Whenever foreign-exchange difficulties or war made importing expensive or difficult, an impetus was given to the development of domestic industries. During the First World War the number of factories in Brazil doubled. The largest sectors of Brazilian industry, which in 1920 employed 30,000, were textiles and food-processing. The Depression and the Second World War led to further industrial expansion, adding chemical and pharmaceutical industries, the manufacture of tools and machinery and an integrated iron-and-steel industry. Since the Second World War industrial development has proceeded at a rapid rate, and in terms of industrial output Brazil now ranks first in Latin America. Manufactured goods now account for about one-third of the value of Brazilian exports.

Not all parts of the country have shared in this industrial growth. Certain industries, such as the processing of foodstuffs, brewing and the manufacture of textiles, can be found in most of the larger cities, but heavy industry is much more localized and is concentrated in the south-eastern parts of the country. The state of São Paulo alone accounts for 55 per cent of Brazil's industrial production. In fact, there is emerging in Brazil one large industrial region, roughly delimited by the cities of São Paulo, Rio de Janeiro and Belo Horizonte (*Figure 8.13*), as

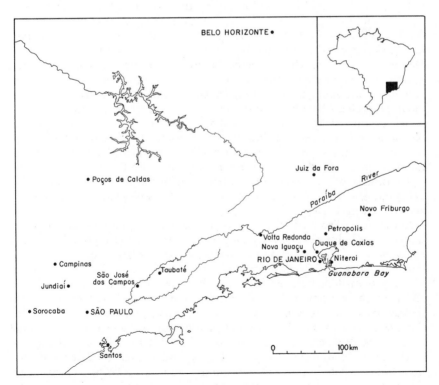

Figure 8.13 The industrial south-east of Brazil

well as several smaller nodes of industrial activity in other parts of the country.

The reasons for this concentration of industry in the south-east are partly historical. Towards the end of the nineteenth century it was in São Paulo, and to a lesser extent in Rio de Janeiro, that the capital existed for investment in industry as a result of the 'boom' in coffee. For the types of industry first established – the manufacture of consumer goods – ease of access to the market was an important locational factor, and in the above two cities there was a concentration of buying power without rival in the rest of the country. The lack of coal in the south-east has become much less of a handicap to industry with the development of hydroelectric power, and the region does contain such basic raw materials as iron ore. The south-east was the first part of Brazil to develop an adequate transportation system, a result of the coffee era and of the need to link the two most important cities of Brazil. Continuing improvements in transportation and the fact that during the course of this century proximity to the market has become an increasingly important factor in the location of industry have confirmed the south-east as the most attractive region in Brazil in which to establish new industries.

The industry of the south-east is centred upon the cities of Rio de Janeiro and São Paulo and the iron mines of Minas Gerais. Rio, which was the capital of Brazil and still contains a large part of the government bureaucracy, is a major commercial and financial centre, port and resort as well as an industrial city. The first industries to be established in Rio during the second half of the nineteenth century were food-processing, the manufacture of light consumer goods and textiles. Since the First World War heavy industry such as metal-working, foundries and ship-building has been gradually added, and the oil refinery at Duque de Caxais, on the northern margins of the city, has encouraged the growth of a petrochemical industry. The rate of industrial growth in Rio, however, is now beginning to slow down. The city occupies a constricted site between mountains, Guanabara Bay and the sea. Flat land is scarce and expensive and, moreover, the city has suffered from chronic shortages of electricity and water. New industry has increasingly begun to locate itself immediately to the north of the city and in towns even further inland. Niteroi, across the bay from the city of Rio, Petropolis, the old imperial summer capital, Novo Friburgo, once a Swiss agricultural colony – these are but a few of the towns in the state of Rio now being industrialized. Volta Redonda in the Paraíba valley has been turned into an iron-and-steel town.

In the city of São Paulo there has been a similar progression from light consumer goods industries and the manufacture of textiles to the establishment of heavy industry. The city of São Paulo now contains the largest single concentration of industry in all Brazil. As around Rio, there has been a decentralization of industry to smaller towns, particularly in the eastern part of the state. Campinas and Sorocaba are now industrial centres. Jundiaí, once a railhead, now has railway workshops as well as textiles and electrical industries; São José dos Campos is a centre for chemical and pharmaceutical industries. Other small towns of the Paraíba valley between São Paulo and Rio are acquiring industries.

The history of Minas Gerais has revolved around mining; gold, diamonds and iron. Industry here began with iron-working, and this remains a fundamental sector. Poços de Caldas is a centre for the smelting of aluminium. There is also a long tradition of textile-working in Minas; some of the earliest mills in Brazil were built at Juiz da Fora. The industry of Minas is now becoming more diversified with general manufacturing.

This trend towards the dispersal of industry from the three original centres and the resulting industrialization of the small towns in the south-east has been greatly helped by improvements in transportation. Railways and good roads now connect these towns with the major consuming centres and with the ports. In addition, the intensification of agriculture within the south-east to supply the huge urban population has led to the establishment of numerous food-processing factories. This dissemination of industry through the region can be expected to continue.

Possibly the most striking developments in the industrialization of Brazil since the Second World War have been the establishment of an integrated iron-and-steel industry and an automobile industry. Both these industries are concentrated in the south-east. Brazil now produces more iron and steel than any other country in Latin America, and its automobile industry is the largest on the continent and ranks among the more important in the world. In the 1930s Brazil was manufacturing about 60,000 tonnes of iron and steel a year, in small foundries and mills, using charcoal for smelting. This amount fell far short of the country's requirements. The difficulties of importing iron and steel during the Second World War, coupled with a desire to expand the national industrial base, led to the establishment of a fully integrated iron-and-steel works. The site selected was the small town of Volta Redonda in the Paraíba Valley, some 80 km from Rio de Janeiro. This inland location reflects in part the strategic considerations of a now distant war. There are advantages to the site: ease of communications with the largest markets in Brazil – São Paulo and Rio de Janeiro – as well as proximity to the sources of iron ore, limestone and manganese in Minas Gerais. Coal is brought from Santa Catarina and abroad. Construction of the plant began in 1942 and in 1946 the first steel was produced. By the mid-1970s, annual production was of the order of 2.5 million tonnes with plans to reach approximately 4.6 million tonnes by 1980.

Table 8.1 Brazil: production of iron and steel 1955–79

	1955	1960	1965	1970	1975	1976	1979
			('000 tonnes)				
Pig iron	1069	1750	2355	4205	7053	8162	11594
Steel ingots	1162	1843	2896	5390	8307	9169	13893
Finished rolled products	982	1358	2022	4150	6738	7525	10854
	Index numbers (1955 = 100)						
Pig iron	100	164	220	393	660	764	1085
Steel ingots	100	159	249	464	712	789	1196
Finished rolled products	100	138	206	423	686	766	1105

Sources: Sinopse Estatística do Brasil and *Economic Survey of Latin America,* 1976 (Economic Commission for Latin America, United Nations, Santiago, Chile, 1977), *Anuário Estatística do Brasil,* 1980.

This increase in the capacity of Volta Redonda plant is but one aspect of a continuing expansion of the Brazilian iron-and-steel industry. The decision taken in 1956 by Juscelino Kubitschek (President, 1956–61) to establish an automobile industry in Brazil, meant that much more iron and steel would be required. The government directly encouraged the iron-and-steel industry by investing in the companies and by permitting them to import equipment at favourable foreign-exchange rates. In Minas Gerais the capacity of a number of small integrated steel works was expanded and new plants built. In 1956 a German company began to manufacture steel tubes in the *Cidade Industrial* of Belo Horizonte. This plant uses local iron and smelts with electricity. During subsequent expansion another blast furnace was added, which uses coking coal brought from Germany in return for iron ore, and is 'backloaded' along the Rio Doce railway. At Ipatinga, to the east of Belo Horizonte, an integrated iron-and-steel works has been built, designed to produce 2 million tonnes a year. Ipatinga also relies on Minas ores and uses imported coal for smelting. There are also a large number of charcoal-fired blast furnaces each producing less than 30,000 tonnes of pig iron a year. Most of these iron works are in small towns to the west of Belo Horizonte, but there are also some at Governador Valadares in the Rio Doce valley of Espírito Santo. There is a new integrated iron-and-steel mill at Cubatão at the foot of the Serra do Mar, near Santos, designed to produce 2.3 million tonnes a year. The raw materials can be brought by sea and the finished products distributed by sea to the large coastal cities of Brazil as well as by rail and road to the city of São Paulo. In late 1978 contracts were signed for the construction of a steel plant at Tubarão, near Vitória, on the coast of Espírito Santo north or Rio. This is a joint Brazilian–Japanese–Italian venture, and is to have an annual capacity of 3 million tonnes by 1982. The far south began to contribute to the iron-and-steel industry in a very modest way with the opening of a small plant in Rio Grande do Sul in 1973. The National Iron and Steel Programme has guided this expansion of the industry with the aim of raising Brazil to a position among the ten most important iron-and-steel producers in the world, and to turn the country from a net importer to a net exporter of iron-and-steel products.

The automobile industry is the showpiece of the industrialization of Brazil. Before 1956 only a few cars were assembled in Brazil and there was one government-owned factory manufacturing trucks. Foreign companies were invited by the Kubitschek government to establish branch plants in Brazil, and pressed to manufacture more and more of the parts for their vehicles in Brazil. By the late 1970s the minimum locally manufactured content of the vehicles was set at 85 per cent. Some models are now designed in Brazil. The industry now manufactures about a million vehicles a year, and exported vehicles and parts to the value of US $843 million in 1979, a substantial increase from the US $3.9 million of 12 years before.

The industry has had to overcome problems of a relatively small home market, a reflection of the generally low standard of living in Brazil and high costs of production. While economies of scale have been difficult to achieve, there has been some rationalization as companies have succumbed to competition, with Volkswagen emerging in a prominent position, responsible for about half the annual production of vehicles. The industry was originally concentrated in and around the city of São Paulo and especially in the suburb of São Bernardino. Ease of assembly of parts and raw materials, local capital and entrepreneurial skill, as well as the fact that São Paulo, the most prosperous state of Brazil, is the main market, explain this location. There has, in recent years, been a modest decentralization, but with the exception of a plant at Jaboatão, near Recife, for the manufacture of jeeps and pick-up trucks, the industry has not spread beyond the confines of the prosperous south-east of the country. Mercedes-Benz has opened a new truck factory in Campinas; Fiat manufactures trucks in Rio de Janeiro and cars in Belo Horizonte; Volvo has built a factory for buses and heavy trucks at Curitiba (*Table 8.2*).

Table 8.2 Brazil: production of motor vehicles 1957–81

Year	Units	Year	Units
1957	30,700	1974	858,479
1960	133,078	1975	914,971
1963	174,126	1976	985,369
1966	224,575	1977	919,242
1969	352,192	1978	1,060,000
1970	416,394	1979	1,128,000
1971	515,000	1980	1,165,000
1972	608,000	1981	780,000
1973	729,136		

Sources: Brazil Today, Instituto Brasileiro de Estatística, Rio de Janeiro, 1967; *Sinopse Estatística do Brasil, Latin America: Economic Reports; Regional Reports*, Brazil.

The advantages of locating in the south-east are so great that it is difficult to persuade industry to select sites in other parts of the country and especially in the north-east and north. In the north-east it is not the lack of people – there are about 35 million – but the scarcity of natural resources, the poor infrastructure, and the great poverty that hinder industrial development. The total purchasing power in the north-east is small, not more than that of the city of São Paulo. In the north there is not only poverty but a sparse population. Until recently industry in the north-east was limited to textile mills, many of them dating from the nineteenth century, as well as some food-processing and manufacturing of a limited range of consumer goods. SUDENE has attempted to attract industry to the north-east by offering substantial tax concessions to companies, whether foreign or Brazilian, willing to invest in approved projects. There have been failures, most notably, perhaps, that of the synthetic rubber factory at Cabo, just to the south of Recife, as well as some notable successes, of which the jeep factory at Jaboatão is an example. The Recôncavo oilfield has helped promote industry in Salvador, and recently the state government of Bahia, with the aid of SUDENE, has established an industrial park at Aratu on the outskirts of Salvador. To encourage the economic development of the north, the government has copied the SUDENE model and has set up SUDAM, the Superintendency for the Development of the Amazon. In an effort to industrialize Manaus, the city has been declared a free port. Industries set up in the free-port area are able to import raw materials and export manufactured goods without paying import and export duties.

The industry of the three southern states of Brazil is based on the manufacture of consumer goods for a relatively prosperous population, on the processing of agricultural products and on the exploitation of forests. Paraná is the most important state for lumbering and woodworking. Porto Alegre, the capital of Rio Grande do Sul, is a centre for meat-processing, tanning and the manufacture of wine and cigarettes – all based on the products of the agricultural hinterland. Blumenau and Joinville, in Santa Catarina, are now small but growing industrial towns which owe many of their factories to the enterprise of the German immigrants.

While industry in the three southern states and the north-east is expanding, and both regions are acquiring metal-working industries, the south will probably move ahead more rapidly than the north-east; but in the foreseeable future the heartland of industrial Brazil will remain that area roughly delimited by Rio de Janeiro, São Paulo and Belo Horizonte.

Transportation
A revolution in transportation has accompanied industrialization. The difficulties presented by the sheer size of the country and the handicaps of its physical geography have in part been overcome, and old and inadequate means of transport replaced by more efficient ones.

The rivers are still little used, except in the Amazon basin, for they either flow towards the interior of the country or are blocked by falls. The use of hovercraft may lead to a re-evaluation of the rivers as lines of communication. Coastal navigation remains important, but in recent years it has been made extremely expensive because of the high wages the unionized workers have managed to obtain. The railways do not provide the country with an integrated transport system. Construction of railways began at the middle of the nineteenth century, but in 1900 there were only about 12,000 km of track, which by the mid-twentieth century had been little more than doubled. These railways were built by several companies, a number of them British, primarily to carry commodities from regions of production to ports. The linking of these separate railway networks has been undertaken by the federal and some state governments, which have gradually assumed ownership of the railways. The fact that the railway lines were built at different gauges – there are six in use in Brazil today – has made this work of unification more difficult. Much of the railway equipment is old fashioned and run-down, and some wood-burning locomotives are still in use. Except for some lines in the south-east and the Rio Doce railway, rail transport in Brazil is slow and unreliable.

Air transport and roads have finally provided the country with efficient transport. Commercial airlines serve all parts of the country, but though fares are low by international standards, air travel remains expensive for Brazilians. A striking advance in the development of transport has been achieved since the mid-1950s through an extensive road-building programme and the extraordinary development of the automobile industry. There has been a great increase in the mileage of paved and improved dirt roads. One result of the location of the new capital in the interior has been an improvement in internal communications through the linking of Brasília with state capitals by all-weather roads. The government has undertaken a programme of road-building in the

Amazon basin (see *Figure 8.12*). The trucking industry is now of growing importance in the distribution of goods within Brazil, and buses and trucks carry a large
percentage of the Brazilian travelling public. Today it is possible to travel from
Belém to Uruguay by scheduled buses, and the bus station is a major focus of
activity in any Brazilian town or village.

Urbanization
The urbanization of Brazil is one of the most striking developments in the
country's geography this century (*Figure 8.14*). Until late in the nineteenth
century the cities of Brazil were comparatively small and growing only slowly.
Since the 1870s the growth of São Paulo has been spectacular, from 30,000 to
approximately 12 million in 1980. São Paulo and Rio de Janeiro were the first
cities in Brazil to reach a million, but now there are nine cities already over the
million mark, and in 1980 some 28 per cent of the population of Brazil was contained in the ten largest metropolitan areas (*Table 8.3*). Part of this growth can
be attributed to the natural increase of the urban population and to immigration
from overseas, but in large measure it is the result of migration from the countryside on a massive scale. Millions of people have already left the countryside to try
their luck in the cities, yet so high is the rural birth-rate that this exodus has not

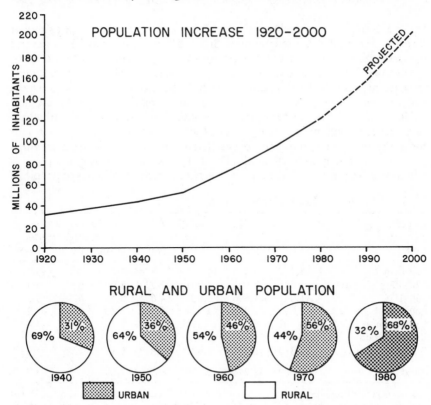

Figure 8.14 Brazil: the pattern of urbanization

Table 8.3 Brazil: population, enumerated and projected, of the largest metropolitan areas ('000)

	1950	1960	1970	1980[3]	1990	2000
Brazil	51,944	70,021	93,215	121,512	157,658	202,437
Belem	242	377	606	1,016	1,299	1,820
Fortaleza	251	490	864	1,616	1,982	2,797
Recife	647	1,028	1,630	2,399	3,170	4,179
Salvador	396	652	1,067	1,795	2,209	2,999
Belo Horizonte	409	778	1,505	2,585	3,208	4,252
Rio de Janeiro[1]	3,044	4,574	6,847	9,154	13,210	17,649
São Paulo[2]	2,336	3,950	7,838	12,709	17,824	24,659
Curitiba	157	381	647	1,472	1,747	2,688
Porto Alegre	464	749	1,409	2,284	3,077	4,213
Brasilia (Federal District)	–	140	538	1,203	1,761	2,735

Source: R.W. Fox (1975) *Urban Population Trends in Latin America*, Washington, DC, Inter-American Development Bank.
Notes: 1 The former state of Guanabara and adjoining municipalities in the state of Rio de Janeiro.
    2 City of São Paulo and adjoining municipalities.
    3 Preliminary figures from the 1980 census, reported in *Anuário Estatística do Brasil*, 1980.

led to a decline in the rural population, but merely to a lowering of its rate of increase. The rural–urban migration in Brazil still continues, with the larger and medium-sized towns attracting the population in recent years (*Table 8.4*). Small-town Brazil either has no appeal to the peasant or else it is but an intermediate stage along the migration route from backlands to metropolis, where people spend a few years before moving on, or seeing their children move on.

   Like so much of the under-developed world, Brazil is undergoing urbanization without adequate industrialization. Industry cannot provide enough jobs for all the newcomers, and there are even large cities, such as Fortaleza, in which there is scarcely any industry at all. These cities without industry are ports or administrative centres and they perform service functions. Millions of people in

Table 8.4 Brazil: the growth of cities 1950–80

City-size category ('000)	No. of cities (1970)	Population ('000)				Annual average percentage increase		
		1950	1960	1970	1980	50–60	60–70	70–80
Less than 20	–	5,791	10,294	14,522	19,378	7.8	4.1	3.3
20–50	160	1,641	2,948	4,891	7,523	8.0	6.6	5.4
50–100	46	1,047	1,896	3,084	4,694	8.1	6.3	5.2
100–250	27	1,350	2,376	4,004	6,102	7.6	6.9	5.2
250 and more	16	3,491	6,172	11,014	16,963	7.7	7.8	5.4
Rio de Janeiro and São Paulo	2	5,380	8,524	14,685	21,892	5.8	7.2	4.9
Total		18,700	32,210	52,200	76,552	7.2	6.2	4.7

Sources: Compiled from *Anuário Estatística do Brasil* and R.W. Fox (1975) *Urban Population Trends in Latin America*, Washington, DC, Inter-American Development Bank, p. 21. 1980 figures are estimates.

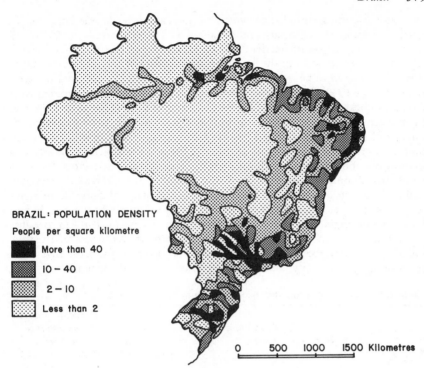

BRAZIL: POPULATION DENSITY

People per square kilometre

More than 40

10 – 40

2 – 10

Less than 2

0   500   1000   1500 Kilometres

Figure 8.15   Population density

the cities eke out a living from badly paid jobs in the lower ranks of a swollen civil service, or as clerks or servants in commercial establishments. Though labour is cheap, it is used lavishly; far more people are employed than are really needed for a particular job, and many find no employment at all. The cities have grown so rapidly that the city administrators have been unable to provide the basic services of water, sewage systems, lighting, paved streets, schooling, medicine or even adequate public transport for much of the population. Each city has its new suburbs of shacks and cardboard houses such as the famous *favelas* of Rio and the *mucambos* of Recife. Conditions of life in these settlements are grim, and life itself all-too-often nasty, disease-ridden and short. The cities also have wealthy suburbs, luxury shops and impressive office buildings. Nowhere, perhaps, is the contrast between rich and poor so great and so clearly seen as in the large cities. The 'urban problem', with all its social and economic ramifications is a major one in Brazil and will take decades to solve. Its solution is bound up with the resolution of the other major problems: the curbing of the rate of population increase, agricultural reform in the broadest sense, the slackening of the rural–urban migration and the increase in the rate of industrialization.

Brazilian cities in colonial times and in the nineteenth century were, with few exceptions, unplanned, and so was the great expansion of the cities this century. Nevertheless, Brazilians have developed a considerable flair for urban planning, which has found expression less in tackling the problems of long-established cities as in founding entirely new ones. Examples are such towns as Londrina, in northern Paraná, state capitals, and Brasilia. Londrina dates from the 1930s and now has a population of nearly 300,000. Minas Gerais began to build a new capital, Belo Horizonte, in the 1890s because the site of the old capital, Ouro Preto, was judged to be too constricted for the expected expansion of the state capital. In the 1930s Goiânia was built to replace the old and remote gold town of Goiás as capital of the state of Goiás. The building of Brasília began in 1956, and the building still goes on. One of the purposes of Brasília was to help develop the interior of Brazil by providing a market for agriculture and new lines of communication; another was to create a symbol of the new Brazil, with a city startling in its layout and architecture. Brasília has its defenders and detractors. It cost an enormous amount of money. It is a monument to its creator, President Kubitschek. The layout is novel, but not always very practical. Some of the architecture is elegant, but many of the buildings are of purely functional design. Hardly surprisingly, it is a far less comfortable city to live in than Rio. Brasília, too, has its slums, in 'satellite cities' a few kilometres from the centre. The total population is now about 1 million. Brasília is a success in the sense that it is a functioning, lived-in city, and it has helped to open up the interior.

Table 8.5 Brazil: population and income by region

Region[1]	% National area	Population per km² 1970	Population (% national total)			Per capita income (% national average)		
			1950	1960	1970	1949	1959	1970
North	42.0	1.01	3.6	3.7	3.9	47	54	51
North-east	18.2	18.23	34.6	31.6	30.2	41	44	40
South-east	10.9	43.38	43.6	43.8	42.8	153	146	151
South	6.8	29.35	15.1	16.7	17.7	105	105	99
Central West	22.1	2.70	3.3	4.2	5.4	54	57	70

Source: Robock, 1975, p. 76.
Note: 1 These regions have been established by the government of Brazil as follows:
North: Acre, Amazonas, Pará, Amapa, Rondônia, Roraima.
North-east: Maranhão, Piauí, Ceará, Rio Grande do Norte, Paraíba, Pernambuco, Alagoas, Sergipe, Bahia, Fernando de Noronha.
South-east: Minas Gerais, Espírito Santo, Rio de Janeiro, São Paulo.
South: Paraná, Santa Catarina, Rio de Grande do Sul.
Central West: Mato Grosso, Mato Grosso do Sul, Goiás, Federal District (Brasília).

During the last 25 years Brazil has made great strides to emerge from the category of 'under-developed country', and has achieved a partial success. Brazil now belongs to a new grouping of countries, characterized by a high degree of development in some sectors of their economies and extreme backwardness in others. It is an important industrial country, the most important industrial country in Latin America, and an exporter of manufactured goods. Industry is predominantly in the south-east and south of the country, where agriculture is also most productive, and the standards of living are comparable with those of southern Europe. Elsewhere in the country the situation is very different. In the north-east a poor, traditional, agricultural society struggles to achieve reforms, to attract industry and to earn a higher standard of living; to the north and west, in the huge frontier of Amazonia, Goiás and Mato Grosso, investors in 'agri-business', peasant settlers and Indian tribes compete for land and resources. These vast regions still belong to the undeveloped world. The country's concern in the coming decades must be to extend the range of modernity and development within its borders, a process in which economic advance will be only one criterion of success; others will be the attention given to the interests of the weaker members of society and to the protection of the physical environment.

Notes

[1] The word probably derives from *bandeira* (flag). The *bandeirantes* marched with a flag at the head of the column.

[2] A word frequently used in Brazil, which may be roughly translated as 'backlands'. One who lives there is a *sertanejo*.

[3] The Amazon remained closed to foreign shipping until 1867.

[4] The Republic, established in 1889, turned over the responsibility of promoting immigration to the states.

[5] This rescript was not withdrawn until 1896 and then only for the southern states.

[6] For example, the FAO reports the following figures for 1980 in Kg/ha: Brazil 56,189:1, Australia 82,082, South Africa 65,182, USA 86,043, Barbados 76,071, Cuba 52,308. (FAO *Production Yearbook*, Food and Agricultural Organization of the United Nations, Rome, 1980, pp. 167–8.) North-eastern production is lower than the Brazilian average: Pernambuco 49,016 and São Paulo 67,074 in 1979.

Bibliography

For discussions of Brazil from different points of view see:
FREYRE, G. (1964) *The Masters and the Slaves*, New York.
FURTADO, C. (1963) *The Economic Growth of Brazil: A Survey from Colonial to Modern Times*, Berkeley.
MOOG, V. (1964) *Bandeirantes and Pioneers*, New York.
POPPINO, R.E. (1968) *Brazil: The Land and the People*, New York.
SMITH, T.L. (1963) *Brazil: People and Institutions*, Baton Rouge.

Physical geography
BROOKS, R.H. (1971) 'Human response to recurrent drought in north-eastern Brazil', *The Professional Geographer*, 23, 40–4.
COLE, M. (1958) 'The distribution and origin of savanna vegetation with particular reference to the Campos Cerrados of Brazil', *Comptes Rendus, XVIII Congrès International de Géographie* (Rio de Janeiro), 1, 339–45.
HUECK, K. (1957) 'Sobre a origem dos Campos Cerrados do Brasil', *Revista Brasileira de Geografia*, 19, 67–81.
JAMES, P.J. (1952) 'Observations on the physical geography of north-east Brazil', *Annals Assoc. Am. Geog.*, 42, 153–76.

Colonial period
ALDEN, D. (1963) 'The population of Brazil in the late eighteenth century: a preliminary study', *Hisp. Am. Hist. Rev.*, 43, 173–205.
ALDEN, D. (1973) *Colonial Roots of Modern Brazil*, Berkeley.
BOXER, C.R. (1962) *The Golden Age of Brazil, 1695-1750*, Berkeley.
HEMMING, J. (1978) *Red Gold: The Conquest of the Brazilian Indians*, London.
PRADO, C., Jr (1967) *The Colonial Background of Modern Brazil*, Berkeley.

The nineteenth century
ANDRADE, M.C. de (1980) *The Land and People of Northeast Brazil*, translated by Dennis V. Johnson, Albuquerque.

LEFF, N.H. (1972) 'Economic development and regional inequality: origins of the Brazilian case', *The Quarterly Journal of Economics*, 86, 243–62.

MONBEIG, P. (1952) *Pionniers et planteurs de São Paulo*, Paris.

MORSE, R.M. (1958) *From Community to Metropolis: A Biography of São Paulo, Brazil*, Gainesville.

ROCHE, J. (1959) *La colonisation allemande et Le Rio Grande do Sul*, Institut des Hautes Etudes de L'Amérique Latine.

STEIN, S.J. (1957) *Vassouras: A Brazilian Coffee County 1850–1900*, Cambridge, Mass.

STEIN, S.J. (1957) *Brazilian Cotton Manufacture: Textile Enterprise in an Underdeveloped Area*, Cambridge, Mass.

The twentieth century

ANUARIO ESTATISTICA DO BRASIL (1980) (and earlier).

AUGELLI, J.P. (1958) 'Cultural and economic changes of Bastos Japanese colony on Brazil's Paulista frontier', *Annals Assoc. Am. Geog.*, 48, 3–19.

AZEVEDO, A. de (ed.) (1958) *A Cidade de São Paulo: estudos de geografia urbana*, 4 vols, São Paulo, Companhia Editora Nacional.

BAER, W. (1965) *Industrialization and Economic Development in Brazil*, Holmewood, Ill.

BAER, W. (1964) 'Regional inequality and economic growth in Brazil', *Economic Development and Cultural Change*, 12, 268–85.

BRET, B. (1975) 'Données et réflexions sur l'agriculture brésilienne', *Annales de Géographie*, 84, 557–88.

DAVIS, S.H. (1977) *Victims of the Miracle: Development and the Indians of Brazil*, Cambridge.

DEAN, W. (1969) *The Industrialization of São Paulo, 1880–1945*, Austin.

DICKENSON, J.P. (1967) 'The iron and steel industry in Minas Gerais, Brazil, 1695–1965', in STEEL, R.W. and LAWTON, R. (eds) *Liverpool Essays on Geography – A Jubilee Collection*, London.

FOWERAKER, J. (1981) *The Struggle for Land. A Political Economy of the Pioneer Frontier in Brazil from 1930 to the Present Day*, Cambridge.

FOX, R.W. (1975) *Urban Population Growth Trends in Latin America*, Washington, DC, Inter-American Development Bank.

GALLOWAY, J.H. (1982) 'Geography in Brazil during the 1970s: debates and research', *Luso-Brazilian Review*, 19, 1, 1–21.

HIRSCHMAN, A.O. (1963) 'Brazil's north-east', in *Journeys Towards Progress*, New York.

KATZMAN, M.T. (1977) *Cities and Frontiers in Brazil*, Cambridge, Mass., and London.

KLEINPENNING, J.M.G. (1975) *The Integration and Colonisation of the Brazilian Portion of the Amazon Basin*, Nijmegen, Katholieke Universiteit, Nijmeegse Geografiska Cahiers, 4.

LATIN AMERICA: *Weekly Reports*; *Regional Reports – Brazil*; *Commodities Report*, London, Latin American Newletters Ltd.

PEBAYLE, R. (1967) 'La vie rurale dans la Campanha Rio-Grandense', *Les Cahiers d'Outre Mer*, 20, 345–66.

PFEIFER, G. (1967) 'Kontraste in Rio Grande do Sul: Campanha und Alto Uruguai', *Geographische Zeitschrift*, 55, 163–206.

ROBOCK, S.H. (1975) *Brazil: A Study in Development Progress*, Lexington, Mass.

SINOPSE ESTATISTICA DO BRASIL, various years.

SMITH, N.J.H. (1978) 'Agricultural productivity along Brazil's Transamazon Highway', *Agro-Ecosystems*, 4, 415–32.

SMITH, N.J.H. (1982) *Rainforest Corridors. The Transamazon Colonization Scheme*, Berkeley.

WAGLEY, C. (1977) *Welcome of Tears. The Tapirapé Indians of Central Brazil*, New York.

9 · *The River Plate countries*[1]

J. Colin Crossley

From being commercial producers of subsidiary significance in the economy of colonial Spanish America, and of no consequence at all outside Latin America, the River Plate lands passed through a period of trial and error and rose during the late nineteenth century to the first rank of world producers of agricultural commodities. Migration and immigration brought about a complete reversal in the relative commercial and demographic importance of different subregions; the Pampas of temperate Argentina and Uruguay came to dominate the scene, and the subtropical Argentine north-west and Paraguay became a backwater of colonial survivals. Such changes did not occur without conflict and difficulty. Nor has a feeling of national unity been able to develop to the desired degree under such adverse conditions. Especially since the Great Depression, Argentina and Uruguay have consciously sought to decrease their vulnerability to world economic vicissitudes by increasing their self-sufficiency, whereas Paraguay has sought a fuller share of world trade. In each case established patterns of land-ownership, production and communications have seriously impeded these efforts and future directions are far from clear.

The margins of empire 1536–1852

Colonial period from 1536 to 1776
When Pizarro was conquering the Inca empire via the Pacific (1532–5), Pedro de Mendoza was attempting the settlement of the Plate from the Atlantic, founding Buenos Aires in 1536. But after 5 years of privation in the face of hostile Indians the settlement was abandoned in favour of Asunción. The assessment of the New World by Oviedo, the Spanish chronicler, that 'the Indies are worth nothing without the Indians', needed slight modification in the context of the Plate. There, lands inhabited by nomadic Indian hunters who could fight or flee at will were useless, whilst lands with indigenous agricultural communities offered a labour force which the superior European weaponry could easily control and direct for the conquerors' ends; and this was particularly true of irrigated areas from which flight was impossible.

Both the process and the pattern of colonial occupation in the Plate thus reflected the pre-Hispanic distribution of settlement, which in turn represented the appraisals by essentially Stone Age peoples of the potential of the physical environment. The forest-dwellers of hilly, subtropical eastern Paraguay and Misiones, perhaps numbering 50,000, were accustomed to shifting agriculture with beans, manioc and pumpkins, and a plentiful water supply. At the foot of the Andes and in the intermontane basins of the arid north-west and west, ease of irrigation had permitted long-established sedentary maize agriculture; here villages with a combined population of about a quarter of a million formed out-posts of the Inca empire. In between and to the south lay the vast plains of the Chaco, the Pampas and Patagonia, hot in the north, cold in the south, covered with grass or scrub forest and everywhere short of permanent water-courses whatever the rainfall; here lived perhaps 90,000 Indians grouped in small bands of hunters obliged to follow the wild guanaco and rhea that afforded their food supply.

In this land of nomads only two cities, Buenos Aires and Santa Fé, were founded successfully during early colonial times, both by settlers from Asunción. To the east, Corrientes was founded among Indian forest farmers, and during the seventeenth century the Jesuits established many *reducciones* or mission villages, into which the scattered Indian populace was organized. In contrast, Spanish official efforts at colonization were concentrated in the north-west and Cuyo, where today's provincial capitals were founded mainly in the late sixteenth century by settlers entering from Peru and Chile respectively (*Figure 9.1*). European crops and animals entering the Plate by the same three routes had also spread throughout the region by the end of the sixteenth century. Indeed, wheat-growing preceded the conquerors in Cuyo, and everywhere the nomadic hunters rapidly acquired the horse.

Despite its agricultural potential, the River Plate remained tied to an economy of industrial and agricultural self-sufficiency with its international commercial activity limited to a marginal role in the mining economy of Peru. Europe of the time had no demand for the temperate produce of other continents – and the subtropical districts of the Plate lay deep in the interior – nor did the region possess any precious metals to exploit. Until 1776 the River Plate formed part of the Viceroyalty of Peru, and so was subject to regulations which gave Lima the monopoly of all trade with Spain. Buenos Aires thus lay at the end of a route to Spain through the Chaco, Lima and Panama, which rendered European goods prohibitively expensive to import legally. Contraband trading with Brazil and with European interlopers and indigenous manufacturing were thus both stimu-lated. Apart from domestic craft activities, regional industrial specialization developed. Tucumán supplied carts to the whole Plate and wheat-flour to the Litoral, and even sent cotton and woollen goods as far afield as Chile and Potosí (a mining city with 160,000 people in the mid-seventeenth century). Mendoza from the earliest times was the centre of production of wine and dried fruits, whilst subtropical Paraguay and Misiones contributed sugar, yerba maté and cotton textiles to the regional markets.

Most food crops were grown for local subsistence, whereas livestock farming was more commercial. The excellence of the natural pastures of the Pampas, Mesopotamia and Uruguay permitted the wild descendants of the cattle and horses introduced in the middle of the sixteenth century to multiply exceedingly; by the end of the century, since cattle were regarded as Crown property, rights were being awarded to individuals to round up and kill so many head of cattle a year within a prescribed area, though for long their only value lay in the hides and to a less extent in the tallow and tongues. Horses were more valuable, as transport on the plains was wholly dependent upon them, and *estancias* were soon formed to oversee the wild herds and tame them. As the wild cattle were reduced in numbers, *estancias* with grants of land became organized for their control, the smallest, of 0.5 × 1.5 leagues (approximately 1875 ha), capable of yielding ninety hides a year. In the absence of fencing, surveillance by gaucho horsemen was necessary to prevent rustling, especially by Indian raiders who could find a profitable market in Chile. There was little interest in sheep, as mutton was despised and wool exports were prohibited.

The most important contribution of the Plate to the Spanish imperial economy, however, lay in its ability to supply sure-footed mules for transport in the mountainous mining areas of Peru. The mules were bred in Mesopotamia and the Pampas, reared in Córdoba and Tucumán, and sold at the annual fair at Salta when 60,000 mules would be purchased for use in Upper and Lower Peru.

Population remained concentrated during colonial times in the areas of indigenous permanent settlement. In the early years, with the cultural shock of the Conquest, native numbers declined and Spanish colonists were too few to compensate. The 'civilized' population by the early seventeenth century is estimated to have been 162,000, of whom 75,000 lived in the north-west, 25,000 in Córdoba (effectively part of the north-west in colonial times), 12,000 in Cuyo, 28,000 in Mesopotamia, and only 22,000 in the Pampas. In contrast the Indian *reducciones* of Paraguay and Misiones, begun in 1610, prospered under the protective rule of the Jesuits, and their population rose to 126,000 in 1733. The cities of the period, with the exception of craft centres like Tucumán, had at best tertiary functions such as administration and commerce, and some were little more than centres of residence for absentee landowners.

But Buenos Aires was not destined to remain at the dead end of the imperial highway system. Contraband trade was already thriving when, by the Treaty of Utrecht (1713), Britain gained limited rights to introduce merchandise and negro slaves directly into the Plate in exchange for hides. By the mid-eighteenth century annual hide exports had risen to 150,000 and negro slaves were widely employed as domestic servants and in the embryonic sugar plantations of Tucumán, where they formed a majority of the population. By the time Buenos Aires became the capital of the new Viceroyalty of the River Plate in 1776, its population, in excess of 20,000, made it by far the largest town in the Plate, and the region's total had probably recovered its pre-Hispanic level of a third of a million (*Figure 9.1*). This figure, however, includes perhaps 100,000 Indians who survived the colonial period in independent occupance of Patagonia, the Chaco and a large part of the Pampas.

The period of transition 1776–1852

The accession of the French Bourbons to the Spanish throne in 1700 began a period of colonial reform, culminating in the creation of the Viceroyalty of the River Plate in 1776 and the 'free trade' regulations of 1777 and 1778. Fear of the growing strength of Portuguese Brazil was another factor leading to the detachment of what are now Argentina, Uruguay, Paraguay and Bolivia from the Viceroyalty of Peru, and to the laws which now allowed Spanish ships to trade directly with the Plate, and the Plate to trade with other Spanish colonies. Montevideo, founded in 1724, benefited as the official port of call for ships

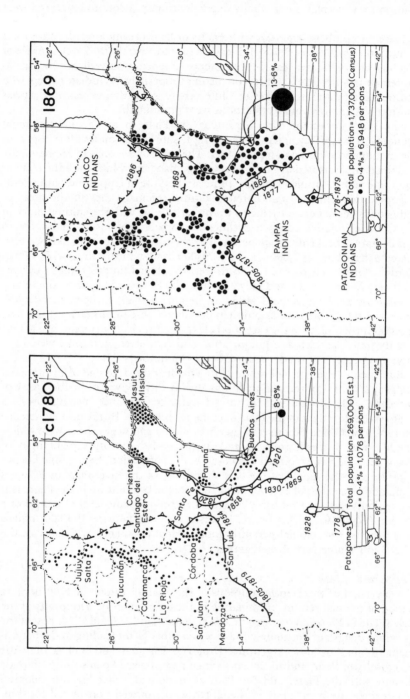

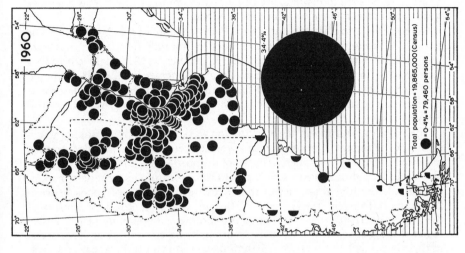

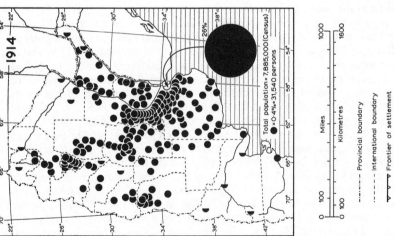

Figure 9.1 Argentina: population distribution and settlement frontiers 1780–1960

sailing to the Pacific, while Buenos Aires exported the new Viceroyalty's produce of silver and hides. The mining centre of Potosí, a 4 month journey from Lima, but only 2 from Buenos Aires, could now get Spanish cloth at one-sixth of its former price. Viceregal action also encouraged the growing of wheat and the salting of meat. New freedoms and fresh developments were very welcome to the merchants of the Plate, but the foreigner was still excluded, and the merchants' desire to trade with industrializing Britain still went unsatisfied. This grievance was a major factor in the wars for independence.

Argentina was the first territory in the Plate to secure its freedom (1810). The rest of the Viceroyalty obtained independence, not only from Spain, but from Buenos Aires as well: Paraguay ousted the Argentines in 1811, Bolivia became a separate state in 1825, and 3 years later Uruguay emerged as a British-sponsored buffer state between Argentina and Brazil, thus ending their long contest for its possession. The stage was set for the region to move to the margins of the British Empire.

The economic disruption of emancipation was succeeded by a reorientation in favour of the Litoral. The creation of a frontier between Bolivia and Argentina narrowed the primary commercial outlet of the north-west, and of Salta in particular. Internal political events served also to isolate it from its secondary market, the Litoral: the liberal policies of the 1820s freed the flow of goods between Argentina (effectively, the Litoral) and Britain, whilst the Rosas dictatorship of the 1830s and 1840s further increased the commercial isolation of the interior, either through persistent civil strife or through the erection of interprovincial tariff barriers. For the first half of the nineteenth century Paraguay, too, was isolated from the world by the whims either of its own or Argentina's dictator.

Throughout the period hides constituted the leading commercial product and provided over half the exports of the Plate, the annual number rising rapidly to 1.4 million in 1783 and to 2.9 million from Argentina alone by 1849. More dramatic, however, was the rise in meat exports: apart from the hides, cattle had been utilized for little more than their tallow, tongues and a certain amount of sun-dried meat. Lack of salt had previously restricted production of salt meat, but in the 1780s, with salt coming from Patagonia, an increased number of cattle and a growing demand for meat from tropical plantations, meat-salting factories – *saladeros* – were established with Viceregal backing. The first was at Colonia, in Uruguay, and by 1804 exports from the eight Uruguayan *saladeros* exceeded 3200 tonnes, but still the meat of over half the animals slaughtered for their hides was being wasted. Following rapidly on the declaration of independence came the first *saladero* on the Argentine shore, established at Ensenada in 1811 by Staples and McNeil, two of the British merchants who flocked into the republic; other factories followed in the next few years on the outskirts of Buenos Aires and in Entre Ríos. By 1829 Argentine exports of salt meat topped 7500 tonnes and by 1851 were running at about 20,000 tonnes a year.

The prosperity of the salt meat industry, for which the *criollo* breed was ideal, gave little encouragement to the breeding of finer cattle although each beast yielded only 75 kg of meat when slaughtered at 5 years. Nevertheless, Miller, a British *saladerista* and *estanciero*, imported the first Shorthorn bull in 1823, and its cross-bred progeny began to spread among the herds of those who sought to develop a local dairying industry. Breeding did not extend, however, to the creation of a single herd of pedigree stock.

It was the sheep-rearing industry that first saw the need for improvement. The typical native sheep yielded a wool not worth scouring, a mutton that was scorned and a carcass fit only for fuel; its principal value lay in the use of fleeces as a comfortable if ungainly kind of saddle. After independence the European demand for long-stapled fine wools could not be satisfied by the native animal and, since Spain forbade the export of her fine merino stock, supplies had to be obtained elsewhere. In the 1820s and 1830s the *estancieros* Gibson and Newton imported stud flocks of Southdowns, Lincolns and Romney Marshes, whilst Hannah and Stegmann acquired German merinos (*Negretes*). In the 1840s merinos were also introduced into Uruguay, but breeding still interested only a few native and foreign *estancieros*. Nevertheless, Argentine wool exports rose steadily, from under 200 tonnes in 1822 to 2100 tonnes in 1837 and almost 7700 by 1850.

Successful breeding, like successful crop farming, depended very much on the farmer's ability to restrict the movements of his animals and, therefore, on the fencing of the boundless Pampas. Olivera tried hedges in 1838 and in 1845 Newton introduced the idea of wire fencing from Wentworth Deer Park in Yorkshire, though his aim was to protect crops, rather than to ring-fence the *estancia* or subdivide the pastures.

Another innovation was the *invernada* or fattening pasture established in the vicinity of port-situated *saladeros*, where cattle could recuperate after long journeys on foot.

Despite the encouragement of wheat-growing and flour-milling by the viceroys and by the early national governments, crop production on the Litoral did not prosper because fenceless farms were defenceless farms, livestock enterprises were much preferred, a labour shortage hindered the intensification of of agriculture, and imported American flour competed too successfully.

But the interior suffered more than the coast: by 1850 over half of Argentine imports – sugar, yerba, tobacco, beverages, flour and textiles – competed directly with its agricultural-processing industries, at least in the markets of the Litoral; its produce was dearer and inferior, and transport costs to the coast were often higher. Conversely the same high internal freight rates helped to protect its economic diversity: Córdoba continued as a textile centre, Tucumán developed its sugar and tobacco industries and Cuyo its wines and brandy, but political difficulties prevented the interior from developing its own integrated exchange economy.

The increasing economic isolation of the interior was not offset by any great move to link up the effectively settled areas of the Litoral and the interior. While Mesopotamia and Uruguay had been successfully, though sparsely, settled in colonial times, west-bank settlement especially in Santa Fé remained a narrow strip, although Rosas's desert campaign of the 1830s pushed the Indian frontier south-westwards in the province of Buenos Aires (*Figure 9.1a*). The 'corridor' between the Litoral and interior, from Rosario to Córdoba, thus remained throughout the period an unsettled territory where only forts protected the route from Indian attacks.

For political and economic reasons early attempts to organize crop-farming colonies in the Litoral, such as Robertson's Scots colony at Monte Grande in 1823, did not succeed. Despite the desire of the national government under Bernardino Rivadavia in the 1820s to see the land occupied by small and medium farmers, and despite the facilities for buying land on instalments, the failure to impose an areal limit on concessions soon led to new lands falling easily into few hands. In any case, large-scale livestock farming needing little labour was the only viable agricultural economy. Thus, by 1840 over 85,000 km² had been sold to under 300 people. Developments in Uruguay were along similar lines, and virtually the whole national territory rapidly became the property of a small number of people.

During the transitional period the total 'civilized' population of Argentina (i.e. excluding nomadic Indian tribes) quadrupled, from 269,000 around 1780, to 1,107,000 in 1855, being about half a million at independence (*Figure 9.1*). The provinces of the interior (the north-west, Cuyo and, at that time, Córdoba) retained their share (57 per cent); among the Litoral provinces, Corrientes declined with the dispersal of the mission Indians after the 1767 expulsion of the Jesuits, whilst the province of Buenos Aires doubled its share to 17 per cent by 1855. Together with the 91,000 inhabitants of the city, this gave the city and province of Buenos Aires exactly one-quarter of the national population; Santa Fé province with barely 40,000 inhabitants still ranked eleventh (out of fourteen provinces). Where racial mixing between whites and Indians had proceeded for generations, i.e. in the north-west and north-east, the population at the end of the period was predominantly mestizo, whereas on the Pampas hostility had kept the civilized whites and the untamed Indians apart. European-born foreigners were important only in the province of Buenos Aires where they constituted almost half the population; a further one-seventh were negroes and mulattos, for the import of slaves continued until 1825 and slavery was abolished only in 1853. Between 1828 and 1852 neighbouring Uruguay doubled its population to 132,000, due partly to the immigration of southern Europeans.

The transitional period marked the first phase in the shift of national emphasis from the interior to the Litoral. Demographically the balance between the two remained constant, but the nucleus of Buenos Aires was already setting the pace that the rest of the Pampas was to follow. Economically the balance shifted markedly: 'The policy of free imports blocked any possibility of spreading (to the interior) the dynamic impulse generated by the Litoral's export expansion' (Ferrer, 1967, p. 67). With little capacity to export, the interior was prevented from importing the new technology; its industries survived because 'the major protectionist barrier was still distance' (Ferrer, 1967, p. 72), but they stagnated, thus forcing the rising population into subsistence occupations. By reason of their locations Paraguay and Uruguay found themselves in the same positions as the interior and Litoral respectively.

The integration of Argentina and Uruguay into the world economy 1852–1930

After the fall of the dictator Rosas in 1852 a long period of semi-democratic constitutional government was inaugurated, and the geography of the Argentine state was transformed in the space of 50 years by immigration, by investment of capital and by an agricultural revolution. From being a nation centred on the oases of the Andean fringe, it became the nation of the humid Pampas; from being a people of creole and mestizo character, its inhabitants became a melting-pot of European settlers; and, from producing little but inferior dried meat and wool, its agricultural economy was converted to the production of high-grade meat from carefully bred animals and of vast quantities of cereals, which gave the country a leading place in the world exchange economy.

Politically, since independence, Argentina had been torn between federal and unitary approaches to the constitutional problem. The *caudillos* of both interior and Litoral provinces, other than Buenos Aires, resented the latter's economic ascendancy and favoured a federal solution. On the other hand the Unitarians, merchants and intellectuals, saw a centralist constitution as necessary for the expansion of trade and the spread of Europe's civilizing influence through Buenos Aires over the boorish gaucho. In the event, the Constitution of the Confederation, drawn up in 1853, was a compromise, and owed much to the political thinker, Alberdi. His view that 'to govern is to populate' was reflected in the constitutional encouragement of immigration, colonization, the import of capital, the construction of railways and the establishment of industries. Interprovincial tariffs were to be abolished and customs receipts nationalized. But for 9 years Buenos Aires province remained independent of a thus impoverished Confederation; in 1856 the latter imposed additional duties on all goods entering via Buenos Aires, thus giving the initial stimulus to the rise of Rosario as a rival port. When union was achieved, under a *porteño* president, it was the north-west that offered resistance. Finally, in 1880, the city of Buenos Aires became the federal capital, thus depriving the province of Buenos Aires of most of the coveted port revenues.

Until 1890 change and development were centred on two foci: Rosario and Buenos Aires. Rosario was the port of entry for European immigrants intending to settle in agricultural colonies in Santa Fé province. The territory occupied by Indians was narrowest opposite Rosario and forts still lined the road linking the settled coastal strip to the old north-west. Here was built the first trunk railway, to Córdoba, followed by extensions to Mendoza and Tucumán, which made the whole north-west tributary to Rosario. Only in the late 1880s were the Rosario and Buenos Aires railway systems joined (*Figure 9.3*).

Activity in Buenos Aires province was of a different type: the Indians were ousted from great areas, the land was made over to the rearing of sheep and cattle, and railway tentacles reached out from Buenos Aires. Exports of wool and salt beef increased and imported wheat was replaced by the produce of the Santa Fé colonies. With the arrival of the *frigorífico*, frozen mutton added to the wealth of the *estancieros*.

British capital, much of it speculative, played a major role in land improvements, railway and port construction and urban redevelopment. Issues of paper currency mounted as the ruling class of *estancieros* saw in depreciation the opportunity both to pay less in wages and to reduce the value of their outstanding debts. Expansion may have laid the foundations for a prosperous future, but the speculative bubble burst in the crisis of 1890, causing the ruin of many.

After the crisis development proceeded more cautiously. With the conquest of the Pampas already achieved, no lands of first-class quality and location remained for settlement, and occupance of the land could only advance towards more arid or more tropical areas. Fortunately, growth was now possible through intensification of land use. Technological advance allowed the shipping of frozen beef to the mass markets of Britain. This stimulated the upgrading of Argentine cattle, which now required more than natural pastures for fattening. The sowing of artificial pastures became a necessity, but *estanciero* attitudes eschewed the tilling of the soil as work fit only for peasants. The preferred solution was to lease blocks of land for a short period to immigrants, who would contract to take three or four crops of wheat and flax, and then return the land laid down to alfalfa. With land values high and public lands suitable for colonization almost exhausted, impermanent tenancy was the only form of land tenure open to the majority. As a by-product of the beef industry, therefore, land use was intensified, cereal exports were expanded, the railways became prosperous with the increased flow of freight, and new lines radiated from Buenos Aires to tap the whole of the Pampas. Cuyo, and the north-west too, were joined directly to Buenos Aires by lines by-passing Rosario. From being an ellipse focused on Rosario and Buenos Aires, the growth area of Argentina was now a circle centred on the capital.

Uruguay, physically well endowed, endured political disorder until the early twentieth century, discouraging both immigration and economic development, although a railway network radiating from Montevideo had been built between 1860 and 1911 in part to aid the suppression of revolutions. With the election of José Batlle y Ordóñez (President 1903-7 and 1911-15), the government began to guide economic development. Immigrants stamped their European character on the country as they did on the Pampas, but with superior natural pastures livestock farming did not produce an influx of tenant colonists.

Paraguay, lying 1200 km upstream, continued to suffer: from isolation, like the Argentine interior; from the war of 1865-70, which reduced her estimated population to a quarter of a million, of whom fewer than 29,000 were men; and from chronic political disorder. Not surprisingly, her attempts to integrate into the world economy, to attract immigrants, to commercialize agriculture and to develop her natural resources met with only limited success, and both economy and society remained those of a peasant nation.

The expansion of settlement
The economic growth of the period was made possible by the world situation and by the existence around the Plate of an area of high-quality land more than twice the size of the United Kingdom, yet utilized almost wholly for hunting by nomadic Indians or for the extensive rearing of inferior cattle and sheep. Argentina also benefited from changes of boundary between the republics of the Paraná basin. As a result of the war of 1865-70 Argentina gained definitive possession of Misiones and Formosa from Paraguay, which also lost territory to Brazil. In 1851 Uruguay had ceded the northern half of its territory to Brazil, and thus ceased in practice to be the 'Republic of the East Bank of the Uruguay' (República de la Banda Oriental del Uruguay) – though it still retains that name officially.

The march of settlement took different steps in different areas. In the Pampas two courses were pursued: limited arable colonization had already begun in the 1840s under *caudillo* Justo José de Urquiza in Entre Ríos, where pastoral farming had also been developed early. This province of rolling parkland was, together with Uruguay, the only part of the Litoral which possessed readily available surface water. West of the Paraná the provision of drinking water for man and beast depended on wells and wind pumps. Yet Santa Fé province was to take the lead in officially organizing colonies for immigrant settlers on a massive scale, big landowners, Rosario merchants and colonization companies sharing the work with the provincial government. Just west of Santa Fé city, Esperanza was founded in 1856, the settlers receiving 33-ha lots. But further developments were slow until after 1865, when colonization promotion laws were enacted by the province. The first spurt came in 1870 with the colonies of the Central Argentine Land Company laid out beside the Rosario–Córdoba railway on land conceded to the Central Argentine Railway Company for this purpose. Although enjoying the unique advantage of location beside Santa Fé's only railway, the colonists did not prosper for a decade or more. Inexperience, drought, locusts, frosts and Indian attacks were burdens alleviated only by financial assistance from the company. Private colonization ventures multiplied: one landowner, M. Cabal, initiated regular shipments of wheat to Europe in 1874 in order to help his colonists.

The boom of the 1880s brought prosperity to existing colonies and the establishment of many new ones: of 361 colonies (covering 3.7 million ha) founded in Santa Fé in the forty years 1856–95, two-thirds were created in the decade 1884–93. Hundred-hectare lots now became common, for mechanization led to greater efficiency. But exhaustion of the supply of good land in the public domain, coupled with the big owners' growing preference for retaining control of their estates, caused a sharp decline in colonization in Santa Fé, as in neighbouring provinces after 1895.

In Córdoba province the Central Argentine Land Company took the lead in colonization, thrusting its colonies along the railway into Indian territory, and thus linking up with the old settled area of the irrigated valleys to the north-west. Of the 139 colonies (covering 1.4 million ha) whose foundation dates are recorded in the 1895 census, only seven were established in 1870–84; 128 came in the decade 1885–94 and indeed half the total were founded after the crisis in 1892–4; another 122 were established in 1896–1901. The delay in colonizing Córdoba is also to be explained by the increasing aridity of the plains, a problem partly overcome after 1888 with the construction of the San Roque dam.

The spearhead of settlement thus became a broad wedge, occupying the central parts of Córdoba as it did in Santa Fé. Córdoba city, alone among the colonial towns of the north-west, now became a colonist centre of the Pampas.

To the north the Indians were pushed back, but slowly. The officials of the Santa Fé Land Company, formed in 1883 to develop the north-western one-sixth of that province, were harassed by Indians until the late 1880s and Resistencia, founded in 1878 opposite Corrientes city, remained for a time merely a fortified enclave. The lower quality of the land and an absence of rail links further restricted northwards colonization. Indeed, settlement in northern Santa Fé and neighbouring Chaco advanced only at the turn of the century, when the extraction of tannin from the *quebracho* forests proved a profitable venture.

Across the river Paraná, Entre Ríos was second only to Santa Fé, with 191 colonies established by 1895, covering 0.8 million ha. After a precocious start under Urquiza further colonies were not founded until 1871; thereafter only fifty-two were created up to 1885, mostly within easy reach of Paraná and

Concepción cities or of navigable highways. The period 1886–93 saw 137 established, with more than half the total founded in 1888–91; at the same time the Paraná–Concepción railway was built to serve them. Thus, with the exception of a few river bank settlements, the northern half of Entre Ríos and adjoining Corrientes saw no colonization ventures, remaining a wooded pastoral landscape. Among the Entre Ríos colonies several were created for German and Jewish refugees from Tsarist persecution.

Whilst colonization dominated the march of settlement across the land between the 31st and 33rd parallels in Entre Ríos, Santa Fé and Córdoba, it played only a minor role in taming lands to the south. In Buenos Aires province during the 1850s and 1860s the Indians were slowly pushed back from the frontier along the Salado river, but during the 1870s groups surviving in the well-watered Tandil and Ventana hills would sally forth and cause such damage to the scattered sheep *estancias* that only through a war of extermination – the 'Conquest of the Desert' (1879–83) – could the hazard be overcome (*Figure 9.1*). The results of this military action were unprecedented: the area under the effective control of the state was doubled and the whole of Patagonia, together with the arid lands south of Cuyo, were opened up. But, more important, the area of the Pampas was doubled. These newly won lands were divided up, though not without dispute, between the provinces of Córdoba, Santa Fé and Buenos Aires, and the western rump became the core of La Pampa National Territory.

In 1876 public land legislation was enacted for the laying out of land in sections of 400 km² each; every section was to be subdivided into 400 lots of 100 ha each, the four central ones being reserved for the creation of a town; all manner of official, private and joint colonization schemes were allowed. In practice, disposal of most lands followed a different course, and in a few years most of the land had become the property of a small number. By 1914 8 per cent of the farms held 80 per cent of the farmland. The 'Conquest of the Desert' had been financed by bonds costing 400 silver pesos, repayable later in a square league of public land, while other lands were sold by public auction. The greater part was given to the soldiers in reward for their services, but, having neither resources nor inclination to develop their estates, most of them rapidly sold out to speculators, livestock farmers and genuine colony promoters. One such was the South American Land Company formed in 1881 with lands in north-eastern La Pampa. Buoyed up by hopes of rapid railway development, a few colonists bought lots and, in the expectation of successful business, the company rented much land to graziers on short lease and set up its own pastoral activities on the rest. But the railways did not arrive in most of the new territories for 15 to 25 years, and neither did the colonists; only large-scale pastoral farming was viable, and the company sold out. Another British-owned venture, the Santa Fé and Córdoba Great Southern Land Company (1888), with lands 175 km west-south-west of Rosario, benefited from proximity to a railway in its early success with colonies at Arias and Venado Tuerto. But the failure of colonists to arrive during the 1890s drove it to the same course that hundreds of individual *estancieros* were already adopting – the installation of wells and wind pumps, fencing the land, stocking it with high grade Shorthorns and converting natural grassland to alfalfa pasture as and when tenant farmers offered themselves. When the immigration boom of

1904–12 occurred, colonization by private enterprise was no longer a profitable venture, and the government no longer possessed public lands to colonize in the Pampas.

With the ousting of the Indians, the whole of the Pampas became available for occupation and improvement, and by the end of the nineteenth century the wave of settlement had engulfed it all. Beyond lay lands either too dry for general improvement or too forested for wholesale clearance. Here settlement was sporadic, like that of the colonial interior, though with one important difference: for the self-sufficient settler of earlier times isolation was less important than the physical attraction of site, whereas for the modern commercial farmer the latter was useless without access to a market.

The expansion of settlement outside the Pampas had begun before 1890 and was associated with the endeavour to find crops able to transcend the barrier of distance. In Patagonia modern settlement dates from the foundation in 1865 of the Welsh colony in the lower Chubut valley. Having been led to believe that the district was a well-watered land of forest and lush meadow, the 153 original settlers (only three of four of whom were farmers) arrived to find a canyon incised in arid scrubland, its pasture-covered floor littered with the driftwood of a river in spate. After many vicissitudes the settlers rediscovered the techniques of flood-farming and canal irrigation, and the colony began to prosper, or at least became self-supporting. But transport costs were high. Wheat cost more to ship to Buenos Aires than from Buenos Aires to Liverpool; 30 years after the foundation of the colony wheat sales were only profitable in seasons of poor harvest on the Pampas. But the colonists were as much the prisoners of their cultural environment as of their physical circumstances.

Within 6 years of the defeat of the Indians in northern Patagonia the territory had been populated by more than 500,000 sheep and cattle. As the pastures of Buenos Aires became the scene of cattle-fattening enterprises, so the production of wool there declined, and the Patagonian wastes acquired a viable role in the Argentine economy. The government, conscious of its dispute with Chile over the possession of much of Patagonia, was anxious to encourage rapid occupation of the land, conceding estates of colossal size from the public domain, often without security of title. Shepherds, frequently Scottish, took up isolated residence either on their own behalf or as managers for absentee owners, both private and corporate. Fortunes were invested in the stocking and fencing of the land and in paying the running costs of enterprises from which no great returns could be expected for years. Britons, Germans, Austro-Hungarians, Spaniards and Chileans provided the capital and Chileans much of the labour. Argentines were either not welcomed or not interested in the development of their Deep South. By 1930 virtually all of Patagonia from the Río Negro to Tierra del Fuego had been carved up into sheep farms of undoubted prosperity.

On the northern margins of Patagonia fertility and accessibility favoured intensive colonization in the irrigated Río Negro valley near the confluence of the Limay and Neuquén. The railway arrived in 1899, and river-control works undertaken in the 1910s brought 23,000 ha into cultivation by 1919. Similar developments took place at San Rafael in southern Mendoza.

At the opposite end of Argentina lay the forested territories of Chaco, Formosa[2] and Misiones, where colonization began with a few riverside settlements in the 1880s, but grew rapidly only after 1918 with the increasing demand for subtropical industrial crops. As already mentioned, the first notable influx of people into the Chaco came with the development of the *quebracho* industry: *quebracho* logs were first exported in 1888 whilst the great *quebracho*-extract factories of Guillermina and Puerto Tirol were opened in the first decade of the present century. As the *quebracho* forests stretched in a belt parallel to but some 80 km west of the river Paraná, there was a need for railways, which were later able to serve the agricultural colonization of the cutover land. After 1910 other lines were constructed from Resistencia and Formosa directly into the interior to stimulate colonization, which was encouraged by the government in 1923 at a time of high world cotton prices. Holdings of cropland were limited to 100 ha and their operators expected to become owners. Many settlers were from neighbouring Paraguay and others were immigrants from Germanic and Slavonic lands; between 1914 and 1937 the population of Chaco Territory rose from 46,000 to 335,000 and the number of small farms from 290 to 14,940.

For much of the nineteenth century Misiones had been the subject of dispute, especially with Paraguay. When the latter finally renounced all claims in 1876, the area came under the control of Corrientes province, which rapidly sold most of it to private individuals. The holdings were defined by the length of their frontage on to the rivers Paraná or Uruguay and by the depth of their penetration of the interior. Re-survey at the turn of the century revealed that nearly a third of this thickly forested territory, mainly along the watershed, still remained public. New colonization laws favouring the small settler and the arrival of the Buenos Aires–Posadas railway along the southern border in 1912 facilitated massive immigration, largely of intelligent and skilful German-speaking peoples, after 1920. For the large landowner no extensive pastoral economy was feasible, and after deforestation only intensive plantation farming or colonization schemes could provide a satisfactory return. Both required considerable management, and the former in addition depended on the availability of manual labour. Neighbouring Paraguay furnished a continuing, if erratic, supply whilst European colonists were also willing to do plantation work while establishing their own farms. Hence plantations and colonies, both of them devoted to tree crop production, were often developed in association, especially along the east bank of the Paraná. The population of Misiones rose only slightly between 1895 and 1914, from 33,000 to 54,000, but by 1947 over 246,000 persons lived in the Territory.

In Uruguay an absence of public lands precluded official colonization, but several private ventures were undertaken, leading to extensive wheat cultivation.

Demographic changes 1850–1930
Over the whole period the total Argentine population rose from about 1 million to almost 12 million and the net immigration exceeded 4 million (*Figure 9.2b*). Between 1855 and 1914 no province except Catamarca failed to double its population, but the proportion living in the north-west (excluding Córdoba) fell from 32 per cent to 13 per cent; despite immigration, the share of Cuyo, Córdoba and Mesopotamia also fell from 40 per cent to 26 per cent. Of the provinces showing an increased share, those that were non-existent in 1855 (The Chaco, Misiones, La Pampa and Patagonia) still only accounted for 4 per cent. The great

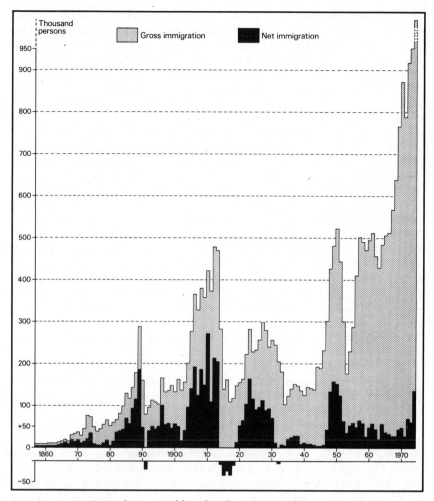

Figure 9.2 (a) Argentina: annual immigration 1857–1974

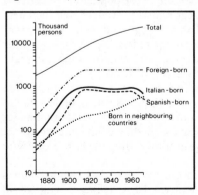

(b) Argentina: contribution of major immigrant groups to total population 1869–1970

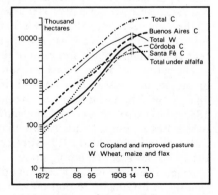

(c) Argentina: expansion of the cultivated area

increase came in the core of the Pampas: the share of Santa Fé rose from under 4 per cent to over 11 per cent, that of Buenos Aires province from under 17 per cent to over 20 per cent, and that of Greater Buenos Aires[3] from 8 per cent to nearly 26 per cent; their absolute population rose from 316,000 to 4,543,000 (*Figure 9.1*).

The role of internal migration in these major regional changes was not very great: although many of the older provinces had a fifth of their sons living elsewhere in 1914, most had simply moved to neighbouring provinces, especially to Mendoza, Tucumán and Córdoba. Of the capital's million and a half residents only 141,000 were born elsewhere in Argentina, and of these only 37,000 did not come from the adjacent provinces of Buenos Aires, Santa Fé and Entre Ríos.

The proportion of urban[4] population to total population increased from 28.6 per cent to 52.7 per cent during the period 1869–1914. Although agriculture was booming it used labour efficiently: beef cattle in fenced pastures required little supervision, arable farming was mechanized, and the climate rendered unnecessary the expenditure of labour on spreading fertilizers or making hay. The growing urban share was not simply due to the expansion of Greater Buenos Aires (whose share rose from 13.6 to 25.8 per cent), for the other towns grew equally fast (from 15.0 per cent to 26.9 per cent). Salient among these was Rosario, whose population rose from 23,000 to 223,000. It had already surpassed the provincial capital in 1847, and still retains the distinction of being the only large city in Latin America which is bigger than the capital of the province in which it is situated.

The part played by immigration was unique in the western hemisphere. During the decades 1881–90 and 1901–10 an annual net inflow equal to between 2 and 3 per cent of the existing population gave Argentina the highest intensity of immigration ever recorded in the New World (*Figure 9.2a*). This is reflected in the percentage of foreign-born inhabitants, which rose from 12 in 1869 to 26 in 1895 and reached a peak of 30 in 1914.

Among the immigrants a distinction must be drawn between those from adjacent South American republics and those from Europe (*Figure 9.2b*). The immigrant neighbours increased steadily in number, but in the period 1869–1914 their share of the total foreign-born population fell from 20 to 8 per cent. Over three-quarters of them simply moved into those Argentine provinces bordering upon their native land, and in some cases they played an important role in provincial growth: thus in 1914, 70 per cent of the population of Formosa was Paraguayan, 48 per cent of Tierra del Fuego and 41 per cent of Neuquén were Chilean, 38 per cent of Misiones was Brazilian or Paraguayan, and 20 per cent of Jujuy was Bolivian. In contrast, few migrated to the federal capital (scarcely 5 per cent of its foreign-born) and most of these came from neighbouring Uruguay.

The pattern of immigration from Europe was very different: 3.4 million immigrants from overseas entered and remained in Argentina in the period 1857–1930. The net inflow rose sharply in the 1880s to a peak of 178,000 in 1889. The crisis of 1890 caused a net outflow in 1891, and reduced movement for over a decade, but the early twentieth century witnessed an all-time record net inflow of over 200,000 in 1910. After a large outflow during the First World War the 1920s saw renewed immigration.

Argentina's role in European emigration during the period 1850–1930 was (jointly with Canada) second only to that of the USA, taking a tenth of all emigrants to the latter's six-tenths, but her importance varied with time. In the mid-nineteenth century most emigrants were Britons and Germans destined chiefly for English-speaking lands. By the turn of the century 30 per cent were from Italy – 17 million left in 1876–1926 – almost equal numbers going to the United States, Argentina and Brazil. Spaniards emigrated almost exclusively to Argentina, and she was also favoured by many refugees from the political disturbances of eastern Europe, especially Jews and Poles. Throughout the period Italians and Spaniards (in the ratio of five to three) represented about 80 per cent of all immigrants into Argentina (*Figure 9.2b*). Many other Italians provided harvest labour in Argentina during the southern summer and in Italy, France and Germany in the following half-year. Between 1890 and 1914 this *golondrina* (swallow) migration was facilitated by return fares as low as £10. Although immigration was officially encouraged, the local attitude can be illustrated by the competition held in 1896 to design a ship able to carry live cattle to Europe and immigrants on the return.

The agriculturally developing Pampa provinces (Buenos Aires, Santa Fé, Entre Ríos, Córdoba and La Pampa) received 57 per cent of the 2.1 million European-born inhabitants of Argentina in 1914, and another 33 per cent stayed in the capital city. In contrast, the long-settled provinces of the north-west, Cuyo and Corrientes (with almost a quarter of the nation's population) attracted only 7.5 per cent and the newly colonized Territories under 2 per cent. Nevertheless, more than a third of southern Patagonia's people were born in Europe.

In addition most Europeans preferred to remain in the towns: in 1914 when 53 per cent of the total population was urban, 78 per cent of the Spanish and 69 per cent of the Italian immigrants lived in towns; for them emigration was the only way of moving from the countryside to the towns. Throughout the period nearly half the Europeans stayed in Greater Buenos Aires and nearly half its population was European-born.

In part the distribution of immigrants reflects the economic opportunities open to them. In 1914, when 30 per cent of the inhabitants were foreign-born, only 22 per cent of the owners but 34 per cent of the tenants of livestock farms were foreign; with crop farms 41 per cent of the owners and 72 per cent of the tenants were foreign. In the towns over two-thirds of all industrial and commercial firms were foreign-owned, and over half their personnel were immigrants, whereas traditional craft industries, such as clothing, and the public services were dominated by native-born staff. It is often argued that this urban concentration of immigrants reflects the decreasing opportunities for landownership after 1890; on the other hand, since migration normally responds rapidly to economic and other pressures, one might suggest that, as the peak inflow occurred just before the First World War, there existed even then a basic satisfaction with the opportunities available. Similarly the percentage returning to Europe in any decade between 1861 and 1910 was lowest in 1881–90 (26 per cent) and next lowest in 1901–10 (38 per cent) – despite the contribution of *golondrina* migration to the latter.

Demographic developments[5] in Uruguay followed a similar pattern, but the scale was much smaller. The total population rose from 132,000 in 1852 to 500,000 30 years later; by 1908 it exceeded 1 million and had almost doubled again by 1930. Immigrants, mainly Italian and Spanish, totalled 650,000 between 1836 and 1926, and a quarter of these arrived in 1904–13. At the 1908 census 181,000 or 17 per cent of the national population – and 30 per cent of the population of Montevideo – were foreign-born. The regional distribution has changed but little, for the departments bordering the river Plate (and to a less extent the river Uruguay) have always been the most attractive to native and foreigner alike. Montevideo has long contained a quarter to a third of the population, and the truly rural population has constituted but a minority.

Capital investment in the economic infrastructure

Capital investment on an unprecedented scale afforded the means by which immigrant labour could be harnessed to the Argentine soil to produce the agricultural revolution. By 1913 75 per cent of foreign capital invested in Argentina was in the nation's infrastructure: in railways, port installations, public utilities and irrigation works. Another 20 per cent was in trade and finance, in banking and import/export houses, merchanting services, loan agencies and processing industries, while only 5 per cent was in agriculture itself. In a word, one-twelfth of the world's foreign investments (to a great extent British) was directed solely towards the lubricating of the machine of Argentine agricultural production. In 1857 Britain's investments were worth under £3 million, but by 1890 they had risen to £175 million and by 1910 exceeded £290 million.

The development of the railway network, mostly British-owned, at first went hand in hand with the advance of the frontier, Rosario and Buenos Aires providing separate and unconnected termini for most routes (*Figure 9.3a*). By the late 1880s Mendoza, Córdoba and the Santa Fé colonies were linked to Rosario, and Bahía Blanca to Buenos Aires, all by broad-gauge lines; Tucumán and Santiago del Estero, alone of the old centres of the north-west, were linked to Córdoba, but only by a narrow-gauge line. During the next 30 years the Buenos Aires and Pacific Railway soon gave southern Córdoba and Mendoza a direct route to the capital, and Rosario was also quickly connected. More gradually the Buenos Aires Pampas became saturated with lines, for a cart-haul of more than 16 km to a railway station made wheat production unprofitable (*Figures 9.3b, c*). Rivalry with Chile over Patagonia also led to the early extension of the Southern system to Neuquén (1899), but the government line from Bahía Blanca to Bariloche took until the 1930s to complete. The north-west was slowly incorporated into a government-built narrow-gauge system, but no through route to the capital was provided until after 1910; similarly the skeletal standard-gauge network of Mesopotamia was not connected to Buenos Aires until 1908, and even then a ferry across the Paraná substituted for a bridge. Thus, by virtue of timing and gauge differences, the growth of Argentine railways worked to the disadvantage of the north-west, Patagonia and Mesopotamia and to the clear benefit of the city and province of Buenos Aires.

Although this pattern may have reflected a realistic appraisal of the economic potential of Argentina's different regions, railway construction by no means led automatically to a profitable flow of goods and passengers. The leading railway companies would buy estates, subdivide them and sell lots on instalments; in the absence at that time of adequate government services they ran agricultural advisory departments, and even established experimental farms to discover the best varieties of crops and make them available to growers. The Buenos Aires

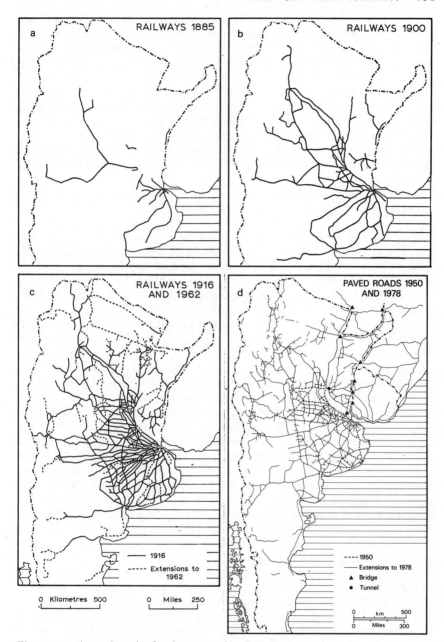

Figure 9.3 Argentina: the development of roads and railways

Great Southern Railway Company carried out irrigation works for the govern-ment along the Río Negro and then, for lack of any local enterprise, set up an organization to advise the fruit-growers, to handle collection and storage and to market the produce in Europe. Other companies independent of the railways also provided the farmer with a wide range of commercial, technical and financial services. 'Agency' firms, often owned and staffed by Britons, offered mortgage loans, handled farmers' legal and commercial business, and even provided managers for *estancias*; wool and cereal producers were served in many ways by produce-purchasing firms, and every town had its distributors of agricultural machinery and medical requisites. As with other developments, these service firms tend to date from two periods: of the forty-three British firms established between 1860 and 1960 in part to provide such services, eighteen (especially the colonization companies) were founded in the 1880s and another fifteen in 1906–12; the four founded in the intervening years were all meat-packing companies.

In Uruguay foreign capital was invested on a less extensive scale, but in similar fields, with several British companies operating in both republics.

The agricultural revolution
This began with the upgrading of the flocks of sheep, and the breed most favoured was the Rambouillet or French merino. For 30 years, until the mid-1880s, the general flocks of the Pampas provinces became progressively more merino, and their fine wools were exported in increasing quantities. From 7700 tonnes in 1850, wool exports rose to 55,000 tonnes by 1865, and 128,000 tons by 1885. One function of the less important cattle was to graze the natural pastures and, by consuming the coarser species, allow the finer grasses preferred by the sheep to dominate. The fencing of the Pampas was also vital to the process of upgrading. In 1876, 5000 tonnes of wire were imported (enough for an equivalent number of kilometres of fence); in 1892, 42,000 tonnes; and in 1907, 84,000 tonnes – by which time over 1 million km of fencing had been erected.

But the course of events changed in the mid-1880s. The arrival in 1876 of *Le Frigorifique*, the French ship with freezing chambers, though hailed by farsighted *estancieros*, was not followed up by France, and development of the frozen meat industry awaited the interest of English firms which by 1880 were initiating the industry in Australia. In 1883 The River Plate Fresh Meat Company built a *frigorífico*, or meat-freezing plant, at Campana, and by 1907 another seven were in operation; except for one at Bahía Blanca all were situated on the south bank of the Paraná–Plate between La Plata and the Buenos Aires/Santa Fé border. At first sheep were preferred because the early equipment could handle the smaller carcass of the sheep more easily, and because the quality of merino mutton was superior to that of *criollo* beef.

But the merino was not really a dual-purpose animal, nor had it ever been a healthy success on the wetter pastures of the older parts of Buenos Aires and of Entre Ríos. The coming of the *frigorífico* and the opening-up of Patagonia now allowed the drier Patagonian pastures to assume production of wool sheep, and the older areas near to the plants converted their flocks to larger dual-purpose breeds more suited to the physical and economic conditions. Lincolns, Romney Marsh and the Downland breeds were now in demand. The annual import of pedigree merinos fell from 380 head in the early 1880s to 37 by 1895, whilst the import of the English breeds rose from 550 to 4550. Between 1886 and 1907, 63,000 of the 67,000 pedigree sheep imported came from the United Kingdom.

Upgrading, however, was not proceeding at the same pace even on the Litoral. By 1907 less than 9 per cent of the Buenos Aires sheep were *criollo*, the rest being cross-bred or pedigree, but *criollos* still accounted for 41 per cent of the flocks of Entre Ríos and 84 per cent of those of Corrientes. Thirty years later, under 1 per cent of the sheep of Buenos Aires and Entre Ríos and less than 3 per cent of those of Corrientes were *criollo*, but the national average was still 6 per cent, and in the five provinces of the Andean north-west, 83 per cent of the sheep were *criollo*.

Although the importance of sheep was initially enhanced by the coming of the *frigorífico*, beef was destined to replace mutton. Total flocks of sheep rose from 58 million in 1875 to 74 million in 1895, but thereafter declined rapidly to 43 million in 1914; since then numbers have usually ranged from 40 million to 50 million. The Pampa provinces were largely responsible for this fall: the flocks of Buenos Aires alone declined from 53 million in 1895 to 19 million in 1914, and then to 14 million in 1930, and those of the four adjacent provinces fell from 16 million to 9 million, and then to 7 million. Conversely, Patagonia's flocks rose from nothing to nearly 2 million in 1895, 11 million in 1914, and 16 million in 1930. Northern Patagonia accounted for most in 1895 but since 1900 the settlement of southern Patagonia has made it Argentina's principal sheep region. By the end of the period a clear regional pattern of improved breeds had emerged (*Figure 9.4b*).

Given such considerable changes, it is not surprising that exports of wool ceased to rise steadily, as they had up to 1885, and began to fluctuate between 100,000 and 200,000 tonnes a year. In 1930, 140,000 tonnes were exported, and as yet little was consumed at home.

In Uruguay livestock-farming followed a similar sequence of developments, but both timing and emphasis were different. Again, the importance of sheep – and in particular the merino – began to grow after 1850, numbers rising from 800,000 in 1852 to a peak of over 26 million in 1908. A sharp decline to 11 million in 1916 then occurred, but thereafter numbers rose again to 21 million by 1930. Once more the advent of the *frigorífico* led to the rise of big, coarse-woolled Lincolns in preference to merino and merino-crosses, although after the great European demand for frozen mutton during the First World War had fallen off, dual-purpose animals of the Corriedale and Corriedale-cross types came into fashion (*Figure 9.4c*).

The development of the Argentine and Uruguayan cattle industry is a curiously protracted one. Upgrading began in earnest in the 1850s with the establishment of the first pedigree Shorthorn herds on the estates of a few leading Argentine *estancieros*. But it was an activity which bore little economic fruit, for *saladeros* continued to provide the main outlet, and slaves consumed most of the salt meat. Even then the carcasses of many animals slaughtered for their hides continued to be wasted. Between the 1820s and 1890s Argentina's annual salt-meat output remained constant at 30,000 to 40,000 tonnes, the product of over half a million head of cattle. In Uruguay the steep rise in cattle numbers from 1.9 million to 6.6 million between 1852 and 1860 gave added impetus to new uses; Giebert's successful attempt at producing meat extract according to the formula of Baron von Liebig led to the formation in London in 1865 of Liebig's Extract of Meat Company[6] and to the conversion of the *saladero* at Fray Bentos to the new process. By the mid-1880s Liebig's was handling almost a fifth of the 780,000

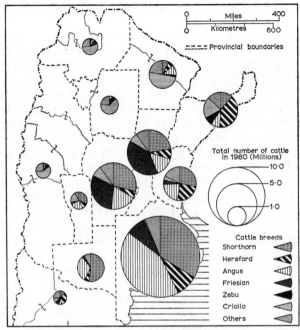

Figure 9.4 (a) Argentina: cattle 1960, provincial distribution and breeds

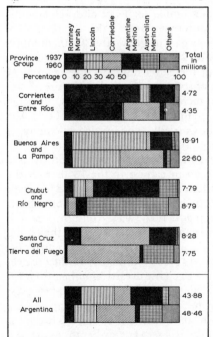

(b) Argentina: sheep 1937 and 1960, regional distribution and breeds

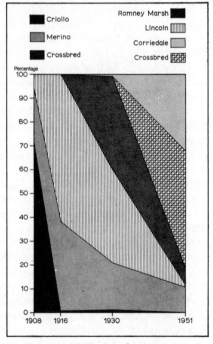

(c) Uruguay: sheep breeds 1908–51, percentage change

cattle which annually passed through the Uruguayan *saladeros*, some of which benefited from the installation of meat-canning equipment. Of the score or so Uruguayan *saladeros* about half were located in Montevideo, and the rest, like Fray Bentos, on the east bank of the river Uruguay.

Both governments sought to expand the long-established export of live cattle to neighbouring countries, and Brazilian *saladeros* took increased numbers of both Argentine and Uruguayan beasts. The disappointment consequent upon the failure of *frigoríficos* to be interested in cattle also led to trials in 1889 with the export of live animals to Europe. Despite the costs of keep, the loss in weight (up to 150 kg) and the hazards of climate and disease, this proved a success, at least for Argentina. From an average export in the 1880s of under 100,000 head, bound chiefly for adjacent lands, numbers rose to over 200,000 by 1893 and averaged well over 300,000 for the rest of the decade, nearly 100,000 going to the United Kingdom. The need for fat, tame animals for this trade stimulated both breeding and fattening activities.

Exports were halved in 1900 when an Argentine outbreak of foot-and-mouth disease, introduced from France, led to the closing of British ports to live cattle. At the same time the French wool textile crisis caused a rapid decline in the profitability of sheep, whilst the Boer War afforded a ready outlet for frozen beef. Hence, by virtue of a series of misfortunes, the Argentine frozen-beef trade began. Exports leaped from 9000 tonnes in 1899 to 25,000 tonnes in 1900 and 150,000 tonnes five years later. *Frigorífico* owners could afford to select only the best animals, and so there suddenly arose a demand for high-grade cattle and for alfalfa pastures on which the steers could be fattened. By 1907 frozen beef had replaced all other meat products in the pattern of Argentine exports (*Table 9.1*) and Argentina had replaced the USA as the largest provider of frozen beef to the British market.

Table 9.1 Value of Argentine livestock exports 1887, 1897 and 1907

Category	1887 (%)	1897 (%)	1907 (%)
Salt beef	48	22	4
Live cattle	28	43	7
Live sheep	1	13	1
Meat extract	2	2	7
Meat flour	<1	<1	6
Frozen mutton	19	17	20
Frozen beef	<1	1	51
Preserved and other meats	1	2	3
Total value (million gold pesos)	5	12	27

The introduction of the chilling process in 1908 gave a further fillip to pasture improvement. Whereas cattle for frozen beef could be bought and killed at the cheapest time of the year and then stored until the market was favourable, chilling animals could only be killed 40 to 45 days before consumption in Britain. A steady demand meant a constant killing and a year-round supply of good quality animals, which could be produced only on artificial pastures. Whereas *saladero* fattening pastures had to be located near the riverside factory, *frigorífico* pastures, thanks to the network of railways, could now develop wherever physical conditions were best. North-western Buenos Aires, southern and eastern Córdoba and adjacent parts of La Pampa and Santa Fé, together with southern Entre Ríos, came to fill this role.

In Uruguay the first *frigorífico*, a locally financed one at Montevideo, was not installed until 1904, and growth came only with the construction of two more US-owned plants in 1911 and 1917. Chilling did not begin until 1921.

Nor did chilling become important in Argentina until the 1920s: during the 1910s exports of frozen beef averaged 350,000 tonnes per annum and chilled beef only 20,000. By the late 1920s, however, chilled beef exports had risen to 400,000 tonnes and frozen beef had fallen to 200,000 tonnes. By this time, too, both Argentina and Uruguay consumed more meat than they exported.

The roles of cattle and sheep in the livestock economy thus changed greatly after 1900. But, as with sheep, changes in the character of cattle-farming were essentially regional. By 1907 55 per cent of Argentina's 29 million cattle (but 91 per cent of the cattle of Buenos Aires province) were cross-bred or pedigree animals, whereas 41 per cent of the cattle of Entre Ríos were still *criollo*, as were 57 per cent in Santa Fé and 84 per cent in Corrientes. These differences were reflected in the later history of the *saladeros*: those in Buenos Aires had almost ceased by 1905, and plants in Entre Ríos were slowly declining, but Uruguayan *saladeros* maintained production until *c*. 1910. There thus developed the two patterns of Argentine cattle movement which have survived until recent years: within the Pampas, especially south and west of the river Paraná-Plate, improved cattle (chiefly Shorthorns) were reared, moved to fattening pastures and then to the eight riverside *frigoríficos*, while in Mesopotamia and the Chaco poorer animals were reared, moved south to better pastures and then into the *saladeros* of Entre Ríos or Uruguay. Although a few *saladeros* continued to operate for several decades, most of the Argentine trade began to concentrate on those two which were converted just before the First World War into multi-purpose canning and extract factories: Colón, Liebigs' factory on the Uruguay, drew its animals from eastern Mesopotamia, whilst Santa Elena, Bovril's plant north of Paraná city, tapped the resources of both shores of the Paraná.

In Uruguay upgrading began in earnest only in 1904, herds becoming dominantly Hereford (as in Mesopotamia) rather than Shorthorn, and numbers rose slowly to about 8 million. The hillier centre of the country became the rearing district, and the loamy and loessic soils flanking the rivers Uruguay and Plate provided the best fattening pastures. The value of upgrading, apart from the quality of the meat, may be judged from the fact that *criollo* steers took 5 or 6 years to reach a weight of half a tonne and then yielded only 40–50 per cent meat, whereas high-grade animals took only 3 years and yielded 50–55 per cent meat.

The evolution of the beef-cattle industry described above may appear simply to reflect the historical peculiarities of the introduction of new processing techniques. Yet the spatial patterns of beef processing and the types of cattle farming which came into existence by the mid-1920s conform to an industrial location model for beef processing developed by the author (Crossley, 1976; Crossley and Greenhill, 1977). The model, analogous to Von Thünen's and Dunn's agricultural location models with their well-known concentric zonation, recognizes that cattle may be destined (like land by the farmer) for a variety of uses,

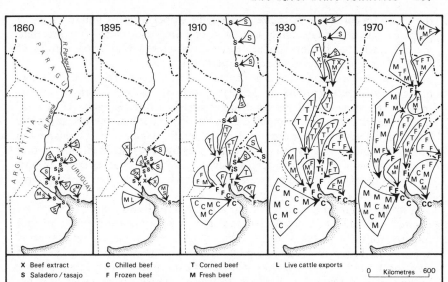

Figure 9.5 River Plate countries: beef product industries and cattle flows 1860–1970

and that that use which yields the highest returns per unit of cattle at a given location will be preferred. In ascending order of market values (gross returns) these uses are: beef extract, salt, corned, frozen, chilled and fresh beef. Since fresh beef is also the raw material for the other processed products, an essential feature of those industries is that they *lose* value by manufacturing. Given such an absurdity, none would exist if some constraint did not prevent the slaughter of all cattle for the fresh-beef market. Such constraints are perishability and costs of transport to the market. With only surface transport available, fresh beef with a short 'life' can be moved only limited distances; chilled beef, with a 'life' of 40 days, can be shipped to European markets from the Plate (but not Australasia); the other products last indefinitely. Perishability thus determines the zonation of production of fresh/chilled/frozen beef about the central world market. The special refrigerated transport required for chilled and frozen beef is more costly than the transport for other products; additionally, until a decade or so ago, refrigerated beef was shipped as quarter carcasses, which are heavier than the boneless beef employed for canning or salting; finally the extract derived from a carcase was miniscule. Thus the relative costs of transporting the further products that may alternatively be derived from a unit of cattle determined the frozen/corned/salt/extract zonation of remoter locations.

Additionally, each superior beef product requires progressively finer animals, and the manufacturer is able to pay more for them. The cattle farmer is thus induced to improve his herds to the highest level for which he can find an outlet. By the mid-1920s a near-perfect zonation had emerged for supplying the chief world market, Britain: *frigoríficos* beside the Plate Estuary, using high-grade cattle from the Pampas, supplied chilled beef; *frigoríficos* along the Lower Paraná produced frozen beef from the moderately good cattle of Entre Ríos and Santa Fé; factories in Entre Ríos drew in the mediocre animals of Corrientes and the Chaco for corned beef; *saladeros* along the Upper Uruguay and in Paraguay prepared salt beef from barely improved or pure *criollo* herds (for Brazilian markets), and production of beef extract was relegated to a few corners of Paraguay and Mato Grosso where no other outlets existed.

The conversion of the Pampas from an almost purely pastoral landscape to one of the world's leading producers of cereals was also the work of less than 50 years. In 1872 Argentina, over eleven times larger than the United Kingdom, had a cultivated area that was smaller than Lincolnshire. By 1900, thanks especially to the efforts of colonists in the northern Pampas, an area equal to half England was under cultivation, and by 1914, under the pressure of livestock farmers further south to plough up the land for improved grazing and fodder crops, almost 250,000 km² of land, an area greater than the whole of the United Kingdom, had come under the plough (*Figure 9.2c*).

From the beginning of colonization wheat was the leading crop, occupying a quarter to a half of all cultivated land, and it was grown initially to replace imports from USA, Chile and Australia; by 1875 regular wheat imports ceased and exports began, exceeding 100,000 tonnes for the first time in 1884 and reaching 1 million tonnes in 1893. Since 1904 exports have rarely fallen below 2 million tonnes. Despite the growth in wheat exports, an equal amount has usually been consumed at home, thanks partly to the growth in population, but also to the changes in diet with Italian immigration, and during the 1880s wheat consumption per head almost doubled. Flour was still a costly luxury in 1870, and in 1875 milling had to be protected for a time by tariffs to encourage its growth, but by 1895, 259 steam-driven mills were in operation, half of them in Santa Fé and Entre Ríos and more than a quarter in Buenos Aires. By 1914, however, Buenos Aires produced most of Argentina's wheat and flour. Flour exports regularly exceeded 1000 tonnes after 1878 and 10,000 tonnes after 1890, but the quality of the product, competition from other countries and tariff protection of foreign mills all helped to restrict Argentine exports to around 100,000 tonnes, first reached in 1904, with Brazil the leading purchaser.

In area, maize was the second crop, usually occupying half as much land as wheat. Eventually it came to dominate the cereal lands of southern Santa Fé and adjacent parts of Buenos Aires, when wheat moved to drier lands in Córdoba, La Pampa and southern Buenos Aires. Nevertheless, exports of maize were greater than those of wheat up to 1900 and closely paralleled them thereafter, Argentina becoming the world's leading exporter of the crop. As with wheat, half the maize produced played a role in the domestic economy, but as fodder rather than flour.

Before 1900 flax was of minor importance, but as livestock farmers sought to improve their pastures through introducing tenants to the land on short 3 to 5-year contracts, flax proved especially suitable as a pioneer crop and as a final crop to be sown with the alfalfa which the *estancieros* so ardently desired. Throughout the period it was grown largely for export – Argentina being world leader – and for its linseed rather than its fibres. Although exports exceeded 100,000 tonnes for the first time only in 1894, they rose sharply to almost 900,000 tonnes by 1904, but not until 1920 did they regularly exceed 1 million tonnes.

The dominance of these three cereals in the pattern of Argentine agricultural production – in 1914 13 out of 22 million ha of cultivated land were dedicated to them (*Figure 9.2c*) – is partly to be explained by the rigid system of land tenure. Tenancy contracts frequently stipulated that only these crops should be grown, that rents were to be paid in kind – usually a third, but occasionally a half of the harvest – and that the rest was to be sold through the landlord or his agent. The administration of tenanted lands was thus greatly simplified, but at a considerable cost, economically as well as socially: at harvest time railways and ports became overloaded, delays ensued and inadequately stored cereals suffered from the rains. Other factors also contributed towards monoculture of cereals: farm labour on the Pampas was always in short supply, and cereal-growing was easily mechanized. Furthermore, immigrant farmers with a little capital could not afford both machinery and land, so tenant-farming with machinery, a mobile investment, afforded a convenient opportunity. At times a shortage of tenants obliged the owner to become a true partner, providing the tenants with work animals, seed and credit for the acquisition of equipment, and himself investing in the fixed assets of fencing, wind pumps and buildings.

But the ultimate aim of tenant-farming was usually the creation of artificial pastures of alfalfa on which the upgraded steers could be fattened for the *frigorífico* (*Figure 9.2c*). In 1872 alfalfa was a crop grown chiefly in San Juan for feeding to cattle destined for export on foot to Chile. By 1922, the peak year, 8.5 million ha, or 40 per cent of the cultivated area was devoted to this single plant and, with the needs of the livestock farmers satisfied, cereal production was on the decline.

As arable farming developed so largely as a by-product of the livestock industry, it is ironical that crops came to dominate the nation's export economy: their share of export values rose from 1 per cent in 1870, to 20 per cent in 1890, and 53 per cent by 1910 (*Table 9.2*), though it had fallen back to 42 per cent by 1919.

Table 9.2 Percentage composition of Argentine exports, 1899 and 1910

	1899	1910
Livestock products	62	43
including: live animals	5	1
meat	3	10
hides	13	11
wool	39	16
Crop products	35	53
including cereals	32	50

The revolution in agricultural exports was made possible by the long-term character of the many kinds of credit which Argentina received – government loans, railway investments made years before routes became profitable, mortgages granted to farmers for fencing their lands, repeated postponements of debts owed by settlers to colonization companies or in the form of profits ploughed back by farming companies. Although some companies eventually reaped handsome profits, for 10 to 20 years many declared no dividends at all or a mere 0.5 per cent.

Uruguay, in contrast, did not experience the same cereal revolution, for no symbiotic development of cattle-raising and cereal-farming took place. With a higher rainfall and more nutritious natural grazing than the Pampas, Uruguay did not need alfalfa for fattening her cattle, though oats were often grown as winter forage. Subdivision of fields for the better control of breeding and grazing led to the redundancy of many herdsmen at the beginning of the present century and hence to a rural exodus.

Nevertheless, the flat, fertile lands adjacent to the river Plate gradually became the scene of more intensive agricultural activities: of wheat-growing for the domestic market and of dairying near Montevideo. Flax-growing also expanded rapidly from 1000 ha in 1899 to 150,000 ha in 1930, linseed becoming the third export after meat and wool. Many farmers were sharecroppers, but their 4-year tenancy contracts were renewable, making for a greater stability of the rural population.

Like Uruguay, Argentina's development in the period 1853–1930 was not geared solely to production of meat and cereals for world markets. Agricultural and industrial production for domestic markets was growing, and other crops were beginning to find a place among the exports. Outstanding were the sugar and wine industries. In 1855 the provinces of Tucumán and Salta had only 223 ha under sugar cane and produced only 400 tonnes of sugar.

The arrival of the railway and the ability of the sugar-mill owners to pressure the government into imposing protective tariffs together afforded the necessary stimulus. By the mid-1870s about 2500 ha were under cultivation and sugar production touched 3000 tonnes; by 1914 over 100,000 ha were in production and output exceeded 250,000 tonnes, most of it from Tucumán. With high production costs in a marginal climate exports did not succeed, but the home market was satisfied.

The expansion of the Argentine wine industry, centred in Mendoza since early colonial times, similarly owed much to tariff protection and also to the influx of Italian settlers. From 6400 ha in 1890, Mendoza's area under vines grew to 45,000 ha by 1910 and exceeded 100,000 ha by 1934.

The production of cotton, although native to the country, expanded only with the rise in world prices after the First World War and the colonization of the Chaco. The 400 ha under cultivation in 1863 were increased to only 3000 in 1917 but jumped by 1925 to over 100,000, and Argentina became the world's seventh-largest cotton producer.

Industry in expanding Argentina was, with the exception of meat-packing, typically small scale, and geared to the processing of local agricultural produce and the manufacture of simple consumer goods to satisfy domestic demand. Some branches were stimulated by tariffs – e.g. the corrugating and galvanizing of imported iron sheets – but most were the work of enterprising immigrants who had no power to urge protection. By 1913 71 per cent of goods consumed in Argentina were manufactured within the country: 91 per cent of foodstuffs, 88 per cent of clothing and 86 per cent of printed matter were locally produced, but only 38 per cent of chemicals and 33 per cent of metallurgical products; only 23 per cent of the cloth consumed was of national manufacture.

The Paraguayan exception
While sporadic attempts were made to develop the Paraguayan economy before

the war of 1865–70, that catastrophe effectively halted them. Before the war most people were tenants of state land, living in the central zone, which stretches for 100 km east and south of the capital; after the war most of the private land also reverted to the state through the inability of its occupiers to prove title to it. In the closing years of the century, however, huge tracts of the public domain were sold or leased to mainly foreign enterprises, which sought to exploit the forestry resources, develop livestock-farming or, less often, establish colonies; thus the Casado Company of Argentina acquired a portion of the Chaco as big as Belgium. Sometimes former tenants were dispossessed, more often they were welcome to remain as largely self-sufficient tenants or squatters who supplied casual labour to the new owners. Most important of the new enterprises was the exploitation of the *quebracho* forests along the west bank of the river Paraguay. As in Argentina, the tree was initially felled for its very resistant timber (*quebracho* = axe-breaker), but at the turn of the century it became increasingly valued for its tannin content. Railways were built – especially by the Casado Company – into the Chaco forests, and extract factories were constructed at their river terminals north of Concepción. To the east the forests of the Paraná plateau were exploited by collectors of wild yerba, who gathered 10,000 tonnes a year in the 1880s, but from 1909 supplies came increasingly from large plantations.

Further south, in the often marshy lowlands of the Asunción–Encarnación–Corrientes triangle, great cattle *estancias* were established. Here, Mulhall reported in 1885, *criollo* cattle fattened better than on the more frost- and drought-prone lands of Argentina, but to little purpose, for no meat-processing factory was established and many beasts were exported, often illegally, to Argentina or Brazil. Nevertheless, hide exports ranked third in the 1880s. With the turn of the century *saladeros* were established along the River Paraguay, and some briefly responded with an inferior product in response to the high First World War demand for corned beef. In the postwar period *saladeros* revived, and Liebigs relegated their extract production from Argentina to a small factory at Asunción.

The development of commercial crop-farming and of associated processing industries depended largely on the efforts of immigrant settlers, and these were few. Between 1880 and 1958 only 55,000 foreigners settled in rural Paraguay and a further 12,000 in the towns. Chief among them were Germans who founded colonies near Asunción and Concepción in the 1880s and near Encarnación in 1900. Some colonists deliberately sought physical isolation in which to develop their political or religious Utopias, among them the Australian socialists who settled near Villarrica in 1893, and the Mennonites who, in 1927, acquired over 100,000 ha of Casado land in the middle of the Chaco. Permanence of settlement at first depended upon subsistence farming, but appraisals of commercial opportunities led to the development of specialized crops. The settlers of the Paraná forests, like their cousins in nearby Misiones, chose to grow yerba mate and, later, tung. Settlers nearer to Asunción preferred to specialize in market-gardening for the capital.

For the small farmer, whether immigrant or native, commercial success depended on his living near railway or river; the rest, including the majority of the native population, were restricted to at best a peasant economy catering for tiny country towns. Even the traditional growing and manufacture of cotton declined when cheaper, mass-produced imports competed in the urban market. And the towns, in any case, still accounted for only a fifth of Paraguay's population of under half a million in 1900.

New directions: 1930–60

By 1930 the rail network had been finished, the agricultural area was expanding only in the remote north-east, the creation of the alfalfa pastures had achieved the first stage in the intensification of land use on the Pampas, and a pattern of regional specialization of agricultural production had been established. With the depression of the 1930s overseas markets declined; multi-lateral trading declined, industrial countries raised tariff barriers to protect their own agriculture, and the growth of cheap Argentine exports, made possible by the technologies of freezing and chilling and the mechanized production of cereals, was at an end. With poor prospects for material advancement, the flow of immigrants dried up.

Although the old conditions for continued progress had now vanished, new conditions favoured industrialization. The prices of manufactured imports fell, but earnings from agricultural exports declined still more. Argentina's large urban population was unable to import the consumer goods it desired, and native manufacturing was thus encouraged to expand.

During the Second World War the belligerents' demand for meat and raw materials boosted export earnings and the nation's reserves grew, from $400 million before the war to $1700 million in 1946. The shortage of manufactured imports stimulated domestic industrial expansion. Most industrial raw materials came from the interior, not the Pampas, and a rural exodus began from its depressed areas, providing labour for the manufacturing industries of the nearby towns, especially Buenos Aires. After the war cereals were again in high demand, but Europe could not yet satisfy Argentina's needs for industrial goods. Nor did the traditional agriculture of the Pampas recover, for the rise to power of Perón in the mid-1940s was supported by the urban proletariat and food prices were held down for their benefit. The government also retained much of agriculture's export earnings to subsidize the growing imports of fuel and raw materials required by industry.

But much of the industrial expansion was inefficient, under-capitalized and labour-intensive, conducted mainly in thousands of tiny workshops and without benefit of the best capital goods from the North Atlantic countries. The results of these economic changes were reflected by the early 1950s in the new structure of the GDP and of employment (*Tables 9.3* and *9.8*): the decline in agricultural production was balanced in part by the rise in manufacturing, but more by the increased share of the transport and service sectors, especially the over-manned railways and bureaucracy.

But by the early 1950s a critical point had been reached. The national market for consumer goods was now satisfied, at least for those which did not require heavy capital investment. Demand for imported fuels and semi-manufactured goods continued to rise, but agricultural export earnings declined and high-cost manufactures could find no export market. The decline in agricultural exports

Table 9.3 Argentina: structure of GDP 1925–9 to 1975–7 (%)

	1925–9	1950–4	1960–4	1965–9	1970–4	1975–7
Farming, etc.	27.1	18.8	16.4	15.0	12.3	12.5
Mining	0.2	0.6	1.4	1.6	1.6	1.5
Manufacturing	23.8	27.8	31.3	34.2	37.3	36.8
Utilities	0.5	1.0	1.5	2.0	2.5	2.9
Construction	3.9	4.5	3.7	3.6	4.3	3.7
Commerce	23.3	18.8	18.7	17.9	18.0	17.8
Transport	6.3	8.3	7.8	7.6	7.3	7.1
Finance	3.6	4.2	3.9	3.7	3.5	3.8
Services	11.3	16.0	15.3	14.4	13.2	13.8

resulted from the failure of the Pampas to increase their productivity, and from the rising proportion of agricultural produce which was destined for the well-paid urban groups that were now among the best-fed in the world. During the 1950s Perón and succeeding governments sought by various means to correct the imbalance and halt the inflation which had caused the cost of living to rise sixteenfold.

Changing patterns of population
Between 1914 and 1960 the Argentine population rose by 8 million to 20 million. In contrast to the previous period the only provinces to increase their share of total population were those of the northern frontier and of the western frontier from San Juan southwards (from 11 per cent to 17 per cent), and Greater Buenos Aires, which rose to 35 per cent. All the rest, and particularly the Pampa provinces, declined relatively, though no province had an absolute decline (*Figure 9.1d*).

By 1960 town-dwellers numbered 14.8 million, or 74 per cent of the population, a rise of 4.8 million in 13 years and 10.6 million in 46 years. Yet depopulation of the countryside began only in the latter years of the period: between 1914 and 1947 not a single province recorded an absolute fall in its rural population, but between 1947 and 1960 rural population fell from 5,960,000 to 5,250,000, mainly in the provinces of Buenos Aires, Santa Fé and Córdoba. In contrast, the rural areas of the northern frontier provinces and of Tucumán, Mendoza and San Juan actually gained 227,000 people.

As always in Argentina, population movements account very largely for the demographic changes observed, but internal migration became more important than immigration from abroad. In 1914 30 per cent of the population had been born abroad; by 1960 only 13 per cent were immigrants, whereas 17 per cent had moved from another province and many of the 70 per cent still in the province of their birth had moved from countryside to town. As in the nineteenth century, most interprovincial migration was for only short distances: almost half the migrants moved only to the next province, with Mendoza, the provinces of the northern frontier and those of the Pampa heartland gaining at the expense of their neighbours. A further third of the migrants moved to or from Greater Buenos Aires, the inflow of 1,640,000 far exceeding the outflow of 460,000; as before, most of the incomers hailed from the nearby Pampa provinces.

The flow of immigrants changed greatly in both number and origin: it declined (*Figure 9.2a*) during the Depression and war years, rose thereafter with the arrival of refugees, only to decline once more. In 1960 34 per cent of the foreign-born population were from Italy and 27 per cent from Spain, but neighbouring countries had doubled their share since 1914 to 17 per cent or nearly half a million (*Figure 9.2b*). Many more came seasonally as harvest labourers, often illegally. Italians and Spaniards now preferred to migrate to western Europe or to English-speaking lands outside Europe rather than to South America.

Whatever their origin, immigrants did not move far from their point of entry. In 1960 64 per cent of the European-born lived in Greater Buenos Aires and 26 per cent in the three main Pampa provinces; in contrast, only 26 per cent of Latin American immigrants lived in the metropolis, whereas 62 per cent were in provinces contiguous to their homelands. Even in areas well known for their recent colonization by Europeans, such as Misiones, the Chaco and Patagonia, most immigrants came from adjacent lands.

In brief, the regional pattern of Argentina's population movements in 1960 presented new features as well as old: Greater Buenos Aires and Mendoza continued to attract people from near and far at both national and international levels; sparsely settled Patagonia continued to depend heavily on incomers from Chile and all parts of Argentina; in the old north-west, Catamarca, La Rioja and Santiago del Estero suffered a continuing and serious exodus. Other northwestern provinces, however, (Salta and Jujuy) experienced a rejuvenation, and like Formosa, Chaco and Misiones, attracted people from adjacent provinces and nations to their countryside. Elsewhere, the core of the Pampas witnessed a tremendous rural exodus coupled with an even greater urban growth, whilst along the periphery of the Pampas only the rural loss took place.

Agricultural reorientation

The basic problem of Argentina's economy in this period was agriculture's failure to keep pace with the dual demand of satisfying a rising domestic consumption and of providing substantial foreign earnings. By 1955–9 the volume of agricultural production *per capita* had fallen by 21 per cent since 1920, consumption had increased by 33 per cent and *per capita* exports had fallen by 65 per cent. Argentina's share of world trade in agricultural commodities dwindled alarmingly. In the mid-1930s Argentina was responsible for 70 per cent of the maize exports of the world, 63 per cent of the meat, 21 per cent of the wheat and 12 per cent of the wool; by 1960 these shares had fallen to 23, 29, 8 and 8 per cent respectively.

Growth before 1930 had been largely achieved by the effective occupation of more land and by the introduction of more labour. The area of improved land rose from 7.3 million to 27.2 million ha between 1900 and 1930, but by only 200,000 ha in the next 30 years to 1960. The only areas available for new settlement were subtropical forests or semi-arid scrub where successful farming depends upon costly soil conservation or irrigation works. In Chaco province the continuing expansion of settlement for the cultivation of cotton carried the population from 335,000 in 1937 to 549,000 in 1960, but by then growth was levelling out as squatters began to abandon exhausted lands. Further east, in Misiones, on the moderately fertile basaltic soils colonies of small farms continued to grow. The population reached 379,000 in 1960, but three-quarters of the land was still covered by natural forest, most of it in huge estates. The only publicly owned forested areas in Argentina still awaiting clearance and colonization were 1.5 million ha of tropical lowland east of Salta, not far from the Tucumán to Santa Cruz (Bolivia) railway.

Nevertheless, in 1960 the five Pampa provinces of Buenos Aires, La Pampa, Córdoba, Santa Fé and Entre Ríos still accounted for about 90 per cent of the 27.4 million ha of improved land in Argentina, though the trend since 1930 was towards the replacement of crop production by more extensive pastoral farming (*Table 9.4*).

Table 9.4 Argentina: use of improved land in the five Pampa provinces (million ha)

	1933–4	1959–60
Fodder crops and grass	4.9	13.3
Other crops	20.2	11.3

During the 1920s and 1930s cereal farming had been in the ascendant, record areas being devoted to wheat in 1928–9 (9.2 million ha) and 1938–9 (8.6 million ha), to maize in 1935–6 (7.6 million ha) and 1939–40 (7.2 million ha), and to flax in 1936–7 (3.5 million ha), whereas cattle numbers in the Pampas had declined from a peak of 28 million in 1922 to around 24 million in 1937. During and immediately after the war cereal production fell and livestock-farming expanded; despite encouragement during the 1950s, the area devoted to wheat remained static at 5–6 million ha; maize declined to 2–3 million ha and flax to only 1 million ha. The area under maize in the famous district west of and within 150 km of Rosario was more than halved, and its population fell by 7 per cent between 1947 and 1960.

In contrast, the expansion of livestock-farming carried the cattle herds of the Pampas up to 35 million by 1960 (the rest of the country accounting for 9 million head). The weight of cattle slaughtered rose from about 1.7 million tonnes in the late 1930s to 2.4 million tonnes 20 years later. But the density of cattle and sheep fell, from 1.0 cattle units per ha in 1937 to 0.9/ha in 1960 (1 head of cattle = 5 sheep).

Despite the general extensification of agriculture in the period 1930–60, a number of improvements did take place: the quality of the livestock was progressively raised, mechanization increased labour productivity, and intensively grown 'industrial' crops rose in importance compared with cereals. Much of this change took place *outside* the Pampas.

Improving the quality of the cattle benefits the farmer in at least two ways: better prices are obtained and steers are ready for slaughter at an earlier age, thus indirectly increasing the productivity of the pastures. Evidence of the steady improvement of herds comes from the declining importance of *criollo* beasts, which accounted for 35 per cent of all cattle in 1914, 20 per cent in 1937 and 10 per cent in 1960. Again, of the cattle killed in the decades 1935–44 and 1955–64, the percentage of 1-year-old to 2-year-old steers rose from 6.2 to 10.9, whereas that of older steers fell from 47.6 to 41.2. The quality of cattle in the five Pampa provinces was already high (only 5 per cent were *criollo* in 1937), and breeding improvements were most marked elsewhere: in the plains provinces to the north and west the *criollo* percentage fell between 1937 and 1960 from 53 to 33 (*Figure 9.4a*). Further proof of improvement on the northern plains comes from the experience of the two firms, Bovril and Liebig, whose Entre Ríos canning factories earlier in the century provided competing outlets for the low-grade animals reared in the north. Upgrading of the herds, encouraged by the companies' own *estancias*, progressively reduced the supply of canning-grade animals to such an extent that in 1965 both factories were converted into *frigoríficos* capable of chilling and freezing as well as canning (*Table 9.5*).

Table 9.5 Steers graded in Entre Ríos and Chaco factories

	Year	Chilled (%)	Frozen continental (%)	Canning (%)
Entre Ríos factories	1943	4.3	18.9	76.8
	1953	14.0	66.4	19.6
	1963	21.4	65.9	12.7
	1977	92.4	6.5	0.2
Chaco factory	1964	6.8	56.7	35.1
	1977	59.7	39.2	0.6

Yet the northern provinces were still the source of most lower-grade animals slaughtered in the *frigoríficos* and factories, as can be seen from *Table 9.6*.

Unfortunately, as the *criollo* strain of the general herds was progressively replaced by that of temperate beef breeds, the animals' resistance to the tick and fly-borne diseases encountered in the humid, subtropical conditions of the extreme north-east declined. In 1938 the manager of Liebigs' Garruchos *estancia*, Finch, introduced a Zebu strain to the Hereford herds, leading to improved health without loss of quality; subsequently the Zebu-cross has spread slowly south and westwards as farmers who once favoured pure-breds came to appreciate the merits of hybrid vigour (*Figure 9.4a*).

A more important change of breeds since the war affected the Pampa herds whose dominantly shorthorn characteristics were being bred out with the spread of Aberdeen Angus: the latter is an earlier maturing producer of the lean beef that became preferred to the fatter meat of the shorthorn.

Other cattle improvements were achieved through vigorous government campaigns to control and eradicate tick fever and foot-and-mouth disease, but brucellosis and bovine tuberculosis remained to be tackled.

Sheep-farming, too, underwent changes, not so much through general upgrading, for even in 1937 only 6 per cent of the sheep were of *criollo* strain, but by the adoption of superior breeds more suited to the regional environments (*Figure 9.4b*). Among the dual-purpose breeds, Lincolns and Romney Marshes declined in favour of Corriedales, whose wool is fine cross-bred rather than coarse cross-bred, and whose less fatty mutton found a readier market. Similarly, the Argentine merinos yielded place to the finer-fleeced Australian merinos. The dominance of dual-purpose breeds in three of the sheep-rearing areas reflected their accessibility to coastal *frigoríficos*, whereas merinos predominated in the semi-arid hills of the remote Pre-Cordilleras of north-west Patagonia. Mutton

Table 9.6 Provincial origin of steers purchased by *frigoríficos* and factories 1964 (%)

Grade	Chilled			Frozen conti- nental	Canning			Chilled			Frozen conti- nental	Canning	
Province	1	2	3	4	5	6		1	2	3	4	5	6
Buenos Aires	39	37	20	4	–	–	100	63	53	32	9	1	3
La Pampa	38	39	20	3	–	–	100	9	8	4	1	–	–
Córdoba	25	39	29	7	–	–	100	5	7	6	2	1	1
Entre Ríos	13	24	40	22	1	–	100	13	21	39	31	13	14
Santa Fé	20	24	24	30	2	–	100	11	11	13	23	21	20
Corrientes	–	1	25	68	5	1	100	–	–	6	25	27	13
Chaco	–	–	9	74	13	4	100	–	–	–	4	11	17
Formosa	–	–	1	67	26	6	100	–	–	–	5	28	32
								100	100	100	100	100	100
Million kg of meat:								47.6	54.1	48.7	34.0	2.2	0.5

and lamb, however, were neither popular at home nor important to the export trade – in 1964 only $9 million-worth of mutton and lamb were exported, compared with $129 million-worth of wool and $228 million-worth of beef.

A second, well-established trend in Argentine agriculture was the mechanization of production. Despite inducements from successive governments to farmers to invest more capital in fixed, rather than mobile assets – in land productivity rather than labour productivity – the latter were for various reasons preferred: the rural exodus obliged many farmers to mechanize, owners often chose to cut their labour force for political reasons, and insecure tenants feared the loss of fixed capital. The number of tractors in use rose from 29,000 in 1947 to around 90,000 by 1959. Yet mechanization did not merit the same priority as other reforms, for even in 1952–4 the output per rural inhabitant in Argentina was on a level with Britain's and higher than France's (though wages hardly reflected this).

Most striking of the changes was the rise of 'industrial' farm produce, fruit and vegetables (*Table 9.7*).

Table 9.7 Argentina: structure of crop production 1935–9 to 1980 (by value)

	Cereals and flax	*Industrial crops (incl. oilseeds)*	*Fruits*	*Vegetables*	*Total*
1935–9	68.0	23.2	2.9	5.9	100.0
1950–4	35.7	38.6	7.1	18.7	100.0
1975–9	46.5	38.9	5.1	9.5	100.0
1980	42.1	44.1	5.8	8.0	100.0

Although the Pampas participated in this development through the rise of oilseed production and of urban-oriented market-gardening and dairying, the greater changes, in part for climatic reasons, took place elsewhere. Some developments represented an import-substitution agriculture associated with processing industries catering for the domestic market; others provided new export crops; virtually all led to more intensive forms of land use. By the mid-1950s the rest of the country with only 11 per cent of the improved land was accounting for 32 per cent by value of the agricultural output, but only 8 per cent of the consumption, whereas the Pampa provinces were consuming more than they produced.

On the Pampas dairying became more important and more intensive. 'Holando-Argentino' (Friesian) cattle increased from 2 per cent of Argentine herds in 1937 to 14 per cent in 1960, and milk products tripled in volume. More than two-thirds of the industry was to be found in the old colony area of Santa Fé and Córdoba, where dairying partly replaced grain production (*Figure 9.4a*).

The production of edible oils from sunflower seeds, groundnuts and olives increased greatly in the cereal-growing districts, where wheat, maize and flax all declined in area without any compensating increase in yields. The sunflower could be cultivated with the same machinery as cereals, and could be grown as a summer crop after the December harvest of wheat or flax, or after the winter grazing of fodder crops, thus making fuller use of the permanent labour force.

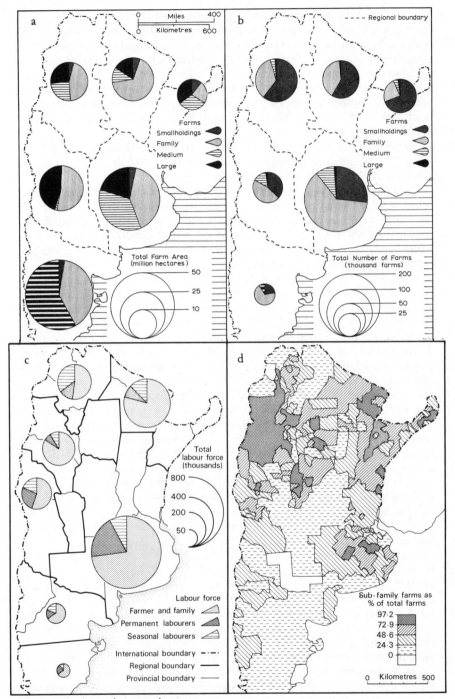

Figure 9.6 Argentina: agrarian structure
(a and b) Scale of enterprise 1960
(c) Agricultural labour force 1969
(d) Distribution of smallholdings 1969

The area devoted to the sunflower first surpassed 100,000 ha in 1935 and reached a peak of 1.8 million ha in 1948; thereafter disease led to reductions in both yield and area and to its partial replacement by groundnuts. The decline of flax-growing was also reversed, for the oil-pressing factories could handle equally sunflower seed, groundnuts, linseed and olives. Four-fifths of the olive trees, however, were grown in Cuyo rather than on the climatically marginal northern Pampas. Cultivation expanded greatly after 1932, when official encouragement began and fewer than 50,000 trees existed; by 1960 4.6 million trees were in production.

The level of production of edible oils thus rose tremendously from 1926–30, when 21,000 tonnes of oil were produced yearly and 48,000 tons imported, to reach 280,000 tonnes in 1960 when almost one-quarter was exported.

Outside the Pampas intensive crop production was concentrated in three broad areas: the old oases of Cuyo and the north-west, the new forest colonies of the Chaco and Misiones, and the newly irrigated lands of northern Patagonia and southern Mendoza. Certain crops were associated particularly, though not exclusively, with one or other of these regions, hence production problems and regional problems frequently coincided.

Long before 1930 the vineyards of Cuyo, together with the canefields of Tucumán, led the way in the development of a protected import-substitution agriculture; nevertheless, growth continued, and by 1960 over 235,000 ha were devoted to vines (75 per cent in Mendoza, 15 per cent in San Juan).

The sugar-cane area also grew, from 190,000 ha in 1941 to 310,000 in 1956. Average yield rose too, from 28t/ha in 1941/45 to 39t/ha in 1956–60, partly through the adoption of better varieties and partly through the establishment of new plantations in Jujuy and Salta, where climatic conditions are more favourable than in Tucumán, the traditional area of production.

Cotton, the leading product of Chaco Province, was first grown in the First World War and occupied 337,000 ha in 1941. With government help, the area sown expanded further to 732,000 ha in 1958. Since the harvest season in the Chaco is wet, yields and quality were poor, yet other crops were unable to compete, resulting in monoculture with all its disadvantages.

Misiones, in contrast, enjoyed the development of a more diversified pattern of farming on its small farms. In 1960 over half grew yerba maté, maize and manioc, and kept cattle, pigs and poultry; and over a quarter produced tung nuts, oranges and tea. Yerba maté, grown on 12,000 of the 19,000 farms, was the leading crop in 1960. Government encouragement from 1926 provided the economic foundation of the growing colonies, and the area planted rose from 9000 ha in 1925 to 63,000 ha in 1936, and to 100,000 ha by 1960. The second product was Chinese wood oil (tung oil), used in high-quality paints and obtained from the nuts of the tung tree, which was introduced by Liebigs in 1929. American reluctance to trade with Communist China, previously the world's chief supplier, encouraged expansion to 50,000 ha in 1958. Misiones' climatic similarity to southern China also favoured the establishment of tea-growing, whose extent rose from 390 ha in 1944 to 30,000 ha by 1960.

Far to the south-west, the twentieth-century irrigation schemes around San Rafael in Mendoza province and along the Rio Negro in northern Patagonia were associated with the rapid expansion of temperate fruit and vegetable production.

Once the static domestic demand for about 90 kg of fresh fruit per head was satisfied, further expansion depended largely upon processing for consumption at home and abroad and upon direct exportation. By 1960 scarcely 1 per cent of the oranges, only a tenth of the apples but half the plums went to the factory; another 40 per cent of the apples were packed for export and these constituted three-quarters of all fruit exports. The San Rafael district grew most of Argentina's peaches, quinces and damsons and many of her pears, apples and plums. Distance made development dependent upon industrialization, upon the canning and drying of fruits and the manufacture of cider, sweets and jams. The Río Negro valley accounted for 70 per cent of Argentine apple production, which rose from 89,000 tonnes in 1941 to 431,000 tonnes in 1960.

The growth of urban occupations 1930–60

In Argentina the process of urbanization has *not* been associated primarily with industrialization, least of all during the Perón era when the development of import-substitution industries was a major feature of government policy. Between 1925–9 and 1940–4, of 1,230,000 new jobs created 440,000 were in commerce and services, 420,000 in manufacturing, 300,000 in agriculture and 70,000 in other sectors. Between 1940–4 and 1955, 1,830,000 further jobs were created, only 350,000 of them in manufacturing and 80,000 in agriculture, but 970,000 were in commerce and services (see *Table 9.8*). During this latter period the proportion of the labour force that was engaged in manufacturing actually fell.

In 1960 Greater Buenos Aires employed 45 per cent of the workers in manufacturing and the provinces of Buenos Aires, Córdoba, Santa Fé and La Pampa a further 35 per cent. In commerce and the services Greater Buenos Aires was more dominant with 52 per cent and 49 per cent respectively to the 28 per cent and 26 per cent of the Pampas.

In the 1930s and the war years the growth industries were those using local agricultural produce – foodstuffs, tobacco, textiles and leather goods – and self-sufficiency had been achieved in most of these branches by 1950 (*Table 9.10*). The location of such activities varied: where processing resulted in greatly reduced bulk – as in the production of sugar, wine, edible oils, cotton fibre and *quebracho* extract – factories arose in the country districts, frequently affording the only non-agricultural employment, albeit seasonal. Such establishments, characterized as 'plantationist' by di Tella (1965), were associated with strong social cleavage between a paternalist management and an unskilled labour force. Other processing industries, such as textiles, clothing, footwear and brewing, were largely concentrated in Greater Buenos Aires, as were such import-based industries as the manufacture of rayon-acetate, synthetic rubber and pharmaceuticals. The Depression years were weathered with unemployment rates rarely reaching 5 per cent, and by 1944 the value of manufactures exceeded that of agricultural produce.

During the Perón regime the 1947–51 five-year plan sought to develop, largely with imported parts, the production of consumer durables (cookers, refrigerators, radios), demand for which was stimulated by low food prices and high wage rates, the wage share of the national income rising from 45 per cent to 56 per cent. At first the necessary capital goods, fuels and intermediate products

Table 9.8 Argentina: occupational structure 1900–4 to 1965–9

Sector	1900–4 ('000)	(%)	1925–9 ('000)	(%)	1940–4 ('000)	(%)	1955 ('000)	(%)	1960 ('000)	(%)	1965–9 ('000)	(%)
Agriculture	783	39.2	1539	35.9	1838	33.3	1916	26.1	1496	18.6	1358	16.8
Manufacturing	396	19.8	890	20.8	1310	23.7	1655	22.5	2166	26.9	1951	24.2
Commerce and services	616	30.9	1377	32.1	1821	33.0	2786	37.9	3205	39.7	4195	52.0
Transport	92	4.6	218	5.1	248	4.5	434	5.9	572	7.1		
Utilities	15	0.8	52	1.2	85	1.5	142	1.9	75	0.9		
Mining and construction	94	4.7	212	4.9	215	3.9	415	5.6	546	6.8	569	7.0
	1996	100	4288	100	5517	100	7348	100	8059	100	8073	100

Table 9.9 Annual output of leading industrial products for selected 5-year periods 1934–8 to 1972–6

Product	Unit	1934–8	1946–50	1954–8	1962–6	1972–6
Beef and veal	Million tons	1.61	1.94	2.29	2.34	2.35
Wheat flour	Million tons	1.5	1.9	2.2	2.2	2.4
Sugar	Million tons	0.41	0.6	0.77	1.03	1.30
Wine	Million hl	6.8	9.6	14.3	19.9	na
Cotton yarn	1000 tons	21.7[1]	69.4	95.6	86.2	na
Wood pulp, mechanical	1000 tons	–	9.0	16.0	27.0	34.0
Cement	Million tons	0.9	1.4	2.1	3.0	5.3
Pig iron	1000 tons	–	17.0	33.0	518.0	1019.0
Crude steel	1000 tons	–	131.0	226.0	1095.0	2055.0
Tyres	Million units	c.0.5	0.76	1.11	2.47	8.31
Motor cars	1000 units	–	–	15[2]	114	179
TV sets	1000 units	–	–	69[2]	133	233
Petrol	Million tons	0.75	1.22	1.73	3.0	na
Electricity, public	Million GWh	2.1[1]	3.9	6.3	10.2	23.0

Note: Averages for other periods [1]1935–8, [2]1957–8.

were imported at especially favourable exchange rates, made possible by obliging agricultural exporters to sell at less-than-world prices through IAPI (the Argentine Trade Promotion Institute). By the early 1950s falling commodity prices forced IAPI to support exporters, and capital from abroad was now sought.

By 1954 nearly all consumer goods were made in Argentina, by a million workers in 150,000 factories. The majority of plants were under-capitalized, inefficient and small. In only two industries, petroleum products and cigarette-making, did the average establishment employ over twenty-five workers; indeed, in the clothing industry factories accounted for only 15 per cent of output, the rest being domestically produced.

By location, industry was heavily concentrated along the Santa Fé–La Plata axis. In 1960 83 per cent of the industrial output and 77 per cent of the industrial labour force was contributed by the capital and the provinces of Buenos Aires, Santa Fé and Entre Ríos. Greater Buenos Aires alone was responsible for 60 per cent of national output. The provinces of Córdoba, Mendoza and Tucumán contributed a further 5, 3 and 2 per cent respectively.

The locational dominance of the metropolis may be attributed to many factors: its labour supply and consumer market, its focal position in the rail net-work, its primacy as a port for imports and its unrivalled facilities for the supply of energy, as well as the intangible advantages of superior health and educational services and of proximity to the centre of government and administration.

One result of Perón's policy was to reduce the importance of the so-called 'vegetative' and relatively self-contained industries in favour of 'dynamic' branches with their potential for stimulating complexes of linked activities (*Table 9.10*). But neither private nor public capital was forthcoming for the adequate development of the necessary, but less profitable, capital-goods industries, and growth was unbalanced.

Investment in energy, roads, and railways also fell short of industry's needs.

Table 9.10 Argentina: industrial structure, 1930–4 to 1977 (% of total value)

Period	'Vegetative' industries			'Dynamic' industries		
	Total	Selected branches		Total	Selected branches	
		Food and tobacco	Textiles, clothing and leather		Chemicals and rubber	Metals and machinery
1930–4	70.4	26.2	14.3	29.6	9.9	8.0
1935–9	67.1	28.8	16.7	32.9	8.2	16.2
1940–4	67.8	27.6	20.6	32.2	9.7	13.7
1945–9	65.1	24.1	21.3	34.9	7.8	17.9
1950–4	57.8	24.8	23.6	42.2	13.1	19.8
1955–9	50.5	22.5	19.9	49.5	14.1	25.9
1960–4	42.3	20.3	15.2	57.7	15.9	33.2
1965–9	38.3	18.7	13.4	61.7	17.5	35.5
1970–1	33.5	16.7	12.1	66.5	19.8	37.9
1977	na	15.3	11.4	na	20.0	40.9

Energy consumption increased (*Table 9.11*), but largely by way of oil imports, which grew from 9 per cent of all imports before the war to 22 per cent in the 1950s. Domestic oil production doubled, but its share in consumption fell from 79 to 50 per cent. Hydroelectricity remained unimportant, accounting for only 11 per cent of electricity output by 1960, although the 1947 programme for developing the resources of the Sierra de Córdoba and the Andes of Mendoza aided the industrial growth of those provinces. The installed public electricity capacity grew only from 1300 MW in 1943 to 2300 MW in 1960, and 30 per cent of that increase was attributable to the San Nicolas oil-fired station, opened in 1957. With public supplies lagging behind needs, private sources had to expand, accounting for 1200 MW of capacity in 1960 and one-quarter of the total production of 10,500 GWh.

Table 9.11 Argentina: energy consumption 1935–9 to 1970

Period	Total (m.t.o.e.)[1]	Solid fuel (%)	Oil (%)	Natural gas (%)	Vegetable matter[2] (%)	Hydro (%)
1935–9	8.5	23.6	45.8	4.1	26.0	0.4
1940–4	9.8	8.9	43.0	4.1	43.5	0.5
1945–9	10.8	8.9	56.4	3.6	30.5	0.6
1950–4	13.0	9.1	70.9	4.2	15.2	0.6
1955–9	16.3	6.1	77.3	4.3	11.1	1.2
1960–4	20.0	4.6	71.5	12.3	10.0	1.6
1965–9	25.3	3.0	70.2	16.9	8.4	1.5
1970	29.4	3.0	70.1	17.8	7.1	1.3

Notes: 1 Million tonnes of oil equivalent.
2 Wood and crop residues.

For many decades the railway network was steadily decapitalized. By 1960 60 per cent of the track was over 40 years old and a further 20 per cent more than 20 (25 years is the maximum for high-density traffic); 96 per cent of the network was single-tracked and on only 5 per cent were speeds in excess of 100 km/h

permissible. A continuing handicap to the regional exchange of industrial products lay in the three different gauges – broad, standard and narrow – whose
lines served, respectively, the south and west, Mesopotamia, and the north and
north-west. The Santa Fé–La Plata axis was the only district served by all three
systems. Until 1932 roads had been built simply as feeders to the railway, and in
the following 30 years annual construction of paved roads averaged only 165 km,
and of other improved roads only 800 km. Whereas the terrain of the Pampas
favours railway-building, highway construction is handicapped by the absence of
road metal (*Figure 9.3d*).

The move toward balanced development from 1960

During the 1950s there had been a marked decline in the apparent importance of
imports, from 37 per cent of GDP in 1925–9 to 11 per cent in 1955–9. Yet economic dependence on imported fuels and capital goods had greatly increased.
Only through massive investment in energy supplies, transport and capital-
goods industries could a sounder modern economy be created. Under the
programme initiated by the Frondizi government in 1958, the highest priority
was given to the expansion of the national petroleum and natural-gas industry
and to the ancillary development of refining and petrochemicals. Next came the
iron-and-steel industry, road transport improvements (through highway construction and the expansion of vehicle building) and increased electricity production. Lastly, encouragement was to be given to the manufacturing of cellulose,
paper, cement and machinery. Participation of foreign capital in authorized
projects was encouraged by guaranteeing it equal rights with domestic capital.

Such policies, however, failed adequately to recognize that unless and until
industry could replace agriculture as the nation's chief source of foreign earnings,
the economy would continue to depend on that sector for both the servicing of
debts and the purchase of capital goods. Yet policies until 1976 continued to
deter rather than to encourage agriculture, and its progress has resulted more
from improvements in the terms of trade during the 1970s and from increased
domestic demand. The Pampas have been more able to respond to these new
opportunities than have the interior regions. The latter, with some exceptions,
are now in decline, and agricultural colonization, so long central to Argentine
development, is almost at an end.

Energy

Despite the discovery of new oilfields, Comodoro Rivadavia has furnished over
half of the Argentine production since its discovery in 1907, but its share has
fallen in recent years to 36 per cent in 1976, compared with 28 per cent from the
Neuquén–Río Negro–La Pampa field, 25 per cent from Mendoza and 5 per cent
each from Salta and Tierra del Fuego. With dwindling on-shore reserves attention is turning to the prospects of the continental shelf which stretches to the
Falkland Islands. Although each oilfield has its refinery, four-fifths of the
capacity is located between San Lorenzo and La Plata, which are equally accessible to imported, coastwise and piped supplies (*Figure 9.8c*).

With the help of foreign contractors oil production has risen; averaging 24
million m^3 in the 1970s it has normally contributed most, if not all, of national

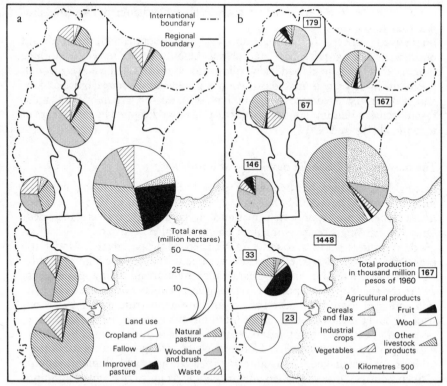

Figure 9.7 Argentina: land use and agricultural production
(a) Main land use types 1969
(b) Main types of agricultural production 1980

needs. The expansion of oil and natural-gas production and the construction of pipelines have also facilitated the use of both fuels for electricity generation, especially along the Litoral axis but also in the areas of extraction (e.g. Mendoza, Salta) and along the pipeline routes (e.g. Córdoba, Tucumán). Since 1960 massive investment, helped by international loans, has increased the public sector generating capacity from 2300 to 7900 MW in 1977, with a corresponding rise in production from 7900 to 26,300 GWh. Private production has levelled off since 1970 at about 4900 GWh.

The greatest single contribution to electricity supplies now comes from the 1200-MW Chocón hydroelectric scheme on the River Limay in North Patagonia. Coming on stream in 1973 it provided 3000 GWh in 1977 or 12 per cent of the nation's public supply. Under construction are the 1000-MW Alicura project further up the Limay, as well as the joint schemes with Uruguay and Paraguay, on the River Uruguay at Salto Grande (1600 MW) and the River Paraná at Yaciretá (2700 MW) respectively. When all schemes under construction in 1977 are completed, total public capacity should rise to 17,400 MW, 56 per cent of it being hydroelectric.

Transport
A major programme of paving arterial roads had by 1977 created a 41,000 km network of highways, which not only duplicates the 39,000 km rail system, but also offers the speed and flexibility of which the latter is incapable (*Figure 9.3d*). Bridges constructed in the 1970s across the River Paraná at Zárate and Resistencia and a tunnel at Santa Fé, for the first time allow rapid transport of perishable produce from Mesopotamia to the major markets. Other bridges across the River Uruguay to Fray Bentos, Paysandú and Salto afford new international road links.

The number of motor vehicles rose greatly from 866,000 in 1960 to 2,912,000 in 1974. Road transport has superseded rail in the long-distance movement of both freight and passengers. In 1971 rail freight accounted for 14,000 million t/km and road freight for 33,000 million. In 1975 inter-provincial motor coaches carried 67 million passengers, but long-distance trains only 34 million (and planes 5 million).

Population changes 1960-70: the rural exodus
By 1970 the total population had risen to 23.6 million, of whom 68 per cent lived in the capital and the three main Pampa provinces. Since 1960 the annual increase has averaged only 1.4 per cent, among the lowest in Latin America. With a birth-rate of 22.0 per thousand, a death rate of 8.6 per thousand and male and female life-expectancies of 64 and 71 years, Argentina has demographic characteristics similar to those of most developed countries.

The urban population reached 18.4 million in 1970 or 79 per cent of the total. About 11 million town-dwellers were concentrated along the south-western bank of the Paraná–Plate in Santa Fé (245,000), Rosario (807,000), Greater Buenos Aires (8,436,000) and La Plata (479,000) and in the smaller centres in between; a further 3 million resided in other Pampa cities including Córdoba (791,000), Mar del Plata (302,000) and Bahía Blanca (182,000). The rest of the country had only two large cities, Mendoza (471,000) and Tucumán (366,000). Although the metropolis overshadowed all other cities and was expected to surpass 10 million in 1980, its share of the urban population continued to decline slightly, from 46.3 per cent in 1947 to 45.7 per cent in 1970, whereas the share of other cities of over 100,000 rose from 20.0 to 24.9 per cent.

During the 1960s the rural population declined by 700,000 to 4,903,000, and rural depopulation spread from the Pampa provinces to all but Formosa and Misiones. Internal migration also showed new trends: of the 4.9 million migrants recorded in 1970, 60 per cent had moved to or from Greater Buenos Aires, the influx of 2.6 million far exceeding the outflow of 316,000. Though the Pampa provinces remained the chief source of migrants to the metropolis, even remote Chaco and Misiones were now losing to it more than they gained from elsewhere. In contrast, migrants to adjacent provinces now totalled only 31 per cent of all migrants.

The immigrant share of the population continued to fall, to 9 per cent by 1970. Italy remained the country of origin of the largest group, with 29 per cent, but Argentina's neighbours now accounted for 24 per cent of the foreign-born, and indeed supplied 76 per cent of the decade's inflow. By 1970 212,000 Paraguayans, 134,000 Chileans, 93,000 Bolivians, 51,000 Uruguayans and 45,000 Brazilians resided in Argentina. No longer were the majority of South

American immigrants to be found in provinces adjacent to their homelands, for the metropolis and other big cities had become their main destination.

Industrial changes

As a result of the government's promotion of technologically sophisticated industries with high capital investment, major changes have taken place in the structure of industry. Manufacturing's share of the GDP grew from 27.9 per cent in 1950 to 36.8 per cent in 1975-7, whereas its share of total employment fell from 27.9 per cent to 25.3 per cent (see also *Tables 9.3* and *9.8*). In detail the heavier 'dynamic' industries have replaced the lighter 'vegetative' industries as the dominant component (*Table 9.10*). The balance between non-durable and durable consumer goods, intermediate and capital goods industries had by 1970 already come to resemble that of the most developed nations (*Table 9.12*).

Table 9.12 Argentina and selected developed nations: industrial structure (% of industrial product)

	1950 Argentina	1970 Argentina	Mid-1960s W.Germany	USA	Japan
Non-durable consumer goods[1]	63	37	35	31	36
Intermediate goods[2]	21	32	29	29	32
Capital and durable consumer goods[3]	16	31	36	40	32

Notes: 1 Including: food and drink, textiles, furniture, printing.
   2 Including: paper, chemicals and petrochemicals, non-metallic minerals, basic metals.
   3 Including: metal products, machinery, equipment.

Foreign capital has played an important part in the establishment of new industries – between 1958 and 1971 foreign investment was heavily concentrated in a few branches: 29 per cent went into chemicals and petrochemicals, 26 per cent into the vehicle and auto-parts industries and 9 per cent into other machinery. By origin, 39 per cent was from the USA and most of the remainder from Switzerland, The Netherlands, West Germany, UK, Italy and France. Of the 393 firms established, 55 per cent chose locations in Greater Buenos Aires.

   With their need for large greenfield sites the new heavy industries have shunned the congested federal capital, preferring its suburbs or sites on the right bank of the Paraná or in the inland city of Córdoba; rarely have they ventured far from the metropolis. Thus, in 1974 the capital and the three Pampa provinces of Buenos Aires, Córdoba and Santa Fé accounted for 92 per cent of all employment in heavy industry, but only 79 per cent of light industry's labour force (compared with 69 per cent of the population). Similarly, 87 per cent of the 2300 factories which employed over 100 workers in 1978 were located in the same area (*Table 9.13*). In the rest of the nation opportunities for employment in such large-scale factories were still mainly confined to agricultural-processing industries with little growth potential. Whereas output between 1960 and 1977 had tripled in the chemicals, basic metals and machinery industries, it had grown by only 50 per cent in the older food, textiles and clothing industries.

   The Pampa provinces have thus been almost the sole beneficiaries of the recent spread of industry from the metropolitan growth pole (*Table 9.13*).

Table 9.13 Argentina: distribution of industrial employment, 1974

Group of provinces	Industrial employees ('000)	Percentage distribution by industrial sector (international industrial classification group)[1]								Percentage distribution by size of firm[1] (Number of employees in firm)		
		31	32	33	34	35	36	37	38	1–5	6–100	>100
Federal capital, Buenos Aires	1096	15	20	5	5	13	5	5	30	12	37	51
Córdoba, Santa Fé	277	24	8	5	3	6	9	6	38	20	34	46
Entre Ríos, Corrientes	34	56	8	9	3	1	13	1	8	26	35	39
Misiones, Chaco, Formosa, Santiago del Estero	42	32	11	27	4	10	13	1	5	32	39	29
Salta, Jujuy, Tucumán	70	53	9	8	4	2	5	7	11	11	22	67
Catamarca, La Rioja	4	55	4	13	5	0	16	1	5	42	55	3
Mendoza, San Juan, San Luis, La Pampa	63	50	3	7	3	8	13	1	14	25	42	33
Neuquén, Río Negro, Chubut, Santa Cruz, Tierra del Fuego	25	24	18	21	3	6	10	1	14	18	42	40
Argentina	1610	21	17	6	5	10	7	5	29	14	36	49

Key to industrial groups: 31 food, drink, tobacco; 32 textiles, clothing, leather; 33 woodworking; 34 paper, printing; 35 chemical products; 36 non-metallic mineral products (bricks, glass, ceramics); 37 basic metal industries; 38 metal products, machinery, equipment.

Note: 1 Data for 1978.

The petrochemical industry has grown rapidly since the Decree of 1961, encouraging the establishment and expansion of basic plants and integrated by-product plants. Synthetic fibres, plastics, pvc, rubber, fertilizers and pesticides of wholly national origin are now in production. The new plants have been located outside the major industrial cities, chiefly along the Santa Fé–La Plata axis (*Figure 9.8b*). Large complexes at San Lorenzo and Campana lie adjacent to refineries and astride the Salta–Buenos Aires natural-gas pipeline, which is also available to lesser plants at Zárate and Pilár. A third complex is associated with the refineries of Ensenada and La Plata. Beyond this axis two plants at Río Tercero near Córdoba are fed by an offshoot of the gas line. In 1977 the chemical industries as a whole employed over 150,000.

The growing iron-and-steel industry is located mainly on the same axis. The San Nicolas works began production in 1960, and other plants have been built at Villa Constitución and Ensenada. Production of crude steel (0.3 million tonnes in 1960) rose to 2.4 million tonnes in 1976, meeting 85 per cent of needs. But until the development of large deposits at Sierra Grande in northern Patagonia, the industry rests heavily on imported iron, iron-ore and scrap.

A large proportion of the sheet steel is destined for the motor-vehicle industry. Although a jeep factory was built at Córdoba in the early 1950s, major growth awaited the Promotion Laws of 1959. By 1972 some 269,000 vehicles a year were being produced, almost three-quarters of them cars and the rest commercial vehicles. At first twenty-one factories were established by such companies as Fiat, Renault, Ford and General Motors, each with a limited and therefore costly production. Rationalization by the early 1970s had reduced their number to 10, located in Buenos Aires, Córdoba and Santa Fé provinces; yet profitability remained elusive, and by 1976 total vehicle production had declined to 196,000.

In recent years the annual production of household durables has levelled off, with domestic demand largely satisfied. In 1971–6 annual production averaged 227,000 refrigerators, 178,000 washing machines, 230,000 TV sets and 69,000 sewing-machines. Tariff and other barriers protecting such industries have subsequently been lowered, with serious consequences.

As Argentina enters the 1980s her industrial structure is far better balanced, yet many problems remain. Greater rationalization is needed if efficient and competitive enterprises are to emerge. In the more advanced branches the home market is too small to absorb output and exporting becomes a necessity. Happily, exports from the newer heavier sectors tripled between 1972 and 1977 from US $250 million to 779 million, though their share of total exports rose only marginally from 13 to 14 per cent. Exports of textiles and leather similarly tripled from US $225 to 694 million. Such developments are necessary if Argentina is to service its debts to foreign investors.

Agricultural reappraisals
Since the 1960s the need to reassess the role of agriculture has become clearer. The vital importance of agricultural exports began to be recognized again by government in the early 1950s, but it was not until the early 1960s that it confronted the problem of the over-valued peso, which both reduced the competitiveness of Argentine exports and made the home market more attractive. Political instability has also unfortunately precluded any coherent long-term

policy, but the rise in world commodity prices in the early 1970s afforded a needed external boost to agricultural recovery, at least in those regions capable of producing for export.

Conversely, production of crops for national industries has levelled off as the supply of those industries' products has attained or even exceeded the modest level of internal demand. Stagnation in the production of such crops, largely grown in the northern interior, has in turn highlighted the social problems inherent in the agrarian structures of those districts. Doubts must arise as to whether large numbers of rural people can – or should – continue to be supported at low levels of income through the use of labour-intensive methods of production on tiny holdings in a nation where 80 per cent live in towns at higher standards of living. Argentina's policy-makers, perhaps uniquely in Latin America, enjoy the options inherent in the EEC's Common Agricultural Policy paradox. Where the non-agricultural population becomes ever more dominant, it becomes increasingly capable of artificially supporting the residual rural poor, yet the social need for such support diminishes as urban opportunities expand. At present there is no sign that the welfare option is favoured.

The Pampas have now reasserted their paramount position thanks to their ability to respond flexibly to changing circumstances. That flexibility allows livestock rearing and fattening to be interchanged, or livestock and crop farming to be substituted the one for the other, or the crops grown to be selected from a wide range. Such versatility derives from the nature of climate and soil, from the ease and low cost of access to markets and ports (now greatly enhanced by the new, paved-highway network), and from farm sizes which are large enough not to require highly intensive and therefore costly modes of production. Valuing the opportunity of revising their production plans annually, few farmers in the Pampas are condemned to produce tree crops for little or no reward, since more profitable enterprises may usually be easily substituted.

Land use
In 1969, at the last available census, 210 million ha, or 75 per cent of continental Argentina, lay in farms. Of the farmland 44 per cent was devoted to natural pasture and 25 per cent to woodland; 15 per cent was waste or otherwise not in use. Only 17 per cent was relatively intensively used, 9 per cent being under improved pasture and a mere 8 per cent under crops. Physical conditions largely explained the great regional diversity: one-third of arid Cuyo was unused, and under 4 per cent was intensively used, with the aid of irrigation; 79 per cent of cold, semi-arid southern Patagonia was in natural pasture, and less than 1 per cent intensively used; in all the northern and less fertile regions nearly half the land remained under forest. Only in the Pampas had the greater part been improved, for 23 per cent was under sown pastures, 19 per cent under crops and of the 30 per cent under 'natural' pasture most had, at one time or another, been ploughed. Even the 28 per cent under wood or unused was chiefly concentrated on either the wetter north-eastern or drier south-western peripheries. The Pampas still possessed the major part of the nation's improved land: 82 per cent of its cropland and 93 per cent of its improved pasture.

Yet the Pampas accounted in 1980 for only 70 per cent by value of the nation's crop production and 85 per cent of livestock production, for farming in the peripheral areas continued to concentrate on more intensively grown products of greater unit value. Thus, other regions accounted for 58 per cent of the vegetables, 72 per cent of the industrial crops and 80 per cent of the fruit grown, whereas the Pampas produced 92 per cent of the less intensively cultivated cereals.

The agricultural economy

Farm production as a whole was in 1980 valued at 2064 million pesos (of 1960), equally divided between crops and livestock. Having resumed its growth in the late 1950s, output rose to 2280 million pesos in 1970–4, levelled off at 2250 million in 1975–9, before falling back with renewed over-valuation of the currency. In the same period the structure of crop production has to some extent moved back towards that of pre-war days (*Table 9.7*), with a resurgence of cereals and a decline in fruits and vegetables. Industrial crops (including oilseeds) have, however, continued to increase their share, which exceeded that of cereals for the first time in 1980.

The Pampas have benefited most from recent changes not only with cereals but also with new industrial crops, while production has stagnated or declined in some peripheral regions. The area of wheat harvested shows no long-term trend, but yields have risen consistently in their decennial averages from 1143 kg/ha in the 1940s to 1545 kg/ha in the 1970s (*Table 9.14*). More impressively, the area of maize harvested increased by a half between the 1950s and 1970s, and yields rose by 63 per cent in the same period from 1674 to 2722 kg/ha. Most outstanding of the changes to cereal production has been the spread of grain sorghum: insignificant in 1950, it rose to third place in area harvested by 1964; its yields rose from 1088 to 3033 kg/ha between 1951 and 1979. As an animal feeding stuff it enjoys a high demand at home and abroad. Resistant to drought, though susceptible to frost, it has been widely adopted in the maize area where the provinces of Buenos Aires, Córdoba and Santa Fé meet. It is also used as a grazing crop instead of alfalfa.

The most spectacular development yet witnessed has been that of soya beans. Long a minor crop which still occupied under 1000 ha in 1961 and under 100,000 ha in 1972, it covered over 2 million ha by 1980, its yield meanwhile doubling to 2 tonnes/ha. In value it became in 1979 the third most important crop after grapes and wheat, and the most valuable export after beef, worth US $718 million. Its wide use as a protein-rich human and animal foodstuff, is well-known. Unlike most field crops its cultivation also enriches the soil, making it ideal for rotations. Production is concentrated in southern Santa Fé, where it often replaces maize. Soya beans also promise a valuable alternative to the troubled crops of the warmer northern provinces: indeed, it is already significant in Tucumán and in Misiones, where its cultivation began.

Outside the Pampas, the Río Negro valley apple industry has continued to advance, with national production more than doubling since 1960 to reach almost 1 million tonnes in 1979 and 1980. The small farmers of the valley use modern techniques, and its packing stations are especially quality-conscious, reserving the best for export. In 1980 over 250,000 tonnes, valued at over US $200 million were exported, constituting the sixth most valuable export crop after wheat, soya, maize, sugar and sorghum. Brazil receives half and EEC one-third. Exports had, however, already achieved the 250,000 tonne level by 1965, suggesting a world market near saturation.

The area under vines has also continued to expand since 1960, levelling off in the late 1970s at about 340,000 ha, 94 per cent of them in Mendoza and San Juan. The grapes are destined almost wholly for wine, whose production has stabilized at about 22 million hectolitres, making Argentina fourth in the world. As yet only about 2 per cent is exported.

Further north, the fortunes of the sugar industry have fluctuated greatly. From 277,000 ha in 1956–60 the area under cane fell to 202,000 ha in 1966–70, though yields rose from 39 to 50 tonnes/ha. With only 11 per cent of sugar production finding a profitable export market, surpluses accumulated, and many mills in Tucumán sold off their cane lands to small growers in an endeavour to reduce costs. Thanks to some diversification, Tucumán's share of cane production fell from 67 per cent in 1960 to 60 per cent in 1966–70, whereas the share of Jujuy and Salta rose from 27 to 35 per cent. The world sugar shortage of the early 1970s and Argentina's subsequent participation in the quotas of the World Sugar Agreement have enabled renewed expansion to a record 360,000 ha under cane in 1977, and to the export of 34 per cent produced by the mills in 1973–7. Tucumán has shared fully in this expansion, its share declining only marginally to 59 per cent in 1980, though that of Jujuy and Salta has advanced to 39 per cent with the decline of minor producers like Santa Fé and Chaco. Thanks to irrigation, a warmer climate and greater adoption of technological improvements, yields in Salta and Jujuy have advanced to over 70 tonnes/ha compared with Tucumán's 40-plus tonnes/ha.

The second industrial crop of both Salta and Jujuy is tobacco, these two provinces accounting for 67 per cent of Argentina's total production in 1980 and for almost all the lighter tobaccos grown. Having doubled between 1941–5 and 1955–60, the total area cultivated grew even more rapidly to 93,000 ha in 1975. The subsequent sharp fall is entirely explained by the collapse in demand for the darker varieties, whose cultivation is concentrated in the north-eastern provinces.

Climate allows the same provinces to dominate in the production of early potatoes (Tucumán) and tomatoes (Salta and Jujuy), though the main crops come from the Pampas. The higher prices fetched by earlies offset the greater transport costs. A similar compensation may be seen in the distribution of citrus-growing: Tucumán, Salta and Jujuy account for 81 per cent of the lemons, 27 per cent of the grapefruit and 20 per cent of the oranges, compared with 11, 46 and 64 per cent respectively grown further south in Mesopotamia and the Paraná delta, where the cost of warding off frosts offsets the lower freights. Tucumán's domination of lemon production, however, also reflects a fourfold expansion since the mid-1960s through planned diversification away from sugar cane.

Cotton production, confined to Chaco and adjoining provinces, suffered a major decline with the ending of guaranteed prices in 1959, the area falling from a 1958 peak of 732,000 ha to 307,000 in 1968. Problems of the wet harvest season, soil exhaustion and erosion and crop disease led to a long decline in yields until the late 1950s. Subsequently yields have risen, initially through contraction to optimal locations; since 1968 the area has also increased, production achieving a record 700,000 tonnes of raw cotton in 1978; the spread of improved varieties and techniques and success with exports following the removal of duties explain this healthier position.

In neighbouring, but more tropical, Formosa, bananas have afforded a major alternative to cotton. Until 1960 Argentina's consumption had been satisfied almost entirely by the annual importation of about 200,000 tonnes. By 1977 Formosa alone was producing 224,000 tonnes and a further 109,000 tonnes were grown in Salta. By 1977 Formosa and Chaco had also diversified in other directions: in rice production the former ranked third and the latter fifth among producing provinces; more importantly, Chaco had become the fourth producer of sunflower and grain sorghum.

Despite its greater diversity of crops, Misiones suffers from the inflexibility inherent in the predominantly tree crops which its uniquely humid subtropical climate favours. Until recently it also suffered the restraints of poor accessibility, its unpaved highways feeding the costly and inefficient rail link to Buenos Aires. As a result high unit-value products, often achieved by local processing industries were the only ones that could yield a profit in distant markets. Indeed, in 1977, 78 per cent of the cultivated area of 450,000 ha was devoted to industrial crops. Relaxation in the 1950s of restraints on the planting of yerba, the province's main crop, led to an expansion to 147,000 ha in 1964. The foolishness of this step is reflected in the growing discrepancy between the area under cultivation and the area which the farmers were permitted to harvest, as new attempts were made to prevent over-production (*Table 9.14*). Nevertheless, a record production of 150,000 tonnes was recorded in 1973. In the late 1970s a respite from the secular decline in demand for yerba was afforded by the high world coffee and tea prices which led to renewed interest in drinking yerba maté.

Uncertainty has also characterized Misiones' tung-oil fortunes, with increased competition in world markets from other producers and from synthetic and cheaper substitutes. In consequence, the province's second industrial crop is now tea, the area harvested reaching 40,000 ha in the late 1970s, and the yield averaging about 4 tonnes/ha. Self-sufficiency had been achieved by 1960, and 80 per cent is now exported. High labour charges are offset by the mechanized methods of production for which many of the plantations were designed, permitting Argentina now to rank fifth among tea-exporting nations.

The beef industry

Since 1960 the beef industry has undergone revolutionary changes. The old pattern consisted of higher-grade steers moving to a few huge *frigoríficos* at locations accessible to ships from which chilled beef quarters were exported to the British market. Local slaughter-houses provided most of the domestically consumed beef, employing chiefly cows and inferior steers, though some also came from the *frigoríficos*. Additionally, up-country factories processed about 3 per cent of the cattle for corned beef. Today the old *frigoríficos* have all closed down, and new ones, more widely dispersed, have been established in response to new technologies, modes of transport and markets, with the slaughter-houses still serving traditional local needs.

The 1950s had seen the steady emergence of new *frigoríficos* of modest size catering to the home demand of the growing number of families with refrigerators, but in 1960 these still accounted for only 6 per cent of cattle slaughtered. Meanwhile the old *frigoríficos*' share had fallen between 1955 and 1960 from 53 to 35 per cent, whereas that of slaughter-houses had risen from 40 to 57 per cent

Table 9.14 'Argentina: crop areas and yields

Crop	1941–5	1946–50	1951–5	1956–60	1961–5	1966–70	1971–5	1976–80
				Area harvested ('000 ha)				
Wheat	5575	4627	4804	4694	4720	5331	4230	5224[3]
Maize	3308	2156	1956	2284	2836	3535	3467	2678[3]
Sunflower	824	1205	773	1063	920	1171	1226	1651
Flax	2055	1218	670	941	1149	804	533	668[3]
Grain sorghum	–	–	21	233	617	1173	2009	2171[3]
Soya beans	2	1	1	1	11	21	192	686[3]
Cotton[2]	363	439	564	650	599	421	487	577
Sugar cane	201	219	270	277	227	202	272	345[3]
Tobacco	17	24	35	34	44	58	76	73[3]
Tea	0	0	1	10	19	29	33	40[4]
Yerba maté	65	64	64	62	88	49	55	na
Vines[1]	142	165	195	227	253	286	324	342
Apples[1]	17	21	26	30	36	41	na	na
Oranges[1]	57	73	93	112	98	71	na	na
				Crop yields (tonnes/ha)				
Wheat	1.1	1.1	1.2	1.3	1.5	1.2	1.4	1.6[3]
Maize	2.0	1.7	1.6	1.8	1.8	2.2	2.5	2.9[3]
Grain sorghum	–	–	1.3	1.8	1.7	2.0	2.2	2.9[3]
Soya beans	0.9	1.0	1.0	1.0	1.1	1.0	1.5	2.0[3]
Cotton	0.8	0.7	0.7	0.7	0.8	0.9	0.9	1.0
Sugar cane	27.8	33.9	33.9	39.1	49.5	49.9	53.8	46.3[3]

Notes: 1 Area cultivated.
2 Area sown.
3 Average for 1976–8.
4 Average for 1976–9.

Table 9.15 Structural changes in cattle slaughter, 1956–78

	Total slaughter (million head)	Old frigorificos	New frigorificos	Slaughter-houses (% of total)	Regional factories
1956	11.2	53.1	5.4	38.0	3.5
1957	11.5	43.9	5.7	46.5	3.9
1958	11.9	38.4	6.4	51.7	3.5
1959	8.7	36.8	7.6	52.2	3.4
1960	8.5	34.6	6.1	56.9	2.4
1961	9.8	30.1	7.4	59.5	3.0
1962	11.4	26.9	15.9	54.5	2.7
1963	12.6	30.7	18.1	47.9	3.3
1964	9.0	25.8	22.6	48.5	3.1
1965	8.8	24.6	26.0	47.0	2.4
1966	10.7	26.0	23.1	48.0	2.9
1967	12.2	25.7	24.6	46.8	2.9
1968	12.5	23.9	27.9	45.4	2.8
1969	13.5	23.1	24.4	50.0	2.5
1970	12.6	17.3	29.4	50.6	2.7
1971	9.1	12.5	33.6	51.2	2.7
1972	9.7	14.8	37.1	44.8	3.3
1973	9.5	12.9	37.3	47.3	2.5
1974	9.8	13.5	37.3	46.8	2.4
1975	11.8	13.0	38.1	46.7	2.2
1976	13.5	10.7	36.7	49.7	2.9
1977	14.4	8.2	39.5	49.1	3.2
1978	16.1	8.6	42.6	45.9	2.9

(Table 9.15). Faced additionally by declining total slaughters, the government sought to release more beef for export by decreeing beefless weeks at home. This device simply boosted sales of domestic refrigerators and freezers and encouraged the creation and expansion of the new *frigoríficos*. By 1965 they accounted for 26 per cent of the kill, with the slaughter-houses falling back to 47 per cent and the old *frigoríficos* declining still further to 25 per cent.

In 1968 a severe outbreak of foot-and-mouth disease in Britain led to the banning of imports of beef on the bone from the River Plate. As in 1900 (p. 405), Argentina responded rapidly by substituting an acceptable product of higher technology – in this case vacuum-packed, boneless beef cuts – and the nation benefited from an export with higher added value *(Table 9.16)*. But the old *frigoríficos*, each designed simply to slaughter and refrigerate the carcases of a quarter-million and more beasts, and to export them to few destinations, were incapable of adjusting. The preparation of beef cuts required new facilities, and the growing world demand for hamburger beef required new kinds of processing. The new *frigoríficos* proved able rapidly to respond, and by 1972 were responsible for 37 per cent of cattle killed compared with the old *frigoríficos'* 15 per cent.

Table 9.16 The Argentine beef industry: total slaughter and types of beef exports

	1965–7	1970–2	1976–8
Total slaughter			
'000 head	10,910	10,801	15,030
'000 tonnes carcasse wt	2,279	2,272	2,974
Exported			
'000 tonnes carcasse wt	595	623	617
(%)	26	27	21
Beef exports			
'000 tonnes actual wt	488	451	438
Refrigerated actual wt	399	348	306
On the bone, % by wt	61	29	13
Cuts, % by wt	7	31	35
% chilled, by wt	38	24	6
Chilled exports			
% to UK by wt	65	32	9

Accession of Britain to the EEC served further to disadvantage the old plants, as well as to reduce the export market for Argentina's prime export, chilled beef. By the late 1970s chilled exports fell to insignificant levels, and the British share of the chilled market also declined in favour of West Germany and Switzerland. Greece and Spain emerged as the only major and, apparently, reliable purchasers of frozen beef, but the actual or impending accession of those countries to the EEC underscores the uncertainty of markets already upset by the disposal of surplus EEC beef at artificially low prices. The emergence in 1980 and 1981 of the USSR as the leading buyer of Argentine refrigerated beef (and wheat) has created a partnership of political as well as economic importance to both nations.

Despite uncertainties in the overseas markets, Argentina has succeeded in keeping up the level of its total beef exports, helped by the increased productivity of its cattle farmers, which has permitted both home and export demands to be met and all-time record production to be achieved. The new *frigoríficos* by 1978 had come to handle 43 per cent of all cattle killed, and the old, soon to disappear, took under 9 per cent.

Most new *frigoríficos* are located within Greater Buenos Aires, the market for most of their beef. Many others have been established in the rest of Buenos Aires Province, in Santa Fé and Córdoba, with few provinces still lacking at least one. Most strikingly, those establishments which cater chiefly for export markets have shifted into the interior, with Santa Fé, Córdoba and Entre Ríos *frigoríficos* being mostly export-oriented, and even some in San Luis and Corrientes (*Table 9.17*). This spatial shift in processing for export may be attributed partly to the expansion of frozen beef relative to chilled, and partly to the ease and speed of the road transport now widely available. The long-established factories in the interior have meanwhile also undergone further adjustments to circumstances. Cattle suitable only for canning have now disappeared even from the most northerly areas (*Table 9.5*), and corned beef must therefore come from inferior parts of better animals. Liebigs' Colón *frigorífico* also found it increasingly difficult to acquire the heavy freezing-grade animals preferred in their British market, as farmers gained more from selling lighter, leaner and hence more rapidly finished beasts to the home market. The bridging of the Paraná at Zárate, speeding livestock haulage to the capital, gave the *coup de grâce* to Colón, which closed down in 1980.

Table 9.17 Argentina: spatial changes in the refrigerated beef industry

		No. of frigoríficos	Slaughter ('000 head)	Slaughter ('000 tonnes carcasse wt)	Exports (% by wt)
Greater Buenos Aires	1958	4	1632	365	58
	1978	19	3690	635	34
Buenos Aires Province	1958	5	1250	290	67
(rest of)	1978	21	2130	337	36
Santa Fé	1958	1	421	91	82
	1978	10	1196	213	72
Entre Ríos	1958	4	565	126	95
	1978	7	530	96	98
Córdoba	1958	0	–	–	–
	1978	5	295	58	61
Corrientes, Chaco	1958	0	–	–	–
and Formosa	1978	3	204	34	28
San Luis	1958	0	–	–	–
	1978	3	327	62	46
Total	1958	14	3868	872	69
	1978	71	8500	1511	44

Sheep-farming
In the late 1960s and early 1970s the fall in wool prices led farmers to reduce their flocks from 48.5 million in 1960 to 34.7 million in 1974. Of the 14 million reduction, Buenos Aires accounted for 8 million, La Pampa for over 2 million and

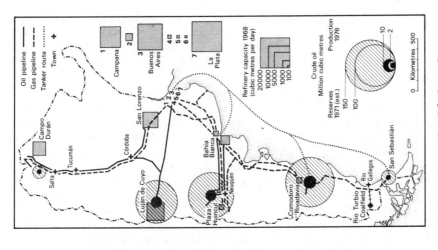

(c) Argentina: national fuel supply

(b) Argentina: litoral industrial axis

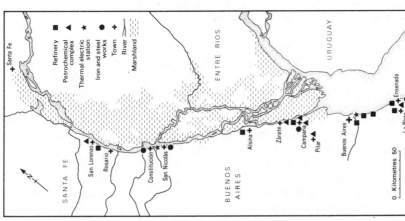

Figure 9.8 (a) Argentina: illiteracy 1977

other Pampa provinces for a further 1.5 million. Farmers in those provinces could often readily shift to more rewarding kinds of livestock or crop-farming. Yet even the flocks of the Patagonian farmers, who had no alternative livelihood, fell by over 1 million head. Overall, the shift southwards of Argentine flocks continued, with Buenos Aires's share falling from 39 per cent in 1960 to 31 per cent in 1974 (and 28 per cent in 1977), while Patagonia's share rose from 36 per cent in 1960 to 46 per cent in 1974 (and 49 per cent in 1977). Nevertheless, mutton and lamb production continues to be dominated by Buenos Aires, with its heavy, dual-purpose Lincolns accounting for 51 per cent of the sheep slaughtered in 1977.

The major consequence of this spatial shift has been the halving of production of the lower-value coarse wools, whereas the clip of finer grades has remained nearly constant for 30 years. Furthermore, as so often before, Argentina's response to the problem of declining world wool prices has been the intelligent one of opting for processes which added greater value before export. Thus, between 1968 and 1977 the proportion of unscoured and scoured wool in the total wool-based exports (measured by the original weight of wool) fell from 97 to 75 per cent, while combed tops increased from 3 to 19 per cent, and wool yarn and cloth rose from almost nothing to 6 per cent.

The adoption of innovations
As the 1960s and 1970s have advanced, the number of pointers to recovery, especially of the Pampas, has increased. Higher yields of major crops, dramatic in the case of maize, sorghum and soya beans, more modest for wheat, are evidence of the widespread adoption of improved varieties (*Table 9.14*). Overall, the productivity of the Pampas between 1960 and 1975 grew at an annual rate of about 2.5 per cent, similar to that of the prosperous early decades of the century and well ahead of the 0.4 per cent rate recorded between 1938 and 1960.

Livestock-carrying capacity depends upon the provision of adequate nourishment during winter cold and summer drought. In the cattle-fattening zone stretching south-west from Rosario into a Pampa, the traditional practice was to supplement natural grazing with alfalfa pastures in summer and cereal-grazing in winter. The so-called natural pastures of the Pampas are not rough grazings, but in fact comprise old alfalfa and stubble fields containing clover and ryegrass, which through neglect have become infested with thistles and other weeds. Recently, in this pattern of management, alfalfa has been replaced by sorghum and other summer cereals. It is possible that the record cattle slaughters of the late 1970s are in part attributable to a greater intensity of production, though increased consumer preference for younger, leaner beef has clearly played its part.

Mechanization may not directly increase yields, but the opportunity to carry out more and deeper cultivations and to harvest speedily at the right time indirectly increases productivity. Between 1959 and 1977 the number of tractors in use rose from 90,000 to 159,000, while their average power rose from under 50 to almost 80 hp. Encouraged by government concessions, the tractor-building industry, established in 1957, had by 1977 produced about 300,000 units. Their purchase has often been financed since 1963 by IDB credit, available at negative real interest rates; a one-third fall in real prices has also boosted sales. Much of the machinery is, in fact, operated by contractors who in 1972 handled half the crops harvested. Studies show that it is more economic for the farmer with under 300–400 ha to hire contractors than to use his own equipment inefficiently.

Improvements to Argentine agriculture have been greatly stimulated by the national agricultural research and advisory service (INTA) founded in 1957 and, locally, by progressive farmers' clubs (CREAs), which combine the local knowledge of farmers with the technical expertise of agronomists in carrying out large-scale trials and demonstrations of innovations.

Soil management, however, still remains remarkably under-developed. Soil erosion is serious: along the western edge of the humid Pampas 16 million out of 61 million ha are affected by wind erosion, a major factor in the rural exodus from La Pampa. In the north water erosion is serious on the deeply weathered soils of small farms in Misiones and the Chaco. The adoption of improved soil-management techniques is, in general, still in its infancy: deep-ploughing trials have been shown to retain moisture and greatly increase yields throughout the Pampas. The use of NPK (nitrogen, phosphate, potash) chemical fertilizers is still minimal in comparison with other countries, despite a doubling from 0.9 kg/ cultivated ha in 1961–5 to 2.1 kg in 1976. Thus New Zealand used 644 kg/ cultivated ha, UK 274, USA 107, Brazil 63; even countries with a similar kind of agriculture used far more: Canada 32, Uruguay 33 and Australia 24 kg. Furthermore, 73 per cent of the straight N and 78 per cent of the compound fertilizers were, in fact, used for intensively cultivated crops, such as cane, vines, fruit trees and vegetables, which occupy only 8 per cent of the sown area. In the most recent survey of fertilizer use in Argentina (*Información Económica*, No. 15, 1980), several reasons are adduced for the very low use of fertilizers on extensive crops: the incorporation of leguminous leys and soya beans in crop rotations provides adequate residual nitrogen for succeeding crops; the cost of nitrogenous fertilizers in relation to crop prices is two to three times higher than in other countries; the protective tariffs on Argentina's infant fertilizer industry, inflating prices, compare ill with the subsidies often received by farmers elsewhere; falling crop prices and dry weather, both of which can wipe out the expected gains from applying fertilizers, constitute too great a risk. These same factors also explain the wide use of fertilizers on intensive crops, which often enjoy abundant moisture from rainfall or irrigation, command high unit prices, and which frequently exhaust the soil through monoculture. However, it is still the case that adequately executed NPK – response studies remain too few, and that their findings certainly do not amount to an unqualified recommendation for fertilizer use: in several wheat areas the application of nitrogen will not repay the cost one year in two; in only one wheat area is the use of phosphates advisable. Only a widely available soil-diagnostic service capable of providing advice specific to precise locations could lead to a greater and rational use of fertilizers. Given the rising cost of chemical fertilizers and the high cost which a diagnostic service would incur, it remains debatable whether the average Pampas farmer in this respect is behind or ahead of the rest of the world.

Agrarian problems

As in many other Latin American countries, the large area occupied by great estates and the large number of under-capitalized smallholdings are salient features of the agrarian structure. In contrast to those countries, it is debatable whether any programme of land redistribution could solve the economic and social problems of the Argentine countryside, especially since the nature and gravity of those problems show great regional variation. Ownership, farm size and economic viability are different though often related aspects of the agrarian scene.

In recent years a major increase has occurred in the proportion of the land operated directly by its owners (*Table 9.18*). The Perón government began in 1948 by automatically extending all tenancy leases, a step repeated at later dates until in 1958 all tenants were granted the option of purchasing their holdings, agreeing to new rent contracts or forfeiting their leases. Inadequate credit facilities reputedly precluded the full implementation of this and succeeding similar laws. Yet census returns for 1947 and 1960 reveal that the percentage of land in owner-occupied holdings grew from 36 to 53 nationally (and from 35 to 62 in Buenos Aires, and from 46 to 70 in Santa Fé), largely through the decline in tenant holdings, but also through the granting of title to lessees of government lands, especially in Patagonia and the Chaco. It is unlikely that this increase in owner-occupation can be attributed simply to large owners recovering direct usage of portions of their estates previously rented off, since the average size of owner-operated holdings grew only slightly from 362 ha to 399 ha nationally, and not at all in the Pampas. Between 1960 and 1969 an even greater expansion of owner-occupation took place by the further granting of titles to state lands and by the conversion of mixed forms of tenure (part-owned, part-tenanted holdings) to simple ownership. However, tenancy and more insecure forms of occupation continued to account for 13.4 per cent of the farmland and, given that such holdings tend to be small, for a larger proportion of actual farms.

Table 9.18 Argentina: land-tenure changes 1947–69

	1947		1960		1969
Type of tenure	Area in farms (%)	No. of farms (%)	Area in farms (%)	No. of farms (%)	Area in farms (%)
Owner-occupation	36.0	36.7	53.1	49.5	71.6
Tenancy	22.2 ⎫	37.3	9.7 ⎫	16.5	9.9
Sharecropping	1.5 ⎭		0.9 ⎭		1.3
Squatting	2.9	3.4	2.4	3.4	2.2
Govt leasehold	22.4	9.3	15.8	8.6	8.9
Mixed types	15.0	13.4	18.1	22.0	6.2
	100.0	100.0	100.0	100.0	100.0
Area (million ha)	173		175		211
No. of farms ('000)		457		465	

In contrast, the period 1952–69 experienced virtually no change in the relative importance of different sizes of holdings (*Table 9.19*). At both dates about 68 per cent of farms were under 100 ha in area and accounted for only 5 per cent of farmland, whereas 0.5 per cent of farms were over 10,000 ha and occupied 33 per cent of the land. The median size of farm was under 25 ha in all provinces west and north of the Pampas, excepting Salta, Jujuy and Chaco, and exceeded 25 ha in all provinces of the Pampas and Patagonia, excepting Río Negro. Size alone, however, is an inadequate criterion for the scale of enterprise; calculations have therefore been made of the number and regional distribution of farms according to the number of families each is capable of supporting (CIDA, using 1960 data; Caracciolo de Basco and Rodriguez Sanchez, using 1969 data). At both dates over 200,000 or about 44 per cent of all farms were estimated to be incapable of providing two man-years of work and hence of supporting a family. Such *minifundia* are totally absent from parts of Patagonia, whereas in parts of the north-west and north-east they constitute well over three-quarters of all farms. Nor are the Pampas devoid of such smallholdings, but the method of estimation, it is admitted, over-estimates their significance in the Pampas where many profitable capital-intensive one-man farms exist, and under-estimates their importance in the north where so-called family farms supporting two to four full-time workers often have low *per capita* productivity (*Figure 9.6d*).

Table 9.19 Argentina: changes in farm size 1952–69

	1952		1960		1969	
Size	Area in farms (%)	No. of farms (%)	Area in farms (%)	No. of farms (%)	Area in farms (%)	No. of farms (%)
≤25 ha	1.0	43.2	1.0	39.7	0.9	41.2
25.1–100 ha	4.2	26.2	4.4	27.8	3.9	26.2
100.1–1000 ha	19.9	25.1	20.1	26.7	20.3	26.5
1000.1–10,000 ha	42.0	5.0	41.3	5.2	41.0	5.5
>10,000 ha	32.9	0.5	33.2	0.6	33.9	0.6
	100.0	100.0	100.0	100.0	100.0	100.0

Nevertheless, the problem of the small farmer with little education and little access to credit is far from absent in the Pampas. A careful investigation by Gaignard in La Pampa province (CNRS, 1968) revealed that smallholdings are typically derived from the lots created by the organized colonization schemes of 50 to 100 years ago. With rising standards of living, a once adequate lot of 75–150 ha no longer affords an acceptable livelihood and, with mechanization, occupies but a third to a half of a farmer's time. If colonies of smallholdings border huge estates – as in Guatraché department, where unviable holdings of under 350 ha account for 79 per cent of the farms but only 25 per cent of the land, and where 32 per cent of the land is occupied by the nine farms of over 5000 ha – social and political tension can easily develop, and the obvious solution may well be pressed. But local solutions can only be palliatives in such cases as Trenel department, where 11 per cent of the land is occupied by four big farms, but over 50 per cent is already taken up by the 611 farms of under 350 ha.

Intensification would seem to offer an answer. Yet elsewhere in the Pampas, in the colonies of the enterprising Jewish Colonization Association (which introduced the sunflower and co-operative marketing to Argentina), attempts to make 75-ha lots profitable have failed, and rural depopulation and the amalgamation of holdings have taken place. According to Winsberg (1964, p. 501), 'the Jewish population is likely to decrease until the size of the holdings of [those] who remain is large enough to provide an income comparable to what they could earn . . . in the cities.'

The concept of an ideal farm size is elusive. Some studies showed that family farms in 1960 devoted over two and a half times as much of their land to crops as did large farms, whereas the latter had two-thirds as much land again in natural pasture as had the former (*Table 9.20*). Data for value added per hectare showed a progressive decline in productivity with increasing scale of operation. Such relationships are often attributed to the satisfaction of large operators with ample returns based on low yields from a large area, or to the unwillingness of absentee owners to devote the extra managerial effort required by intensive farming. Such assumptions do not, however, explain the even stronger positive correlation between scale of operation and productivity per man. Indeed, another study, in the cattle-rearing area of Ayacucho (Buenos Aires Province), revealed statistically significant positive correlations between farm size and the adoption of innovations, the desire for profit maximization, and the decision to intensify production (Scheinkerman de Obschatko, 1971). Although farms of over 2200 ha were shown to invest less and use less labour per hectare than farms of 200–1000 ha, they actually achieved more than twice the income per hectare and three times the rate of return on invested capital. The suggested chain of explanation of the observed correlations begins with large farm size and high investment leading to large incomes, which allow higher levels of education and absentee residence in the big cities; these in turn facilitate contacts with sources of information about innovations and access to credit, thus favouring innovatory attitudes and the desire for profit maximization. Such findings echo those of developed world studies of the diffusion of innovations. They lend support to the view that large farmers will serve themselves and the nation best until, as a matter of social policy, the state decides to finance the wider dissemination of information and credit.

The problems of the small farms of the north and west are however more intractable. They begin with the need to select products which yield high returns per hectare, thus excluding readily exportable cattle, sheep, cereal and oilseed products – cereals, for example, typically yield only one-seventh to one-tenth the gross returns per hectare of industrial crops. In addition, the more northerly provinces are most apt for subtropical crops such as sugar cane, tobacco, tea and cotton, which are chiefly exported by nations enjoying cheap labour. Argentina, therefore, cannot export its surpluses competitively without subsidies; even the ceiling on production imposed by the modest size of the home market is lowered when imports are permitted. Furthermore, a good number of intensively grown crops are perennials – citrus, yerba, tung, tea, vines – which cannot be rapidly substituted for alternative crops when prices fall; many, too, are associated with contract arrangements for supplying specialized processing plants. Such plants frequently enjoy oligopsonist power – six sugar mills process 56 per cent of all cane, seven ginneries handle 48 per cent of cotton, six wineries process 44 per cent of San Juan's grapes, and five cigarette manufacturers take 99.5 per cent of the tobacco grown – and can therefore pass on to the grower the full cost of any reductions in market prices.

The problems of many crops are also compounded by the inflexibility of regional specialization. Thus, on the one hand 90 per cent of San Juan's agricul-

Table 9.20 The Pampas: land use 1960 and productivity by scale of enterprise

Scale	No. of workers	Farm area (%)	Farms (%)	% of land in				Value added, per	
				Crops	Improved pasture	Natural pasture	Other	Hectare (pesos)	Man ('000 pesos)
Smallholdings	1–2	3.2	26.8	31.7	19.5	43.6	5.2	4002	63
Family farms	2–4	40.7	62.7	34.9	29.7	31.0	4.4	1946	111
Medium farms	4–12	36.0	9.6	22.1	22.6	47.5	7.8	1387	250
Large farms	>12	20.1	0.9	13.4	27.8	49.0	9.8	1150	345
		100.0	100.0						

tural output by value consists of vines, 75 per cent of Tucumán's comes from sugar cane and 60 per cent of Chaco's from cotton. On the other hand Misiones produces over 90 per cent of Argentina's yerba, tea and tung, Chaco and Formosa 70 per cent of the cotton, Mendoza and San Juan 91 per cent of the wine grapes, and La Rioja 56 per cent of the olives for bottling. Furthermore, most farms specializing in such crops are very small: 78 per cent of Chaco's cotton-growers occupy 25 ha or less, 77 per cent of Misiones's yerba growers have 10 ha or less, 77 per cent of San Juan's vineyards cover 4.5 ha or less, 52 per cent of Formosa's cotton-growers have up to 5 ha and 73 per cent of Corrientes's tobacco producers have no more than 9 ha.

The opportunity to adopt high-yielding varieties is rare, for research has concentrated on more extensive crops such as maize, sorghum and sunflower. Thus, it is the medium and large farmers who by adopting better varieties have been able to maintain their incomes despite falling prices. The small cane-growers of climatically marginal Tucumán have, however, benefited from the introduction of frost-resistant varieties. Cotton-growers, too, have gained from better varieties.

Where farms are large enough and the crops lend themselves to it, mechanization often offers no real answer to the problem of increasing *per capita* income, since such farms frequently already suffer from under-employment of the existing labour force, composed as it is of the members of the farming families. In general, labour forces exceed 200 persons per 1000 ha, and in some tobacco and vine-growing areas rise above 500. In contrast, on the Pampas densities rarely exceed 25. Not surprisingly, productivity per man is low: on Formosan cotton farms it is only 7 per cent of that achieved on Buenos Aires wheat farms, on north-eastern tobacco farms only 6 per cent, and on Misiones yerba farms 18 per cent. Significantly, it reaches 50 per cent or more on the typically larger tobacco farms and cane plantations of Jujuy and Salta, which enjoy the capital resources and scale of operations to permit the adoption of technological packages of biological, chemical and mechanical innovations.

In all, studies showed that Formosa cotton-farmers with 5 ha of the crop in 1969 enjoyed a net annual income of $670 and Misiones yerba growers with 5 ha only $590. In contrast even the permanent farm labourer averaged $700 and the unskilled industrial labourer $1400. The conclusion seems inescapable that in such areas far too many people depend on agriculture, and that increased *per capita* incomes can only be achieved by encouraging rural emigration and concomitantly amalgamating smallholdings in order to create viable family farms. By the early 1970s, however, so many crops, at least in Misiones, were in crisis, occasioned by lack of demand, low crop prices, crop disease or the rising wages of a unionized labour force, that few settlers had the resources to expand their holdings, even if that could lead to a profitable solution (Eriksen and Reisch, 1974). Much larger enterprises, however, had long recognized the good prospects afforded by reafforestation under climatic conditions almost uniquely favourable for rapid tree growth in a country that was very largely dependent on imports of timber and its products. By 1972 such companies, mainly located beside the River Paraná where they once concentrated on plantation crops and on exploiting the natural forests, had over 80,000 ha under rapidly growing pines and eucalyptus, and had expanded by the purchase of 61,500 ha of former agricultural holdings. A factory was producing 40,000 tonnes a year of chemical pulp. Calculations showed that, with its low labour costs, afforestation could yield net annual returns of $5600 from 100 ha – but only in the long term. By 1971 over 3500 settlers in Misiones had succeeded in afforesting part of their lands, but without long-term credit few could afford to forgo the dwindling income that would be lost if traditional crops were wholly replaced by trees.

In general, it might be argued, in Argentina as elsewhere, that a more equal distribution of land would provide the necessary number of viable farms for the existing rural population. Such action would inevitably lead to the general intensification of production. In much of interior Argentina, however, this would create an excess of supply to a limited domestic market (Ras, 1977). Indeed, in many recent multi-purpose hydroelectric and irrigation schemes the irrigation works have already been deferred because the ensuing extra total production would cause gluts on the domestic market, and thus render such works unprofitable.

In the nineteenth century peopling the land was sound government policy, to increase exports; today, peopling the cities with consumers would appear to be the prime need if the agricultural resources of the interior are to be further developed.

The problem of Uruguay

Although the smaller River Plate republic has a population scarcely one-eighth that of Argentina, and enjoyed uninterrupted democratic government for many years, the course of her economic development, the role of her agriculture and the growth of her industry have been remarkably similar to Argentina's.

Exports derived from agriculture still account for 97 per cent of all exports, with wool representing at least half, and beef and arable crops the rest. Thanks to high world prices for wool, Uruguay's export earnings remained high until the early 1950s but, as in Argentina, exports represented a declining proportion of total production and production itself, of both crops and livestock, was almost stagnant. Thus in 1935–7, 23 per cent of crops and 54 per cent of livestock products were exported, but these shares had fallen to 12 and 38 per cent by 1959–63, due not only to rising domestic demand but also to the disincentive of export duties.

The patterns and problems of Uruguayan agriculture most clearly resemble those of the Argentine Pampas: in 1963, 16.8 million ha, or 91 per cent of the national territory, was in agricultural use; no fewer than 15 million ha were devoted to natural pastures and a further 500,000 ha to improved pasture and fodder crops, leaving only 1.3 million ha for cash crops (although a further 4 million ha, then under pasture, were capable of annual cropping). Crop-farming and dairying are to be found particularly on the erosion-prone Tertiary and more recent deposits which flank the Plate estuary west of Montevideo; livestock-rearing is concentrated on the northern hills and plains, and fattening further south. As on the Pampas, the major traditional crops (maize, wheat and flax) were supplemented by the sunflower during the war in order to obviate the need for olive-oil imports. Edible oils have now become an important export. The yield of none of these four crops, however, had risen perceptibly by 1960.

The yield of beef per hectare, as of the leading crops, is not only stagnant but far lower than that obtained in Argentina. Nor is the meat from her 23 million sheep and 8 million cattle subjected to rigorous grading. Livestock-breeding is the only continuing improvement in the agricultural field, but it receives little encouragement from the pricing system, which operates to keep domestic prices of beef almost as low as those of bread and potatoes. Because of inadequate feeding practices, fatstock is not available in the winter and spring, and the *frigoríficos* remain closed. Some, indeed, have been forced by a shortage of supplies for export to close down altogether. Improved techniques could quadruple the yield of beef per hectare and triple that of wool and mutton. The Agricultural Development Plan of 1961 has made a contribution with the aid of international funds for the extension of artificial pastures, fencing and the use of fertilizers.

Much of the blame for agricultural stagnation is attributed to the same two defects in the agrarian structure that are held responsible in Argentina – insecurity of land tenure and excessive size of holdings. In 1961 39 per cent of farmland was operated by tenant farmers, a fall of 4 per cent since 1951, whilst owner-operated farmland had risen by 3 per cent to 58 per cent. The tenant farmer enjoyed no compensation for improvements and even the 5-year contracts were subject to rent revision every 2 years. Yet a study of the 1961 survey (Plottier and Notaro, 1966) reveals that, contrary to expectation, tenant farmers are generally the more efficient in the use of land, labour and capital. The proportion of owner and tenant farmers' capital devoted to fixed investment (buildings, fencing, water supplies) was identical at 26.6 per cent; tenants devoted somewhat more than owners to machinery but less to orchards. The yields of tenant-produced wheat, maize, sugar beet, fruit and animal products per hectare were all slightly higher, and only their yields of sunflower, flax and apples were lower. Both labour productivity and returns on capital were a fifth to a quarter higher on tenant farms; the study even concludes that declining agricultural production may be attributable to the fall in areas tenanted! It is possible, however, that inadequate attention has been paid to the somewhat above-average importance of tenanted (and smaller) farms in the more fertile arable districts near the Plate estuary.

Less evidence is available on the relationship of farm size to productivity. In 1961 46 per cent of farms were under 20 ha and occupied under 2 per cent of the land, whereas the 4 per cent of the farms which had over 1000 ha occupied 57 per cent of the land. According to a report on the 1951 census farms under 200 ha had a labour productivity that was less than a third that on larger farms, but output per hectare was three times as great. Such global figures may conceal the effect of important intrinsic differences in soil and location between large and small farms. A more detailed study in Paysandú department in 1960 showed that incomes per hectare generally decreased with increasing farm size. This reflected the importance on all farms of under 100 ha of intensively grown vines, cotton or vegetables, crops that were completely absent from the larger holdings.

Official efforts to increase the number of small farmers have been limited by the shortage of money and personnel. Between 1948 and 1967 less than half a million hectares were distributed by the Colonization Institute to some 3100 settlers, and much, therefore, remains to be achieved.

As in the rural sphere, Uruguay's urban and industrial development since 1930 has closely paralleled that of Argentina and the Pampa provinces.

Of the 1963 population of 2.6 million, only 18 per cent are classed as rural, 72 per cent live in towns of over 5000 and 46 per cent live in the capital city, Montevideo. Between the two censuses of 1908 and 1963 Montevideo grew at an annual rate of 2.5 per cent compared with 1.2 per cent in the rest of the country. The age structure of the population is similar to that of the most developed countries: 11 per cent receive social security pensions, which absorb 60 per cent of public expenditure; in addition, over half the active population are employed in the tertiary sector, the public services alone accounting for 20 per cent. Some 17 per cent work on the land and 22 per cent in manufacturing.

Import substitution had already proceeded far before the Second World War, when over half Uruguay's manufacturing needs were produced at home (and over three-quarters of the foodstuffs, clothing and leather goods). The growth of manufacturing continued at over 8 per cent per year in the decade following the war, only to encounter the same problems of stagnation and import-financing as Argentina; with a smaller market and smaller factories, however, Uruguay's inefficiency is more serious.

In certain respects Uruguay's development has differed from Argentina's. With a tiny market, her prospects for heavy industry based on economies of scale are limited – her steel consumption is only 100,000 tonnes a year – yet a few branches of industry (petroleum-refining, cement and alcohol production) have existed under a state monopoly for over 30 years. New agricultural processing industries (edible oils and wool combing) have also expanded with the aim of increasing export earnings.

Nor does Uruguay suffer the same transport and energy problems as her neighbour. Both road and rail networks cover the whole country, although many improvements are required, and the number of vehicles in use has more than doubled since 1953 to over 196,000 in 1965. Three-quarters of the electricity produced comes from installations on the Río Negro, and the country's energy capacity always exceeds demand.

Lastly, Uruguay obtains 17 per cent of her foreign-exchange earnings from tourism (which ranks third, after wool and meat, in the 'export' table). The visitors come overwhelmingly from Argentina, for Uruguay's south coast resorts of Montevideo, Punta del Este and Piriápolis, afford the nearest places of escape from the summer heat of Buenos Aires. But Mar del Plata alone attracts ten times as many Argentine visitors, and Uruguay has little to offer the tourist from other nations.

The integration of Paraguay

During the first 30 years of Argentina's agricultural and industrial reorientation, no permanent new trend disturbed the Paraguayan scene. The Chaco War with Bolivia (1932–5) brought her new territories at a high cost in human resources, and in the Second World War her beef-canning industry enjoyed a brief boom which depleted her cattle stocks, but the disadvantages of political instability and physical isolation persisted. Indeed, in 1955 the country possessed only 1100 km of roads, of which scarcely half were surfaced.

The countryside remained a land of huge estates, squatters and peasant smallholdings. In 1961 a third of Paraguay, which occupies 41 million ha, was covered by twenty-five properties, each over 100,000 ha, and almost a quarter was still public; in contrast 142,000 properties, each under 50 ha, covered but one-fortieth part. At the same time, according to a partial survey of tenurial systems, owner-operators held 37 per cent of the farms but 81 per cent of the land, tenants held 8 per cent of the farms and 4 per cent of the land, whereas squatters ran 42 per cent of the farms on only 7 per cent of the land. Another partial survey in 1956 revealed that 98 per cent of the farms in the relatively densely settled central district were under 50 ha, yet a third of the land still lay in huge farms over 5000 ha (*Table 9.21*).

Table 9.21 Paraguay: farm sizes 1956

Selected regions	Selected classes		
	0–5 ha	5–50 ha	Over 5000 ha
Nation			
% of farms	45.9	48.9	0.3
% of area	1.0	5.3	73.5
Central 5 departments			
% of farms	55.3	42.6	0.05
% of area	6.7	23.4	32.9
Chaco departments			
% of farms	22.7	24.3	9.0
% of area	0.02	0.1	85.6

In brief, the major agrarian problem lay in the insecurely held smallholdings of the central area, where 77,000 farms devoted to subsistence and peasant agriculture had a median size of only 3.5 ha. Yet, with 60 per cent of the nation in forest and 34 per cent under natural pasture, there is no real shortage of land.

Indeed, under-population was really a more serious problem than any shortage of land. International migration had done little to help. In 1950 Paraguay, over three times the size of England, had only 1.4 million inhabitants, of whom only 47,000 were foreign-born. Although most immigrants had been born in Argentina or Brazil and only 3000 had come from Germany, a large proportion of the former were descendants of German settlers. With few immigrants or native migrants new land settlement did not assume the character of a continuous pioneer fringe, though Itapúa department (capital Encarnación) facing Argentine Misiones was the most popular area of sporadic Germanic settlement. The strong social cohesion within the different European colonies helped greatly in the establishment of permanent settlements, but this exclusiveness, coupled with physical and economic isolation as well as the linguistic barrier of Guaraní, the language of the mestizo countryside, retarded the assimilation of immigrants into the national society.

Nevertheless, the foreigner continued to dominate the production of tree crops for the export market, whereas the native farmer still practised a peasant economy, selling locally his surplus of food staples, and growing chiefly cotton and tobacco as cash crops for export.

By the late 1950s new trends were at last beginning to emerge and with them the prospect of integration into the world economy. A basic part has been played by the stable government of General Stroessner, who gained power in 1954. Though severe in its treatment of political opponents, it has pursued moderately enlightened policies in some directions. Thus, the Agrarian Reform Law of 1963 sought to alleviate agrarian insecurity and inequality. By 1967 33,000 farmers had had their titles confirmed to some 1.25 million ha, and another 11,000 had received 500,000 ha of land expropriated or purchased from their owners. More importantly, between 1956 and 1966 about 85,000 families, involving perhaps half a million people, were resettled in new colonies on forested lands exceeding 2.6 million ha in area. Few of these colonies lie near the new paved highways, however, and no technical and financial help was provided. The economic consequence for many was thus merely to shift the location of their subsistence activities.

In contrast, a highly organized programme of Japanese immigration between 1955 and 1968 brought over 7000 colonists to Paraguay, especially to the areas around Encarnación and to a lesser extent Puerto Presidente Stroessner. The 100-ha lots offered afforded better prospects than the 30-ha lots accorded to Paraguayan settlers; rural roads and other aid further enhanced their prospects. Finally, by 1979 300,000 Brazilians had settled spontaneously, and often illegally, in the frontier areas beside the River Paraná and in the north-east near P.J. Caballero, the Paraguayan town adjoining Ponta Pora in Mato Grosso. The foreign-born population continues to constitute but a small proportion of the total population, estimated to have reached 3 million in 1980, and numerically does not compensate for the economic and political refugees from Paraguay, estimated to exceed half a million, who now live in Brazil and Argentina.

During the 1960s and early 1970s Paraguay's economic prospects were greatly improved by the construction of three paved highways from Asunción: eastwards, across the River Paraná and to the free port of Paranaguá; south-eastwards to Encarnación which is linked by ferry to Posadas, now enjoying both rail and paved road connections to Buenos Aires; and finally southwards, via the Paraguay bridge, to the Argentine highway running beside the Paraguay/Paraná river to Buenos Aires. Additionally, new but unpaved roads join Asunción and Concepción to P.J. Caballero and the Brazilian system. The trans-Chaco highway is destined ultimately to reach Santa Cruz in Bolivia.

In the 1960s the cattle industry became the most dynamic sector of the economy. With canning-grade cattle disappearing from Argentina, Paraguay was able to expand its production and exports of corned beef. By value these rose from under $0.5 million in 1952 to an average of $14 million in 1962–8, representing a quarter to a third of all exports. Paraguay's cattle, numbering about 6 million, are still mostly lean, bony animals derived from the poor *criollos* of Corrientes or zebus of Mato Grosso, imported after the decimation of herds in 1865–70. Herd improvement, begun over 30 years ago with the introduction by the US Agricultural Mission and by Liebigs of high-grade zebu-type animals (Santa Gertrudis and Brahman), was further stimulated by World Bank credits in the mid-1960s. By 1971, with the completion of the paved highway direct to Buenos Aires, cattle were commanding higher prices for exportation live to Argentina. In an endeavour to compete, Liebigs' Asunción factory began to produce frozen beef, a higher-value product than corned beef, but by the end of the 1970s all three beef-processing factories in Asunción had closed down as ranchers discovered that the fresh-beef markets of Brazil, Argentina – and Asunción itself – could absorb virtually all the animals on offer and pay higher prices.

The 1970s have also witnessed the expansion of export-crop production. Salient among these has been cotton. Averaging 48,000 ha in 1962–71, the area rose to 93,000 ha in 1972–6, reaching 312,000 ha in 1978. It is cultivated by the Paraguayan peasant on both old and new lands, the average farm growing no more than 1 ha with family labour. Expansion owes much to the establishment of government buying-stations offering fixed prices, and yields have risen thanks to the introduction by French advisers of disease-resistant varieties.

In the wetter forest lands tree crops have typically been preferred by the immigrant colonists: around P.J. Caballero the coffee zone of Brazil has spilled over into Paraguay, while near Encarnación the Japanese colonists rapidly adopted the yerba and tung of neighbouring Argentine Misiones as their chief cash crops. Nevertheless, they also brought with them silkworm rearing and the soya bean. With the uncertainties of the tung-oil and yerba markets the latter has rapidly spread as a field crop, over half a million hectares being devoted to it in 1980, four-fifths of the area being located in the forest colony provinces. Other colonists have adopted a corn-hog economy for the exportation of pig-meat to neighbouring Brazil and Argentina, while others have seized the opening of the paved roads to Asunción as the opportunity for producing sugar, manioc and maize for the capital city.

Currently the Paraguayan economy is being revolutionized by the construction of the Itaipú HEP project (the world's largest) on the Upper Paraná. Destined to produce 12,600 MW when completed in 1988, the scheme will make electricity the country's biggest export. Meanwhile the tens of thousands of construction workers constitute a ready market for foodstuffs, and all branches of the construction industry have been stimulated. It is expected that as construction nears completion the construction boom will be resumed by the work of building the Yacyretá HEP scheme shared with Argentina downstream.

For nearly 20 years the boost to Paraguayan development has come chiefly from Brazil, as its people, capital and demand have crossed the frontier and as Brazilian markets have grown, creating more lucrative markets than those afforded by Asunción or Buenos Aires. Concepción, Paraguay's second city, with 20,000 inhabitants and once the port for shipping cattle, timber and tannin extract downstream, now sends its cotton to P.J. Caballero and Brazil by truck, and receives in return the manufactured goods of that country.

Yet the attractiveness to the Paraguayan peasant of supplying the Brazilian market is relative – relative to the low *per capita* income previously obtained as a subsistence farmer isolated from wider economies. How far the new HEP scheme will act as a catalyst to permanent occupations affording incomes on a par with those of metropolitan Buenos Aires or São Paulo, or will merely provide frustratingly temporary, well-paid employment in construction, remains to be seen. The only certainty is that the disturbance will be profound.

Notes

[1] 'The River Plate' and 'The Plate' are commonly used phrases to refer to Argentina, Uruguay and Paraguay. 'Mesopotamia' refers to Entre Ríos, Corrientes and Misiones, the Argentine provinces lying between the rivers Paraná and Uruguay. 'Litoral' refers to the older provinces bordering the Paraná–Plate estuary, namely Corrientes, Entre Ríos, Santa Fé and Buenos Aires (and even Uruguay).

[2] Formosa was part of Chaco Territory until 1884. They are still known together as The Chaco.

[3] Greater Buenos Aires includes the federal capital and the contiguous urban divisions of Buenos Aires province.

[4] A town is defined as a population centre with over 2000 inhabitants.

[5] Uruguayan censuses are very defective, and there was none between 1908 and 1963!

[6] Better known by its brand names, Lemco, Oxo and Fray Bentos.

Bibliography

ARNOLDS, A. (1963) *Geografía económica argentina*, Buenos Aires.

ASOCIACIÓN ARGENTINA DE CONSORCIOS REGIONALES DE EXPERIMEN-TACIÓN AGRÍCOLA/BANCO DE LA NACIÓN/FUNDACIÓN BANCO DE LA PRO-VINCIA DE BUENOS AIRES (1978–81) *Información Económica*, various numbers, Buenos Aires.

BANCO GANADERO ARGENTINO (1980) *La producción rural argentina en 1971 . . . . 1980*, Buenos Aires.

BANCO GANADERO ARGENTINO (1974) *Temas de economía argentina. El sector agropecuario 1964/73*, Buenos Aires.

BOURDE, G. (1974) *Urbanisation et immigration en Amerique Latine: Buenos Aires*, Paris.

BRANNON, R.H. (1968) *The Agricultural Development of Uruguay*, London.

CARACCIOLO DE BASCO, M. and RODRIGUEZ SANCHEZ, C. (1978) *El minifundio en la Argentina*, Pt 1 (Secretaria de Estado de Agricultura y Ganadería. Servicio Nacional de Economía y Sociología Rural. ESR 11/78), Buenos Aires.

CARLEVARI, I.J.F. (1979) *La Argentina. Geografía humana y economía*, 5th edn, Buenos Aires.

CENTRE NATIONAL DE LA RECHERCHE SCIENTIFIQUE (1968) *Les problèmes agraires des Amériques Latines*, Contributions by R. Cortés Conde, J.C. Crossley, R. Gaignard, H. Giberti, J.A. Martínez de Hoz, Paris.

CENTRO DE ESTUDIANTES DE CIENCIAS ECONÓMICAS Y ADMINISTRACIÓN (1966) *Plan nacional de desarrollo económico y social 1965–1974 (Uruguay)*, 2 vols, Montevideo.

CIDA (Comité Interamericano de Desarrollo Agrícola) (1965) *Land Tenure Conditions and Socio-economic Development of the Agricultural Sector, (Argentina)*, Washington, D.C.

CIRIO, F.N. *et al.* (1980) 'Aspectos económicos del empleo de fertilizantes en el agro, *Información Económica*, 15, Buenos Aires.

CIRIO, F.N. *et al.* (1981) 'Productividad, eficiencia, y problema energético en la agricultura, *Información Económica*, 20, Buenos Aires.

CONSEJO FEDERAL DE INVERSIONES (1962–5) *Programa conjunto para el desarrollo agropecuario e industrial*. 1er, 2do, 3er, 40 Informes, 15 vols, Buenos Aires.

CONSEJO FEDERAL DE INVERSIONES (1964) *Tenencia de la tierra*, 4 vols, Buenos Aires.

CROSSLEY, J.C. (1976) 'The Location of beef processing', *Annals Assoc. Am. Geog.*, 66.

CROSSLEY, J.C. and GREENHILL, R. (1977) 'The River Plate beef trade', in PLATT, D.C.M. (ed.) *Business Imperialism, 1840-1930*, Oxford.

DE APARICIO, F. and DIFRIERI, H.A. (eds) (1959-61) *La Argentina. Suma de Geografía*, vols 4, 6, 7, 8, Buenos Aires.

DIRECCIÓN NACIONAL DE ESTADÍSTICA Y CENSOS (1960a) *Censo nacional agropecuario*, 3 vols, Buenos Aires.

DIRECCIÓN NACIONAL DE ESTADÍSTICA Y CENSOS (1960b) *Censo Nacional de población*, 9 vols, Buenos Aires.

DI TELLA, T.S. (1965) *La teoría del primer impacto del crecimiento económico*.

ECONOMIC COMMISSION FOR LATIN AMERICA (CEPAL) OFICINA EN BUENOS AIRES (1976) *Desarrollo regional argentino: la agricultura*, 2 parts, Buenos Aires.

EIDT, R.C. (1971) *Pioneer Settlement in North-East Argentina*, Madison.

ERIKSEN, W. and REISCH, E. (1974) 'Agrarkrise und Aufforstung in Misiones', *Die Erde*, 105.

FERNS, H.S. (1960) *Britain and Argentina in the 19th Century*, Oxford.

FERNS, H.S. (1969) *Argentina*, London.

FERRER, A. (1967) *The Argentine Economy*, Berkeley and Los Angeles.

FIAT/CONCORD (1976) *Paraguay*, Buenos Aires.

FILLOL, T.R. (1961) *Social Factors in Economic Development: The Argentine Case*, Cambridge, Mass.

GAIGNARD, R. (1968) 'Sous-développement et déséquilibres régionaux au Paraguay', *Revta Geogr.*, 69.

GAIGNARD, R. (1972) 'Les villes du sous-développement: le cas du Paraguay', *Revue Géographique des Pyrénées et du Sud-Ouest*, 43.

GIBERTI, H. (1961) *Historia económica de la ganadería Argentina*, Buenos Aires.

GIBERTI, H. (1964) *El desarrollo agrario Argentino*, Buenos Aires.

GIBERTI, H. *et al.* (1965) *Sociedad, economía y reforma agraria*, Buenos Aires.

INSTITUTO NACIONAL DE ESTADÍSTICA Y CENSOS (1973) *Censo Nacional de Población, Familias y Viviendas, 1970. Resultados provisionales. Localidades con 1000 y mas habitantes*, Buenos Aires.

INSTITUTO NACIONAL DE ESTADÍSTICA Y CENSOS (1975) *La población de Argentina*, Buenos Aires.

INSTITUTO NACIONAL DE ESTADÍSTICA Y CENSOS (1979) *Anuario estadístico de la Republica Argentina, 1978*, Buenos Aires.

INSTITUTO NACIONAL DE ESTADÍSTICA Y CENSOS (n.d.) *La migración interna en la Argentina, 1960/70*, Buenos Aires.

JUNTA NACIONAL DE CARNES *Reseña 1957* and *Reseña 1964*, Buenos Aires.

JUNTA NACIONAL DE CARNES (1979) *Sintesis estadística 1978*, Buenos Aires.

KATZ, J. and GALLO, E. (1968) 'The industrialization of Argentina', in VELIZ, C. (ed.) *Latin America: A Handbook*, London.

MARTINEZ RODRIGUEZ, I. (1962) *Apuntes de geografía del Uruguay*, Montevideo.

MINISTERIO DE AGRICULTURA Y GANADERÍA. DIRECCIÓN NACIONAL DE ECONOMÍA Y SOCIOLOGÍA RURAL (1972) *Series estadísticas retrospectivas de las especies agrícolas cultivadas en la República Argentina 1940/41-70/71*. ESR 34/72, Buenos Aires.

MINISTERIO DE ECONOMÍA. SECRETARIA DE ESTADO DE PROGRAMACIÓN Y COORDINACIÓN ECONÓMICA. INSTITUTO DE PLANIFICACIÓN ECONÓMICA *et al.* (n.d.) *Sector agropecuario argentino. Situación y perspectivas 1978-79*, Buenos Aires.

MINISTRY OF ECONOMY. COORDINATION AND ECONOMIC PLANNING SECRETARIAT (1978-9) *Economic Information on Argentina*, monthly issues, Buenos Aires.

OFICINA DE ESTUDIOS PARA LA COLABORACIÓN ECONÓMICA INTERNACIONAL (1966 and 1973), *Argentina económica y financiera*, Buenos Aires.

PENDLE, G. (1967) *Paraguay: A Riverside Nation*, London.

PINCUS, J. (1968) *The Economy of Paraguay*, New York.

PLOTTIER, L. and NOTARO, J. (1966) *El arrendamiento rural en Uruguay*, Montevideo.

RAS, N. (1977) Participación del sector agropecuario en el desarrollo de las economías regionales. *La producción rural argentina en 1976*, Buenos Aires.

RECA, L.G. (1971) 'La producción agropecuaria y los precios en 1923–65', *La producción rural argentina en 1971, 1er semestre*, Buenos Aires.

SCHEINKERMAN DE OBSCHATKO, E. (1971) 'Factores limitantes al cambio tecnológico en el sector agropecuario', *La producción rural argentina en 1971, 1er semestre*, Buenos Aires.

SCOBIE, J.R. (1964a) *Argentina. A City and A Nation*, New York.

SCOBIE, J.R. (1964b) *Revolution on the Pampas*, Austin.

SCOBIE, J.R. (1975) *Buenos Aires. Plaza to Suburb 1870–1910*, New York.

STEWART, N.R. (1967) *Japanese Colonization in Eastern Paraguay*, Washington, DC.

TAYLOR, C.C. (1948) *Rural Life in Argentina*, Baton Rouge.

WHITE, D. *et al.* (1979) Analisis económico de la maquinaria agrícola. *Información Económica*, 6, Buenos Aires.

WILHELMY, H. and ROHMEDER, W. (1963) *Die La Plata-Länder*, Braunschweig.

WINSBERG, M.D. (1964) 'Jewish colonization in Argentina', *Geogr. Rev.*, 54.

WINSBERG, M.D. (1970) 'Introduction and diffusion of the Aberdeen Angus breed in Argentina', *Geography*, 55.

10 · Chile

Harold Blakemore

In recent years few countries in the world have attracted more sustained international attention than Chile, and the paradox is enhanced when one considers the country's modest size, comparatively small population and, far from least, its peripheral geopolitical position in an ideologically divided world. The reasons, of course, are primarily political. An abrupt change occurred in Chile in 1973, when the freely elected socialist government of Salvador Allende was overthrown by a military *coup*; the subsequent authoritarian order, apparently completely contrary to Chile's long-standing constitutional system became, throughout the world, a political symbol, much as Spain was in the 1930s. Since 1973 interest in Chile has grown rather than diminished, as that symbolic importance has been given economic, social and international dimensions, deriving from the policies adopted by the military government. For, as in politics so in economics, the government since 1973 has deliberately turned its back on half a century of economic development in which the state was increasingly the motor, and has adopted a model which, if successfully pursued, can make Chile one of the most free-enterprise economies in the world. These developments – to which we shall return – have profound implications for the future of the republic: they also have some significance for the rest of Latin America and, indeed, the wider world, increasingly polarized as it becomes between rich and poor nations. Models may not be strictly exportable, but example can be infectious, and there is already no doubt that Chilean experience in the last decade or so has exercised some influence beyond the country's boundaries. Yet whatever changes are wrought in Chile as a result of these events, they take place against a geographical and historical background which has made Chile what it is, and many of the characteristics of that environment will not be eradicated by political change, though they may be modified by it. The geographical features of Chile have played a major part in determining its historical evolution, and they will continue to do so.

Some basic characteristics

The best-known popular work in Chilean geography, by Benjamin Subercaseaux, is entitled *Chile, o una loca geografía* (Chile, or a Crazy Geography), and the adjective could hardly be more apt. For how else should one characterize a country over 4200 km in length – excluding Chilean Antarctica – but nowhere more than 400 km wide, and with an average width of less than half that figure; its entire western frontier an ocean shoreline, more clearly defined but no less natural than its eastern counterpart of mountain ranges, with varieties of physiography and climate extending from absolute desert in the north to rain-drenched forest in the south? Much more than many of her larger neighbours, Chile is a microcosm of the continent in which they all lie, with its great physical diversity and regional contrasts, reflecting not only basic geographical and economic differences but also a distinctive historical experience. That Chile exists as a unitary state at all is, in effect, the result of the organic expansion north and south from a central core of long-standing human occupance, and the incorporation of frontier regions of strikingly different character into a national polity.

The unity in diversity that Chile exhibits is a unity imposed from the centre to the periphery, as Chile expanded from its colonial base in the heart of the modern republic, when economic and strategic considerations during the nineteenth century determined the march of her frontiers north and south. This expansion, however, has entailed disadvantages not apparent 100 years ago, notably the growth of disparities of economic and social well-being between the centre and outlying regions, which has been manifestly aided by increased centralization of decision-making. How to reduce those disparities has been an increasing preoccupation of successive governments in modern times, no matter what their political complexion.

In Chile these features – common to most Latin American states – are particularly striking for several reasons. First, Chile's economic growth has depended in the past, and still depends today, though perhaps to a lesser extent, on the exploitation of resources located in the outlying regions of the country, notably minerals in the northern provinces and oil and wool in the far south. Secondly, the rapid growth of cities in recent years, most notably of the capital Santiago, has meant an increased concentration of population and of the services it requires in areas that were already favoured recipients of national expenditure. Not surprisingly, therefore, regional feeling against the metropolis, itself a feature of the colonial period, should still be strong today, and that many should feel that just as in the colonial era Chile was no less a colony of Peru than Peru was of Spain, so today Antofagasta and Magallanes are colonies of central Chile, and central Chile a colony of Santiago.

National governments have long been aware of these tensions and of what, unchecked, they might signify for the republic as a whole; and they have been no less conscious of the gross internal disequilibria *within* provinces – between, for example, town and country. During the past 40 years in particular, attempts have been made to ameliorate such imbalances through planning and investment mechanisms, and how far they have worked will be considered in due course. Nevertheless, the fact remains that for over 400 years central Chile has dominated the country, and modern attempts to change that balance can only be appreciated within the framework of Chile's evolution as a whole. It is an evolution in which the political tradition of strong, centralist government has almost always obtained. From the foundation of Santiago by Pedro de Valdivia in 1541 to the achievement of independence from Spain in 1818, the core region was firmly established between the rim of the northern desert at Copiapó and the edge of the southern forest area at the line of the river Bío-Bío: more than that, within these confines were laid down over nearly three centuries the specific racial, socio-economic and even quasi-national characteristics which made Chile what it is today as a national state. Subsequently, in the second half of the nineteenth century, when Chile grew physically by three times its original size, the economic structures which were established were highly dependent on export-orientated resources located outside the core region, and the pattern was largely preserved: what nitrates meant to Chile at the turn of the century, copper still means today. Hence, no true understanding of contemporary Chile can be achieved without some attention to *regional* evolution in particular, for the patterns of the past, though modified, are persistent. In that evolution geographical factors have played a vital role, and it is necessary to sketch a description of the land and the people before turning to their interaction through time.

The physical framework
Chile extends from 18°S to 56°S, a length equivalent to the distance between central Norway and Morocco or, to take a New World comparison, to approximately the distance from New York to San Francisco. Chile's area, some 740,000 km², is roughly three times the size of the United Kingdom, but few areas of comparable size have such striking climatological and physical diversity. The climatic division is a latitudinal one, in contrast to the longitudinal character of landscape, which we will consider first.

There are three basic features of the Chilean landscape, which run from north to south in roughly parallel lines: the Andean cordillera to the east, decreasing in height towards the south of the country; the much lower and more fragmented *cordillera de la costa*, fronting the Pacific and becoming an archipelago, partly submerged, from about 42°S; and, between the two highland chains, a central depression. Each of these longitudinal features has quite markedly different characteristics from north to south, with the clearest definition of each in the central portion of Chile.

In the north the eastern highlands are, in effect, a continuation of the Peruvian–Bolivian *puna*, the high plateau, entirely over 3600 m above sea-level with surrounding peaks at 6000 m or more. Composed of a large number of interior basins, separated by recent volcanic-flows, this cold and windswept highland, with its salt lakes (*salares*) and dramatic scenery, has not changed much since Bowman wrote his classic description of it (1924, pp. 257 – 342).

From about 27°S to 38°S the orographic structure in the east is of high parallel ranges with steep narrow valleys between them. Here are to be found Chile's highest peaks, with Aconcagua on the Argentine side of the frontier at 7500 m. In this region the Andean ranges occupy from a third to a half of the width of Chile, their peaks snow-covered and the lower slopes heavily timbered. Near Santiago, on latitude 33°S, the snow-line at a height of between 4000 m and 4500 m gives the capital one of the most superb backdrops of any city in Latin America. But from latitude 38°S the mountain chains are lower and, whereas in the central section none of the passes over the Andes to Argentina is less than 3000 m, here a profusion of lakes, rivers and passes at lower altitudes, combined with a series of magnificent volcanoes skirting the western edge of the range, produces a more varied and beautiful landscape, albeit a less awesome one. Further south still, the Chilean Andes confront the Pacific on a fjord coastline, with extensive icefields inland, the mountains curving south-east to the Straits of Magellan, where they submerge to reappear in the southern part of Tierra del Fuego and the islands of the Beagle Channel.

The coastal range also exhibits variety from north to south, though, because of its more broken character and lower elevation, this is somewhat less striking than with the inland mountain chain. Geologically much younger and considerably more eroded, the *cordillera de la costa* begins south of Arica as an abrupt elevation from the Pacific shoreline, attaining varying heights between 550 m and 850 m north of Iquique, but rising in altitude further south to over 1800 m south of the river Loa, the boundary between the provinces of Tarapacá and Antofagasta. The highest point of the *cordillera* in the desert regions is the Sierra Vicuña Mackenna south of Antofagasta, almost 3000 m above sea-level. The proximity of the coastal range to the Pacific coast gives the latter a wall-like appearance from the sea for over 960 km, broken only by marine terraces at infrequent intervals: on these terraces stand the few ports and portlets – Pisagua, Iquique, Tocopilla, Mejillones, Antofagasta and Taltal – but they lack the protection of natural bays and harbours on what is a fairly uniform coastline. Behind them the coastal range might be best described as a plateau, having an average depth of over 50 km and lacking on its eastern side the abrupt descent that characterizes its coastal aspect.

South of Chañaral the *cordillera de la costa* virtually disappears, since from here to the river Aconcagua Andean spurs run westwards, and transverse river valleys – such as the Copiapó, Huasco and Elqui – coupled with broad marine terraces, 40 km or more in width, are the characteristic features. The range reappears, however, south of the Aconcagua, and in the provinces of Valparaiso and Santiago attains virtually Andean elevations of over 2500 m. But though the identity of the coastal range is here re-established, greater areas of coastal plain are also apparent at the mouths of the more abundant rivers. South of the river Maule, the coastal range divides into two chains, one along the coast and the other some 16 km inland, with fertile basins between them, but this feature disappears near the river Itata and the mountains only once again reach a height of 700 km as the Cordillera de Nahuelbuta, south of the river Bío-Bío. Thereafter, a succession of high hills rather than mountains runs to the channel of Chacao, which separates the large island of Chiloé from the mainland of Chile, and the coastal range re-emerges on the island as the Cordillera Piuche, at over 300 m, and also in the Guaitecas islands, to disappear finally in the peninsula of Taitao.

The third longitudinal feature of Chile and, from the point of view of human settlement by far the most important, is the great alluviated central depression. Again there is a strongly contrasting picture from north to south. In the desert north, huge alluvial fans mark the foot of the Andes and a series of dry basins (*bolsones*), some 80 km wide and over 600 m above sea-level, stretch to the coastal plateau. Formerly lakes, these basins are rich in salts, especially nitrates, the export of which provided Chile with approximately half her government revenue from the 1880s to the First World War. The basins are crossed by deep, ravine-like valleys (*quebradas*) made by intermittent rivers, such as the Aroma, Tarapacá and Quisma, but surface water supplies are very meagre and between 18°S and 27°S only the river Loa reaches the Pacific.

The latter latitude is approximately the limit of a transitional zone which extends to 33°S. In its northern part, in the province of Atacama (Region III),[1] permanent streams begin to appear crossing the longitudinal depression – the rivers Copiapó and Huasco. As one moves south, the river valleys, separated by transverse Andean spurs, widen, and in the province of Coquimbo the rivers Los Chorros, Elqui, Limari and Choapa have formed more extensive plains than exist in the north, as have the Ligua, Petorca and Aconcagua in the province of Aconcagua. South of the river Aconcagua lies the last great Andean spur to cut across the central valley for some 960 km, the Chacabuco ridge, which separates, at a height of 730 m, the Aconcagua valley – 'the Vale of Chile' – from its southern neighbour, the valley of the Mapocho where Santiago stands. From here, the central valley extends south to Puerto Montt at latitude 41° 30'S.

The central valley is divided by the river Bío-Bío into two quite distinct sections. North of that line it is a continuous series of river basins – the Maipo, Mapocho, Rapel, Mataquito, Maule and Itata – with the valley floor sloping from east to west and with a declining elevation from north to south. These rivers, fed by rain in winter and by the melting snows of the Andes in summer, have built up large alluvial fans, and the slope of the valley floor permits irrigation of the rich soils by gravity flow. Near Santiago, which stands at 530 m above sea-level, the eastern and western valley margins lie at approximately 700 m and 330 m, but in the region of the Bío-Bío they are much lower – 300 m and 100 m respectively. Only with the latter river, however, does the broad valley floor reach to the Pacific; the rivers further north cross the central valley more or less at right angles and break through canyons in the coastal ranges to the Pacific.

South of the Bío-Bío the alluvial fans on the eastern border of the northern part of the central valley give way to moraines and lakes, as deep-flowing rivers – the Imperial, Toltén, Valdivia, Bueno, Maullín and Petrohué – and the reappearance of Andean spurs gives a more fragmented appearance to the landscape. At the Gulf of Reloncaví, where Puerto Montt stands, the great longitudinal depression disappears beneath the sea, to emerge again for a short distance as the Isthmus of Ofqui, before finally being submerged in the Golfo de Peñas.

Climate and its effects

While the relief of Chile has clearly been of fundamental importance in historical settlement patterns and must remain a critical factor in the exploitation of natural resources and other economic activity, no less significant is the influence of climate and, above all, the country's rainfall pattern. With a latitudinal extent of 38°, Chile experiences a climatic range no less striking than the variety of its physical features and at least as important in creating the basic character of each of Chile's regions.

Chile's temperatures range from subtropical to subarctic, though extremes are tempered by the proximity of the sea to all parts of the country, by the effects of the cold Humboldt current, which are felt north of latitude 40°S, and by the onshore character of the prevailing westerly winds. Moreover, the Andean cordillera operates as an effective barrier to continental influences. These factors, taken in conjunction with the topographical and, of course, linked to the global patterns of air circulation, account for the remarkable variations in rainfall throughout the country. Rainfall increases from north to south, being higher everywhere on the coast than at equivalent latitudes inland, but lower than on the higher ranges of the Andes. Temperatures steadily decline from north to south, the diurnal and seasonal range being less marked on the coast than inland, while in the Andes it is, naturally, altitude rather than latitude that determines temperature.

Table 10.1 Chile: annual rainfall and mean temperatures at selected locations

Region	Place	Latitude south	Rainfall (*mm*)	Mean temperature (°C) Annual	Hottest month	Coldest month
Desert	Iquique	20°21'	2.1	17.9	21	14.7
	Antofagasta	23°39'	9.0	16.6	25.5	13.3
	Copiapó	27°21'	28.0	16.0	25.7	12.3
	La Serena	29°54'	133.3	14.8	16.4	11.8
Mediter-	Los Andes	32°50'	307.0	15.3	21.2	9.3
ranean	Viña del Mar	33°0'	482.0	14.7	16.1	11.7
	Santiago	33°27'	362.0	14.0	20.0	8.1
	Talca	35°26'	716.3	14.8	22.1	8.5
	Concepción	36°40'	1292.8	11.6	18.0	9.6
Forest	Temuco	38°45'	1360.0	12.0	19.0	7.6
	Valdivia	39°48'	2488.7	11.9	17.0	7.7
Archipelagic	Puerto Aisén	45°28'	2865.0	9.0	13.2	4.6
	Evangelistas	52°24'	2569.7	6.4	8.8	4.4
Atlantic	Punta Arenas	53°1'	437.1	6.7	11.7	2.5
	Punta Dungeness	52°24'	253.8	7.1	11.4	2.5

Sources: Bohan and Pomerantz (1960, p. 36); CORFO (1967, pp.123–49).

The natural environments produced by the interaction of land form and climate are in striking contrast, and the names commonly given to the natural regions of Chile indicate their basic characteristics. From north to south, these are Desert Chile (latitude 17.5°S to 30°S), Mediterranean Chile (30°S to 37.5°S), Forest Chile (37.5°S to 41.5°S), Archipelagic Chile (41.5°S to 56°S) and Atlantic Chile (from about 44°S to 54.5°S).[2] *Table 10.1* illustrates the temperature and rainfall of selected locations within these regions from north to south.

Desert Chile
Known best to Chileans as the Norte Grande, Desert Chile, composed of the provinces of Tarapacá and Antofagasta (Regions I and II) occupies approximately one-quarter of the land surface of Chile, excluding Antarctica. It is one of the driest regions on earth, and includes places where no rain has ever been recorded, and others, again, where extraordinary rainfall in a short space of time makes up a very low average rainfall over a long period of years, as in 1911 (Bowman, 1924, pp. 42–3). On the coast there is a higher relative humidity than inland, more uniform seasonal and diurnal temperatures and, from north of Antofagasta city, the phenomenon known as the *camanchaca*. This is a greyish mist,

formed as follows: cold upwelling water, landward of the Humboldt current, cools the stable air, leading to condensation, and the prevailing temperature inversion prevents vertical air movement and the formation of precipitation. Iquique has almost a third of the year under the *camanchaca*, particularly during May to August, and enjoys cloudless skies only for about 60 days, but the heavy dews along the coast allow xerophytic and herbaceous plants to grow. In the desert hinterland, in contrast, the air is dry, the sky cloudlessly blue, and vast areas are quite devoid of vegetation. However, in *quebradas* shrubs may grow, and dwarf trees can survive in certain areas where tap roots reach underground water, as in the extensive Pampa de Tamarugal. The short-lasting but intensive, and highly infrequent, rainfalls that may occur in desert regions cause them, proverbially, to blossom as the rose for a short while before resuming their customary barren appearance.

On the eastern fringes of the desert important oasis towns and villages, such as San Pedro de Atacama, Toconao and Peine, have survived through centuries, while today large networks of pipelines bring water from the wetter eastern highlands as far as the coast.

Mediterranean Chile

The southern half of the province of Atacama (Region III) and the province of Coquimbo (Region IV), together with the northern part of Aconcagua, is a transitional zone between the Desert and Mediterranean Chile proper. Precipitation increases as one moves south, particularly on the coast; the winter rainfall regime is short in duration and little rain falls, but it is adequate to support typical *matorral* cover, outside the cultivated river valleys. Aconcagua province marks the end of the transitional zone. From the valley of that name to the river Bío-Bío, the climate is truly Mediterranean, with cool to cold rainy winters, mild springs and autumns, and warm, dry summers. Since the *cordillera de la costa* acts as a rain shield south of the Chacabuco ridge, the rainfall and humidity figures for the central valley are lower than on the coast, and temperature ranges are consistently higher. This region, with its rich soils and equable climate, is generally regarded as one of the most suitable for human settlement in the world; its climatological and vegetational similarities to Mediterranean Spain made it immediately attractive for early settlement (Encina and Castedo, 1964, I, pp. 75–7), and, indeed, this was one of the factors that made Chile a distinctive colony of Spain from a very early period, leading as it did to genuine and permanent establishment on the land.

Forest Chile

The clearly defined seasons of Mediterranean Chile, and notably the dry summers in contrast to a winter rainfall regime, begin to lose their distinctive character in the southern part of the region. Linares, for example, near latitude 36°S and approximately the same distance inland as is Santiago on latitude 33° 27'S, has an average annual rainfall nearly three times as great, and the rainy season is a month longer. But it is south of the Bío-Bío that the striking change occurs as migratory depressions from the west make their influence felt in cloud and rain throughout the year, and the vegetation cover markedly reflects the higher precipitation.

Forest Chile is truly named. With cool, wet summers and stormy winters, dense forests replace the scrub-type forest of Mediterranean Chile, and the *espino* as a characteristic tree gives way to the much taller beech, pine and laurel. The province of Valdivia (Region X) is particularly densely forested, with a multiplicity of species.

The region is a frontier region, for, although towns such as Valdivia and Osorno in the respective provinces of the same names go back in time to the colonial period, they were always essentially outposts of the Kingdom of Chile, and not until the mid-nineteenth century was settlement more widespread and permanent.

Archipelagic Chile

If Mediterranean Chile has always attracted man, Archipelagic Chile is quite the opposite, particularly south of latitude 44°S. Only in areas protected from the persistent force of the westerly storms, such as eastern Chiloé and the mainland of Aisén, is settlement possible in this cold, wet region. Puerto Aisén, at latitude 45° 28'S, has an annual rainfall of over 2865 mm, and an annual mean temperature of 10°C, but further south still up to 5100 mm of rain annually has been recorded for some locations. The region, with its maze of channels and islands, is virtually uninhabited in the west, and the quality and extent of the vegetational cover, unlike Forest Chile, is limited by the severity of the climate.

Historically, the region has been of minor significance and, with its paucity of resources and climatic deterrence to man, it seems destined to remain so, unless current exploration for oil in off-shore waters proves successful.

Atlantic Chile

As its name suggests, this region is untypically Chilean in its location. It lies east of the high Andes, on both sides of the Straits of Magellan, and, with a character and climate markedly different from Archipelagic Chile, it is in many respects more an extension of southern Argentina than of Chile proper. Rainfall, again, is the key determinant. Lying in the rain-shadow of the Andes, the region avoids the heavy rainfall of the west and, indeed, comparative figures for approximately similar latitudinal locations on both sides of the Andes are very striking (see *Table 10.1*).

Low, undulating terrain with extensive grasslands characterizes northern Tierra del Fuego, the mainland opposite and, much further north, scattered areas near the Palena, Cisnes and Simpson rivers in the province of Aisén, which is the attenuated northern finger of this region. The wetter and higher areas – southern Tierra del Fuego, eastern Aisén, the *cordilleran* zone of Magallanes – are clothed with evergreen forest.

The initial reasons for settlement in the Magallanic zone during the nineteenth century were as much strategic and political as economic, but late nineteenth and twentieth-century growth of pastoralism and of the oil industry have given Atlantic Chile considerable importance to the country as a whole.

Population

Chile's last national census was held in 1982: it revealed a population of 11,275,440, an increase of over 2,390,000 since the previous census of 1970.[3] In its rate of growth over the decade, at about 1.7 per cent, Chile lagged behind the Latin American average of 2.5 per cent but it shared with most other Latin American states demographic characteristics which account for the very rapid increase of human beings in the continent, chief of which are declining mortality-rates with only slight falls in birth-rates. The death-rate has fallen consistently from 15.7 per 1000 in 1950 to 6.7 in the 1980s, infantile mortality falling dramatically from 153 per 1000 live births to 36.3 over the same period. Birth-rates, though falling considerably by Latin American standards, do not show such a dramatic decline: between 1950 and 1982 they fell from 34 per 1000 to 21.4. One consequence is that, between the censuses of 1970 and 1982, the population over twelve years of age rose by no less than 45 per cent, with obvious implications for employment, education and housing. A complicating factor in assessing demographic change in Chile in the 1970s relates to the number of Chilean refugees who have left the country, voluntarily or otherwise, in disapproval of the military regime: estimates vary from around 300,000 to over one million, though the latter figure seems unreasonably high and probably includes the 300,000 Chileans who habitually work in Argentine Patagonia.

The preliminary results of the 1982 census confirmed trends in distribution as well as growth which had been obvious from previous decennial counts. Among these were a continuing fall in rural population; a staggering increase in urban population (defined as those living in towns of 20,000 and more) by over 20 per cent, and the irresistible rise of the capital, Santiago, and the metropolitan region as a whole, where the population rose from 3,154,765 to 4,294,938 (36.1 per cent). Expressed another way, in the last intercensal period, the proportion of the national population living in the metropolitan region rose from 36 per cent to 39 per cent.

It is a commonplace of twentieth-century Latin America that the other side of the demographic coin to a large increase in numbers is their uneven distribution, both between regions and within regions, with large-scale urbanization and heavy migration from the rural areas to the cities as characteristic modern phenomena. In Chile's case – as in those of Argentina and Uruguay – these movements have considerable antecedents, and the present volume of these changes may be regarded as much as a dramatic quickening of long-established tendencies as something appertaining principally to the twentieth century.

The population of Chile has always been heavily concentrated in the provinces of the central valley from Aconcagua to Ñuble, containing not only the two major cities of the republic, Santiago and Valparaiso, but also what has been the richest agricultural region since the colonial period. This is one of those regions, of which Latin America has few examples, characterized by James (1959, p. 236) as a zone of expanding settlement in which the rate of population growth has been sufficient to allow an expansion of its frontiers without lessening the density of the nucleus. Nevertheless, the total proportion of the national population living in this region has fluctuated over the last 100 years as the northern and southern extremities of the republic were effectively incorporated into Chile during the nineteenth and twentieth centuries. In 1854, the region contained approximately 70 per cent of Chile's 1.5 million people (Encina and Castedo, 1964, II, p. 1140). Thirty years later, not long after the northern desert region and the southern forest area had been made part of the republic, the proportion had fallen to 60 per cent and by 1920 stood at 55 per cent (*Censos, Provincias*, 1964). Not until 1940 was this downward trend arrested: the proportion of Chileans in the central provinces was then 56 per cent and it rose to 59 per cent by the census of 1952, and was held at that figure to 1960 (*Censos, Provincias*, 1964). By 1982, it had fallen but slightly, to 58. 5 per cent.

Provincial variations within the region indicate clearly that the modern return to the picture of 100 years ago is due predominantly to the growth of the conurbations of Santiago and Valparaiso. Between 1885 and 1960 the population of Santiago province rose sixfold and that of Valparaiso province threefold; that of O'Higgins doubled and that of Talca almost doubled, while the rate of growth in the other provinces – Aconcagua, Colchagua, Curicó, Linares and Ñuble – has been much slower, and in Maule the population in 1960 was, in fact, about one tenth less what it was in 1885 (*Censos, Provincias*, 1964). The proportion of people in Santiago province living in the capital city in 1960 was no less than 82.2 per cent (*Censo*, 1960, Santiago, p. 17), while the corresponding figure for Valparaiso was about 33 per cent (*Censo*, 1960, Valparaiso, p. 7).

By the mid-1970s, however, it was estimated that the percentage for Santiago had risen to over 90 and that for Valparaiso-Viña del Mar to over 61. (Cunill, 1977). The grouping of provinces into regions, since 1974, makes exact comparison with previous figures impossible, since census data is based on the new administrative structure, but Table 10.2 indicates the variations in provincial/regional population between 1885 and 1982: what stands out is the disproportionate increase in the total for the regions containing the major cities – Santiago, Valparaiso-Viña del Mar and Concepción.

Certain regions have always been predominantly urban for obvious reasons. The desert and semi-desert areas of the north, for example, have population clusters related to mining and port activities, and over 90 per cent of the population of Tarapacá and Antofagasta was classified as urban in 1960 (CORFO, 1967, p. 379): 20 years later, it was 91 per cent for Tarapacá and 97 per cent for Antofagasta. In the central valley, however, rural exodus as well as natural increase has been a most significant factor in the growth of cities, and the process has been in train for 100 years, increasing in momentum with the passage of time. For the whole country, as it then was, rural population accounted for 73 per cent of the total in 1875; by 1920, however, the percentage had fallen to 54 and by 1960 to 31. Indeed, between the censuses of 1952 and 1960, urban population had an annual growth-rate of 3.9 per cent, compared with one of only 0.7 per cent for rural population (Herrick, 1965, p. 53). By 1980 urban population totalled 9,006,448 compared with rural at 2,097,485, or 80.7 per cent, compared with 73.7 per cent in 1975. In short, the growth of urban population is vertiginous: on present trends, by the year 2000 urban dwellers will outnumber rural nine to one.

Racially, Chile's population has a high degree of homogeneity, with some 65 per cent reckoned as mestizo, 30 per cent white and the rest basically Indian. The mestizo element is more European than Indian, and 97 per cent of the total population is Chilean by origin. The formation of the Chilean people is the product of historical circumstances from the establishment of the colony by Spaniards in the sixteenth century, and to those circumstances we now turn.

Table 10.2 Chile: population by provinces 1885–1980 ('000 – to nearest 1000)

Modern region	Province	1885	1895	1907	1920	1930	1940	1952	1960	1970	1982
I	Tarapacá	54	101	120	119	115	105	109	127	175	273
II	Antofagasta	33	46	113	175	182	146	196	225	252	341
III	Atacama	64	62	64	49	62	85	85	119	153	183
IV	Coquimbo	190	182	190	179	202	248	278	318	340	419
V	Aconcagua	128	102	112	102	105	119	136	145 ⎫	900	1205
	Valparaiso	203	230	281	326	367	428	529	638 ⎭		
Metropolitan	Santiago	364	464	550	735	990	1278	1862	2525	3231	4295
VI	O'Higgins	131	137	133	161	174	202	238	269 ⎫	475	585
	Colchagua	110	115	113	121	123	133	148	164 ⎭		
VII	Curicó	63	71	73	76	77	82	95	111 ⎫	619	723
	Talca	122	123	123	128	145	158	184	214		
	Maule	91	94	81	85	76	71	77	82		
	Linares	110	105	108	121	125	136	155	177 ⎭		
VIII	Ñuble	210	215	218	228	236	245	267	296 ⎫	936	1517
	Concepción	163	177	199	233	273	311	437	559		
	Arauco	70	62	62	63	62	67	77	93		
	Bío-Bío	99	91	96	106	115	128	147	174 ⎭		
IX	Malleco	57	98	109	123	138	155	169	181 ⎫	600	693
	Cautín	53	93	176	255	321	378	387	409 ⎭		
X	Valdivia	39	55	81	121	152	193	247	265 ⎫	749	843
	Osorno	30	41	52	67	89	108	131	150		
	Llanquihué	33	41	54	71	94	118	149	172		
	Chiloé	72	80	87	111	94	103	107	103 ⎭		
XI	Aisén	–	–	–	2	9	17	28	39	49	65
XII	Magallanes	2	5	17	29	39	49	59	76	89	132
	Total	2492	2790	3213	3785	4385	5063	6295	7628	8885	11,275

Source: El Mercurio (Santiago), International Edition, 13–19 May, 1982.

General historical factors

It is a historical commonplace that the Spanish empire in America was never a
unity, despite strong attempts to centralize and control political and economic
life. The hard facts of physical geography, coupled with a scattered pattern of
European settlement, tended inevitably to the growth of regionalism, and in
Chile it was a unique combination of distinctive geographical, racial, political
and economic circumstances that set the colony apart from other regions in much
more than the mere spatial sense.

First, Chile was one of the most isolated of Spain's ultramarine possessions,
the journey from Europe being attended by constantly dangerous factors
– whether one travelled around Cape Horn by sea, or took the overland route
from Buenos Aires. For the former, timing was critical in those storm-tormented
seas; for the latter, apart from the Atlantic crossing and its threat of corsairs, there
were the hostile Indians of the Pampa in Río de la Plata, and the crossing of the
Andean cordillera. This physical isolation from Spain was reinforced by comparative isolation in America, certainly for most inhabitants of Chile. Throughout the colonial period normal intercourse with Peru was by sea, along the Pacific
coastline, and in the national period it was control of the sea that determined
control of the land in Chilean conflicts with Peru. For commerce and political
contact, the sea route was preferable to that by land, which involved crossing the
Atacama desert to the Bolivian *altiplano* and thence to Peru. Though the desert
trails of Atacama provided a hard but negotiable route for the conquest of Chile
from the north by both the Incas in the fifteenth century and the Spaniards in the
sixteenth, the hazards were enough to prevent the establishment of a permanent
way and, for all practical purposes, the northern desert served Chile as an effective natural, though ill-defined, boundary.

The Andean cordillera, likewise, was a visible wall serving the same purpose,
though, again, its passes carried considerable commerce throughout the colonial
period, and, indeed, the Cuyo region of what became Argentina was administered from the sixteenth to the late eighteenth century by Santiago. But if the
Andes were by no means a barrier, neither were they a highway: the closure of the
passes for several months a year by snow, the general inadequacy of mountain
routes and the periodic nature of the traffic again served to reinforce rather than
relieve Chile's sense of isolation.

Another geographical circumstance of Chile militated against the forcible
breaking down of barriers of space and time – namely its comparative poverty in
the precious metals. While the early *conquistadores* of Chile felt the lure of El
Dorado, and some sizeable fortunes were founded on gold extraction north of
the central valley, north-west in the coastal range and south of the Bío-Bío, these
avenues were virtually closed by 1600. The northern deposits were exhausted,
and the southern possibilities cut short by Indian resistance.

Nevertheless, the search for gold predominated as an economic motive in the
sixteenth century, with agriculture and stock-raising as little more than subsistence activities. By 1600, however, Chile's expansion was checked at the line of
the river Bío-Bío by the fierce resistance of the Araucanian Indians.
Consequently, Chile was effectively confined to that part of the central valley
between the Andes and the Pacific on longitudinal lines, and bounded on the
north by the transitional zone to the desert and on the south by the Bío-Bío. As

already noted, this region, some 560 km in length and 160 km wide, is delimited by relief and rainfall alike: its climate is akin to that of parts of Mediterranean Spain, and the combination of fertile soil and adequate water made it attractive for settlement. The boundaries of the colonial Kingdom of Chile, thus established, remained almost static for nearly 300 years. Not until 1882 did a strong military force finally subdue the Araucanians of Cautín and bring the forest regions under the national flag, and it was only in the same decade, after the War of the Pacific against Bolivia and Peru, that the northern desert provinces were also incorporated. The point is reinforced by *Figure 10.1*, which plots the foundation of over 100 Chilean urban settlements: those of the first three centuries to 1800 are almost exclusively in the central region and its marches; those of the last two reflect the territorial and economic expansion that followed these events.

Another consequence of the Indian resistance south of the Bío-Bío was the racial formation of the Chilean people. The Araucanians generally were agriculturalists, cultivating maize, potatoes, squashes and beans, while fish and shellfish were important to the coastal peoples. Terracing was practised north of the Choapa and irrigation north of the river Rapel; the higher rainfall further south and its more even annual distribution made irrigation unnecessary there. Communal cultivation in open glades or in clearings where the forest cover had been burnt off was the norm, the ground being left fallow for a year or more after one season.

At the time of the Spanish Conquest the area north of the Itata river was much less heavily populated than that south of it, and since, as elsewhere in the Americas, the early *conquistadores* brought few women with them, miscegenation began early and continued in the colonial period. A negro element was also introduced subsequently as slave labour. During the sixteenth century two factors affected the original distribution of the Indian races: the first was a heavy demographic decline through war, European diseases and plain exploitation of the natives; the second, the transfer of Indians northwards across the Bío-Bío as labour for the Spaniards. But the rapacity of the conquerors sparked off Indian resistance from the Mapuche – as the Araucanians south of the Bío-Bío were known – and turned the region south of the Itata into a battleground. By the end of the sixteenth century the Spaniards had been obliged to retire north of the Bío-Bío, which became, in effect, the Indian frontier. Thereafter, until the nineteenth century, the Mapuche largely maintained their territory intact, together with much of their culture, though they were subject to missionary penetration and Spanish slave-raiding expeditions, attentions which they periodically repaid by marauding raids into the Spanish lands.

The racial consequences of these events were highly significant. The mixing of the white and Indian races, the process of *mestizaje*, took place in a limited area and, since the Indians who did not submit were beyond a recognized frontier, Chile did not have by the end of the colonial period the threefold structure of whites, mestizos and Indians so characteristic of many other parts of the Spanish empire. The negro element was also virtually completely absorbed in the process. Whereas in 1540 there were 154 whites, 10 negroes and perhaps 1 million Indians of all types in Chile, by 1620 there were 15,000 whites, 40,000 mestizos who were predominantly white, 22,000 mestizos who were more Indian and

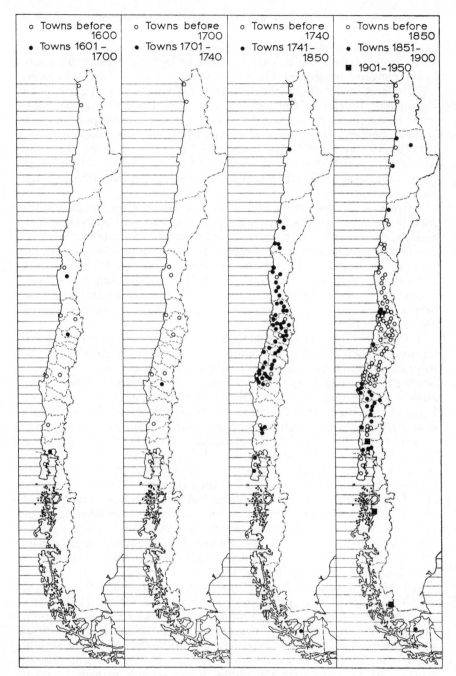

Figure 10.1 Chile: the growth of settlement

negro than white, 230,000 peaceful Indians in the central valley and some 250,000 hostile Indians beyond it (CORFO, 1967, p. 342). By the end of the eighteenth century the best estimate we have gives a total population of 600,000, of whom about 150,000 were white, 350,000 were undifferentiated mestizos, and 100,000 were unassimilated Indians beyond the Bío-Bío (CORFO, 1967, p. 347). While some pure-blood Indians remained within the central valley, they were numerically insignificant compared to whites and mestizos, and the negroes likewise had virtually disappeared as a significant, separate racial element. The mixing of races proceeded more rapidly in the seventeenth and eighteenth centuries than in the sixteenth, one of its characteristics being the large-scale adoption by all races of Spanish plant foods and domesticated animals.[4] In effect, by 1800 Chile was a mestizo and white colony, having a racial composition much less complex than elsewhere in the empire, and having also a cultural character of higher uniformity.

Two other consequences of Mapuche resistance should be noted briefly for their persistent impact on Chilean evolution. With the exception of the *provincias internas* of New Spain (Mexico), where Indian problems were also military ones, Chile was the only part of the Spanish American empire with a permanent military frontier, necessitating a standing army and posing a problem of imperial defence. Spain and Peru had to bear some of the liability, and the standing threat to Chile itself made that geographically compact colony even more cohesive. Secondly, when the Mapuche decisively defeated the Spaniards on the frontier at Curalavá in 1598, they stopped the Spanish search for gold in the south. The settlers were obliged to turn to the exploitation of the real wealth of Chile's central region, its rich agricultural and pastoral land, and in this development there emerged systems of land tenure and exploitation, as well as economic and social relationships, that not only survived to the twentieth century but also gave a particular stamp to the development of the whole nation.

The heartland

Land and labour in the central valley: the evolution of the great estate
As elsewhere in the Spanish empire, the basic instrument of settlement in Chile was the *encomienda*, an institution whereby natives were obliged to pay tribute to the Spaniards in return for the latter's assumption of obligations to Christianize and civilize them. It was a grant of Indians, but not of their land, at least in theory, and land for the conquerors was assigned on other principles. Each member of a new community was entitled to a piece of land commensurate with his rank, as well as to the use of lands held to be common; the former were *peonías* or *caballerías*, reflecting the status of foot-soldiers and cavalry respectively. Additionally, early governors and town councils, and later royal officials, distributed land in legal title to the more prominent as *mercedes*. These were usually huge in size, and the smallest plots, the *peonías*, tended to be rare since few would accept the inferior ranking they implied, the name *peon* being applied only to Indians, mestizos and very poor whites. Labour, in fact, was no less important than land, as no self-respecting Spaniard would work his own property, and the demand for *encomiendas* to provide this, and for siting

mercedes close to Indian villages, soon ran up against the intractable fact of a serious labour shortage in Chile. It was to remedy this by finding more manpower, as much as the desire for gold, that prompted the early conquerors to move south. Their ultimate failure by 1600 did not have such serious effects on the early colonial economy as might have been expected. For economic and demographic change coincided: had gold continued to be of predominant importance, with stock-raising and agriculture of secondary significance, Chile would have faced a critical labour problem at a time when demographic decline was well under way. As it was, the forced change of economic direction at the end of the sixteenth century from mining to stock-raising, requiring fewer hands, salvaged something for Chile out of what contemporaries regarded as an unmitigated disaster.

Though agriculture and stock-raising came second to mining as economic activities in sixteenth-century Chile, they were not simply subsistence activities after the earliest years. The pastoral industry was particularly important in the Araucanian war for the supply of horses, leather and provisions for the frontier forces. Agriculture was chiefly concerned with wheat cultivation, taken to Chile by Valdivia in 1540, where it soon supplanted the Indian staple, maize, in a region particularly suited to its cultivation. In the first 50 years the major wheat zone lay south of the Bío-Bío, with stock-raising predominant in the central valley, but thereafter, with the Indian rebellion, it was transferred to the area between the Bío-Bío and the Maule, and by 1600 a small market in Chilean wheat existed in Peru (Sepúlveda, 1959, p. 13).

The Viceroyalty of Peru, however, was much more significant in the growth of the Chilean pastoral industry, the predominant economic activity of the colony for most of the seventeenth century. The dramatic growth of Peruvian mining, not least in Upper Peru (Bolivia) created a large demand for leather and tallow – for candles and soap – commodities that Chile could supply. The consequences of this development had a profound effect on Chilean land and society: the great estate, destined from the earliest days of the colony when *mercedes* were granted to be a characteristic feature of the countryside, now grew greater, as the Peruvian demand for pastoral products put a premium on extending pastures. Although *encomienda* carried no juridical title to land, it was an instrument of land colonization, as *encomenderos* successfully sought outright grants near Indian villages and equally successfully circumvented the persistent efforts of the Spanish crown and the Spanish church to protect the natives from exploitation. With the rise of stock-raising there occurred an accelerated takeover by landowners both of Indian and of common lands. This was assisted by the increasing process of *mestizaje*, since the incorporation of Indian villages into private estates became easier as mestizos drifted in as squatters and as the erosion of the Indian character of the villages deprived the Indians of legal protection.[5]

But the rise of stock-raising in seventeenth-century Chile not only firmly established the hacienda – now usually called *estancia* – as the dominant feature of the rural landscape; it also called into being another institution which, like the great estate itself, survived into modern times. This was the basic system of rural labour, known in Chile as *inquilinaje*, the development of which was intimately related to changes in the land economy of the colony.

By the mid-seventeenth century, the better parts of the central valley had been divided up into land grants to the most powerful and influential, and later changes in landownership came from sales, gifts, inheritances and donations – all private forms of transfer. Those excluded from the circle of landowners could not now acquire large holdings from the state, though smallholdings might still be acquired from undistributed state domain, as when new towns were founded and *chacras*, lots for personal cultivation, were allocated. There arose a form of tenancy on the *estancias*, commonly called *arriendo* or *préstamo*, initially the letting of a piece of land to a natural son, a kinsman, or at least someone with whom the owner had a personal tie, in return for some specified payment. Examples date back to the early seventeenth century (Góngora, 1960), and the rise of stock-raising accelerated the process. *Estancieros* were now looking for trustworthy tenants to guard their boundaries and oversee stock, and since estates were vast and there were plenty of applicants – not least retired officers from the Araucanian wars seeking to maintain their status – fairly large *préstamos* were granted on which tenants could graze their own stock separate from the owner's, in return for duties performed for him. A decisive factor at this time was still the personal relationship: clearly, with unbranded stock mutual confidence was at a premium and, while *préstamos* varied enormously in detail, two common traits stand out – the temporary tenancy of property on agreed terms, and the tenant's obligation to provide some form of service.

While, in the seventeenth century, some *préstamos* were agricultural rather than pastoral, the decisive change here, and in the nature of the contract, came in the eighteenth century with the rise of wheat cultivation to predominance. A key date for the capture of the Peruvian market by Chilean wheat was 1687; the fortuitous occurrence then of an earthquake in Peru, which disrupted local production and necessitated large imports from Chile, enabled the superior quality of Chilean wheat to establish itself, so that thenceforth Peru became dependent on Chilean supplies (Sepúlveda, 1959).

By this time, along with the continuing process of *mestizaje*, population levels were recovering from the sharp decline caused by the shock of European conquest. Wheat cultivation needed more intensive labour than stock-raising and the hands were now more readily available. The eighteenth century saw not only a large increase in the tenancy agreements previously described as *préstamos*, but now called *arriendos*; it also saw a number of changes in the character of the contracts. Whereas under the stock-raising economy landowners were looking for tenants to guard their herds and flocks and assist at round-ups and at related activities, with cereal cultivation they were now looking for a more fixed labour force to work the land. In the seventeenth and early eighteenth-century tenancy agreements, cash payments often figured as part of the tenant's obligation to the owner, in addition to services, but the latter increased during the course of the eighteenth century at the expense of the cash element. As time passed this tendency increased: for example, about 1760 the new obligation of tenants to provide the landowner with a *peon* for house duties appeared, and from then on personal service obligations became a common feature of contracts. Another increasing tendency, apart from the depression in the status of the tenant, was for tenancies to become hereditary, and by the nineteenth century the rural labour system associated with the great estate had virtually settled into the form in which it was to remain to the present day.

A classic modern description of the twentieth-century *fundo* (hacienda) and of *inquilinaje* (McBride, 1936, pp. 3–14, 146–70), supplemented by more recent research (Bauer, 1975, *passim*) differs little in essentials from the picture of 100 years before (Gay, 1844). Owners of great estates allowed tenants the usufruct of a small piece of land, in return for labour service. The contract between them was a free one, usually verbal, varying in its terms not only from area to area but also from *fundo* to *fundo*, and there was also considerable variety between the labourers themselves in terms of possessions and standards of living. In theory the tenant had complete liberty of movement, and was not bound to the soil as a serf; in practice, by the nineteenth century, *inquilinos* were inheriting their fathers' status and obligations to the family that owned the *fundo*. Their duties included not only the provision of labour in agriculture, stock-raising and general work, such as clearing irrigation channels, but also personal service at the master's side or in his house, though all work done in excess of the agreement was paid for by the master, sometimes in cash, much more often in provisions. The tenant would till the land he lived on on his own account, and he might also own stock and reach agreement with his master about pasture. If the contract ended, the tenant took the harvest of his own land, but got nothing for any improvements he might have effected. At the beginning of the nineteenth century, *inquilinaje* was the predominant form of rural labour in Chile; it dominated the central valley, though it was weak in its northern and southern extremities, and it had been elaborated by the growth of a hierarchy of workers on the estates, contracted in the same way, from the lowest hand to the *mayordomo* with supervisory duties over agricultural workers and the *capataz*, overseer of the pastoral sector. These persons also held their own land and stock, but their contract with the owner revolved more around the cash nexus than did that of the mere worker.

The great estate, the *fundo*, and the system of *inquilinaje* proved outstandingly stable elements over more than two centuries, with profound consequences for Chile's political and economic evolution. Moreover, they reflected the racial structure of the country. The landed, white aristocracy ruled the national life in all its branches, while an illiterate, mestizo peasantry obeyed, and such crucial historical events as the securing of political independence from Spain early in the nineteenth century had little or no effect on Chile's internal social structure. This is not to say that fragmentation of land and extension of ownership were not continuing processes in the central valley from the seventeenth century to today, nor that *inquilinaje* was by any means the sole system of rural labour in Chile: as will be seen shortly, other forms of landholding and of labour have their own significance. Nevertheless, the *fundo* remained the characteristic feature of the Chilean countryside. In 1925, in the fourteen provinces of Chile from Coquimbo to Bío-Bío, 5396 estates over 200 ha in size accounted for 89 per cent of all farmland, with over 76,000 other properties accounting for the other 11 per cent (McBride, 1936, p. 124). But, within these figures, 1507 properties were larger than 1000 ha each and accounted for 73 per cent of all farmland, with a number of estates of more than 100,000 ha (McBride, 1936, pp. 125 ff.). More than 30 years later, 4709 properties accounted for no less than 87.9 per cent of all farmland (CIDA, 1966, p. 48), and indeed, in Chile as a whole, the concentration of ownership was such that 73 per cent of land in farms were in holdings of over 1000 ha, representing only 2.2 per cent of the total number of farms (Warriner, 1969, p. 326).

But, while the traditional *fundo* survived, *inquilinaje* had been a declining institution for over a century. Comment on the depressed conditions of the *inquilinos* was widespread in both the nineteenth and in the twentieth centuries, though often balanced by the view that, in comparison with the share-cropper or vagrant hired labourer, the *inquilino* had, in his plot of land and social relationships, at least a degree of stability and certainty. Conditions varied enormously from province to province and from estate to estate, depending greatly on the individual character of the landowner. Significant changes in *fundo* ownership did occur in the mid-nineteenth century when adverse economic conditions caused many of the old aristocracy of the land to sell out to *nouveaux riches*, men grown wealthy in commerce, banking and mining: the change does seem to have affected the old paternalism of the *fundo* to the detriment of the *inquilino*. Absenteeism of the landlord from his land became more pronounced, and many estates were run down since a number of new owners were less interested in land as an economic investment than as a sign of arrival in the ranks of the Chilean aristocracy. The upward rise of the demographic curve paralleled this change, and pressure of numbers on the land began to be relieved by peasants leaving it, a rural exodus shown by regional study to have been well under way in the later nineteenth century (Borde and Góngora, 1956, I, pp. 113 ff.). For however much local circumstances prevented him from exercising it, the *inquilino* always had legal freedom and, where conditions became intolerable, an avenue out of farming could be found in the mining regions of the north or the expanding frontier to the south. Thus, in 1929 it was reckoned that there were about 256,000 *inquilinos* in Chile, out of a rural labour force of 386,000 (McBride, 1936, p. 164); in 1955, however, by which time the modern city and industry had greatly developed in Chile, of a total farm population of 664,000, there were only 82,000 *inquilinos* (Warriner, 1969, p. 327), and a study of land division in central Chile made a little earlier revealed how attenuated *inquilinaje* had become by then (Martin, 1960).

Minifundia and landless labour: the historical context
The origins of the *minifundia*, of which perhaps the most workable definition is that of farms too small to provide a living for their owners from agriculture alone (Smole, 1963, p. 2), like those of the *latifundia*, lie in the colonial past, and the major factor in their creation was subdivision of larger properties, mainly through inheritance, in the colonial period and after. In Putaendo in the province of Aconcagua, for example, a certain Pedro de Silva had built up an *estancia* in the mid-seventeenth century but on his death four lots — *hijuelas* or 'little daughters' — passed to three sons and a daughter. One son died in 1700, leaving his land to be divided between seven sons, and one of these lots was itself divided in 1731 into six properties. By 1821, of the four original parcels of land left by Pedro de Silva, one had been subdivided into eighty properties (Baraona, Aranda and Santana, 1961, p. 159). It is true that inheritors of great estates generally sought to maintain properties intact by family compacts, such as providing for many sons in other ways than through land so that one heir might hold his father's property, and a number of *hacendados* received from the crown

in the colonial period the grant of *mayorazgo*, the entailed estate that could *not* be broken up as ordinary Spanish inheritance laws demanded. Nevertheless, in the first two centuries of the colonial period estates were subdivided, sold and given away, reconstituted and, in some few cases, maintained in their original size. Thus, of the twenty-five hacienda proprietors in Puangue in Santiago province in the seventeenth and eighteenth centuries, only one family property survived to the middle of the nineteenth century with the holding largely intact (Borde and Góngora, 1956, I, p. 60). The two processes of the maintenance of the great estate on the one hand, and its subdivision on the other, went on side by side in the eighteenth 'and nineteenth centuries: hence *latifundia* and *minifundia* in Chile have basically a common origin.

During the nineteenth and twentieth centuries, however, another factor besides inheritance created the division of the larger properties, namely the commercial opportunity offered to great landowners to sell a portion of their holding to classes and individuals having no traditional connection with the land, and not really wishing to acquire it for commercial farming, but seeking rather a hedge against the inflation that has been a feature of the Chilean economy since the latter part of the nineteenth century (Martin, 1960, p. 38). Not surprisingly, lots acquired in this way changed hands frequently, but more characteristic types of the small estate are those of owner-cultivators, who have a genuine attachment to the land, and those of community holdings. Included in the latter are Mapuche reservation lands, particularly in the province of Cautín, held in common but with the usufruct individually apportioned (Faron, 1968, pp. 15–21), and peasant communities, notably in the northern provinces, especially Coquimbo, where individual holdings are complemented by communal lands, usually grazing grounds and unirrigated areas (CIDA, 1966, p. 128).

It is no less impossible to generalize about owner-cultivators and *minifundia* in Chile than it is to do so for great landowners and their estates. In the 1960s Smole's study of owner-cultivatorship (1963), however, in a carefully selected region of Middle Chile, where both *minifundia* and *latifundia* were found, suggested certain important characteristics of this type. Allowing for enormous variations in farm size, wealth of natural resources (notably availability of water), tenure and labour supply in the study area, owner-occupied farms generally functioned as self-sufficient units, and only to a very limited degree were they market-orientated. They lacked capital resources and access to credit, and their physical isolation and limited resource-base further limited opportunities for commercial development. Labour was drawn chiefly from the immediate family and friends, and, while share-cropping arrangements and use of casual labour were not neglected where they were needed, the farmers preferred to rely on traditional and family relationships in their work. The sense of community was strong, and while the growth of some markets created an empirical balance between production for use and production for gain, the latter decidedly came second in motivation. In short, these properties represented a way of life as much as a way of earning a living. The increasing impact of agents of change – improved communications, radios as well as roads, visiting emigrants, the operation of selective military service, drawing young men out of their environment for a spell – was breaking down the isolation of such owner-occupied areas, but doing little to modify farm technology, though cases existed of a successful shift to more commercial production where its advantages had been clearly demonstrated. In fact, the new market opportunities created by increased urbanization in Chile were largely taken up by great estates and medium-sized properties, which could operate on the required scale, in contrast to these more self-sufficient units with their conservative ethos and traditional practices.

Owner-occupiers, however, with an adequate resource-base for self-sufficiency in basic needs and, perhaps, a small marketable surplus, were much better off than the smaller *minifundistas* and the landless labourers. In 1955, of a total of 329,000 of small landowners, 70,000 had farms too small to support a family, and worked part-time on other properties, while of 335,000 wage-paid workers, including *inquilinos* and administrators, there were 27,000 sharecroppers and 180,000 seasonal labourers (Warriner, 1969, p. 327). The landless were permanent labourers (*voluntarios*) or migrants (*afuerinos*), the latter created by past growth of commercial agriculture and by the fact that Chile's wide climatic range and variety of crops created a peak demand for labour in different regions at different times; hence, well-paid but transient jobs were available to migrants. Sharecroppers (*medieros*) could also be migratory, usually contributing all their labour, half the seeds, fertilizer and other operating costs, while the owner, apart from the land, contributed the rest of the inputs, living expenses and credits necessary until the harvest, which was divided equally (Thiesenhusen, 1966, pp. 18–19).

Land reform, revolution and reconstitution: 1960–80
The deeply entrenched forms of land tenure in Chile were challenged in the 1960s by demands for reform, part of a continental movement but with specific implications. The basic economic arguments for change entailing the break-up of the great estate were that it was not sufficiently productive and, more important, that it perpetuated the extreme inequalities of income characteristic of Chile as a whole. On the first point it was argued that the *fundo* system inhibited higher productivity and prevented the agricultural sector from producing what it could and should, given the country's climate and soils (e.g. CIDA, 1966, pp. 203–6; Thiesenhusen, 1966, pp. 30–1). Indisputably, agricultural production did not match population increase: in the 1960s Chile was importing some two-thirds of her food requirements, temperate-zone foodstuffs – cereals, meat and dairy products prominent among them – with a significant effect on her balance of payments. The inelasticity of agricultural production, it was alleged, played a major role in one of the country's endemic economic weaknesses, heavy inflation. The *fundo*, it was argued, manifestly failed to supply Chile with food because much land was under-utilized, too much irrigated arable was under pasture, there was low investment in technical improvement and labour was wasteful. The assumption of reformers was that, if large estates were broken up and a great expansion of owner-occupance resulted, coupled with adequate credit and technical inputs, inefficiency would be overcome, since land would belong to those who worked it and this would provide its own incentive. A greater equalization of rural income would result, leading not only to a more just society, but also stimulating the industrial and commercial sectors of the economy by creating more of a mass market.

Opponents of land redistribution, however, and notably the SNA (Sociedad Nacional de Agricultura), the landowners' professional organization, also had strong counter-arguments. If the *fundo* had failed to feed the people, it was alleged, the fault lay less in any inherent structural defects than in policies pursued by successive governments, which had the effect, if not, indeed, the intention, of penalizing the rural sector to support the industrial through pricing policies for agricultural and pastoral produce. Urban consumers benefited at the expense of rural producers (Mamalakis, 1965, pp. 117–48). Under-capitalization on the land was thus the result of lack of incentive: food imports might have been reduced and local production increased by raising farm prices, and tariffs on imports might have helped reduce balance-of-payments problems much more quickly than land redistribution, itself necessarily a process of dislocation as well as one needing large inputs of capital and technology if production were not to fall further.

The polemic for and against redistribution of land dominated rural affairs in the 1960s, and the proponents of the measure argued insistently that to diminish existing gross inequalities of income distribution it was necessary to redistribute capital in the form of land. They assumed that the high cost on new inputs to owner-occupied and collective properties could be met, that managerial bottlenecks could be surmounted, and that the major incentive of ownership by peasants would overcome supply problems. The expression of their success was the Agrarian Reform Law of 1967, passed under the Christian Democratic Government of President Eduardo Frei (1964–70), which succeeded a law of 1962 generally regarded as ineffective. Under the 1967 law properties larger than 80 ha of basic irrigated land could be expropriated, but efficient owners of larger properties (up to 320 ha) could claim exemption. Norms of efficiency were clearly laid down, with latitude allowed to a new National Agrarian Council to decide questions of expropriation. This Council (CORA) was responsible for running expropriated estates and distribution to peasants, and due regard was paid to compensation provisions for affected landowners. A useful summary is in Warriner, 1969, pp. 340–5.

President Frei had expressed his desire to create 100,000 new farms for peasants (who would be given the option after 2 or 3 years working collectively with CORA to choose collective or individual ownership thereafter); in fact, by July 1970 less than one-tenth of the 3.5 million ha expropriated was irrigated land, and less than 37,000 families had benefited. There were many reasons for this slow progress: intense political opposition on varied grounds to the law, which took nearly 2 years to get through Congress, a disastrous drought which afflicted Chile for much of 1967 and 1968, and the complexity of the law itself. By the end of Frei's term in 1970, three effects were apparent: a much higher political consciousness in the countryside, expressed in the swift unionization of rural workers; a higher standard of living for beneficiaries of the reform, and, more significantly, the rapid creation of expectations among the peasantry which the reforming government itself could not satisfy. Its successor reaped the reward in the presidential elections of 1970, when the rural vote was a significant factor in the election of the Marxist coalition of Salvador Allende, part of whose programme was a promise to accelerate rapidly the redistribution of land and the destruction of the great estates.

Under Allende (1970–3), the process was certainly speeded up, but with mixed consequences. Throughout, the Minister of Agriculture Jacques Chonchol was under pressure from the left-wing, quasi-revolutionary movements, and notably the Movement of the Revolutionary Left (MIR), which fomented peasant seizures of land as part of its strategy to radicalize the government's programme. While frequently condemning such activities verbally, the government, in effect, took no positive steps to curb them, while itself accelerating land distribution under Frei's law of reform. Under that law, previously expropriated lands had constituted *asentamientos*, in which the peasants worked the land as collective units under the management of CORA; it was a strong criticism of the new Popular Unity (UP) government of Allende that the system was paternalistic, did not permit real peasant participation, and was intended to create a new class of peasant proprietors who would form a new rural bourgeoisie (Chonchol, 1980, pp. 68–70). Under these arrangements peasants made agreements for 3 years with CORA, whereby they provided their labour and modest amounts of capital, while CORA provided land, water, credit, technical and administrative experience, though shortage of suitably trained personnel was always a bottleneck. The Agrarian Reform Societies (SARAS) thus formed were to endure for a maximum of 5 years, after which peasants were to exercise their option to go for individual ownership of land, maintain the SARA or choose a co-operative structure among themselves. Nine hundred *asentamientos* were formed under Frei, of which 109 completed the transition period, 95 choosing co-operative ownership and 14 the mixed form (World Bank, 1980, p. 68). With a stronger ideological commitment to state supervision, if not control, the UP government allowed SARAS to function at the local level, but also created new regional co-operative structures, known as Agrarian Reform Centres (CERAS), and state farms (CEPROS) where large capital inputs would be required. Moreover, it declined to award individual title to *asentados* who wished it and had completed the transitional period. (For detailed treatments of the subject see Chonchol, 1980; World Bank, 1980; Kay and Wynne, 1974.)

Moreover, certain UP spokesmen threatened to reduce the basic limit of 80 ha of irrigated land allowed to expropriated landowners to 40 ha and even 20 ha, while unauthorized land seizures, subsequently ratified by the government, created further uncertainty. The outcome, apart from increased bureaucratization of the state-intervened rural sector, was decapitalization. Southern dairy farmers preferred to drive their cattle over the cordillera to Argentina; output fell dramatically, and Chile was obliged to increase its food import bill by large amounts: while, with the land expropriated under Frei, the UP took over 10 million ha, almost half the country's agricultural land, only about 12 per cent of agricultural labourers received land, with preference given to those already attached to it, and *minifundistas, voluntarios, afuerinos* and indigenous communities derived little or no benefit at all. Not one single land title was issued under the UP, the state retaining overall control of expropriated land. By late 1973 296 CERAS and 76 CEPROS had been created, benefiting, in theory at least, 50,000 families (World Bank, 1980, p. 90). But the consequences for agricultural output and the balance of payments were very serious. Between 1971 and 1973 overall annual output fell by 27 per cent (*Hoy*, 1981). Food shortages were a major factor in the fall of Allende in 1973.

Since then the successor military government has dismantled the apparatus of the UP's agrarian reform. In line with a general free-market philosophy, it stopped expropriations, restored land held to have been seized illegally and confirmed just title to those who had received it under the Agrarian Reform Law. It freed agricultural prices, except on three products, and drastically reduced the role of the state in providing agricultural credit, shifting that responsibility to the commercial banks. It has favoured *asentados* and notably individually owned family farms, as *Table 10.3* indicates:

Table 10.3 Allocation of expropriated land (Sept. 1973–Feb. 1978)

	'000 ha	%	Basic irrigated '000 ha	%
Total expropriated	9,965.9	100	895.7	100
Assigned to reform-sector farmers	3,097.8	31.1	458.0	51.1
Returned to original owners	2,833.0	28.4	237.5	26.5
Transferred to other public sector institutions	671.3	6.7	20.2	2.3
Still held by CORA	3,363.8	33.8	180.0	20.1
CORA reserve	(865.8)	(8.7)	(36.2)	(4.0)
Forest and dry land	(2,498.0)	(25.1)	(143.8)	(16.1)

Source: World Bank, 1980, p. 184.

Such policies, allied to a strong export-oriented thrust in agro-products – notably fruit, vegetables and wine – had two major results in the 1970s: a massive surge in exports in such products and a decline in 'traditional' foodcrops. From 1973 to 1977 food and agricultural exports experienced a sevenfold growth, and a survey of reform-sector farmers at the end of 1976 indicated a stronger commercial, as opposed to subsistence, operation than previously. However, there is no doubt that the new freedom in transferring titles had benefited those with capital and other resources at the expense of the poorer farmers, and that union power – not viewed sympathetically by the present government – had declined appreciably. The traditional *minifundistas* – numbering about 182,000 or 42 per cent of the farm population – appear to have been little affected by these changes, but suffered more, in tight credit conditions, from the lack of financial support and technological inputs; and it may be doubted whether, outside the rich Central Valley where access to markets through middlemen is far easier than elsewhere, they have participated in the general expansion of the economy in recent years. What is not in doubt, however, is the increased commercialization of Chile's farm sector, and the opening to larger-scale operations with easier access to credit and technology: these have undoubtedly boosted the country's exports massively, but at the cost of both subsistence and commercial farming of basic foodstuffs, necessitating increased imports. Hence, while the area planted to fruit for export increased considerably between 1973 and 1980 (in the case of apples and grapes by over 50 per cent), production of staples such as wheat, beans and rice has fallen (see *Hoy*, 1981 for a detailed breakdown). Chilean imports of foodstuffs rose over 150 per cent by value between 1979 and 1980.

More significantly in the long run, the process of reconstituting the Chilean agrarian sector, after the undoubted chaos of the years of the UP, carries considerable implications of a social kind for the future. Unless the *minifundistas* and the poorer sectors of the rural population can secure increased access to credit, they will play an ever-diminishing role in the rural economy, and the divergence between their subsistence activities and those of commercial farming will simply grow, with what consequences cannot yet be foreseen. A further consequence of promoting export crops at the expense of staples will probably be an increased volume of basic food imports: during the 1970s Chile was able to afford this, on the strength of her proved export capacity and expanding markets, fully utilizing the model of comparative advantage in world markets. But this is an unpredictable asset in the likely world trading conditions of the period to the year 2000, whereas Chile's rich, irrigated soils are capable, with appropriate inputs to promote efficiency, of feeding a population twice her current size at least. Modifications in land use and in land tenure are, therefore, likely to re-emerge in the remaining years of this century.

Centralization, urban growth and industrial development
In dealing generally with Chilean population, brief mention was made of the urban concentration of people in the two key provinces of the central valley with their eponymous conurbations, Santiago and Valparaiso. At the 1960 census no fewer than 41.5 per cent of all Chileans lived in these two provinces, which also accounted for 54.5 per cent of the country's total urban population. Moreover, the city of Santiago itself contained 84 per cent of the provincial population, while the conurbation Valparaiso–Viña del Mar held 60 per cent of the total for the province of Valparaiso (calculations from data in *Censos, Provincias*, 1964, Santiago and Valparaiso). Twenty years later, in 1980, the metropolitan region, with over 4 million people, contained more than a third of all Chileans with over 94 per cent of the provincial population within the city boundaries. While some other urban centres showed comparable rates of growth in the period, this has done little to to affect the particular dominance of the Chilean economy and society by the central valley in general and by Santiago in particular.

The process of centralization in Chile has undoubtedly gathered increasing momentum in recent years, but it is as much a product of a long historical evolution as it is of contemporary development. Like other capitals of Latin America, Santiago plays the dominant role today because it has always done so. Colonial Santiago was roughly equidistant from the northern and southern confines of the Kingdom of Chile, and equally favourably placed with regard to the Andean passes to Cuyo; it occupied a key position in the rich central valley, so long the basis of the colonial economy, and it was only 130 km by road from a suitable port on the Pacific, Valparaiso. As the seat of the governor and of the *audiencia*, the religious centre of the colony, and, from the mid-eighteenth century, the site of the only university in Chile, Santiago's administrative and cultural predominance was unchallenged throughout the colonial period, and effectively underlined its economic and strategic importance. After independence a combination of circumstances increased Chile's centralization on Santiago. The political organization of the republic, in contrast to many other Latin American states, took shape in a unitary, rather than a federal, constitution, introduced in 1833 and lasting (though with amendments) until 1925. This constitution concentrated decision-making in the national capital, such local autonomy as remained to the provinces being truly parochial in form and content. Equally significant as centralizing factors were the growth of communications, notably railways, and the emerging economic structure of the country from the mid-nineteenth century.

The railway system in the central valley was state-owned almost from the beginning, and it served the national purpose of binding the outlying provinces to the core region, as the lines ran north and south, east and west from Santiago. The great central trunk line, running today from Zapiga, some 300 km south of Arica in the desert north, to Puerto Montt in Forest Chile – over 3000 km in length – was built from the centre outwards, beginning in the 1850s. By 1859 it had reached Rancagua, 82 km south of Santiago; by 1862, San Fernando, 139 km; by 1893, Temuco, 690 km; and by 1913, Puerto Montt, 1079 km. Branch lines were built during the same period to feed other towns in the central valley and ports on the Pacific: thus, the very important Valparaiso–Santiago line was begun in October 1852, and formally inaugurated in September 1863 (Bohan and Pomerantz, 1960, p. 195). With the northern extensions, however, it was a different story, the network in both the Norte Grande and the Norte Chico (Atacama and Coquimbo) being built mainly by foreign capital and held in foreign ownership until more recent times. The growth of these networks was intimately connected with the exploitation of mineral wealth, nitrates in the Norte Grande and copper in the Norte Chico, during the nineteenth century. Chile's first railway, indeed, and the first railway of any considerable length in South America, was opened in early 1852 to link the copper mines of Copiapó to the port of Caldera, 41 km away.

In these facts lies another basic reason for the persistent centralist character of the Chilean political and economic system. For much of the nineteenth century, and well into the twentieth, a high proportion of government revenue came from export taxes on minerals mined in the northern provinces, but the government in Santiago limited its role in the mining industry predominantly to the collection of these taxes, leaving to private enterprise the ownership and exploitation of the mineral deposits. This policy certainly resulted in sizeable revenues, easy to collect, for government financing of major public works such as railways and port installations; it can also be argued, however, that it left the export-orientated mining sector as something of an economic exclave of the nation, and thus helped to prevent the possibility of the northern mining industry becoming the basis of local growth points within the national economy. Such points might well have come, in time, to challenge the dominance of the core region and the metropolis. A less conjectural result of this development was to make it unnecessary for government to raise essential income from land taxes on the aristocracy, a negative factor in the latter's continued social pre-eminence and political influence, exercised in the national capital and the central valley. This situation was compounded by the social habits and inclinations of the Chilean *nouveaux riches*, men who had risen to prominence through mining, banking and commerce during the nineteenth century. Rather than constituting a separate social force from the traditional aristocracy, they sought above all to emulate and join it through marriage and the purchase of land, and the Chilean upper class showed a remarkable resilience not only in accepting these new elements but also, in fact, in absorbing them. Thus, the emergence of new bases of economic power in mineral exploitation, in contrast to land as the source of wealth in the colonial period and after, did nothing to change either the traditional political structure or the centralist tradition.

At present the overwhelming dominance of Santiago in Chile is a fact, and if its growth may have seemed inevitable historically, its future expansion may well seem irreversible in the light of prevailing circumstances. As *Figure 10.2* indicates, the city grew moderately and regularly during the colonial period around the original nucleus of settlement on the southern side of the Río Mapocho, near the Cerro Santa Lucía. At the time of independence it contained

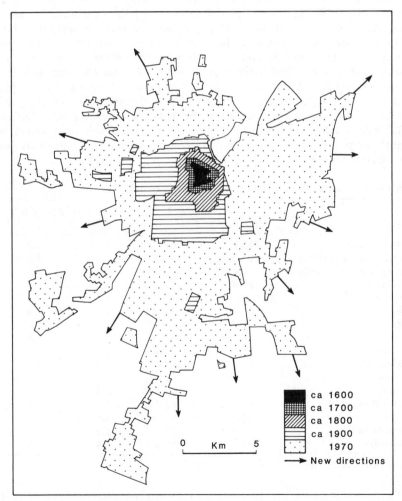

ca 1600
ca 1700
ca 1800
ca 1900
1970
New directions

0 Km 5

Figure 10.2 The growth of Santiago, 1600–1970 (after Cunill, 1977)

some 40,000 to 50,000 people, almost a tenth of the total Chilean population. The nineteenth century, and notably the second half, saw a considerable increase in both physical extent and numbers, not least with the extension of residential suburbs to the west and south, and also north of the river. By 1900 Santiago contained some 300,000 people, still only one-tenth of the national population. In the twentieth century, however, more-or-less regular spatial growth and a population increase more-or-less commensurate with national expansion of numbers has given way, with increasing momentum, to a rapid enlargement of the city boundaries and a vast rise in population. Physically, the city's growth has been most marked on the east, south and west, the eastern side containing the *barrio alto*, the fashionable residential sector. Lower land values in the less attractive southern and western sectors have also attracted residential development, but for the less well-off. But, if Santiago's physical growth has been dramatic, especially in the past 30 years, its population increase has been astounding. Between 1900 and 1925 population doubled – to 600,000 – but by 1952, a census year, it had more than doubled again – to 1,438,000. Between then and the next census, in 1960, it rose by a further 500,000, to contain more than a quarter of the total national population. As *Table 10.4* indicates, the average annual growth-rate of Santiago's population in these years was 4.2 per cent and, while it fell slightly to 3.7 per cent in the next inter-censal period (1960–70), the actual population rose by 878,000, increasing by a further 1,215,000 by 1980, to a total of just over 4 million. The table also shows how primate a city Santiago is (disaggregated figures for cities within regions are not yet available for 1980).

Table 10.4 Chile: population growth for selected cities, 1952–70

City	Population		Annual growth-rates	
	1960	1970	1960–70	1952–70
Gran Santiago	1,983,945	2,861,900	3.7	4.0
Valparaiso–Viña del Mar	384,324	430,000	1.2	1.7
Concepción–Talcahuano	314,412	326,200	0.4	2.2
Antofagasta	87,860	125,100	3.6	4.0
Temuco	72,132	110,300	4.3	4.3
Talca	67,463	94,500	3.4	3.0
Arica	43,334	87,700	7.3	9.1
Chillán	46,774	87,600	2.7	2.9
Rancagua	53,318	86,400	4.9	4.4
Valdivia	61,334	82,400	3.0	3.4
Osorno	54,693	68,800	2.3	3.0
Iquique	50,665	64,500	2.4	2.8
Puerto Montt	41,681	62,700	4.1	4.4
La Serena	40,854	61,900	4.2	2.8
Punta Arenas	51,200	61,900	1.9	3.3
Coquimbo	33,794	50,400	4.1	4.6

Source: World Bank, 1980.

As *Table 10.4* shows, it is the predominantly rural and backward provinces of the central valley which, since the late nineteenth century, have shown the smallest growth in population, and that the provinces containing the largest cities – Santiago, Valparaiso and Concepción – show the highest gain. Nor is this surprising. Provinces such as O'Higgins, Colchagua, Linares, Talca and Ñuble, though now regrouped into regions which makes precise comparison more difficult for the latter years, have found their populations feel increasingly the lure of the city, though Herrick's (1965) careful study of internal migration indicated a predominantly urban origin of recent migrants to Santiago, rather than an immediate rural one. He proposed, in fact, a two-generation model of migration: a first generation movement of people from countryside to towns, usually of less than 50,000 population, and a second generation movement from the latter to Santiago. This process represented, therefore, a smoother accultura-tion of rural population to new ways of life than the bare statistics might suggest. But the increasing momentum of these trends, against a background of rapidly rising population in the past 40 years in particular, has produced in Chile a real megalopolis, which even an abundance of economic resources and a wealth of planning talent would find very difficult to overcome.

The social consequences of Santiago's vast and rapid expansion are obvious in the *callampas* (mushrooms), those squatter settlements which have acquired a permanent character. The provision of amenities such as piped water, electricity and telephones to these deprived environments increased markedly during the Frei and Allende governments (1964–73) and a good deal of low-cost housing was also erected. Nevertheless, in 1981 it was estimated that there was a deficiency of some 700,000 housing units in Santiago (*Ecosurvey*, 6 July 1981) and while the yearly average of housing units completed nationally rose from 19,474 in 1965–70 to 24,400 in 1974–8 (Méndez, 1980, p. 94), this is a rate of growth quite incommensurate with the need.

On the other hand, other urban amenities have greatly increased in the 1970s and early 1980s. Perhaps the major development has been the construction of the Santiago underground transport system, providing an efficient, cheap and rapid method of travel. By 1981 over 25 km of line were in use, with further extensions under construction. Not the least advantage of the metro will be the reduction in the number of buses and the volume of exhaust fumes: Santiago's smog problem has increased dramatically in the last 30 years, and while its topographical situa-tion in the lee of the high Andes is a constant factor, it is the growth of industry and of road transport which have been the major causes of air pollution.

Under the military government successive mayors of the city have, to some extent, remodelled downtown Santiago, creating traffic-free shopping precincts (*paseos*), and renovated many of the old colonial buildings, giving the city a much more pleasant aspect than formerly. The construction boom of the late 1970s also entailed the building of many high-rise office blocks and apartments, and the visitor to Santiago of 20 years ago would be surprised at today's transformation.

Urban growth in the central valley of Chile is, of course, intimately linked to economic development, but it has been economic development of a particular kind. Whereas in western Europe there was a direct and positive connection between the growth of cities and the rise of industry, notably in the nineteenth century, in Chile, as in much of Latin America, the urban tradition predates the industrial age, and city growth as an aspect of industrialization is a fairly recent development. Before the 1930s Santiago and Valparaiso experienced a historical evolution more akin to that of cities of the Europe of the Middle Ages than of the Industrial Revolution: they were centres of trade and commerce, government and administration – in short, service centres – rather than factory-dominated, urban complexes of manufacturing industry. The reasons for this are many and involved, but, stated briefly, they relate to the historical evolution of the country's economic structure.

From the colonial period to the present day the Chilean economy has been based on primary commodity exports from the farm and the mine. In the nineteenth century, and well on into the twentieth, consumer needs were essentially satisfied by imports from manufacturing countries and by domestic, small-scale production of such commodities as foodstuffs and textiles. A comparatively small internal market, in which a heavily rural population lived on mostly self-sufficient great estates, offered little stimulus to the growth of industry, except in particular times and places when conditions were right for its introduction. Thus, for example, the growth of the Chilean wheat trade, at its height during the nineteenth century, was accompanied by the development of the flour-milling industry, first in the southern part of the central valley near Tomé and in the basin of the Maule river, sites with good access – and therefore cheap transportation – to both the wheat fields and to the Pacific outlets. As the export trade in flour grew, the need both for more capital and a better system of marketing concentrated the business in the existing commercial centre of Chile, Valparaiso, where foreign merchant houses performed the functions of credit institutions and consigners for the growers and millers in the south. By the 1860s, however – the real beginning of the modern age in Chile – Santiago was becoming Chile's centre of banking and credit and, assisted by the concentration of railways on the capital and by the growth of the export-orientated wheat agriculture on the estates around it, Santiago province supplanted the south as the focus of the milling industry, and Santiago city supplanted Valparaiso as the commercial centre of the wheat trade. The growth of Santiago itself had some effect on manufacturing, with textiles and food-processing, glass-blowing and sugar-refining appearing as typical small-scale industries. Railway construction was a further stimulus to industrial growth: workshops belonging both to foreign-owned private companies in Valparaiso, and to the state railways in Valparaiso and Santiago, produced many of the locomotives and passenger and freight cars that ran on Chilean lines in the latter part of the century. The War of the Pacific, 1879–83, with Bolivia and Peru, was another stimulus to industry for the provision of military equipment such as munitions and clothing, and the annexation by Chile, after the war, of the nitrate provinces of Tarapacá and Antofagasta saw some demand in Chile itself for equipment to exploit the deposits, though the bulk of this was imported, chiefly from Great Britain.

Such developments cannot be ignored, but with few exceptions, they did not lead to the real growth of heavy industry in Chile, though light industry continued on an upward curve of production thenceforward. Probably the major reason for this (bearing in mind the nature and structure of Chilean society as already described) was the simple fact that windfall wealth, in the shape of export taxes on nitrate and copper, made it unnecessary, in the eyes of its leaders, for Chile to industrialize. Dependence on mineral exports enabled successive

Chilean governments to meet commitments, develop public services to a certain degree and service their foreign debts, without needing to tax the Chilean aristocracy heavily. Demand for luxury and consumer goods, imported from abroad, continued to be met, since here again there seemed no necessity to go in for home production of such goods.

The weakness of this position was exposed time and time again with the inevitable fluctuations in world demand (and therefore exports of the minerals), but the shock was not sustained until the First World War. Then, cut off from their traditional markets for minerals, and from their normal suppliers of necessary imports, Chileans began to realize the fundamental weakness of their position in the world economy. Moreover, the demographic and urban expansions were getting under way. Much more severe, however, was the impact of the Great Depression of the early 1930s, which hit Chile harder than any other country in the world, because of extreme dependence on the two commodities copper and nitrates, which in 1929 amounted to three-quarters of all exports. Recovery in the 1930s was based upon strict government control of the economy, protection for local industries and their promotion with government intervention, characteristics that were accentuated with the coming of the Second World War. Chile's real industrial development, indeed, and the major role played by state organizations in promoting it, dates essentially from these events.

Not surprisingly, import-substitution industries were the first to expand, and the 1930s saw a considerable growth of plants producing textiles, footwear, furniture, construction materials and fittings, and processed foodstuffs. Most of these industries were, in fact, concentrated on Santiago and Valparaiso well before the First World War, and they were the chief locations for these new developments, though Concepción and some other towns outside the central valley (Valdivia, for example) also benefited. It was entirely natural that growth in non-durable consumer goods, having little or nothing to do with exports, should take place in the vicinity of the country's most populous market, and in 1930 Santiago already contained a quarter of Chile's total population. By 1945 virtually the entire national demand in prepared foodstuffs, beverages, tobacco products, shoes, clothing, matches, furniture, window glass, paints and varnishes, paper products and lighting fixtures was being met by national industry (Ellsworth, 1945, p. 179).

It was, however, private enterprise – including such foreign firms as Grace (United States) and Duncan Fox (Great Britain) – that was responsible for this development, assisted by general government economic policies of recovery from the Depression such as tariff controls on imports. A new phase in Chilean industrialization, emphasizing both direct intervention by the state and the promotion of basic industries, began in 1939 with the establishment of the Chilean Development Corporation (CORFO), the first Latin American general development and economic planning authority, endowed with very wide powers to promote economic growth (for a description of these, see Ellsworth, 1945, pp. 85 ff.). Fortuitously, and beneficially, its establishment coincided with the outbreak of the Second World War, which created in Chile a serious shortage of imported manufactured goods, including semi-finished products and capital equipment not then produced in Chile: this gave an enormous boost to Chilean industrialization precisely when the machinery had been created to promote it.

Industrial expansion since the 1940s proceeded to the extent that by 1970, Chilean industry was meeting virtually all internal requirements for durable consumer goods, but this expansion depended on the existing availability of labour, power and transport facilities, highly concentrated on Santiago. Moreover, the durable consumer goods sector relied heavily on imported components and, in this connection, Valparaiso's historic position as Chile's most important port for incoming cargoes was an important factor in underlining the dominance of Santiago and Valparaiso as major locations for industrial plant.

The modern economic history of Chile, and notably the growth of industry, has thus reinforced the historical centralism the country has experienced since its beginnings, and the actual process of economic change, with continued dependence on primary commodity exports, situated by and large outside the central region, has imposed constraints on Chile's economic growth. National industry grew rapidly in the first half of the 1940s, profiting from the Second World War, with a growth-rate of around 10 per cent per annum, but this fell to 4 per cent between 1945 and 1955 as international imports again became more readily available and as the scope for further import-substitution declined. The following decade saw the virtual stagnation of the industrial sector when the severity of inflation necessitated a policy of stabilization, which had a most depressive effect on the economy (Benham and Holley, 1960, p. 109). The basic reasons for the vicissitudes of the Chilean economy and the slow general growth of industrialization were essentially structural; partly because of its link with primary product exports, partly because of the low rate of demand for manufactures in a country where average income, though perhaps good by Latin American standards, is still low compared with genuinely industrialized countries, and partly due to endemic inflation – encouraging speculative investment rather than the accumulation of industrial capital – the industrial sector was not able to grow commensurately with the rise in population. The unemployment, underemployment and inefficient employment – always, of course, more noticeable in cities – characteristic of Santiago and Valparaiso were, in effect, a consequence of Chile's distorted economic structure. Moreover, the extractive industries, which provided the bulk of Chile's foreign-exchange earnings, were capital rather than labour-intensive, and the transfer of technology to Chile, to promote import-substitution, was of the same character. Consequently, while the growth of manufacturing in Santiago and Valparaiso generally drew in migrants looking for work and for higher standards of living, it did not grow enough to absorb them, and could not do so until other economic conditions were met.

Among these were the need to change Chile's backward agricultural sector, notably in the central valley, first to reduce dependence on imported foodstuffs, accounting for a quarter or more of all imports by value in most years in the 1950s and 1960s (a major factor in inflation); and secondly, so to raise rural purchasing power as to make it a more significant stimulus to industrial growth as well as creating greater social equality. Hence, the increasing concentration on government-promoted agrarian reform of the 1960s and early 1970s. Seen as another critical necessity in the same period was the promotion of manufactured exports and their diversification, to strengthen Chile's terms of trade and reduce dependence on primary commodities. Some progress was made here in heavy industry, petrochemicals and cellulose, significantly outside the Santiago –Valparaiso region. Between 1950 and 1973 successive governments sought to use the centralization of decision-making to ameliorate regional imbalances, forging planning and investment mechanisms for the purpose. CORFO had as one of its declared objectives the decentralization of production through the distribution of investments elsewhere than in the core region, though the results were not always happy. Two examples must suffice. Weaver (1968, pp. 141–82) has shown that improvement in transport facilities, notably roads, intended in part to benefit provinces outside the core region, in fact enabled industries in Santiago to penetrate outlying markets and drastically affect small-scale local enterprises producing the same commodities. He concluded that

> the geographically centralising forces inherent in Chile's economic and social structure have not been significantly affected by such government policies as the geographical dispersion of public services, the construction of transportation facilities in outlying areas, or tax incentives to industries established outside the Santiago–Valparaiso area. Only the creation of governmentally owned enterprise appears to have had any lasting impact on balancing the nation's economy spatially. (Weaver, 1968, p. 240)[6]

The second example – the Chilean automobile industry – is a classic case of government-supported, import-substituting industrialization, apparently sensible in its intention but ridiculous in practice. Between 1954 and 1962 the industry developed in the northern free port of Arica purely as an assembly industry, using imported components, as part of a policy to attract industry to regions with little other economic potential. By 1962 some twenty companies were assembling over 6600 vehicles of all kinds a year. In 1962 a law was passed to keep Arica as the privileged base for the auto industry, but imposing strict controls on future multi-national location in Chile and increasing the proportion of locally produced components to be used in assembly. The multi-nationals which entered the market, however, found it difficult to support and invest in a location some distance from the real Chilean market, especially when costs of locally produced components were roughly two and a half times their costs at home. In 1967 the government, responding to this situation, permitted location in the central valley – in the then provinces of Valparaiso, Aconcagua and O'Higgins, where Fiat, Ford and the Chilean licensees of Peugeot built plants, producing by 1969 over 12,000 units a year, 3000 more than the Arica units combined. So investment in Arica fell, also because of inability to meet the law's requirements on locally produced parts. By 1970, with a national market of only 25,000 vehicles a year, ten producers were in operation, seven of them subsidiaries of multi-nationals. The socialist government of Allende sought to reduce this excessive fragmentation by toughening registration rules for foreign companies and forcing them to accept mixed companies in which CORFO would hold 51 per cent of the equity, thus reducing the number of foreign firms to four, with allocations of categories of vehicles to be produced. The conclusion of Gwynne (1977) is irrefutable:

> although the Chilean government originally committed itself to concentrating motor vehicle production in Arica, the multiple and complex forces of industrial centralisa-

tion in an intermediately developed country, together with the pressures exerted by multi-national companies, have meant that the government's commitment has dissipated through time. This paper . . . has pointed to the possible incompatibility of two government programmes favoured by newly industrialising countries – those of import substitution industrialisation and regional industrial development. (pp. 139–40)

The automobile industry in Chile, prior to 1973, illustrated many of the characteristics of national economic development as a whole in the period since the early 1930s and which, despite changes of government and, hence, ruling economic philosophies, showed both remarkable consistency and the increasing role of the state in the economy. One feature was the growth of state supervision of certain economic activities through subsidiaries of CORFO, such as the National Petroleum Corporation (ENAP), the National Electricity Board (ENDESA), the Steel Company of the Pacific (CAP), and many others, entailing a marked increase in the size of the bureaucracy. Import-substitution behind high tariff walls, notably in the 1930s, did promote industrial development, much of it in private hands, but it was also accompanied by the manipulation of exchange rates, whereby they were kept artificially low to allow imports of raw materials and capital goods, encouraging capital-intensive technologies and under-utilizing the national work force, though the latter did also find outlets in the social policies of successive governments through public works and housing programmes. But high tariffs against manufactures resulted in many cases in inefficient, high-cost domestic production catering for a small internal market. Similarly, low tariffs were applied to food imports to benefit urban consumers – politically significant in the electorate – and price controls on domestic production discouraged efficiency on the land. Thus, exports played little part in industrial development, accounting for only about 2.5 per cent of the gross value of industrial output in 1970 and being concentrated in industries based on domestic raw materials – metals, chemicals, pulp and paper. Similarly, at about that time domestic industry contributed over 90 per cent of the total supply of manufactured consumer goods as compared with about only 15 per cent of manufactured capital goods. A high degree of market concentration resulted also from the high protection given to the domestic consumer market. The steady growth of state enterprises meant that by 1970 about 12 per cent of manufacturing output came directly under the state, which also held about 28 per cent of industrial assets (World Bank, 1980, p. 235).

Under Allende these processes were intensified through extensive state take-overs of privately owned assets in banking and credit, mining, distribution, external trade, with a vast extension of central planning and, inevitably, an already top-heavy bureaucracy. Given an opposition majority in Congress likely to block legislation, the government resorted to a variety of means to achieve its aims by by-passing the legislature. These included using CORFO to buy out shareholders in enterprises, under considerable pressure, and resuscitating an old law of the 1930s which gave government powers to intervene in concerns deemed to have failed the public. With a highly organized and unionized labour force generally supporting the government, it was not difficult to create the circumstances which justified intervention. Moreover, the government's immediate aim to redistribute income through raising wages and controlling prices soon led to disinvestment in the private sector, after a short-lived boom as the redistributed income was spent. There was little increased productivity, and large government spending on social measures resulted in massive fiscal deficits which were simply covered by increasing the note issue. By 1972 shortages of supply, massive bottlenecks, government controls and a rampant black market characterized the Chilean economy, and a fall in copper prices (still the major export) and diminution of external credit compounded Chile's economic difficulties, though national mismanagement was perhaps the major factor (for contrasting views see de Vylder, 1976; Sideri, 1979; Sigmund, 1977).

The complex political panorama of the Allende period does not concern us, though there is a wide measure of agreement that economic failure was what most undermined the regime. In the context of Chile's economic evolution, however, there is little doubt that the period 1970–3 was the culminating phase of development policies pursued for some 40 years, though sharply accelerated to the point of virtually dividing the nation. For whatever reasons, the economic collapse of 1973 which preceded the military take-over, marked the end of those policies. But it could hardly be said that prior to Allende, import-substituting industrialization had been anything but a very qualified success anyway, and the increasing role of the state in the running of the economy had not, in general, either produced efficient and cost-effective sectors or laid the basis for sustained growth. For example, by 1970, against an annual rate of population growth of 2.13 per cent in the previous decade, GDP increased by only 4.4 per cent a year, and was actually less than 4 per cent in the period 1966–70. Unemployment fluctuated between 5 and 8 per cent of the labour force, though cushioned by extensive social-security payments; private industrial investment had fallen from 85 per cent of total investment in 1960 to less than 65 per cent in 1970 and both the foreign debt and deficit financing rose rapidly. Increasing paper money issues by the Central Bank to balance the national books, coupled with little or no increase in productivity in major industrial sectors, did nothing to reduce endemic inflation, running at over 25 per cent a year throughout the 1960s. Chile's dependence on one export commodity, copper, increased: as a proportion of exports by value, it represented 68.4 per cent of the total in 1960 and as much as 75.5 per cent in 1970, exposing the economy as a whole to high vulnerability in the world market situation.

In the social sphere, while successive governments had commendable achieve-ments to their credit in such spheres as education, health and housing, Chile's erratic economic performance had not enabled the country to correct massive dis-crepancies in services between the cities on the one hand and the countryside on the other. The national infantile mortality-rate fell, it is true, from 120.2 per 1000 in 1960 – then one of the world's highest – to 79.3 in 1970, but this was still very high for a country of Chile's political sophistication and culture. The country's apparent inability to capitalize effectively on its undoubted resources was, indeed, the basic frustration which had led electorates to vote for a wide variety of governments from the 1950s, ranging from apolitical ones, to free-enterprise administrations, centre-party reformists and, finally, Allende's Marxist coalition in 1970. Whatever their complexion, however, their policies followed broadly similar lines as indicated above.

The most striking characteristic of economic policy since the military assumed power in 1973 has been precisely the abandonment of those traditional policies, in seeking a new economic model for Chile's future growth, based upon the acceptance and encouragement of free-market forces. This entails the dismantling or, rather, the reduction of the role of the state in running the economy; the removal of protective tariffs against imports of manufactures; the opening of the Chilean economy to foreign investment on terms of parity with national capital; the imposition of strict budgetary control on ministries and the end of overmanning; in short, the reduction of the role of the state to that of a neutral, not an active, manager of the economy, though the government remains, of course, responsible for such issues as national security and social welfare, and sets out the parameters of future development. The government's chief role is to set out the rules of the economic game, including fundamental changes in labour laws and social security, but to leave to market forces the determination of production, employment and consumption.

Accordingly, having abolished the massive price-control structure of its predecessor and undertaken a major reorganization of public-sector enterprises – both measures entailing severe hardship for many people – it instructed CORFO to begin hiving back to private hands, through auction, many concerns taken over previously. It retained, however, certain key organizations, such as the Copper Corporation (CODELCO), ENAP and CAP, under state control, though they can no longer expect subsidies and must be cost-effective. On the external side, it reduced the extraordinarily high tariffs against imports of the previous regime to a uniform 10 per cent *ad valorem* duty, the only exception being vehicles, the domestic assembly plants being given a higher degree of protection until 1985.

Given traditional policies, the sharp exposure of Chilean manufacturing to foreign imports was dramatic in the late 1970s: many high-cost, over-manned and inefficient factories went to the wall, some, such as textiles, not least because of dumping by cheap producers abroad. But, while bankruptcies rose dramatically in the late 1970s and early 1980s, the index of industrial production and the rate of investment both rose significantly. Between 1976 and 1979, GDP growth averaged 7.5 per cent, and still reached 6.5 per cent in 1980 when world copper prices fell considerably. Many industries were given a sharper and competitive edge, as the initial shock was absorbed, though it is still too early to say how far Chilean industry can do what it has not done much in the previous decades, namely find export markets and operate competitively in them as a result of higher efficiency.

Another major plank in the new economic policy which has, so far, proved effective has been the diversification of Chile's export pattern away from overall dependence on copper. A deliberate and aggressive policy of export expansion in the so-called 'non-traditional' products, those in which Chile's natural advantages could be fully exploited, had reduced copper's share of exports by value from 75.5 per cent in 1970 to 45.7 per cent in 1980, though copper still remains, of course, the single most significant export item, and will continue to do so. But, whereas in 1973 forestry products accounted for only 2.8 per cent of exports by value, in 1980, after a period of expansion under new and generous afforestation laws, they accounted for 9.7 per cent. Fishing has been another major growth area, raising Chile's world position in terms of annual tonnage catch, from nineteenth in 1975 to fifth place in 1980. Such products, together with fruit, vegetables and wine, have found hitherto unpenetrated markets throughout the world under the new free-enterprise philosophy of the government, and their production has been greatly modernized in the process. Yet they remain primary products, and there must be doubts of the capacity of the revived Chilean economy to expand continually in increasingly competitive world conditions.

Later sections of this chapter, dealing with Chile's regions, will refer to the more recent changes discussed broadly here, but the national framework is necessary to understand them. For, both during the 40 years previous to the military government of 1973 and onwards, and in the last decade, there is scant evidence as yet of any major change in the highly centralist structure of the state, and the dominance of the Santiago–Valparaiso axis spatially in the country. Indeed, authoritarian military government has reinforced that tradition, through its direct control of regional as well as central institutions and, with politics in abeyance for some years yet, a higher degree of regional autonomy is most unlikely. This is not to say that the present government, like its predecessors, is not unaware of regional imbalances, but how it will seek to correct them is not at present clear.

Norte Grande and Norte Chico

Introduction

The northern provinces of Chile, Tarapacá (now Region I) and Antofagasta (Region II), known together as the Norte Grande, and Atacama (Region III) and Coquimbo (Region IV), generally called the Norte Chico, make up both the desert region of Chile and the transitional zone between Desert and Mediterranean Chile, as described in the early part of this chapter. But while the relationship of each of these northern regions to the heartland has been markedly different in some respects, it has been very similar in others. The difference is historical: it lies in the fact that whereas the Norte Grande has not yet celebrated its centenary as Chilean territory, the Norte Chico has been Chilean from the beginning in theory if not precisely in fact. The similarity is economic, deriving from the common character of both regions as zones of mineral exploitation, and the comparable nature of their link with central Chile because of this feature.

The Norte Grande became part of Chile in the 1880s as the prize won in war with Bolivia and Peru: Tarapacá was ceded to Chile by Peru in the Treaty of Ancón, 1883, and 5 years later Bolivian Antofagasta was also incorporated into the southern republic. The Norte Chico, on the other hand, was part of the colonial Kingdom of Chile and, in fact, the town of La Serena in Coquimbo is the second oldest town in the country, dating from 1549:[7] it was built by order of Pedro de Valdivia at the northern confines of the infant colony, both as a centre for gold-mining operations and as a convenient staging post on the land route to Peru. The northern boundary of the Norte Chico was very ill defined throughout the colonial period, and the province of Atacama, unlike that of Coquimbo, saw little Spanish settlement before the latter part of the eighteenth century. *Figure 10.1* indicates the marked increase in the number of towns founded in the Norte

Chico between the mid-eighteenth and the mid-nineteenth centuries, a development intimately connected with the expansion of mining. In the Norte Grande, too, there was a close relationship between urban growth and mineral extraction, though, as the map indicates, this less hospitable region with its serious problems of water and food supply, saw far fewer foundations than the Norte Chico, and some of these occurred, with the rise of nitrates after the mid-nineteenth century, under Peruvian and Bolivian auspices.

It is, indeed, mineral extraction that has always dominated the life of the northern provinces in the past, and it seems destined to do so in the future. In the Norte Chico it is a story of gold, silver, copper and, latterly, iron ore; in the Norte Grande, it is a story of, first, natural nitrates and later of copper. Today, the Norte Grande and the Norte Chico contain little more than a tenth of the total Chilean population, a slightly smaller proportion than 70 years ago, in the same third of the total national territory. Yet, the significance of these regions in the national life is out of all proportion to these numbers because of Chile's dependence, past and present, on their mineral resources. The dominance in the Chilean economy during the nineteenth and twentieth centuries, first of copper, then of nitrates, then of copper again is, perhaps, the most significant single fact in the economic history of the republic. And the evolution of these regions is essentially the story of man's exploitation of mineral riches. It is not, however, the whole story, for there are other aspects of the historical geography of the Norte Grande and the Norte Chico that deserve, at least, some passing mention.

Traditional communities of Desert Chile
Throughout the Norte Grande and the Norte Chico, climate and relief inhibit agriculture, as indicated in the early sections of this chapter, except in the permanent river valleys to the south and in the oases of the Atacama desert. Yet the desert regions themselves have always supported agricultural and pastoral communities, from the pre-Spanish period to the present day, notably in piedmont oases of the Andes, such as Pica, Matilla and Pozo Almonte in Tarapacá, and San Pedro de Atacama, Toconao and Peine in Antofagasta. The classic description of these settlements was made nearly 50 years ago (Bowman, 1924), and in most respects that account still holds. From pre-Spanish days to the mid-twentieth century the *pueblos* of the eastern Atacama, situated north and east of the Salar de Atacama, experienced little change, apart from losing their native language, Kunza, and adopting Spanish as well as the Christian religion, but in much more recent times civilization in its modern form has begun to reach the villages with the advent of schools, motor roads, modern lighting and radio. Another factor bringing these communities more into touch with the outside world today is, perhaps paradoxically, the increasing interest of archaeologists and students of folklore in their culture: the annual fiestas, with their strange blend of Christian and pagan elements, attract increasing numbers of outsiders, though, of course, the settlements are still well off the beaten track of the tourist trade. Yet the ecological base of the *pueblos* has changed little in centuries: irrigation farming, producing corn, wheat, barley, vegetables and a wide range of fruits – apples, pears, grapes, figs, quinces – together with sheep-herding in the upland pastures, today supports a population of about 2000 people at San Pedro de Atacama and its environs, and the next largest *pueblo* of the Antofagasta

cordillera, Toconao, with a similar number scattered over a large number of smaller settlements.

Despite their location, these communities, highly sedentary and self-centred as they have always been, have never been really isolated, for they lie on well-worn trails from both north to south and east to west. These trails were trodden by the Incas in the fifteenth century, and by the Spaniards in the sixteenth. Pedro de Valdivia's expedition from Peru in 1540 travelled from Arica inland to Tarapacá, thence to Calama and San Pedro de Atacama, where it rested some time before proceeding south. Christianity came early in the colonial period, and the bells of the churches in many of these *pueblos* bear dates of the seventeenth and eighteenth centuries. During the whole of the colonial period San Pedro came under the jurisdiction of Peru and, after 1776, Buenos Aires, through the governor of Potosí; at the same time, the maritime link between Chile and Peru took precedence over any land route, but the region remained an important transit zone throughout, between the mountains on the one hand and the desert and Pacific coast on the other (Bowman, 1924, pp. 236–8). This role was enhanced after the region came under Chilean control with the rapid development of the nitrate industry from the 1880s: the supply of fresh meat to the nitrate works (*oficinas*) of Antofagasta, situated north-east and south-east of the port of that name, came from Salta in Argentina, a 2-week cattle drive from San Pedro de Atacama. Having crossed the Puna de Atacama, the cattle rested at San Pedro for a few days, feeding up after their difficult journey, before being driven for the 3 days required to get them to the *oficinas* (for a detailed description of the trade see Bowman, 1924, pp. 218–35). Similarly, the demand of the *oficinas* for mules for haulage purposes was met from the north-west provinces of Argentina, and San Pedro's facilities for accommodating transient droves were well utilized. One consequence of this development for San Pedro itself was the expansion of alfalfa-growing for fodder.

The cattle trade was at its height during the first quarter of this century, and when the world market in natural nitrates succumbed to the competition from synthetic nitrates in the late 1920s, a further lease of life was given to San Pedro's important role on the cattle road by the rise of copper at Chuquicamata. There followed another decline with the opening of the Antofagasta–Salta Railway, routed to the south of the Salar de Atacama (see *Figure 10.3*), and the old cattle trail seemed doomed with this development in 1948. But, for various reasons, the railway has been a failure (see Rudolph, 1963, pp. 33–6), and San Pedro has resumed its role as a transit zone between Argentine Salta and Chilean Antofagasta, with the construction of a road, following the old cattle trail, in the early 1960s. Cattle now travel from Salta to San Pedro in one day, instead of 14, and they go in trucks, not on the hoof.

The changes that increasingly make an impact on the traditional communities of the Atacama desert may well raise the standards of living, but they will also affect the quality of life. New generations seem likely to be less inclined to maintain their forbears' pattern of existence, as education and improvements in communications – radios no less than roads – diminish its cultural individuality. Fishing communities of the Antofagasta coast, such as the Changos of Cobija, whose precarious survival to the modern age was noted by Bowman (1924, p. 59), have now quite disappeared. The desert communities are much

Figure 10.3 Copper-mining in Chile (after World Bank Report, 1980)

more firmly rooted, and they have shown a remarkable capacity to survive through the centuries, but their future cannot be taken for granted simply in the light of their past.

Agriculture in the northern regions
In the Norte Grande, apart from the oasis settlements, it is only in the very few watered valleys of the province of Tarapacá that agriculture has any hold. Here, the Lluta and Azapa rivers – and, to a lesser extent, the Vitor, Camarones and Camina – provide some scope for the growing of tropical produce, as does the Loa, which separates Tarapacá from Antofagasta. Vegetables, fruit, especially raisins, and almonds are the specialities in the far north, but only 2 per cent of the total area of the Norte Grande is arable, and its total proportion of the national territory devoted to fruit-growing is only half that percentage.

The northern part of the Norte Chico – Atacama and northern Coquimbo – have the same, or similar, conditions, though the ribbons of vegetation that follow the course of the rivers Copiapó and Huasco in Atacama are much wider than further north and, while precipitation is not much greater than in the truly arid regions, it is sufficiently so to make a considerable difference. Seen from the air, the irrigated valley floor in which the town of Copiapó stands provides a flash of green amidst its barren surroundings, and it has done so certainly for centuries, if not, indeed, millennia, since the Spanish conquerors found Indian irrigation works and a settled population when they arrived. Yet everything depends upon the *amount* of water available and, since both rainfall and river supply in valleys like that of the Copiapó and the Huasco fluctuate very considerably over the years, the agricultural economy of the Atacama valleys has had a very chequered history (Bowman, 1924, pp. 113–28, and cf. the comments of Charles Darwin (1960, pp. 332–8)).

The agricultural products of these valleys, apart from supplying the highly important local mining industry – to which we shall return – were also exported to the northern desert during the boom years of the nitrate industry, and the fattening of cattle, imported from Argentina and southern Chile, was also an important activity for the same market. Alfalfa is still a significant crop but more important today are products known as *primores*, vegetables such as tomatoes which, due to the climatic conditions, are in season here before they ripen in the central valley. The urban markets of Santiago and Valparaiso are the chief consumers. The Huasco valley also has extensive vineyards, olive groves and orchards. The pastoral sector also takes due advantage of the peculiar climatic situation: outside the irrigated valleys, the *camanchaca* on the coastal range and the higher precipitation on the Andean slopes provide pasturage for cattle, sheep and goats, which are driven there when it is the dry season in the valleys.

The significantly higher rainfall of Coquimbo, in comparison with Atacama, makes its river valleys – the Elqui, Limari and Choapa – much more productive in cereals, vegetables, fruit, sheep and goats. Indeed, agricultural settlement from the time of the Spanish Conquest was very much more pronounced than in the essentially mining province of Atacama. But agricultural exploitation of non-irrigated areas has, historically, been closely linked with the fortunes of mining camps: as veins of metal were worked out, so agricultural villages established to provision them also declined, and permanent concentration of population occurred primarily in the stable river valleys. Today about 30 per cent of cultivated land is dry farmed, and it is areas such as these, of course, that are hit the hardest by the periodic droughts that afflict Chile, the worst of which occurred in the late 1960s.

Wheat, barley and maize are the chief grain crops, but in the Elqui and Limari valleys alfalfa and clover predominate, and La Serena is an important dairying centre. Transhumance is highly significant in the pastoral industry, but the coastal range and the Andean slopes have been seriously eroded by the vast numbers of goats which outnumber cattle, sheep and horses in the province. Stock-raising, in fact, predominates in the rural economy.

Throughout the northern provinces, the dual pattern of *latifundia* and *minifundia* we have observed in the central valley also obtains, but in Coquimbo there also exist quite distinctive communities holding land in common. These derive from the colonial period when large land grants made by crown agents often included common lands, though many of these were subsequently incorporated, by one means or another, into private properties (see pp. 471–5 above). Not all such common land was incorporated, however, and, in addition, some unirrigated and unirrigable lands included in *private* land grants were subsequently felt by heirs to the property to be of such little value that they did not take the trouble to legalize their possession of them, and the lands were then taken over by others. The processes whereby agricultural communities came into existence – with some members holding both individual properties and a share in common lands, and others having only recognized rights of common usufruct – have a very complicated history, often obscure. In Coquimbo, however, such communities are of considerable significance, since they represent a third of the total population and a half of the rural population of the province, and they occupy and work no less than a quarter of the usable land, though this is only 3.5 per cent of the irrigated land in Coquimbo. The organization and government of these agricultural communities, and their use of the land they hold, are extremely diverse, but they are worthy of note since they represent a peculiar hangover from the colonial period, and also an interesting testimony to the variety of systems of land tenure in Chile (see CIDA (1966, pp. 126–42) for a detailed description, and Batallán (1980) for the impact of recent developments).

Notwithstanding the importance of agriculture and stock-raising in the northern provinces, increasing as one moves south, the economy of the Norte Grande and Norte Chico is, and always has been, that of the mine rather than the farm. It is mineral extraction that has determined population movements, the establishment of towns and ports, the siting of communication routes and, indeed, in the War of the Pacific, the expansion of Chile's national frontiers 960 km to the north of their colonial location. Mining itself, as we have already observed, was an important stimulus in regional food production, not only in the northern provinces but also in the central valley, and the commerce associated with northern mining was also a factor in the growth of Valparaiso as well as the ports of the region. The role of mineral exploitation has been, and still is, very important for government revenues and for Chile's balance of payments, and while other regions of Chile also have mineral deposits of great significance, either past or present or both – copper at El Teniente in O'Higgins, coal in Arauco and oil in Magallanes – they do not dominate the economy, the life and the history of their locations as they do in the Norte Grande and the Norte Chico, nor have they yet played the crucial role in the evolution of Chile that they have done, and continue to do, here. Two other general factors should be mentioned: in the exploitation of the nitrates and copper of Chile's northern provinces, foreign capital has predominated, and, secondly, proximity of the mineral deposits to the Pacific has made for comparatively easy shipment to overseas markets.

The rise and fall of the nitrate industry
The significance of natural nitrates in the Chilean economy between 1880 and 1925 is illustrated in *Table 10.5*. Even before 1880, when the nitrate regions belonged to Peru and Bolivia, Chilean labour and capital, with British finance also prominent, played a major role in the exploitation of the deposits. Indeed, as early as 1870, of the 18,000 workers at the nitrate port of Iquique in Peruvian Tarapacá, approximately half were, in fact, Chilean; already it was Valparaiso, rather than a Peruvian commercial centre, that was the real focus of the nitrate trade. Until the 1870s both the Peruvian and Bolivian governments remained apparently indifferent to this state of affairs, confining their intervention in the nitrate business to the collection of export duties on nitrate shipments. In 1875, however, the virtually bankrupt government of Peru decided to expropriate foreign capital in its nitrate areas and, in effect, nationalize the industry. The privately owned producing plants (*oficinas*) were to be paid for in government bonds, made payable to bearer, which were to be redeemed by the government within 2 years but, in fact, they were never cancelled by Peru since her government failed to raise the European loan required to buy out the owners. The late 1870s were depression years for nitrates, partly owing to the great uncertainty created by Peruvian government action, and partly owing to the earthquake of March 1877, which destroyed many of the loading platforms along the coast. The intervention of the War of the Pacific, however, changed the situation dramatically: Chile's seizure of the nitrate provinces and her subsequent decision

Table 10.5 Export of nitrates (tonnes) 1880–1925 (at 5-yearly intervals) and proportion (%) of government revenue derived from export taxes on nitrate

Year	Exports	Revenue (as % of total)	Year	Exports	Revenue (as % of total)
1880	223,974	5.52	1905	1,650,363	56.67
1885	435,988	33.77	1910	2,335,941	55.14
1890	1,063,277	52.06	1915	2,023,321	54.81
1895	1,238,605	66.03	1920	2,794,394	49.65
1900	1,453,707	56.29	1925	2,517,099	37.18

Source: Hernández Cornejo (1930, pp. 174, 177–8).

to return the industry to private ownership was a decisive moment in the history of the nitrate industry. In the first place, Chile secured for herself the export revenues from nitrate that had formerly gone to Peru and Bolivia. Secondly, the devolution of the nitrate grounds to private hands again gave foreign, and notably British, capital, the pre-eminent position in the industry.

This situation arose from the circumstances of Peru's attempted nationalization. The bonds issued by the government in 1875 – in effect, the title deeds to nitrate properties – fell markedly in value in the late 1870s, and the War of the Pacific brought their price down even more rapidly. Speculating on both the chances of a Chilean triumph in the war itself, and on the possibility of the Chilean government subsequently handing back the industry to private control, British entrepreneurs bought up large quantities of the depreciated certificates in the war years, and thus acquired the nitrate grounds very cheaply.[8] In 1878 British capital had controlled about 13 per cent of the nitrate industry of Tarapacá; by 1884 the proportion had risen to 34 per cent, but by 1890, after a period of feverish activity on the London Stock Exchange, 70 per cent of the province's nitrate industry was in British hands.

There were five major areas of nitrate exploitation in the high period of the industry, the most extensive in the province of Tarapacá, the others known from their location in Antofagasta province as El Toco, Antofagasta, Aguas Blancas and Taltal. In Tarapacá the fields ran in a south–south-west direction from the latitude of Pisagua to that of Chucumata, east of the *cordillera de la costa* and on the western edge of the central depression known as the Pampa de Tamarugal, less than 20 km inland in the far north but more than double that distance from the coast in the south. In Antofagasta province the main northerly deposits of the raw material (*caliche*) were to be found some 60 km inland, due east of the port of Tocopilla, in the same central depression which runs from Tarapacá to Atacama: it is the southern part of this nitrate field of El Toco that is the basis of the modern industry, based on María Elena and Pedro de Valdivia. The third field, moving south, was of much less importance than those of Tarapacá and El Toco, lying 80 km north-east of Antofagasta port, but much more significant was the field of Aguas Blancas, a similar distance south-east of the port. Finally, in the southernmost part of the province lay the Taltal deposits, from 40 to 70 km east of the port with the same name.

The rapid growth of the nitrate industry from the 1880s naturally had a significant effect on the economic life of the regions concerned. It encouraged immigration from other parts of Chile and stimulated the growth of ports and the building of railway lines to carry the nitrate from the *oficinas* to the ports. The virtual doubling of the population of Tarapacá between 1885 and 1895 shown in *Table 10.2* reflected, in fact, the growth of its nitrate industry. Its port towns of Pisagua and Iquique had, in this period, the character of real 'boom' towns, Iquique boasting, as it still does today, an opera house built by the French architect Eiffel. In the early period of development water supplies for Iquique came by tanker from Arica and from coastal distillation plants, while the nitrate *oficinas* inland were supplied by mule carts and, later, railways conveying the precious necessity (Bowman, 1924, pp. 76–8). Subsequently, however, pipelines were laid to the Andean cordillera region, and this has remained the major source for both the coastal and inland centres (Bowman, 1924). Reference has already been made to the stimulus provided by the rise of the nitrate industry to the trans-Andean trade in cattle, to stock-raising and agriculture in the Norte Chico and the central valley, and to the production of comestibles for humans and animals alike in the desert oasis settlements. The Chilean coastal shipping trade was affected similarly. But the close interconnection of these segments of the Chilean economy carried with it the danger that circumstances affecting the nitrate industry itself would be transmitted far beyond the nitrate regions, having a multiplier effect of the harmful kind.

Throughout the history of the natural nitrate industry, boom was followed by slump with considerable frequency. The major markets for Chilean nitrates were the farming industries of western Europe and the USA, but their vicissitudes, climatic or otherwise, were a critical factor in nitrate production. Overall, world consumption of nitrates did not keep pace on a regular rising trend with Chile's productive capacity, and over-stocking of the world nitrate market faced producers with the fear of falling prices. Between 1880 and 1909 no fewer than five agreements, or combinations, were made between the various nitrate producers to restrict their output in these circumstances, leading to serious economic difficulties not only for the workers of the *oficinas* and shippers at the nitrate ports but also for the many, distant interests that had become enmeshed in the nitrate trade. The consequences for the Chilean government, deriving half its revenue from export taxes on nitrates between 1890 and 1920, were no less severe.

The periodic booms and slumps of the nitrate industry would, no doubt, have continued had Chile retained its monopoly of nitrate production to the present day, since the industry depended on world market forces over which it had little control; nevertheless, given its monopoly position down to the First World War, falling markets were always followed by rising ones, and recovery was usually rapid. This cycle, however, was fundamentally affected first by the war of 1914–18, and, secondly, by the Great Depression of the early 1930s. Germany was a major market for Chilean nitrates, and the advent of the First World War, when the British naval blockade disrupted German trade, almost immediately affected Tarapacá. Within a few months of the declaration of war, half the nitrate *oficinas* in the province were closed down, and many of the workless labourers, numbering some 33,000, left Iquique and Pisagua for more southern parts of Chile. More significantly in the long run, German scientists turned their attention to producing synthetic nitrates in commercial quantities by techniques that spread quickly to other advanced countries in the post-war period. It is true that between 1913 and 1929 there was no significant real decline in the quantity of nitrates shipped from Chilean ports, but Chile's share in world nitrate production, natural and synthetic, fell in the same period from 90 per cent to about 24 per cent. The Great Depression greatly accelerated the decline by its complete disruption of world trade, and many large *oficinas* that had survived synthetic competition into the 1930s did not last out the decade. The *oficina* Chacabuco, for example, some 80 km north-east of Antofagasta, off the road to Calama, had 7000 people living there as late as 1938; today, its vast leaching tanks turning to rust and the trees of its central square the same colour as the surrounding desert,

Chacabuco is a lifeless testimony to the transient nature of Chile's nitrate age, and of man's temporary conquest of this part of the Atacama, though in the 1970s it was used by the military government as a detention centre for political opponents.

The worldwide development of synthetic nitrate forced the Chilean industry to seek, through reorganization and the application of modern technology, to concentrate production in order to secure, through cost-competitiveness, at least a share in the world market. Expansion and modernization, including a solar evaporation plant at Coya Sur, mechanical loading at the port of Tocopilla, improvements in other equipment and better housing for the labour force were undertaken in the late 1950s and 1960s by the Compañía Salitrera Anglo-Lautaro, an Anglo-American company, then responsible for 85 per cent of total Chilean production at its plants at Maria Elena and Pedro de Valdivia in Tarapacá. But the company was nationalized by the Allende government in 1971, and joined the Empresa Salitrera Victoria, a subsidiary of CORFO, under a new organizational umbrella, the Chilean Chemical and Mining Company (SOQUIMICH), responsible for all operations, including marketing (see *El Mercurio*, special supplement on nitrate, 17 November 1976 and SOQUIMICH, 'Tópicos de interés de sus oficinas salitreras', mimeo, Oct. 1976).

Between 1960 and 1966 an average of 900,000 tonnes annually was exported out of a production figure of some 1,100,000 tonnes, and Chile held about 4 per cent of the world market in natural nitrates. That share has continued to fall, however, and when in October 1978 the last shipment of nitrate from the classic nitrate port of Iquique took place (to China), it marked the symbolic closure of a century-long cycle of Chile's dominance of an industry and trade which had had a more profound effect upon her development than any other economic factor. SOQUIMICH, subjected by the military government of 1973 to the financial disciplines of its economic philosophy, while still remaining a state entity, has sought since then to cultivate specific overseas markets for nitrate, but it is fighting a losing battle against synthetic production, and it seems unlikely that nitrates will ever again emerge as a major Chilean export. The surviving nitrate communities of María Elena, Pedro de Valdivia, Coya Sur and some other minor plants, surrounded as they still are by vast deposits of unexploited mineral, and numbering over 30,000 workers and their families, can only look forward to diminishing activity and reduced employment.

Yet in other respects there are bright prospects for the region. Iodine, occurring in nitrate deposits as iodide and iodates of sodium, calcium and potassium – of over 95 per cent purity – remains the major by-product and, despite competition mainly from Californian and Japanese brines, Chilean iodine can count on rising world demand for a variety of uses – sanitation, metallurgy, livestock feeding and the synthetic organic chemistry industry – to remain a major world supplier.

Other prospects lie in the increasing exploitation of non-metallic minerals in which the region abounds, and for which world market conditions are propitious. For example, in Region II (Antofagasta) lies one of the world's largest deposits of lithium carbonate and other salts, estimated to contain some 4 million tonnes of recoverable mineral, about 40 per cent of known world reserves. To exploit this resource – lithium is widely used in the glass and ceramic industries, in lubricants, pharmaceuticals, batteries and metallurgy generally – and in accordance with its free-market philosophy, the military government encouraged an agreement in 1975 between CORFO and the Foote Mineral Co. of Pennsylvania, followed by the formation in 1980 of the Sociedad Chilena del Litio Ltda, a joint venture in which CORFO owns 45 per cent and Foote Mineral the rest of the stock. Capital cost for infrastructure and extraction plant will be US$26 million, to come on stream by 1984 with an initial annual production of 5500 tonnes of lithium carbonate a year. Other salts in the same massive bed of the Salar de Atacama include (with estimated reserve figures), potassium (53 million tonnes), magnesium (26 million tonnes) and borate (11 million tonnes), and plans are far advanced for their exploitation. The nitrate era may well be over for Chile, but the Atacama desert continues to yield up its riches in natural salts, while its deposits of metallic minerals, and particularly copper, are very far from being exhausted.

Metallic mining in the northern regions
The nitrate industry of Chile was essentially a development of the nineteenth century, but the mining of metallic ores goes back to the beginning of the colonial period and it has had a continuous history since then. In the seventeenth century, for example, Peru's requirements for arms to defend the Viceroyalty against corsairs was a major stimulus to copper-mining in the major known region of deposits, the province of Coquimbo, the copper being shipped from the port of that name to the Peruvian arsenal at Callao (Pederson, 1966, pp. 70 ff.). In the eighteenth century – a period of considerable economic expansion in Chile – it was gold, rather than copper, that came into its own in the Norte Chico. Old centres of gold-mining, dating from the period of the Spanish Conquest, such as Combarbalá and Illapel in southern Coquimbo, and La Serena in the north, once again became the scenes of feverish activity as old workings were reopened and new sources discovered. More significant, however, was the northward march of the mining frontier as prospecting was intensified and extended: new centres were established as focal points both for mining operations themselves and for the variety of economic operations mining brought in its train – agriculture and stock-raising to supply the mining camps with food; the growing of forage to feed the mule trains that carried supplies and ore in that pre-railway age; the provision and transport of timber to supply the mines with supports and the smelters with fuel. Prospectors ranged and honeycombed the hills, the outcrops of Andean diorite that form the coastal batholith between Santiago and Copiapó, and as the expansion of vein-mining got under way so was the landscape transformed. Copiapó itself and Freirina and Vallenar were characteristic foundations of the period, near to mining lodes and lying in river valleys that could also be exploited for agriculture and pastoralism.[9] By the end of the eighteenth century the Norte Chico had become differentiated from the rest of Chile by its specialization in mining, which dominated the life of the region and has continued to do so.

Gold production had passed its peak by the time Chile achieved her independence from Spain in 1818, but already silver and copper were becoming highly significant. The silver strikes of major importance occurred at intervals between 1811 and 1870, creating bonanza conditions at particular times and places, and enabling Chilean silver production to show a constant rise through most of the century. The first, at Agua Amarga, south of Vallenar in 1811, not only quadrupled the population of Vallenar within a few years; it also extended the amount of land under cultivation, in order to supply the miners, and expanded the cutting of irrigation canals. Another major discovery took place in 1825, north-west of La Serena, but this was soon overshadowed by the most significant, and certainly the most romantic, strike of the century, the accidental discovery in 1832 of the silver hill of Chanarcillo, near Copiapó, by the woodcutter Juan Godoy. With this discovery Copiapó grew rapidly from a town of 4000 to one of 12,000 within 3 years – not surprisingly in view of the fact that masses of nearly pure silver were found at Chanarcillo, one piece weighing 2700 kg (Bowman, 1924, p. 170). The greater climatic handicaps of the Copiapó valley, in comparison with those of Vallenar, however, imposed burdens on its inhabitants with regard to supplies and their costs: Darwin (1960, p. 340) described the cost of living in 1835 as 'wonderfully exorbitant'. Another major silver discovery was made south-west of Copiapó in 1848, and the last of the great silver strikes occurred in 1870 at Caracoles, roughly a third of the way along the road from San Pedro de Atacama to Antofagasta, in what was then Bolivian territory in law, if not entirely in fact, given the number of Chilean compared with Bolivian prospectors there (see Vicuña Mackenna, 1882, for a detailed account of the silver discoveries; Bowman, 1924, pp. 169–72, for a brief resumé).

Chilean silver production rose constantly until the 1850s, reaching an annual average of over 123,000 kg in that decade; a slight drop in the following 10 years was reversed with the Caracoles boom, and production reached its peak in the decade 1881–90, with an annual average output of over 157,000 kg. After 1900 a sharp decline set in, and this, in effect, was permanent (Pederson, 1966). In world terms, however, the peak of Chilean production never exceeded 13.8 per cent, and, while its importance in the Chilean economy was profound, the total value of silver produced in Chile during the nineteenth century was approximately half the value of the copper mined in the same period (Pederson, 1966, p. 171).

The production curve for Chilean copper in the last century was remarkably similar to that for silver, though copper was more important as a source of national income, and the history of copper-mining in Chile is not only older than that of silver, but also more continuous. Chile has experienced two quite distinctive phases of copper-mining in the last two centuries. In the first, from the 1820s to the 1870s, production increased rapidly as the rich veins of easily concentrated oxide ore in the provinces of the Norte Chico were opened up and exploited, *pari passu* with silver. At the peak of this first boom an annual average of over 46,000 tonnes was produced, and the importance of copper in the national economy may be gauged from the fact that in 1860 copper exports amounted to well over half of Chile's total export trade by value (Martner, 1923, I, p. 299, II, p. 307). Between 1840 and 1850 Chile's share of world production of copper stood at over 30 per cent, rising to 44 per cent in the 1860s, and holding to 36 per cent in the 1870s. But, as the veins in Coquimbo and Atacama were worked out, as international competition grew – notably from the United States and Spain – and, despite the efforts of a number of Chilean innovators in technology, as the generally crude methods of production in Chile could not keep pace with these developments, Chile's pre-eminence in world markets fell away. In the 1880s Chile's share of world production fell to 16 per cent, and by the 1890s it had fallen as low as 6 per cent. Moreover, copper, like silver, was now over-shadowed by nitrates in the Chilean economy, and could not provide so high a yield on invested capital as other activities, particularly since more efficient, larger-scale, and lower-cost suppliers of the metal had emerged in other countries (Reynolds, 1965, pp. 210–13). By the 1880s, in fact, the world was looking for some other standard for copper than the 'Chile bars' that had predominated for 40 years, and the decade marked the beginning of the end of Chile's first cycle of copper production.

The second great phase of copper exploitation in Chile was closely associated with the revolution in copper technology that occurred in the USA between 1900 and 1910, whereby it became economical with capital-intensive techniques of mining, concentration and smelting to work low-grade porphyry ores, found in abundance in Chile but hitherto unexploited. The visible and vast symbols of this development are enterprises such as the Chuquicamata open-cast mine (the biggest in the world) in Antofagasta province, and El Teniente (the largest underground copper mine in the world) in the province of O'Higgins, in Chile's central valley region. To this development we shall return, but it is important to note that the two great phases of copper-mining in Chile are separate and distinct, not only in terms of mineral extraction but also in their different impacts on the Chilean landscape; what links them together, however, is the critical dependence of the Chilean economy during each phase on the produc-tion and export of copper, a dependence interrupted only, in commodity but not in character, by the nitrate years.

The nineteenth-century phase of copper-mining in Coquimbo and Atacama, and the silver boom with which it was co-existent, was in many respects a con-tinuation of the colonial period in methods and techniques, characterized by a vast number of small mines and a small number of larger ones. It was the latter that, naturally, were furthest from the colonial tradition and could introduce such technological innovations as the use of coal rather than wood for smelter fuel, sink deeper shafts and utilize modern machinery. As late as the 1870s, in the whole of the Norte Chico only 33 out of 788 mines in 111 *minerales*, or mining zones, were using steam engines for the extraction of ore, at a time when there were over 30,000 miners in the two provinces (Pederson, 1966, pp. 191–2). Nevertheless, the fact that such mines existed indicated the advent of modern technology in the Norte Chico, brought about not least by changes in the communications network.

The steamship arrived in Chilean waters in 1840 when the North American, William Wheelwright, founded the Pacific Steam Navigation Company, and the railway was not far behind. As already noted, the Copiapó–Caldera Railway was built between 1849 and 1852; other lines subsequently followed to link inland mineral areas with the ports of Chañaral, Carrizal, Huasco, Coquimbo and Los Vilos, all these lines running east–west, and following, where possible, the valley floors, running transversely to the Pacific coastline. At this period Swansea in south Wales was the world's largest smelting centre for copper, and it was to Swansea that Chilean ore was carried, mostly in British vessels, which in turn carried coal to the Norte Chico for local smelting operations, which had also developed, notably at port sites, to cater for the multiple sources of ore mined in the Norte Chico. In fact, the depredations of the colonial period and the nineteenth century on the resources of wood in the Norte Chico not only created serious erosion problems in parts of the region, but also caused some mine-owners to establish smelting plants south of the central valley where wood was abundant, and where, in Arauco, was Chile's only domestic source of coal. Quantities of coal were also shipped from Lota and Coronel in Arauco to the northern smelters, though imports of British coal were much more significant and of better quality. By the 1880s nine-tenths of Chilean coal imports came from Great Britain. Other far-off economic factors were brought into play by the development of mining in the Norte Chico: the trans-Andean settlements of north-western Argentina supplied cattle on the hoof, driven for three and a half weeks or so from San Juan, Catamarca and Tucumán through the mountain passes to the valley of Copiapó, where they were fattened on green alfalfa before being sold.

The period of high activity, however, was over by the 1870s, and the mining industry of the Norte Chico was in decline by the 1890s. The decline was visible in the many ports that had grown up entirely for mineral shipments and the importation of supplies required by the industry: only Coquimbo survived at anything like its former level of activity while, inland, depletion of the high-grade deposits on which the boom had been based led to the abandonment of many mines. Only a few large operators with adequate capital resources and modern technology, such as the British-owned Copiapó Mining Company, were able to hang on profitably. Population figures also indicate the fluctuations in mining in the Norte Chico: in 1854 the region contained 161,000 people, over a tenth of the national population, and by the time of the next census, in 1865, the figure had risen to 225,000, approximately one-eighth; it fell to one-tenth again by 1885, but thenceforward, while the national population grew, that of the Norte Chico declined, absolutely until the 1920s, and proportionately from 1885 to 1960 (*Censos, Provincias*, 1964).

In the second major phase of copper-mining in Chile, the twentieth-century exploitation of the low-grade but huge porphyry deposits, operations have been, and are, capital rather than labour-intensive, with American money dominating the industry until the very recent past. Apart from the El Teniente mine in O'Higgins province, owned by the Kennecott Copper Corporation until the late 1960s, the major enterprises were at Potrerillos and El Salvador in Atacama and at Chuquicamata in Antofagasta, both belonging to the Anaconda Company (*Figure 10.3*). Hence, the Norte Chico and the Norte Grande retained their interest in copper-mining, though it was of a very different kind from that of the past. In 1916 the Andes Copper Mining Company – a subsidiary of Anaconda – was formed to develop the industry at Potrerillos, some 160 km east of Chañaral in northern Atacama and, although construction started and plant was established, the mine did not go into production until 1926 because of delays caused by the First World War and by the depression in copper prices that followed it (Reynolds, 1965, p. 218). The open-cast mine at Chuquicamata, bought by an American entrepreneur from a British company and sold by him to the Guggenheim interests in 1911, was started in 1913 and began production 2 years later, using power shovels that had been employed in the cutting of the

Panama Canal. El Teniente, much further south, began production earlier, in 1912, but in its second year of operations Chuquicamata already equalled the output of El Teniente – 21,000 tonnes of fine copper.

When Potrerillos was being prepared for production in the 1920s, the population of the Norte Chico was at its lowest point since 1854 and, in fact, the population of Atacama province was actually smaller than it had been in that year. Potrerillos aided the recovery with the growth of a complex of installations, which for the 30 years between 1927 and 1959 created an island of activity in a barren land, producing nearly 2 million tonnes of copper in the period. Some 30 km north-west of Potrerillos is the El Salvador mine, and when in 1959 Potrerillos itself was closed by Anaconda as being no longer economic, some of its installations were retained and developed for El Salvador. A township of over 10,000 people now exists there with many modern facilities, a far cry from the traditional mining camps of nineteenth-century Atacama.

Chuquicamata is even more impressive both as a mining centre and as a town. Over 3000 m above sea-level, and some 230 km by road north-east of Antofagasta and 150 km east of Tocopilla, the town of Chuquicamata has a population of over 30,000 and very modern facilities, including an excellent hospital. The mine, with over twenty benches descending to a depth of over 350 m had produced over 7 million tonnes of copper by 1980 and over 350 million tonnes of oxide ore had been removed by then. The total plant area is enormous, and includes an oxide-ore treatment plant, a sulphide-ore treatment plant and other installations with a total daily production capacity of over 100,000 tonnes of copper. In 1967 work also began on the Exotica mine, a little over 3 km south of Chuquicamata, then estimated to contain over 153 million tonnes of oxide ore of 1.35 per cent copper content.

These great installations of the Norte Grande and the Norte Chico, together with El Teniente and other mines in the central zone of Chile, formed, and are still called, the Gran Minería, accounting in the 1970s and today for well over 80 per cent of Chilean production (see Sutulov, 1975, for the most comprehensive account of Chilean copper). But there exist also the Mediana Minería, medium-scale mining, composed of a number of companies largely located in the Norte Grande and the Norte Chico, including the state corporation, the National Mining Enterprize (ENAMI), and the Pequeña Minería, composed of a large number of small operations, selling ore to ENAMI, but producing only a fraction of national output and employing, probably, less than 5000 workers, compared to over 30,000 by the Gran Minería and about 8000 by the Mediana Minería (see *Table 10.6* for recent production figures). Yet it is the very small-scale operators who still represent the true mining tradition of the Norte Chico and the Norte Grande, the large-size, highly capitalized enterprises being outside that tradition (Pederson, 1966), though it was naturally the latter which dominated the long debate on nationalization from the 1950s onwards.

The nationalization of Chilean copper
In 1970, copper accounted for about 8 per cent of Chile's GDP, 60 per cent of exports by value and 80 per cent of government tax receipts, while Chilean output – then well over 600,000 tonnes – amounted to about 15 per cent of world production, with reserves estimated variously from a fifth to a third. Given

that the Gran Minería dominated the industry and that its mines were largely foreign-owned, and given the crucial significance of copper in the Chilean economy, the growth of national sentiment in favour of larger Chilean owner-ship was hardly surprising. Prior to 1955 the position of the Chilean government with regard to the foreign-owned sector was very ill defined: the American com-panies paid taxes (of crucial importance in government revenue) as the nitrate shippers had done, though it is generally agreed that Chile's share of receipts from the exploitation of her major natural resource was small. In that year, there-fore, the government established the Copper Corporation (CODELCO) to study all matters concerning copper, and other measures followed, by agreement with the companies, to increase refining capacity, thus increasing value-added accruing to the country, and raising tax receipts. But this was not enough, and national pressure for greater Chileanization of the Gran Minería grew. Accordingly, in 1965–6, President Frei carried a bill through Congress which, briefly, laid down terms for specific state participation in each of the large mines, increased investment by the foreign companies (but with substantial tax and other concessions), and target figures for production and refining over the following 5 years. But this was not nationalization: for example, while Chile's agreement with the Kennecott Copper Corporation, owner of El Teniente, allowed for a state share of 51 per cent, that with Anaconda over the Exotica mine allowed for a state stock-holding of only 25 per cent, the rest being held by the American company. Copper issues, consequently, were a major debating point in the presidential election of 1970, and not long after Allende's victory outright nationalization took place in 1971. It is, perhaps, significant that this was one of the very few proposals of the Allende government that was sufficiently uncon-troversial nationally to pass through Congress and secure a large majority.[10] The Gran Minería then passed under CODELCO, but the next two years were marked by labour troubles, incompetent local management and declining output, coupled, it must be said, by clear evidence of inadequate preparation for state take-over and lack of co-operation from the former owners.

The subsequent military regime did not de-nationalize the existing Gran Minería, but it drastically reorganized CODELCO by a law of 1976, imposing norms of planning and cost-effectiveness in accordance with its broad economic philosophies. At the same time, by the promulgation of decree laws relating to foreign investment, it completely reversed the policies of its predecessor with regard to new mining activities, seeking to encourage massive foreign parti-cipation on generous terms. Compensation was agreed for the previous nation-alization, and during the late 1970s investment commitment in Chilean mining grew substantially. At the same time Chilean output and exports grew as shown in *Table 10.6*.

CODELCO's own investment programme for 1981–5, announced at the end of 1980, calls for a total of US$ 1.84 billion to be divided between Chuquicamata El Teniente, Andina and El Salvador (see *Figure 10.3* for locations), with a planned output of 2 million tonnes of fine copper a year by 1990, and 2.5 million tonnes by the year 2000. Emphasis will be laid on expansion of ore-treatment capacity, given the declining copper content of ore from existing mines, and energy-saving technology. But new mines, many of them with heavy foreign-capital participation, are planned to come into operation, particularly in the

Table 10.6 Production and sale of Chilean copper, 1976–80 ('000 tonnes of fine quality)

Year	Commercial production			Exports	
	Total	CODELCO	Other[1]	Total	Refined
1976	1005.2	846.8	158.4	981.9	594.7
1977	1056.2	892.7	163.5	1005.2	617.5
1978	1035.5	876.5	159.0	977.8	702.1
1979	1060.6	910.2	150.4	1003.5	741.8
1980	1072.2	904.4	167.8	1041.1	768.2

Note: 1 Mediana and Pequeña Minería.
Source: *Ecosurvey* (Santiago) 16 February 1981, using CODELCO figures.

northern and central regions of the country: between 1974 and mid-1979, of a projected foreign investment in Chile of over US$4 billion approved by the Committee on Foreign Investments, the Chilean government agency charged with that responsibility, over US$3.7 billion was in mining, predominantly copper.

Chile's reserves of copper are vast, amounting on current estimates to about 25 per cent of total world resources, while present production equals about one-eighth of world demand. The country's exploitation of that resource, and its contribution to national economic development must, of course, turn on world market conditions over which government has no control, and the development of major projects elsewhere in the world, for example in Peru and Panama, in the 1980s and 1990s may have an unpredictable impact. But Chile enjoys comparative advantages of low production costs and rich ores, and must remain a major world-market force in copper to the end of the century. And while copper's share of Chile's export earnings has, healthily, fallen in the 1970s to less than 50 per cent of the total, the mineral and its by-products, such as molybdenum, will remain the single chief national earner of foreign exchange for the foreseeable future.

The development of iron ore in the Norte Chico

The Norte Chico produces almost Chile's total output of iron ore. Mining began only in 1910 at El Tofo in Coquimbo (now Region IV) to supply the new blast furnace at Corral in Valdivia (now part of Region X), but the enterprise failed, and in 1913 the Bethlehem Steel Corporation of Pittsburgh acquired El Tofo to produce ore for shipment to the United States (Pederson, 1966, p. 247). From 1921 to 1957 the mine produced over 48 million tonnes of ore, but declining yields obliged the company to move its major operations to El Romeral, some 25 km north of La Serena, which it acquired in 1936 but did not develop until 1948. It then did so in association with the Chilean state steel organization, the Steel Company of the Pacific (CAP), established in 1942, Guayacan being equipped with mechanical loading systems and storage facilities to serve as the major port of shipment of the ore.

The CAP itself acquired another very large ore deposit in 1959 at Algarrobo, some 35 km south-west of Vallenar, with an estimated 70 million tonnes of reserves. Prior to 1971 the Gran Minería of iron in Chile, represented by

Algarrobo, El Tofo and El Romeral, accounted for about 25 per cent of total national ore output, with the Mediana Minería, composed of a large number of private producers, producing 45 per cent, but in that year, the privately owned mines were nationalized by the Allende government and CAP assumed control of some 96 per cent of production. The successor military government, while retaining CAP as a CORFO affiliate, announced in January 1981 plans for a drastic reorganization, whereby CAP became a holding company divided into seven subsidiaries, completely independent, individually incorporated and each responsible for its own budget.

Chilean iron-ore reserves amount to around 900 million tonnes (proven) though with a further probable 2 billion tonnes, and current production is around 15 million tonnes a year. Hitherto unexploited deposits, such as those at El Laco in Antofagasta (Region II), estimated to contain over 150 million tonnes of ore with a 66 per cent iron content, will probably require heavy foreign investment to exploit them, but that will depend on world market demand, currently (1981) depressed. Chile, however, possesses about 0.8 per cent of proven world reserves of iron ore, sufficient for her own domestic steel industry and capable – as she has been since the 1950s – of supplying overseas demand, notably from Japan, which currently takes about 85 per cent of total Chilean exports. The future of Chilean iron-ore mining, however, as a major contributor to general economic growth is not promising, in current world conditions of supply and demand, though national resources will be more than adequate for domestic demand at least to the end of the century.

Manufacturing industry and other activities
The proximity of mineral deposits to the coast in Chile naturally made for the growth of suitable ports of shipment situated as near as possible to the ore bodies and, as we have seen, the growth and development of places like Iquique, Antofagasta, Chañaral, Huasco and Coquimbo were intimately linked to mineral exploitation. Mines needed equipment and provisions and, in the pre-railway era, a large amount of both came by Chile's major colonial highway, the Pacific coastline, the ports being the link between sources of supply and external markets and the hinterland mines. Apart from the incoming and outgoing traffic through the ports, a number of ancillary activities to mining naturally developed in these distribution centres, such as small industrial establishments in the ports of Tarapacá to provide machinery for the nitrate *oficinas*, and smelters for copper ores in the port towns of Atacama and Coquimbo. There was, therefore, something of an industrial and manufacturing tradition in these places long before Chile deliberately set out on the path of economic diversification. In addition, the Pacific waters off Tarapacá and Antofagasta provinces are part of that immensely rich fishing ground, which, thanks to the ideal conditions for plankton created by the cold Humboldt current, stretches from Antofagasta to the coast of central Peru, and which, from time immemorial, has compensated man for the barren nature of the land opposite. Consequently, while the fishing industry has long been a characteristic economic activity here, recent times have seen the growth of fish-canning and fishmeal-processing, notably at Iquique and Antofagasta, though also at Arica and Pisagua. In 1959 CORFO was requested by the government to prepare a development plan for the fishing industry, and

this was motivated in large part by the decline of the northern nitrate regions, leading to much migration to the south. Thanks to credits made available to establish plant and build modern fishing vessels, Iquique, once the nitrate capital of the world, became in the 1960s Chile's major fishing and fish-processing centre, having by 1965 twenty-three of the thirty-five fishmeal plants in the northern region, and producing 173,000 tonnes of fishmeal. The greater part of this was for export, and by 1965 Chile had become the world's fifth largest exporter of fishmeal. Moreover, as part of its export-diversification programme, the military government since 1973 has encouraged the fishing industry. The northern regions now account for some 90 per cent of national production of fish products, and canning and processing facilities were enlarged in the late 1970s and early 1980s at such ports as Iquique.

The growth of other port cities in the modern period has been linked in part by international as well as national developments. Antofagasta, the largest city of the Norte Grande and the Norte Chico, derives its significance not only from the fact that it is the port of shipment of Chuquicamata's copper, but also from its position as the terminal point of both the railway from Oruro in highland Bolivia – some 1155 km length – and the railway to Salta in north-west Argentina – some 845 km distance. Until the 1960s, in fact, the Antofagasta–Oruro railway was crucial to Bolivia and, as late as 1962, 87 per cent of Bolivian tin – the country's major export – was shipped by this route, though it has since been challenged by the lines from Oruro to Arica and to Peruvian Mollendo. The Antofagasta–Salta line, commissioned finally in 1948, had disappointing results, and was superseded by the motor road built in the early 1960s. Nevertheless, the Antofagasta– Bolivia Railway, still owned on the Chilean side by a British company, is an important link for Bolivia with the outside world, and seems likely to remain so.

Concepción, the Frontera and the Lake District

Introduction
The region between the river Bío-Bío and the gulf of Reloncaví, described in the section on climate in this essay as Forest Chile, is, in its northern part, really a continuation of the central valley and, from the province of Cautín south (Region IX) is usually known as the Lake District. The most appropriate designation, however, is for the region bounded by the provinces of Concepción in the north (Region VIII) and Cautín in the south is *La Frontera* – the frontier – for it was the line of the Bío-Bío that for more than three centuries marked the effective demarcation of Spanish from Indian Chile, and the modern history of this region is essentially the story of the southward march of settlement from that line. The city of Concepción is the metropolis of the south, and so it has been since the discovery and conquest of Chile, despite changes of site since its foundation in 1551. Valdivia, originally founded a year later, is the second most important centre in the region. Most of the other settlements of the period of conquest in the *Frontera* – such as Angol in the province of Malleco (Region IX) – were abandoned because of Indian pressure and only refounded centuries later, after the birth of the republic of Chile.

Though the late eighteenth century saw some expansion of Spanish settlement south of the Bío-Bío, it was mostly in the 1840s and 1850s that a major impetus to colonization far south of the river was given by a law of 1845, amplified in 1851, to attract European immigrants, and particularly Germans, to what became the province of Valdivia. The majority of the new settlers – some 1200 arrived in 1850 in the vicinity of the fort of Corral – were small farmers and peasants, as were those who came in 1853 to settle on the shores of Lake Llanquihué, a colonization which was completed by 1861, and which was marked, among other things, by the foundation of the towns of Puerto Montt and Puerto Varas in 1853 and 1854 respectively (Jefferson, 1921).[11] This movement, in effect, leap-frogged the *Frontera*, which in this period was still Indian territory, held by the Mapuche, and it was not until the beginning of 1883 that the last Indian stronghold fell to Chilean troops, the culmination of a campaign – often desultory – that had been in train since 1862. During the 1880s new immigrants settled in the region between the Bío-Bío and Valdivia – Germans, Swiss, Spanish, French and English as well as a considerable number of Chileans, many of them demobilized soldiers after the War of the Pacific, encouraged to settle in the south by government policy. The provinces of Malleco and Cautín were formally incorporated into Chile in 1887, but they still contain the vast majority of Araucanian Indians in Chile.

The numbers of immigrants from Europe was small, the *Frontera* receiving some 32,000, Valdivia and Llanquihué between 5000 and 6000 by the end of the nineteenth century. In the regions south of the Bío-Bío, however, it was not so much numbers as dynamism that gave the foreigners a predominant position, and towns such as Valdivia, Osorno and Puerto Montt still reveal today in the architecture, the physical features of many of their inhabitants and in the ubiquity of German names the imprint of the colonists of a century ago. It was these pioneers of the frontier, a class of smallholders independent of the traditional *hacendado-inquilino* relationship of the central valley, who tamed the wilderness, built the towns, and by their initiative and enterprise in agriculture, stock-raising and manufacturing had an impact out of all proportion to their numbers, and justified the extension of the communications network longitudinally from the central valley, the railway reaching Puerto Montt in 1913. *Figure 10.1* indicates the southward march of the Chilean frontier in terms of town-foundations in this period.

No region of Chile presents so many contrasting types of human life and activity as that between Concepción and Llanquihué, ranging from the highly urbanized, industrial zone of Concepción itself to the Mapuche reservation lands of Malleco and Cautín.

The Concepción industrial zone
Throughout the colonial period the province of Concepción, then including the area between the rivers Maule and Bío-Bío, was the second major zone of Spanish settlement, containing at the end of the colonial period approximately one-third of Chile's population of some 500,000, excluding unassimilated Indians in Araucania. With the closing of the frontier at the Bío-Bío at the end of the sixteenth century, Concepción became the military province guarding Chile against the Indians, and the city of Concepción was the major port and the major

fort of the south. It was the second city of Chile and, indeed, at times the first, despite changes of location due to Indian depredations and destruction by earthquake. The present site of Concepción, occupying the strategic gateway location where the *cordillera de la costa* is broken by the wide valley of the Bío-Bío, dates from the mid-eighteenth century, though the first Concepción, where the town of Penco now stands, was two centuries older. The city's outlook has always been a maritime one, for, throughout the colonial period and after the essential communications link with Santiago was by sea, through Valparaiso, and even today Concepción's dependence on sea-borne commerce remains considerable. In fact, the partly enclosed bays of Concepción and San Vicente provide the most favoured natural harbours on the Chilean coast, and this physical feature has been well utilized by man throughout Chilean history.

The hinterland region north of the Bío-Bío, as far as the river Maule, was an important one in the essentially pastoral agricultural economy of the colonial period, with its higher and more regular rainfall compared with the more northern parts of the central valley and, in addition, a certain local tradition of small-scale manufacturing developed in response to Concepción's isolation from Santiago. Further impetus to these developments came in the nineteenth century: thus, for example, the growth of Chilean wheat and flour in international markets, and not least with the demands of the Californian and Australian goldfields in the 1840s and 1850s, stimulated the milling industry at Concepción, Penco and Tomé, and the rising port of Talcahuano. More significant in the long run was the opening of the coal mines in the area of Lota and Coronel, south of Concepción in the 1830s, and at Lirquén, north of Concepción, in the 1840s, developments coincidental with the advent of the steamship, followed soon after by the railway. It is here, and in the neighbouring province of Arauco, that Chile's major coal deposits are to be found, sub-bituminous in quality and in irregular seams running beneath the Pacific Ocean. In a continent poorly endowed in coal deposits, Chile possesses fields here that are still among the most important in Latin America, though modest in output and quality compared with those of Europe and the United States. In the nineteenth century, however, the steady expansion of coal-mining in the provinces of Concepción and Arauco made a considerable contribution to such industrial developments as took place in Chile. Moreover, special clays found in association with the coal deposits became the basis of clay-products plant and refractory brick manufacture in the 1860s, particularly near Lota, and other local industries — glass-blowing, copper-smelting (cf. above) and timber mills — utilized this local fuel.

Other industries were established in the Concepción region during the course of the nineteenth and early twentieth centuries — woollen textiles at Tomé, cotton textiles at Chiguayante, sugar-refining and ceramics at Penco, glass products at Lirquén, and fish-processing plants at Talcahuano and San Vicente (Butler, 1960). Many of these were based on locally produced materials and supplied the Chilean domestic market, with the aid of a communications network built both to serve that purpose as well as to bind Chile together. The branch line of the great central railway linking San Rosendo to Concepción and Talcahuano was finished in 1872, and in the 1880s particularly, under a programme of active government promotion, communications and port developments were given special emphasis. The Bío-Bío was bridged at Concepción in 1888 – a significant event for the coal mines of Lota and Coronel – the great dry dock of Talcahuano was completed and port facilities at Coronel and Tomé were improved (for a contemporary description, see Russell, 1890, pp. 33 ff.). Equally significant was the colonization of the *Frontera* and the Lake District, which, while giving Concepción a more central position in Chile as a whole, also reinforced its traditional role as the metropolis of the south.

Chile's economic experience in the first 40 years of the twentieth century, and notably her reaction to the Great Depression, clearly stimulated the types of economic activity to which the province of Concepción was already dedicated – consumer goods production with low capital needs but with high requirements in unskilled labour, and manufacturing dominated by traditional lines such as textiles, ceramics and glass, and not so diversified as the import-substitution manufacturing of Santiago and Valparaiso. The Industrial and Commercial Census of 1937 showed the number of industrial establishments in the province as 1115, employing over 17,000 workers, though, significantly, only twenty-four concerns employed more than 100 hands (Butler, 1960, pp. 13–14).

However, events in the late 1930s presaged a fundamental transformation. In 1938, the Popular Front – a combination of left wing and centre parties committed to a policy of active state intervention in the economy – came to power in Chile, and in 1939 there occurred both the outbreak of the Second World War and a devastating earthquake in south-central Chile, which disrupted life in Concepción. This event led to the creation of the Chilean Development Corporation, which immediately began to plan Chile's economic future. The coincidence of this development with the impact on Chile of the Second World War, during which Chile's import-substitution process was stimulated to produce an average annual rate of manufacturing growth of some 11 per cent, was highly significant, since, in the post-war period, the stage was set for the next phase of Chile's economic growth: the establishment of heavy industry under the aegis of a state organization. The core of this programme was a new, large, integrated steel plant, and the site chosen was Huachipato, on San Vicente Bay. Work began on the site in January 1947, and the steel mill was commissioned in November 1950 (for details of construction and finance, see Butler, 1960, pp. 17–19).

It was this development, above all, that converted the Concepción region into Chile's major zone of heavy industry, and expansion of both the steel industry and a variety of activities associated with it has continued since that time. In 1947 Chile's total production of iron and steel was about 26,000 tonnes, with consumption at 163,000, necessitating large imports to be paid for with scarce foreign exchange. By the mid-1960s, Chile was producing an annual output of around 500,000 tonnes of steel and steel products, and supplying over 80 per cent of national requirements (Soza, 1968, p. 618). The steel complex itself necessarily entailed the growth of other industrial activities – coking plant, sulphuric acid plants, coke oven by-product shops, machine shops and so on. The dependence on sea-going transport to sustain the steel mills naturally led also to mechanization and the development of storage facilities at Huachipato and San Vicente ports. The vast amounts of water required by the various industrial plants came a distance of 5 km, by a complicated intake structure, from the Bío-Bío, and energy was supplied from the hydroelectric station at Abanico, some 160 km east of Concepción on the river Laja, near the lake of the same name.

The impetus given by the rise of Huachipato to ancillary industrial development in Concepción was enormous, and here the same advantages of site also operated – extensive level tracts of land in the lower reaches of the Bío-Bío, a good communications network by road and rail with the major markets further north, proximity to good port facilities and, with the rapid urbanization of Concepción city and of Talcahuano, easy access to the labour force and the usual commercial services a city provides. Companies producing steel-wire products, ferro-alloys, zinc and tin recovered from slag are among enterprises that sprang up in Huachipato's shadow.

Other industrial activities, of increasing importance in Chile's export trade, also developed in the region, again because of favourable physical site factors. The Concepción region and the neighbouring provinces of Forest Chile are immensely rich in timber, notably the pine, *Pinus insignis radiata*, which grows here at a faster rate than anywhere else in the world. In the last two decades of the nineteenth century, private companies began a systematic policy of forest plantation, providing the raw material for the highly important pulp and paper industries, and eventually for the production of cellulose. In 1957 a modern pulp and newsprint plant was built at San Pedro, across the Bío-Bío from Concepción and, at the same time, a cellulose plant was begun at Laja, some 65 km southeast of Concepción. A later industrial development in the region, the petrochemicals industry was initiated in the mid-1960s by the building of a refinery at Concepción, utilizing oil imported from Chile's fields in Magallanes. The plant was inaugurated in April 1967, since when additional plants for the production of ethylene, caustic soda and synthetic resins (to meet the needs of the plastics industry) have been opened.

During the Allende period the industrial region suffered from the vicissitudes of the rest of the economy as already described, and some industries experienced great uncertainty owing to the government's professed aims of extensive nationalization. Some of these aims were as much political as economic: for example, a determined attempt was made to take over Chile's largest manufacturing establishment in wood pulp and paper, the Compañía Manufacturera de Papeles y Cartones, because of its virtually monopolistic position over newsprint, and despite the fact that it had never used that role politically. Supply bottlenecks in many industries were acute, owing to government control policies over prices and materials, and the manipulation of the exchange rate. With the advent of the military government in 1973 and its free market economic policies, however, while the political uncertainties were removed, the economic impact on the region has been very different for different interests.

A prime beneficiary has been the forestry sector and related industries, and, indeed, in the 1970s they experienced spectacular growth. A major cause was the new Afforestation Law of 1974: this exempts from general taxes on agricultural land those areas designated, after survey by the National Forestry Corporation (CONAF), as prime forest; it imposes low taxation on profits from forestry exploitation and, for 20 years from the date of the law, the government will pay a subsidy of 75 per cent of the net costs of plantation. Private enterprise has thus been encouraged to move heavily into the sector, and it has done so: whereas in the period 1970–5 an annual average of 55,000 to 60,000 ha were planted, the average for 1975–80 rose to 80,000 ha. And exports in forestry products rose dramatically, in relation to total exports, during the decade.

Table 10.7 Forestry exports 1973–80 in relation to total exports (US$ million in current dollars)

Year	Total exports	Forestry exports	%
1973	1,310.5	36.4	2.8
1974	2,238.9	127.0	5.7
1975	1,529.6	125.0	8.2
1976	1,990.7	169.1	8.5
1977	2,190.3	180.5	8.2
1978	2,400.6	236.9	9.9
1979	3,763.4	349.5	9.3
1980	4,818.1	468.1	9.7

Source: *Chilean Forestry News* (Santiago), 37, February 1981.

Large investments in the sector in these years – many by Chilean banks – and aggressive marketing policies which had enabled Chile to export wood and wood products to over fifty other countries by 1981, made forestry and wood-processing the most dynamic sector of the Chilean economy in its rate of growth, by 1981 second only to copper in the foreign-earnings table. These developments, largely concentrated on the Concepción region and the Forest Area, are perhaps the classical example of a positive response to the government's free-market policies.

The same, however, cannot be said of other industries in the zone. The policy of drastic tariff reductions and removal of protection from national industries has hit textiles hard, and during the 1970s major textile firms experienced a serious fall in production in the face of cheap foreign competition. Despite mergers to promote economies of scale, and the halving of the work force over the period, this sector of the economy does not seem likely to become sufficiently competitive to survive with a fair share of the national market and, as forestry is a success story, textiles is one of the casualties of the new economic philosophy. Similarly, consumer goods industries have suffered from the import boom of the late 1970s, notably white goods and products of the electronics industries, and Chilean manufacturers' market share has been falling. This is due not only to the low tariffs on foreign imports, but also to the maintenance by the authorities of an over-valued peso in its battle against inflation, which penalizes exporters and removes the incentive to be competitive. Thus, while local industry – much of it based in Concepción – produced over 45,000 refrigerators in 1978 (compared with 24,000 in 1976), imports grew between the 2 years from 6000 to 35,000. And in the latter year domestic production of monochrome television sets stood at 653,000, while imports amounted to 806,500 (*El Mercurio*, Santiago, 25 October 1979).

The differential impact of the military government's economic policies on

Chilean economic activities, and not least manufacturing, cannot yet be fully evaluated, but they have particular significance for the Concepción region. The distinctive position it occupied in Chile as the only province in which industry is the predominant activity, and the only province to offer anything like a serious challenge to the exclusive concentration of advanced industrial development in Santiago and Valparaiso, might well be threatened if current policies are continued without modification. Its growth and development were largely determined by the deliberate, state-planned location of industry there, under the aegis of CORFO, though that choice was also dictated by highly favourable geographical factors and by a previous historical development which seemed highly conducive to future progress. In the far different conditions which obtain at the time of writing, it is an open question as to whether its industrial base will survive international competitive pressures or whether, like the *oficinas* of the Atacama Desert, it will become one of their significant victims. If it does, it will be yet another example of the unintended, but inevitable, consequences of the dominance of centralist direction on regional evolution, and to that extent will represent a historical continuity in Chilean experience.

The farming economy of Forest Chile

While Concepción is the pre-eminent industrial zone of the south and, indeed, of all Chile, other cities of the forest region – Temuco, Valdivia, Osorno, for example – also have important industrial activities located there, and Valdivia in particular produces a wide range of commodities – leather goods, ceramics, wood products, oxygen, nails, refined sugar, freight cars, tugs and lighters. Nevertheless, it is true to say that such manufacturing as takes place in these centres is of greater regional than national importance; the industries rely heavily on local resources of raw materials, and much more significant in the national economy is the contribution the provinces make in agricultural and pastoral products, as they have done since their incorporation into the republic. The colonization of the forest region during the last decades of the nineteenth century had the effect, for example, of moving Chile's major wheat zone from the central valley to the south: whereas in the 1870s the region between Bío-Bío and Valdivia accounted for only 8 per cent of the cultivated area of Chile sown to wheat, by the First World War it accounted for 40 per cent, and wheat production there tripled in volume (Sepúlveda, 1959, pp. 112–14). Today, the region between Bío-Bío and Llanquihué accounts for approximately two-thirds of Chile's cereal production, almost two-thirds of the cattle population, five-sixths of the lumber production, and most of the apple crop. Such figures represent not only the natural-resource endowment of south-central Chile but also the evolution of its exploitation by man, a process in which foreign immigration was significant by virtue of its example rather than by its volume, and one that has given this essentially agricultural region of Chile characteristics which are markedly different from those of central Chile.

One such feature is landownership. While the large hacienda is by no means unusual in the forest region, and neither is the *minifundia*, there is a far greater variety in sizes of holdings than in the central valley. Owing primarily to the far heavier rainfall throughout the year, the region is more devoted to mixed farming than is central Chile, and, while this is much more true of the Lake District than of the *Frontera*, the agricultural economy is far more dynamic. During the late nineteenth and early twentieth centuries, considerable numbers of Chileans from these southern provinces migrated – many temporarily – to the trans-Andean provinces of Argentina (Jefferson, 1921, p. 35), and in recent years what have seemed to be better opportunities there have attracted some Chileans across the frontier. Most migrants, however, head north, not east, to the growing towns and cities of their own country.

With the exception of the province of Concepción, all the provinces of south-central Chile have a larger rural than urban population but the region lacks the feudal atmosphere of the central valley, owing to the quite different nature of settlement and the absence of such *patrón–peón* relationships as *inquilinaje*. Nevertheless, the process of colonization itself had its dark side. One such aspect was the treatment of the Mapuche Indians, to be considered shortly. Another was the treatment of Chileans by their own governments, which consistently favoured the foreign immigrants and their own upper class. In the early 1870s, for example, the government of the day permitted Chilean peasants from the north to clear and till land in the *Frontera* in return for very low rents, and a great deal of land came under cultivation, with considerable Chilean settlement. The renter, however, had no security of tenure, and his land could be simply taken away by government agents, without payment for improvements effected. Public auction of state lands soon followed, and this effectively froze out the Chilean peasantry, who had no capital, and brought in the speculator, notably the wealthy of Santiago. Land grants and other forms of assistance to foreign immigrants were not extended to Chilean peasants who became, in effect, landless labourers for the new colonizers, and evictions of hard-working squatters were common. If it is undoubtedly true that the development of Chile's forest region owes a great deal to the industry and pertinacity of the European immigrant, it is no less the fact that the labour of Chileans was crucial in that development, which was purchased, in part, through policies that consistently militated against Chilean peasants acquiring their own land (Jefferson, 1921).

The life of the towns of Forest Chile is closely integrated with that of the rural landscape. Temuco is a large lumber, livestock and agricultural centre, with many flour mills and tanneries; Osorno has one of the largest dried-milk plants in Latin America, and one of the biggest butter plants in Chile, as well as a modern chilled-beef packing plant, and a number of flour mills; Puerto Montt has utilized its coastal location to become the supply point of over half the shellfish production of Chile, and possesses a number of fish-canning and processing factories. As in the case of Concepción, state-directed developments were also of some significance in the region's growth: the National Sugar Corporation (IANSA, Industria Azucarera Nacional SA), a subsidiary of the CORFO, established sugar beet refineries at Los Angeles in Bío-Bío and at Llanquihué and, indeed, was instrumental in the rapid growth of sugar-beet cultivation, a marked feature of agricultural production in the region in the 1950s and 1960s.

During the Allende period no part of Chile was subject to more political turbulence than this region, affected as it was by politically fomented land seizures by peasants, and severely affected by government pricing policies on foodstuffs. The reaction of landowners was to disinvest in their holdings, either slaughter their cattle or drive them into Argentina, and even organize quasi-militias to resist illegal occupations. Re-organization under the military government was a slow process, but the elimination of uncertainty regarding expropriation and the gradual return to normality enabled farmers and growers to participate in the new export-orientated production of fruit and vegetables, as well as forestry products, already described (pp. 515–16). As noted, however, export opportunities have led to a decline in traditional-foodstuff production for internal consumption, and concentration on exports. As in the Central Valley, the farming economy of Forest Chile has turned more in the 1970s to food processing for export – tinned fruits, jams and dried vegetables, for example – all of which performed well in the late 1970s. Similarly, the coastal regions have benefited from the cultivated growth of fishing and the processing of fish and shellfish products. Domestic meat production, heavily concentrated in the region, increased from 260,000 tonnes in 1977 to 317,000 tonnes in 1980, and there was a considerable increase in cattle production in the provinces of Cautín (Region IX) and Valdivia (Region X) in these years, by over 25 per cent in the former and by over 20 per cent in the latter (*Chile Economic Report*, no. 120, March 1981). Chile is self-sufficient in dairy and meat products, and if these developments continue will have exportable surpluses in the 1980s and 1990s. She is also the only country in South America to have been declared free of foot-and-mouth disease by the Pan-American Health Organization (1981).

The Indian lands
The area between the rivers Bío-Bío and Tolten contain well over 90 per cent of the pure-blood Indian population of Chile, the descendants of those warlike Araucanians, or Mapuche, who withstood the Spanish empire for almost 300 years, and successfully resisted the Chilean republic for another 60. During the nineteenth century, however, the northern part of the *Frontera* was encroached on by Chilean settlers before the last Araucanian resistance crumbled in the 1860s, and in the colonization that ensued, Indian rights were not respected. In the early 1880s, for example, Indian chiefs were induced to accept European notions of property rights, in terms of family, rather than tribal, property, a change that clearly weakened their capacity to prevent expropriation. Detailed, and sometimes benevolent legislation to protect the Indians was often more honoured in the breach than the observance, and the establishment of reservation lands, legally held in title, was subject to provisions permitting a *comunidad* to request division of common land if at least one-third of its members so desired. The laws and their operation were, and are, very complicated: suffice to say that, with ample scope for exploitation by outsiders, and with increasing population pressure within the reservations, expropriation of Mapuche lands was the most pronounced characteristic during the last century, and the growth of *minifundismo* the most striking feature of this one. There are some prosperous Mapuche, but they are very few, and the vast majority, faced with deterioration of their lands as a result of demographic pressure, low technological levels and weak credit facilities, present a growing social and economic problem. Modern Chilean governments have made conscientious efforts to preserve the Araucanian heritage and help the Indian communities, but the task is difficult and today many Mapuche simply migrate from their homelands to other parts of Chile (for detailed discussions see CIDA, 1966, pp. 79–95; Universidad de Chile, Departamento de Extension Cultural, 1956; Faron, 1968).

The Mapuche are a national as well as a regional problem, as is, indeed, the development of the Forest Region itself. But its potential for a future major contribution to national well-being is very large. Two examples will illustrate the point. First, owing to the abundance of rivers in this region of high rainfall, Forest Chile's share of the country's hydroelectric capacity – *per capita* the highest in the world – is large, and already the area between Bío-Bío and Llanquihué is highly significant in national power supply. Under the National Electricity Company (ENDESA), a CORFO affiliate, plants at Abanico in Bío-Bío, Pullinque in Valdivia, Pilmaiquen in Osorno, El Toro in Ñuble and other plants in the region feed their power into the national grid; and ENDESA'S current development plan, both of expansion of existing as well as the opening of new plant, seeks to produce 19.3 billion kWh annually by 1988. Chile's hydroelectric potential, much of it located in Forest Chile, is currently only tapped to 7.5 per cent, and now supplies about 21 per cent of the country's current energy requirements. Several new major plants under construction are located in the region. The second example for growth potential in the region lies in tourism. Already the Lake District is a major attraction to foreign visitors – mostly from neighbouring Latin American countries, particularly Argentina – and here, as elsewhere in the republic, recent years have seen a great improvement in facilities with government support. The region of the lakes, with their superb volcanoes and tree-clad slopes, is an area of outstanding natural beauty, and will undoubtedly contribute heavily to Chile's foreign-exchange earnings in the future. Moreover, domestic tourism is increasing rapidly as a consequence of a higher standard of living generally, and the region is perhaps the most favoured one to benefit from this development.

A pioneer frontier in the nineteenth century, Forest Chile is no longer so, but its future contribution to national well-being will depend no less on the human capacities to overcome obstacles created by geography than they did a hundred years ago.

Chiloé and Atlantic Chile

Chiloé
The province of Chiloé consists of the large island of that name and the mainland opposite, but, while the mainland section partakes of the nature of the frontier, a region of very little settlement, and that in recent times, the island of Chiloé was an object of attention by the Spaniards in the sixteenth century. The reasons were almost entirely strategic: the need to maintain a watch over the southern Pacific against the constant threat of corsairs and maritime rivals to Spain throughout the colonial period. Castro was founded as early as 1567, and Ancud, the island's second town, exactly two centuries later. The population was always small, owing to the island's rigorous climate and scanty resources, and even today its western side is virtually uninhabited and uninhabitable, left to the constant rainfall and impenetrable forest. In the nineteenth century, after the final expulsion of the Spaniards in 1826 (the island was the last part of Chile to be liberated) various attempts at colonization were made, notably towards the end of the century. With generous government assistance, and with the now customary disparage-

ment of the native-born inhabitants, the Chilotes, European immigrants – and not least British and Dutch – were encouraged to settle on the island, but the experiment was not a success (Jefferson, 1921, pp. 40–1). Land was cleared, and potatoes and rye were sown, but the development was essentially for subsistence since market conditions did not exist, even if there had been surpluses to export. Moreover, the traditional type of land tenure here was smallholdings, and as population increased, however slowly, subdivision by inheritance became so acute that the only answer to the problem of subsistence was migration.

The Chilotes are the great migrants of Chile. Today, they are prominent in the Chilean merchant marine, and hundreds journey every year to the vast sheep ranges of Magallanes to help with the wool clip, returning to the island to resume work in the lumber trade, fishing and cultivation, which are its only economic activities, or not returning at all. Apart from Ancud, the provincial capital, and Castro, there is only one other centre on the island of Chiloé with more than 2000 inhabitants, the population being dispersed inland as agriculturalists, and scattered along the coast as fisherfolk. Economically, the region is one of the poorest of Chile. Potatoes are the main crop on the island, with a little cereal cultivation, but the possibilities for an expansion of the formerly active timber trade are now limited by the virtual extinction of the more valuable trees, such as cypress and larch, and by the high costs of exploitation and transport. There are better prospects for timber on the mainland, in the areas of Chaiten, Lake Yelcho and Palena. The Palena basin has also been an area of stock-raising since its first settlers arrived there in the early years of this century, but it is still very remote from the mainstream of Chilean life, and the province of Chiloé as a whole seems destined to slip still further into the backwater of Chilean development, unless current exploration for off-shore oil proves positive.

Atlantic Chile: Magallanes
Throughout the colonial period Chile's boundaries were understood to extend south to Cape Horn, and some of the independent republic's early constitutions specifically said that they did. But this was theory not fact. A number of expeditions south, almost from the very beginnings of the Kingdom of Chile, were complete fiascos, not surprising in view of the extraordinarily fragmented coastline and innumerable islands of the archipelagic region, coupled with a forbidding climate and inhospitable terrain. What, more than anything, resolved the Chilean government in the nineteenth century to make more positive its claim to Magallanes was the advent of the steamship in the 1840s, and the danger, which was real and not merely apparent, that foreign powers might seize possession of the now strategic narrows from the Atlantic to the Pacific. The threat that this might pose to Chile, whose entire western frontier was the Pacific coastline, and whose communications were still largely maritime, was obvious to the government of President Bulnes (1841–51). In 1843 accordingly, an expedition founded Fort Bulnes and claimed the incorporation into Chile of the Straits of Magellan. The early years of the settlement were very arduous, and in 1849 Fort Bulnes was abandoned for a new site, 56 km north, backed by an expansive grassy plain, with beech trees, a permanent watercourse and, higher upstream, coal in placer deposits. Punta Arenas was thus founded for strategic reasons and for national satisfaction; true colonization came later.

Until the decade 1865–75 Punta Arenas was basically part military outpost and part penal settlement, operating an essentially subsistence economy. Attempts were then made to mine both coal and gold, and while success was limited, these developments brought in more people, led to the building of a light railway, a wharf and a store, and were accompanied by the laying of other economic foundations. In 1867 a government decree offered cheap grants of land to families of colonists, who settled in the Brunswick peninsula and began to supply passing vessels with fresh provisions. Population increased from about 200 in 1865 to 1100 ten years later (Butland, 1954, pp. 30–1). In the same period two important events occurred, though their significance only became apparent in the 1880s: in 1868, Punta Arenas was declared a free port, which it has remained to the present day; and in 1866, for the first time, a large ship navigated the channels route between Atlantic and Pacific through the Straits of Magellan.

The decade of the 1880s was a watershed in the development of Magallanes. It was a period of considerable exploration of the southern reaches of America, promoted in part by the growing power of Argentina and Chile, and by the desire of their governments to know what resources their unexplored territories contained. The realization dawned, on both sides of the frontier – still not delimited in many areas – that the extensive grasslands and low, undulating terrain of Magallanes and Patagonia were admirably suited to sheep-rearing. *Estancias* were established on land occupied with government permission, often obtained retroactively, and the population began to grow. By 1885 it had reached a total of over 2000, more than a third of it in Punta Arenas itself, and of the total population over a third was foreign-born, English and Scots emigrés playing a prominent part in the rise of the *estancias* (Butland, 1957, p. 57). The beginnings of colonization on the island of Tierra del Fuego soon followed.

From 1885 to 1907, what was still the territory, and not the province, of Magallanes experienced an economic transformation. Population grew almost eightfold (see *Table 10.2*), and the number of sheep increased by some forty-two times, to nearly 2 million. Colonization spread westward and northward on the mainland, south and east in Tierra del Fuego, so that by the end of the period it had reached Ultima Esperanza in the north and Riesco Island to the west. By the end of this period, it was reported that profits in some of the land companies were over 300 per cent within two years (Scott Elliot, 1907, p. 308), while the number of ships entering and clearing Punta Arenas had risen from an annual figure of less than 300 to more than 1900 (Butland, 1957, p. 58). Punta Arenas itself began to assume the aspect of a real town rather than an isolated outpost of settlement.

Important changes also occurred in this period in the land-tenure system. By the 1880s it was no longer necessary for government to induce colonists to settle in Magallanes by easy grants of land, and laws were passed providing for the lease of property, ownership remaining with the state and the majority of farms in the hands of existing holders. But the leasing system lasted only to 1902; in that year new legislation provided for the sale of enormous tracts of land at public auction, without any reservation of the lands already in farms. The result was that the principal land companies, which had risen with the pastoral boom of the previous 20 years, had the financial power to acquire huge estates, absorbing in the process most of the smaller farms (Butland, 1957, pp. 60–1). In fact, between 1903 and 1906, a mere twenty-nine owners acquired more than 1,600,000 ha (CIDA, 1966, p. 116). Shocking cases of eviction occurred, the most notorious at Ultima Esperanza. This hitherto empty quarter had been colonized in the 1890s by German settlers under licence from the governor of Punta Arenas; they had established highly prosperous sheep farms, built hotels and wharves, cut roads to Punta Arenas and to Gallegos, in Argentina, and even established steamship connections with German transatlantic lines, and by 1906 the colony numbered 600 people. Moreover, when in 1902 a frontier dispute on this region between Chile and Argentina was arbitrated by Great Britain, the presence of these Germans as settlers from Chile was a major factor in the decision to award the territory to that country. Under the new laws, however, much of the land in Ultima Esperanza was acquired by the Tierra del Fuego Exploitation Company, and most of the settlers were obliged to leave (Jefferson, 1921, pp. 46–9).

No less significant were the long-term repercussions of the creation of pastoral *latifundia* in Magallanes. However efficient the large estates might be, this system of land tenure imposed patterns on the territory that may not have been in its best interests. In the first place, it discouraged permanent immigration while establishing the practice of seasonal migration, since sheep-farming required additional hands in the shearing season only. Most of the migrants came, as noted, from Chiloé. Moreover, despite the growth of other economic opportunities in Magallanes, these were not enough to absorb the migrants in permanent employment after the shearing season, and many moved on to Patagonia. The size of estates, and the fact that many were in foreign hands, became a subject of a controversy in Chilean politics which is not yet resolved. Finally, there can be no doubt that lack of opportunities on the land for immigrants has been an important factor in the urbanization of Magallanes and the concentration of population into one town, Punta Arenas. By 1960 it contained almost 70 per cent of the provincial population (*Censo*, 1960, Magallanes, p. 6).

From this period at the turn of the century sheep-farming has been the mainstay of the Magallanes economy, and its progress has been marked less by an increase in numbers of sheep – from an average of 2 million in the first 30 years of the century to about 3 million today – than by improvements in breeds, land utilization and communications, and similar developments to benefit both wool and meat production (for a detailed account, see Butland, 1957, pp. 69, 85–102). Today Magallenes produces over half Chile's output of wool, the average annual amount for the early 1960s being about 20 million kg. The meat-packing industry began to develop at the time of the first pastoral book: by 1905, the *frigoríficos* of Punta Arenas were sending some 75,000 sheep carcasses to Smithfield (Scott Elliot, 1907, p. 309), but by 1929 meat exports, at 20,000 tonnes, were ten times what they were in 1905 (Butland, 1957, p 71). In more recent years, however, both higher demand for meat in Central Chile, and high wool prices, leading to concentration on wool production at the expense of meat, has led to a marked fall in exports of frozen mutton and, indeed, to an overall decline in meat production. A third factor in this fall has been the diversion of Argentine sheep – formerly an important proportion of those processed at Punta Arenas – away from Magallenes to Argentine plants, and in the late 1950s a number of *frigoríficos* in Magallanes were closed (Bohan and Pomerantz, 1960, p. 36).

The growth of the industrial fabric of the province, based on Punta Arenas, was naturally intimately linked in the early days of expansion with sheep-farming. But this was not all, and during the period 1885–1907 other important activities were gold-mining (of placers), a development that naturally attracted a large number of immigrants – Yugoslavs, Spaniards and Italians in particular – a timber industry, and a revival of coal-mining, in view of the vastly increased number of steamships calling at Punta Arenas. Other settlements arose, including the only urban settlement in Chilean Tierra del Fuego, Porvenir, across the strait from Punta Arenas, founded in 1894, and having in 1960 a population of 3000 (*Censo*, 1960, Magallanes, p. 8). Certain of these developments were short lived, and while, even today some gold-prospecting continues, the hectic boom of the mid-1890s seems unlikely to be repeated. But a more general characteristic of the development of Magallanes has been the filling out of structures established three generations ago, with some important changes to which reference should be made.

One of these was a reduction in the size of the great *estancias*, at their peak in the first decade of this century. The basic reason for this was the growing awareness of the Chilean government of the need to promote immigration and to guard against the dangers of monopoly. From the First World War to the Second, as leases expired and as land companies themselves gave up some of their holdings, large areas were subdivided and offered to colonists by government agencies (for details, see CIDA, 1966, pp. 116–17; Butland, 1957, pp. 69–71). The companies still retained enormous estates and, in general, much the best land, but these developments did make the region somewhat less dependent on a few vast enterprises, though, it is interesting to note, among the first candidates for expropriation under the Agrarian Reform Law of 1967 were some of the great *estancias* of Magallanes.

A more significant new development in the province, both for the region and the nation, has been the rise of the oil industry. Again, it was the CORFO that initiated it, through exploration for deposits in the 1940s, resulting in the development of the field at Manantiales between 1946 and 1957. Within a few years, over 500 wells had been sunk in the province, and by 1961 over 1,500,000 m^3 of petroleum were being produced (*Censo*, 1960, Magallanes, p. 7). By 1968, however, petroleum output had reached over 2,177,000 m^3, an increase of over 10 per cent on 1967 (ECLA, 1969, II, p. 145). Development of the highly significant oil industry of Chile – Magallanes is the only province producing oil commercially to date – is under the aegis of a state corporation, ENAP (Empresa Nacional de Petróleo), created in 1950. An increasing proportion of Chile's production of crude oil is refined in Chile at the large refinery at Concón, near Valparaiso, opened in 1954, and at the new refinery at Concepción and it is, indeed, the oil of Magallanes that is the basis for Chile's petrochemicals industry.

Since 1973 the military government has put considerable emphasis on expanding domestic energy production, in the light of world-wide oil price increases in the 1970s, and particularly in association with foreign capital and technology. Chile today (1981) depends on imported oil for over 50 per cent of requirements, and estimated resources are not expected to reduce this proportion to the end of the century; nevertheless, in the 1970s, expansion of the Magallanes fields, with heavy foreign participation, succeeded in raising production to 45 per cent of needs by 1981, and new fields in the Magellan Straits, Ostion and Spiteful, were then providing some 18,000 barrels a day. Perhaps more significant in the long term will be the exploitation of the region's coal-fields, Pecket on the mainland and Isla Riesco in the straits of Magellan, total deposits of some 400 million tonnes, but requiring sizeable investment to bring them on stream. This, in the early 1980s, seemed forthcoming, as foreign investors, no less than the Chilean government, realized the necessity to maximize natural energy resources in the light of the world oil situation. Nevertheless, and despite the oil and coal reserves of Magallanes, and Chile's hydroelectric potential, energy supply seems likely to continue as a restraint on Chile's economic development to the end of the century, unless further exploration, now intensive, results in the discovery of hitherto unsuspected reserves in off-shore Magallanes and the mainland.

Atlantic Chile: Aisén

The province of Aisén is Chile's last frontier, having in 1980 the smallest population, and celebrating its centenary as part of the republic only recently. It was, in fact, in 1870 that Robert Simpson, the British-born vice-admiral of the Chilean navy, first explored the basin of the river that bears his name, following its course from where Puerto Aisén now stands, and crossing the Andes to the valley of the Coyhaique. Others followed in the 1880s, in that rather hectic period of exploration patronized by the Chilean and Argentine governments, and it was these discoveries that made known the other transverse valleys of Aisén. Yet, as late as 1902, when Chile and Argentina concluded an important boundary agreement, there were less than 200 colonizers on the Chilean side, and those who were there had drifted on, not from the west but from the east, crossing into Argentina much further north and coming into Aisén from Neuquén and Chubut (Butland, 1957, p. 77). Numbers grew only slowly, but in 1903 a newly formed land company, the Sociedad Industrial de Aisén, was given a licence by the government to occupy large areas of three river valleys for stock-raising, and its efforts began to have a marked impact on the province. Vast numbers of sheep were introduced, and large tracts of land were fenced; no less important, the company built a road from the Coyhaique region west to Puerto Aisén, which was founded by government at about the same time. Since, however, Magallanes was then by far the major southern area of attraction for would-be colonists, settlement was still slow, and another reason for this was the fact that in Aisén the physical obstacles to settlement were far greater, and the favourable lands much less extensive than in the far south.

Until the 1920s the estates of the Aisén Industrial Company employed a fair proportion of the settlers in the province, though more of them held their own family properties for stock-raising and farming. In the late 1920s, however, in a serious attempt to promote further settlement, the Chilean government drastically reduced the Company's holdings, and enacted other measures to encourage immigration. Thenceforward, Aisén's population and the provincial economy grew more rapidly. By 1940 population had risen to over 17,000 and sheep numbers to 600,000 (Butland, 1954, p. 40), the most important local centre being Coyhaique: founded only in 1931, the town had a population of some

9500 by 1960, a quarter of the total for the province, and having half as many people again as the provincial capital, Puerto Aisén (*Censo*, 1960, Aisén, p. 18).

Nevertheless, and certainly in comparison with the growth of Magallanes, the evolution of Aisén has been slow, and seems likely to continue so. The province is less well endowed in natural resources than its southern neighbour, and the exploitation of its resources are hindered by far more formidable physical obstacles, as the lack of communications clearly indicates. The economic basis of the province is sheep and cattle-raising, though a little copper, lead and zinc are mined in the vicinity of Lago General Carrera, and shipped across the lake to Chile Chico, whence it goes by Argentine transport to market. There is, as yet, little industry, but the economy that exists, unlike that of Chiloé, is a flourishing rather than a declining one, based on the 600,000 sheep and 100,000 cattle of good quality that inhabit the province. Aisén will be a frontier zone for many years to come, but it is a gradually expanding frontier. The closer integration of Aisén into the national life will depend, above all, on improvements in communications, and while the air link with central Chile is well established – there are airports at Coyhaique, Balmaceda and Chile Chico – better roads and more frequent maritime connections would serve the purpose more effectively. But this, in the highly fragmented landscape of Aisén, with its densely forested western region, is more easily said than done.

Conclusion: diagnosis and prognosis

The historical patterns of regional development in Chile have reinforced, rather than weakened, the highly centralized nature of the national life. While the country's shape and physical characteristics might seem to militate against this, and the economic growth of the outlying regions might appear to encourage a practical as well as a theoretical autonomy, the fact remains that the economic and social structure created in the core region over a long period of time has been little affected by changes at the periphery, and drastic changes in the nature of government have done little to affect this historically imposed pattern. One reason for this is undoubtedly the sense of national belonging which links people from Punta Arenas to Arica in a common allegiance to one flag and government, a truly national allegiance, rare, if not unique, in Latin America, which has blunted to a large extent the feeling of regionalism, so disruptive of national integration elsewhere in the continent. Politics, in short, has triumphed over geography in Chile's evolution against all the odds of a country so peculiar in shape as to encourage disintegration, which, given the economic contribution of peripheral regions to national well-being – the minerals of the north, the oil of the south, and so on – might well have been expected. Tensions between the centre and the periphery persist, whatever government may be in power, but Chile is a nation as well as being a state, and this is rare in the continent to which she belongs. Her survival as a cohesive entity, therefore, seems likely to depend upon the capacity of her governors to recognize that unity in diversity can only be attained if local susceptibilities can be accommodated within national plans.

In this respect, Chile's historical advantages -- racial homogeneity, compact physical size and structure, strong natural frontiers and recognizable international identity – are permanent assets which are unlikely to be impaired by the passage of time. What will determine her internal cohesion in the future will turn more on domestic rather than international factors, and here, in so centralized a state government policy will be crucial.

The current National Development Plan, 1979–84, recognizes regional, as well as national, priorities, setting out projects for regional development in some detail (ODEPLAN, 1979, pp. 105–27). But prospects for regional development must turn on national development, and it is here that the major area of controversy arises. Under the military government since 1973, Chile has abandoned a cherished concept of Latin American development fostered by the Economic Commission for Latin America (ECLA), namely that import-substituting industrialization and the increasing role of the state in the economy, coupled with major internal structural alterations, as on the land, were the keys to future growth. Adopting, instead, a free-market philosophy, in which domestic industry and other economic activities are forced by tariff arrangements, the absence of government control over exchange and supply, and the abandonment of a paternalistic attitude for one of crude competition, the current Chilean model of economic growth relies on comparative advantage for Chilean products, the inflow of foreign capital on generous terms to develop natural resources and the assumption that sharp competition with international suppliers will provide domestic manufactures with the opportunity to streamline production techniques and become cost competitive in outlook. In Latin American terms, let alone Chilean, this is a revolutionary approach to developmental problems, but it depends, as did the structuralist emphasis, on economic imponderables.

First is the role of foreign investment and the inevitable growth of the foreign debt. The current national plan calls for overall investment of US$9 billion up to 1984, with a major role for private enterprise in industrial and commercial development. At the end of 1980 foreign debt stood already at over US$11 billion (much of it private, not public), giving Chile one of the highest *per capita* degrees of indebtedness in the world. Her future capacity to service it, based on the continued growth of export earnings as described previously, must remain a major question mark. Secondly, despite a significant increase in domestically produced energy and sizeable scope for expansion, Chile must remain at least 50 per cent dependent on imported fuels, mostly oil, for the foreseeable future, and in volatile world conditions the cost of that dependence cannot be estimated. Thirdly, the major switch on the agricultural front from primary food production to export-earning crops, buoyant in the 1970s, may not be sustainable over the long term, and could lead to sizeable food imports with a consequent impact on the balance of payments.

Nevertheless, as a comparatively modest economy, Chile has the resources, physical and human, to perform as well as any other Latin American state of comparable size in the rest of the twentieth century. Her history to date has been that of a country which, despite extraordinary geographical obstacles, has created a national unity out of very diverse regional structures, in defiance of nature and logic alike. That historical experience alone is considerable ground for confidence in the country's future.

Notes

[1] In 1974, the Chilean government reorganized the administrative structure of the country, replacing historic provincial names by regions numbered I to XII from north to south. Provincial names, however, are still widely used, and are retained here, but see *Table 10.2* for equivalents.

[2] Butland (1956, pp. 5–11) places Archipelagic Chile with Forest Chile but, in view of its much more fragmented physical character, turbulent weather and lack of settlement, there is a good case for treating it as a separate region (Bohan and Pomerantz, 1960, p. 38). In historical terms, certainly, the case is overwhelming.

[3] Detailed results of the 1982 census are not yet available, but sufficient data has been published by the National Institute of Statistics (INE) for broad comparisons with the 1970 census.

[4] A major factor, indeed in the successful Mapuche resistance to Spain was their rapid adoption of horses for cavalry purposes, and of Spanish arms, to such effect that Spanish technological superiority in arms, clearly demonstrated in the conquest of Mexico and Peru was quite nullified in Chile. It was this factor rather than deep forest and high precipitation, that forced the Spaniards back beyond the Bío-Bío.

[5] Under Spanish colonial law the Indians were a special charge of crown and church, and a considerable amount of protective legislation was passed on their behalf. Mestizos had no such status.

[6] The author is indebted to Professor Weaver for permission to quote this passage.

[7] The first foundation in 1543 was destroyed by Indians.

[8] The principal speculator was John Thomas North, known as 'the Nitrate King', who bought many bonds at prices around one-tenth of their nominal value. See Blakemore (1974) for a detailed account of his activities.

[9] Spanish colonists, of course, had long been in occupation of particular areas when such towns were founded, exercising *encomienda* rights and so on. Thus, in the Copiapó valley, Francisco de Aguirre established his estates in the mid-sixteenth century; but it was mining two centuries later that made Copiapó more than an outpost of settlement (Bowman, 1924, pp. 99–106).

[10] Allende refused not only to compensate the foreign owners of the mines but even demanded repayment of alleged excess profits made in the past. This was a major factor in the drying-up of external credit to the Allende regime.

[11] The success of this development owed much to the chief Chilean colonization agent, Vicente Pérez Rosales, who not only founded colonies himself, and set up in Germany the organization to encourage emigration to Chile, but also in his excellent drawings and paintings, and his classic literary work, *Recuerdos del Pasado,* left a detailed and valuable record of this movement of colonization.

Bibliography

BARAONA, R., ARANDA, X, and SANTANA, R. (1961) *Valle de Putaendo: estudio de estructura agraria*, Santiago, Instituto de Geografía de la Universidad de Chile.

BATALLAN, G. (enero-mayo, 1981) 'Las comunidades agrícolas del Norte Chico chileno', *Revista Paraguaya de Sociología*, 18, 50, 89–139.

BAUER, A. (1975) *Chilean Rural Society from the Spanish Conquest to 1930*, Cambridge.

BENHAM, F. and HOLLEY, H.A. (1960) *A Short Introduction to the Economy of Latin America*, London.

BLAKEMORE, H. (1972) 'Chile: continuity and change on the road to socialism', BOLSA *Review*, 6, 61, 2–10.

BLAKEMORE, H. (1973) 'Chile: the critical juncture?' in *The Year Book of World Affairs, 1973*, London, pp. 39–61.

BLAKEMORE, H. (1974) *British Nitrates and Chilean Politics, 1886-1896*, London.
BLAKEMORE, H. (1975) 'Chile: current realities and historical perspectives', BOLSA *Review*, 9, 75, 1-8.
BLAKEMORE, H. (1981) 'Chile: the real revolution', BOLSA *Review*, 15, 81, 1-9.
BOHAN, M.L. and POMERANTZ, M. (1960) *Investment in Chile: Basic Information for US Businessmen*, Washington, DC, US Dept of Commerce and Bureau of Foreign Commerce.
BOLSA (Bank of London and South America) *Review* (cited in text as BOLSA *Review* with relevant dates).
BORDE, J. and GÓNGORA, M. (1956) *Evolución de la propriedad en el Valle del Puangue*, 2 vols, Santiago.
BOWMAN, I. (1924) *Desert Trails of Atacama*, New York, American Geographical Society, Special Publications, no. 5.
BUTLAND, G.J. (1954) 'Changing land occupance in the south Chilean provinces of Aisén and Magallanes', *Geographical Studies*, 1, 1, 27-43.
BUTLAND, G.J. (1956) *Chile: an Outline of its Geography, Economics and Politics*, 3rd edn, London.
BUTLAND, G.J. (1957) *The Human Geography of Southern Chile*, London, George Philip, Institute of British Geographers, publication no. 24.
BUTLER, J.H. (1960) *Manufacturing in the Concepción Region of Chile: Present Position and Prospects for Future Development*, Washington, DC, National Academy of Sciences and National Research Council.
CHILE ECONOMIC REPORT, monthly, New York, CORFO.
CHONCHOL, J. (1980) 'Organización económica y social del sector reformado chileno en el gobierno de la Unidad Popular', *El Sector Agrario en América Latina*, Stockholm, Institute of Latin American Studies, 67-81.
CIDA (Comité Interamericano de Desarrollo Agrícola) (1966) *Chile: tenencia de tierra y desarrollo socio-económico del sector agrícola*, Santiago.
CORFO (Corporación de Fomento de la Producción) (1967) *Geografía económica de Chile*, Santiago, texto refundido.
CUNILL, P. (1978) *Geografía de Chile*, 7th edn, Santiago.
DARWIN, C. (1960) *The Voyage of the Beagle*, London.
DIRECCIÓN DE ESTADÍSTICA Y CENSOS (1960) *XIII Censo de Población*, (cited in text as *Censo*, 1960, followed by name of province to which reference is made), Santiago, serie B, provincias, 25 parts.
DIRECCIÓN DE ESTADÍSTICA Y CENSOS (1964) *Población total por provincias. Chile 1885-1960*, (cited in text as *Censos, Provincias*, 1964), Santiago.
DIRECCIÓN DE ESTADÍSTICA Y CENSOS (1970) *XIV Censo de Población. Total Pais*, Santiago.
ECLA (Economic Commission for Latin America) (1969) *Economic Survey of Latin America 1968*, 2 vols, Santiago, UN Economic and Social Council, General Series E/CN 12/825, mimeographed.
ECOSURVEY, weekly, Santiago.
ELLSWORTH, P.T. (1945) *Chile: an Economy in Transition*, New York.
ENCINA, F.A. and CASTEDO, L. (1964) *Résumen de la historia de Chile*, 3 vols, 5th edn, Santiago.
EYZAGUIRRE, J. (1965) *Historia de Chile*, vol. I, Santiago.
FARON, L.C. (1968) *The Mapuche Indians of Chile*, New York.
GAY, C. (1844-54) *Historia física y política de Chile*, 24 vols, 2 vols on agriculture, Paris.
GONGORA, M. (1960) *Origen de los inquilinos de Chile central*, Santiago.
GWYNNE, R.N. (1978) 'Government planning and the location of the motor vehicle industry in Chile', *Tijds. Econ. Soc. Geografie*, 69, 3, 130-40.

HERNANDEZ CORNEJO, R. (1930) *El salitre*, Valparaiso.

HERRICK, B.H. (1965) *Urban Migration and Economic Development in Chile*, Cambridge, Mass., and London.

HOY (16-22 Sept. 1981), weekly, Santiago.

IADB (subsequently IDB) (Inter-American Development Bank) (1968-81) *Socio-Economic Progress in Latin America*, Washington, DC, Social Progress Trust Fund Annual Reports (cited in text as IADB or IDB with year).

JAMES, P. (1959) *Latin America*, 3rd edn, London.

JEFFERSON, M. (1921) *Recent Colonisation in Chile*, New York, American Geographical Society.

KAY, C. and WYNNE, P. (1974) 'Agrarian reform and rural revolution in Allende's Chile', *J. Latin American Studies*, 6, 1, 135-59.

MCBRIDE, G.M. (1936) *Chile, Land and Society*, New York, American Geographical Society, Research Series, no. 191.

MAMALAKIS, M. (1965) 'Public policy and sectoral developments: a case study of Chile, 1940-58', in MAMALAKIS, M. and REYNOLD, C.W. *Essays on the Chilean Economy*, Holmwood, Ill., pp. 3-206.

MAMALAKIS, M. (1976) *The Growth and Structure of the Chilean Economy*, New Haven, Conn., and London.

MAMALAKIS, M. (1978 and 1980) *Historical Statistics of Chile*, I-National Accounts; II-Demography and Labour Force, Westport, Conn., and London.

MARTIN, G.E. (1960) *La división de la tierra en Chile central*, Santiago.

MARTNER, D. (1923) *Estudio de la política comercial Chilena e historica económica nacional*, 2 vols, Santiago.

MÉNDEZ, J.C. (1980) *Chilean Socioeconomic Overview*, Santiago.

O'BRIEN, P., ROXBOROUGH, I. and RODDICK, J. (1977) *Chile: the State and Revolution*, London.

ODEPLAN (1979) *Plan nacional indicativo de desarrollo, 1979-1984*, Santiago.

PEDERSON, L.R. (1966) *The Mining Industry in the Norte Chico, Chile*, Evanston, Ill., Northwestern University Studies in Geography, no. 11.

REYNOLDS, C.E. (1965) 'Development problems of an export economy: the case of Chile and copper', in MAMALAKIS, M. and REYNOLDS, C.W., *Essays on the Chilean Economy*, Holmewood, Ill., pp. 207-398.

RUDOLPH., W.E. (1963) *Vanishing Trails of Atacama*, New York, American Geographical Society, Research Series, no. 24.

RUSSELL, W.H. (1890) *A Visit to Chile and the Nitrate Fields of Tarapacá*, London.

SCOTT ELLIOT, G.E. (1907) *Chile*, London.

SEPULVEDA, S. (1959) 'El trigo chileno en el mercado mundial', *Informaciones Geográficas*, Santiago, Instituto Geográfico de la Universidad de Chile, año VI (1956), 7-135.

SIDERI, S. (ed.) (1979) *Chile 1970-73: Economic Development and its International Setting*, The Hague, Boston and London.

SIGMUND, P. (1977) *The Overthrow of Allende and the Politics of Chile, 1964-1976*, Pittsburgh, Pa.

SMOLE, W.J. (1963) *Owner-cultivatorship in Middle Chile*, University of Chicago, Dept of Geography, Research Paper, no. 89.

SOZA, H. (1968) 'The industrialization of Chile', in VÉLIZ, C. (ed.) *Latin America and the Caribbean: A Handbook*, London, pp. 614-21.

SUBERCASEAUX, B. (n.d.) *Chile, o una loca geografía*, several edns, Santiago.

SUTULOV, A. (ed.) (1975) *El cobre chileno*, Santiago, Corporacíon del Cobre.

THIESENHUSEN, W.C. (1966) *Chile's Experiments in Agrarian Reform*, Madison and London, Land Economics Monographs, no. 1.

VICUÑA MACKENNA, B. (1882) *El libro de la Plata*, Santiago.

VYLDER, S. de (1976) *Allende's Chile: the Political Economy of the Rise and Fall of the Unidad Popular*, Cambridge.

WARRINER, D. (1969) *Land Reform in Principle and Practice*, London.

WEAVER, F.S. (1968) *Regional Patterns of Economic Change in Chile, 1950–1964*, Ithaca, NY, Cornell University, Latin American Studies Program, Dissertation Series, no. 11, mimeographed.-

WORLD BANK (1980) *Chile: an Economy in Transition*, Washington, DC.

11 · Conclusion: unity and diversity in Latin America

The Editors

The term 'Latin America', like so many commonly accepted designations of geographical regions, is an invention of western Europe, first propounded in France, and gaining currency there in the 1860s when Napoleon III was seeking to establish Maximilian of Austria on a Mexican throne under French tutelage. It is essentially a cultural term, intended to stress the region's historical relationship with the Latin nations of Europe, and also to differentiate the vast area impregnated with Iberian culture from the other, Anglo-Saxon America to the north. The term is a handy, shorthand expression, though Latin Americans themselves have often been chary of using it, and there is, in addition, a large literature of Latin American protest at the pre-emption by the United States of the title 'America', to which, in fact, her southern neighbours have a better historical claim. Nevertheless, 'Latin America' has passed into common usage, and the term is most unlikely to be replaced. Its blanket use, however, does at times encourage popularly held notions, which it has been one of the objectives of this book to dispel – namely, that the twenty republics of Latin America have so much in common that comparisons between them are more revealing than contrasts.

That many such comparisons exist and are perfectly valid has been suggested in the opening chapter of this book: they are, in large measure, the product of a common, or similar, historical experience, and not least a colonial epoch, which lasted some three hundred years. No visitor to Latin America, moving from country to country, could fail to be struck by the patent similarities he would observe, yet, the longer his stay and the more extensive his experience, the more does he come to appreciate the remarkable diversity the continent presents, not only from country to country, but also within each of the countries themselves. The eighteen heirs of the Spanish empire have a common written language, and the Spanish-speaking visitor would certainly have little difficulty in verbal communication, moving from place to place. But he would notice marked changes in pronunciation as he travelled, and wide variations in vocabulary from country to country, as well as markedly different meanings for the same words in many instances.[1] This is an obvious example of diversity within a general unity, which Latin America presents. The Spanish empire itself, stretching at its zenith from California to Cape Horn – in theory if not precisely in fact – was never much of a unity: geography alone imposed intractable barriers to tidy notions of administrative order, and the application of general laws was continually frustrated by the diversity of local conditions. As in the past, so in the present, the Latin American states possess generalized common features in terms of the aspiration for a better future and the obstacles that lie in the way. Politically, the big issues everywhere concern stability in government, the orderly transfer of power and the establishment in fact of those principles of participatory democracy that every state constitution proclaims. Economically, every country in Latin America seeks to secure growth, and not least to reduce its critical dependence on the production and export of few commodities, and socially, every government declares – though not all may be sincere – that it seeks to promote social justice by the eradication of inequalities, inherited or imposed. But these aspirations are not peculiarly Latin American, and while they are common to all the states there, the responses reveal, in their wide variety, great diversities, which spring from the distinctive geographical factors and the unique historical experience that have made each country what it is. That diversity has been explored in the body of this book, and it may be relevant, therefore, to devote its concluding chapter to some discussion of the theme with which it began – Latin America as a recognizable entity in the modern world, and the ways in which the individual states have sought to achieve a wider unity.

Expressions of unity in modern Latin America have usually been short lived. During the wars for independence in the first quarter of the nineteenth century, solidarity of sentiment united many Spanish Americans, and for the great liberators San Martín and Bolívar the cause against Spain was a continental cause. Argentinians helped to liberate Chile, as Chileans helped to liberate Peru, and in the northern half of the continent Bolívar's armies were truly international. Bolívar himself created a union of Venezuela, New Granada (Colombia) and Ecuador, but this Gran Colombia dissolved into its component parts before the Liberator himself, disillusioned and dejected, died in 1830. Bolívar, indeed, had wider visions of unity: in 1826 he had convened at Panama, symbolic bridge between the two Americas, a conference intended to be the first step towards a continental system of co-operation and friendship, but it was a fiasco, though the idea lived on. In Central America independence was achieved by a union of what became the states of Guatemala, Honduras, Nicaragua, El Salvador and Costa Rica, but the union collapsed in 1839 and, despite later attempts to revive it, the states went their own ways. In the late 1830s the abortive confederation of Peru and Bolivia collapsed after defeat in war with Chile, though there is little doubt that even without the war the union would not have survived.

These attempts – and there were others – to eradicate newly established boundaries in Spanish America by voluntary means were doomed to failure by the fundamental fact that a common heritage of language, religion and culture, with similar social structures and political habits, was not strong enough to over-come the local loyalties and regional differences which had been growing for some three centuries. Moreover, after independence the individual states were not yet nations, and their boundaries were often ill defined. The search for national unity predominated, and this essential quest was complicated, from country to country in varying degrees, by political, social, economic and racial divisions. Regional feeling played a major part in the story of political instability, which characterized most states and derived from the fact that the removal of imperial authority created a power vacuum, which many sought to fill. The few exceptions prove the rule. Brazil, the colossus of the continent, already set apart by reason of language, racial composition and a different imperial connection, ratified her distinctiveness in nineteenth-century Latin America by remaining the only monarchy in a continent of republics. By retaining monarchical institutions, Brazil experienced greater administrative and emotional continuity from the colonial past than her Spanish-American neighbours and soon acquired stability, though Brazil, too, passed through a period of regional revolt against the national government before unity was confirmed. Chile, after a decade of turbulence, established a unique constitutional system, which made authority impersonal and its transmission from one president to the next an orderly process. Paraguay enjoyed tranquillity for nearly 30 years under the iron rule of a strong dictator who virtually isolated the country from the outside world. Elsewhere, however, internal strife was the common experience for many decades, interrupted only by the peace of despotism as individual *caudillos* secured enough support to impose their own kind of order.

Internal order and some degree of national unity within the separate states were obvious prerequisites to co-operation between them, but the growth of national unity was a slow process, hindered, as it was, by physical as well as human factors. The sheer size of many states, their scattered settlement pattern, the universal inadequacy of communications before the advent of modern methods of transport – such features compounded political problems, as did the sharp social cleavage that obtained almost everywhere. Racial divisions in countries with large negro or Indian populations confused even further a complicated picture. Moreover, to the difficulties of relations within states were added particular problems of relations between them in consequence of their independence. As in Africa today, the legacy of colonial frontiers in Latin America was a major cause of conflict between the separate countries, and few international wars there have not derived, in greater or lesser degree, from boundary disputes between contiguous states. The lines of separate jurisdictions, drawn for administrative convenience within empires, are often ill recorded, leaving the successor states to settle between themselves the exact delimitation of their inheritance, by reason or force. Conflicts arising from such causes clearly militated against the growth of internationalism in Latin America, and, even today, many of the disputes that exist between countries arise from opposing interpretations both of the original colonial boundaries and of attempts to adjust them in the nineteenth century. Boundary disputes, however, and the wars to which they gave rise served the useful purpose of helping to unify the states internally and strengthen national feeling – necessary steps on the way to wider unities.

There was a further reason why, in nineteenth-century Latin America, continental solidarity took a low second place to the pursuit of individual national interests: it lies in the economic and intellectual relationships that linked the continent to Europe. The level of economic development in all the states throughout the century, and well into this one, and the integration of their national economies into the world system – described in previous pages – naturally caused the countries individually to look abroad, first to Europe and later to the United States, rather than to one another. The lines that linked Latin American markets with European manufacturers, and Latin American producers of primary commodities with European consumers were lines that ran parallel and did not intersect. Supplies of capital and the provision of technology came to the continent, country by country, from European sources and, if London was for long the economic capital of Latin America, Paris was its intellectual home. The role of foreign capital and enterprise in the economic development of Latin America has been discussed in the country surveys which make up the bulk of this volume, though much less has been said of the importation into the continent of foreign ideas and ideologies. Yet they, too, played a highly significant part in changing the Latin American landscape, and often provided the intellectual framework for economic attitudes. Thus, the impact of *laissez-faire*, itself one aspect of European liberalism, was pervasive and profound, and in some countries – Mexico and Brazil, in particular – the role of positivist philosophies was equally significant. The economic and cultural relationships between individual Latin American countries and the nations of western Europe were reinforced by the immigrant streams that flowed to Argentina, Uruguay, southern Brazil and Chile, and it is no exaggeration to say that for these and other Latin American states relations with Europe were much more intimate than relations with one another. Of the states that had formerly belonged to one Spanish empire, with their common legacies of language, religion and culture, a clear-sighted observer remarked, almost 60 years ago: 'the feeling of a common Hispano-American brotherhood is weak' (Bryce, 1912, p. 445). The national evolution of the independent states, including the growth of nationalism itself, separated the states one from another, confirming differences that, although they existed during the time of empire, had been partly overlaid by the structures of empire.

Yet this is not the whole story. While geographical circumstances, historical evolution and economic influences combined to divide, rather than unite, the nations of Latin America, there persisted in the writings of intellectuals and in particular instances the aspiration to solidarity. It was most in evidence in the expression of a specifically Latin American system of inter-state relations, and in attempts to co-ordinate resistance to outside threats. Thus, in the 1860s, when Spain made a last and desperate attempt to coerce Peru and Chile to her will, these two inveterate rivals for hegemony on the Pacific Coast of South America sank their differences in temporary conciliation and, with Bolivia, Colombia and Ecuador, declared their joint resistance to Spanish pretensions.[2] Other instances could be cited of temporary co-operation between the Latin American states in similar circumstances, and there is a considerable literature in Latin America in the nineteenth century and after that testifies to the persistence of Bolívar's unrealized dream of continental unity.

In practice, however, what brought the states of Latin America into formal structures of continental co-operation was the initiative of the United States, and the modern phase of the Pan-American movement began, and has continued, in a system of relations between them. The story, from the date of the First International Conference of American States, held at Washington in 1889–90, to the present day, is extremely complex, and does not concern us here (for a full account, see Connell-Smith, 1966). Suffice it to say that the growing disequilibrium in power and global influence of the United States on the one hand, and the comparative weakness of the Latin American republics on the other, has been the greatest hurdle to effective and genuine acceptance by both parties of a common definition of co-operation. The history of their relations within the

American system during the past 80 years has been one of marked fluctuations between amity and hostility.

Increasingly in the twentieth century, much more significant for most Latin Americans than the formal features of the Inter-American system – now embraced by the Organization of American States – have been, and are, the economic realities of the relationship between the United States and Latin America, a relationship marked by the replacement of Europe by the United States as the dominant external factor. Figures of foreign investment provide an indication of this change. In 1914 the nominal value of foreign investment in Latin America was $8500 million, Britain's share amounting to $3700 million – a quarter of all British overseas investment – compared to $1700 million from the United States, $1200 million from Germany and $900 million from France (IADB, 1966, p. 54).[3] But the cost of two world wars, centred on Europe, and the impact of the Great Depression drastically diminished the European stake in Latin America, and enabled American interests to rise to predominance. The United States became, in fact, Latin America's major market for primary products, its chief supplier of manufactured goods, and its principal source of foreign capital. At the end of 1975 US investments in Latin America totalled almost $15,000 million, compared with about $800 million at the end of 1959, though as a proportion of total US investment abroad this represented, in fact, a fall from 27 per cent to 14 per cent (Grunwald, 1978, p. 251).

The role of foreign investment in developing countries is a subject of great complexity and much controversy, and this is nowhere more true than in Latin America. It cannot be denied that the Latin American affiliates of US companies provide employment for a million and a quarter Latin Americans, make a sizeable contribution to the continent's balance of payments problems, earn substantial sums in foreign exchange through exports and save large sums through import substitution. But what most Latin Americans remember is the fact that investment earnings by American affiliates is generally far more than the size of new investment. More importantly for them, in addition, in a highly nationalistic continent, is the way they regard the mere existence of sizeable foreign investment as a positive reflection of their basic underdevelopment.

Previous chapters in this book have devoted considerable attention to the development problems of Latin America on which there is, of course, a voluminous, and often partisan, literature, impossible to summarize here. In the context of this chapter, however, what is important about debates on developmental issues is the distinctive Latin American contribution to them, a contribution that is derived from Latin American experience but has much wider validity, and one that in many respects may be regarded as a cogent expression of Latin American unity. It is in this context also that relations between Latin America and the United States, including the issue of investment, have particular significance at the present time.

The formulation of distinctive and coherent Latin American views on the relationship between the developed and developing worlds was largely the result of the foundation, in 1948, of the Economic Commission for Latin America (ECLA), an organ of the United Nations Organization with its seat in Santiago de Chile.[4] From 1950 to 1963 ECLA's Executive Secretary was the Argentine economist Dr Raúl Prebisch, and it is his name which has been given, justly, to the particular thesis on Latin American development problems, which has, perhaps, been ECLA's most constructive contribution to the general debate. Summarized very briefly, the thesis demonstrates how the world free-market system for primary commodities penalizes producing countries, subject as they are to fluctuations in demand from industrial countries with high standards of living. At the same time it also shows how manufacturing countries, exporting to primary-producing countries, pass on a large proportion of value added in their high-cost economies, accentuating the deterioration in terms of trade for primary producers. The latter, their import capacity geared to their export performance, itself closely linked to capital formation for their own economic growth, are forced to seek financial aid and development loans, thus increasing their indebtedness to the developed world. In short, according to Prebisch, the gap between the developed and the developing world is continually increasing, and the answer to the problem lies, at least in good measure, in a whole series of policies for Latin American countries, which includes trade protection, control of imports, acceleration of import-substitution, increased foreign aid, wider economic integration and long-term planning (for an excellent short summary see Huelin, 1968, pp. 472-3).

Although specifically derived from Latin American circumstances, and propounded from a Latin American institution, the Prebisch thesis is applicable to the developing world in general, a fact increasingly recognized in the 1950s and 1960s, as its virtual adoption by the United Nations Conference on Trade and Development at Geneva in 1964 indicated. Indeed, for a continent not particularly noted for its originality in political and economic thought, and historically one that has borrowed extensively in these matters from Europe and the United States, Latin America has produced in the Prebisch thesis and other contributions from ECLA on the universal debate on development its most significant expression of originality. Inside Latin America itself, ECLA has been not only a very tangible manifestation of a new kind of continental consciousness, but also a very important agent for the promotion of other instruments of co-operation, and a valuable service agency in the implementation of economic policies. Thus, for example, its role as adviser in the formulation of development plans, its development of subsidiary institutions for various purposes, and, far from least, the impressive range of its own publications, all testify to a dynamic entity serving Latin America as a whole in its search for economic growth and social justice.

Two other aspects of this question deserve some consideration, however briefly. The first concerns finance, the second economic integration, both subjects dear to ECLA's heart and central to the Prebisch thesis. In December 1959 the Latin American republics, with the exception of Cuba, and the United States of America joined together to establish the Inter-American Development Bank as a financial agency to help accelerate the economic development of its member states, individually and collectively (IDB, 1970, p. 503). At that time its ordinary capital resources were fixed at $850 million, $400 million as paid-in capital and $450 million as callable capital. By the end of 1980, however, its total resources stood at US$23,989 million and its authorized loans for that year were as high as US$2,309 million (IDB, *Annual Report, 1980*). Cumulative lending between 1961 and 1980 was US$17,840 million (op. cit). Under the able direction of, first, the Chilean economist Felipe Herrera and, since 1970, of the Mexican Antonio Ortiz Mena, the Bank has been one of the great successes of Latin American economic development in the past 20-odd years, with an extraordinarily wide-ranging programme of financial support to a vast number of projects throughout the continent (for details, see Ortiz Mena, 1980, *passim*). Its membership has been progressively widened to include extra-hemispheric members, now totalling with those of the Americas forty-three countries. The Bank has become, in fact, a major institution of Latin American economic co-operation, in which other nations participate without running the risk of the rancour that direct unilateral investment can often inspire.

The closer integration of the Latin American economies has been, and remains, a major argument of the Inter-American Development Bank, as well as of ECLA, for intensifying industrialization, diversifying the foreign trade of the region and increasing the pace of economic growth in the individual countries (Herrera, 1968, p. 563). The case for Latin American integration has been warmly debated in recent years, and it remains a somewhat controversial subject; in principle, however, there is general acceptance of the theoretical arguments that urge the advantages of group protection for larger-than-national markets, economies of scale, rationalization of investment and greater use of industrial capacity (Dell, 1966, pp. 15–34). Undoubtedly, in addition, the examples of the European Economic Community and the European Free Trade Area were not insignificant in Latin America, though in terms of levels of development the difference between the two regions is, of course, enormous. Beginning in 1951 the governments of the five countries of Central America gradually worked towards the establishment of a Common Market, which came into formal existence in 1958, aiming to establish a free-trade area over a period of 10 years, and to move thereafter to a customs union. In fact, however, the agreement got off to a poor start since the treaty establishing the organization abolished duties only on a limited number of goods, and additions to the list were to be negotiated anew. But early in 1960 Guatemala, El Salvador and Honduras signed a new treaty, which approached the problem of free trade between them by the opposite direction, not listing the duty-free goods as previously, but establishing free trade for all goods produced by any of them, except for a restricted list subject to pre-ferential tariffs or restrictive quotas for 5 years only. This action galvanized the whole movement towards economic integration in Central America, the tripartite agreement of 1960 being converted into a treaty including Nicaragua and Costa Rica as well in 1961. The Central American Common Market (CACM) experienced quite dramatic results in the 1960s: 98 per cent of intra-regional trade was absolutely liberalized, and 98 per cent of import tariff items from outside the region was subject to a common external tariff, while the value of intra-regional trade grew five fold between 1960 and 1966, from $32.7 million to $173.6 million (Calvo, in BOLSA *Review*, 2, 18, June 1968, p. 314). Trade with the rest of the world increased by 58 per cent in the same period.

This substantial early progress, however, slowed down in the later 1960s, partly because rapid expansion earlier was based, in large measure, on utilizing hitherto unused capacity, agricultural and industrial, and also because external factors were favourable (see Wionczek in BOLSA *Review*, 1, 3, March 1967, p. 134). Deterioration in the terms of trade for Central America's primary products and balance-of-payments difficulties dampened the previous euphoria, and the obviously uneven nature of development within the CACM itself put severe strains on the organization (for a detailed discussion see Cable in BOLSA *Review*, 3, 30, June 1969, pp. 336–46). Behind all the intricate economic data lay the fundamental fact that effective integration depended basically on a *political* will to make integration work, and this was lacking. Indeed, in July 1969 two member-nations of the CACM, El Salvador and Honduras, found themselves at war, ostensibly over football matches between national teams in the qualifying rounds of the 1970 World Cup, but actually over the problems involved in the presence of some 300,000 Salvadorean immigrants in Honduras. This war was undoubtedly the greatest setback experienced by the CACM, and although the fighting phase soon ended, tension between the two countries persisted until diplomatic relations were finally restored in 1981. By then, however, the deteriorating internal situation in a number of states in the wake of the Nicaraguan revolution of 1979 seemed likely to prevent the Market's recovery in the 1980s.

At about the same time that the CACM was becoming a reality in 1960, the Latin American Free Trade Area (LAFTA) also came into existence. In February 1960 representatives of Argentina, Brazil, Chile, Mexico, Paraguay, Peru and Uruguay signed the Treaty of Montevideo, which was ratified in May 1961 and came into force soon after (the text of the Treaty may be found in Dell, 1966, pp. 228–56). Colombia and Ecuador joined later that same year, to be followed by Bolivia, and Venezuela in due course. The long-term objective is the establishment of a Latin American Common Market, by progressive reductions of tariff barriers between member nations, and a beginning was made on this in 1962. By 1967 trade within LAFTA had increased substantially, particularly in the early years (though regional trade in 1961, immediately before the 'base' year, was abnormally low, and improvement in the early 1960s may, in fact, have had little to do with LAFTA arrangements). Tariff reductions were made by annual negotiations in which each country was involved – 'national schedules' – and by triennial negotiations involving all member countries to produce 'common schedules', consisting of products for which all trade barriers were to be removed between members completely by 1973. After the first five rounds of annual 'national schedules', some 9000 concessions had been included in the national lists, but the negotiations were extremely hard and members generally were reluctant to lower tariff barriers for domestic products (Eckenstein, 1968, pp. 543–4). Naturally, the highly uneven levels of the individual Latin American economies were recognized, and preferential concessions were extended to the less developed states. Increase in trade, however, occurred mostly in non-manufactured goods in the first 5 years, and widespread disillusionment set in about LAFTA's capacity to achieve a fundamental breakthrough in the fields of tariff reduction on manufactures that compete with domestic products, and planning new industrial development on a regional, rather than a national, basis (BOLSA *Review*, 1, 2, February 1967, pp. 60–70). The problems, enormously complex, plagued LAFTA throughout the 1960s and 1970s and indeed, at the end of 1969 representatives of member states, meeting at Caracas, voted for modifications of the Montevideo Treaty, postponing the scheduled date for the completion of trade liberalization from 1973 to 1980.

Disillusionment with LAFTA's progress did not, however, lead to despair but to more positive attempts, less ambitious in scope, to establish, in the late 1960s, sub-regional groups where complementarity was more likely. The Andean Group – originally consisting of Bolivia, Colombia, Chile, Ecuador, Peru, and later, Venezuela – came into being in 1969 by the Agreement of Cartagena, its aims to abolish gradually internal tariffs between members, plan the location of specific industries, set up a development bank, and so on. The members also adopted common rules on the treatment of foreign investment, but subsequently the military government in Chile found these irksome in the operation of its free-market philosophy and withdrew from the Pact in the mid-1970s. Progress within the Group has been halting, with particular difficulties over agreement on industrial location, but the development bank may be adjudged a success, and despite friction between members in the late 1970s and threats by some to withdraw, none, in fact, did so. The River Plate countries – Argentina, Brazil, Bolivia, Paraguay and Uruguay – also drew together in co-operative arrangements, though they made no pretensions of establishing a common market as the Andean Group did. The *Cuenca del Plata*, as preceding chapters indicate, has fostered practical co-operation, and one testimony to that spirit is the large number of impressive hydroelectric works, utilizing the shared rivers of the region, which has probably been the major constructional effort in the basin of the River Plate in this century.

Perhaps more significant for the future, however, has been the regeneration of the continental ideal of closer economic union, expressed in 1980 with the formation of LAFTA's successor, the Latin American Integration Association (ALADI). It was then universally recognized that LAFTA had failed, though the experience had been very valuable, and had produced a vast amount of expertise, continental contact and a number of useful institutions. But a new impetus was required and, given the remarkable growth of a number of countries in the past two decades and their growing weight in international affairs – Brazil, Mexico and Venezuela, for example – there was sufficient common will to establish ALADI as the phoenix of LAFTA, but with more realistic aims and methods of moving towards integration (for an excellent summary, see *El Exportador Latinoamericano*, no. 31, 1980). Only time will tell whether this fresh attempt to promote Latin American integration will be more successful than its predecessors or, whether as in the past, common continental aspirations will be thwarted by a lack of political will on the part of the individual governments and nationalistic jealousy in the preservation and extension of their specific economic interests.

But, whatever the future of the continent and the very diverse countries that compose it, there can be little doubt that the significant transformations – political, economic and social – which have occurred in the decade since the first edition of this book appeared, have placed Latin America more centrally in the world view. The remarkable growth of Brazil and her increasing ties with other continents as her economy is transformed; the massive oil discoveries in Mexico and their international importance; even the contemporary political turbulence of many Central American republics as conflicting forces and ideologies assume a more than regional interest – these are but three indications that Latin America today claims that share of world attention which the continent has long been denied, but which, because of its significance, can no longer be ignored as it has been in the past. In the obscure future facing the international community for the rest of the twentieth century, one may confidently assume that Latin America will play an increasingly significant role in it.

Notes

[1] The word *guagua* is, variously, the popular name for a bus in Cuba, a baby in Chile, an edible amphibious rodent in Colombia and a fruit-destroying insect in Spain. The scope for memorable *faux pas* even among Latin Americans is thus considerable.

[2] The Monroe Doctrine, declared in 1823 by the President of the United States, affirmed, in effect, that the continental island of America contains a political system different from that of the Old World, and that the extension of European power to the New World was prohibited by the United States. At the time of the Spanish pressure on Peru and Chile, the United States was involved in civil war but, even so, the Monroe Doctrine only became effective when the United States acquired the power to make it so, in the 1890s.

[3] At the end of the 1970s the Bank changed its acronym from IADB to IDB. Hence both references are used subsequently.

[4] At about the same time, the Economic Commission for the Far East and the Economic Commission for Europe were also founded and, like ECLA, lacked permanent status until 1951. It should also be remembered that in the immediate post-war period, a third of the membership of the United Nations was Latin American, and it was the existence of these states, not that of Spain, that ensured that Spanish would be one of the five official languages of the organization.

Bibliography

BOLSA (Bank of London and South America) *Review* (cited in text as BOLSA *Review*, with relevant dates).

BRYCE, J. (1912) *South America: Observations and Impressions*, London.

CABLE, V. (1969) 'Problems in the Central American common market, BOLSA *Review*, 3, 30, June, 336–46.

CALVO, R. ALBERTO (1968) 'Financial aspects of Latin American integration', BOLSA *Review*, 2, 18, June, 312–32.

CONNELL-SMITH, G. (1966) *The Inter-American System*, London.

DELL, S. (1966) *A Latin American Common Market?* London.

ECKENSTEIN, C. (1968) 'The Latin American Free Trade Association', in VÉLIZ, C. (ed.) *Latin America and the Caribbean. A Handbook*, London, pp. 542–50.

GRUNWALD, J. (ed.) (1978) *Latin America and World Economy: A Changing International Order*, Beverly Hills and London.

HERRERA, F. (1968) 'The Inter-American Development Bank', in VÉLIZ, C. (ed.) *Latin America and the Caribbean. A Handbook*, London, pp. 558–75.

HUELIN, D. (1968) 'Latin America: a summary of economic problems', in VÉLIZ, C. (ed.) *Latin America and the Caribbean. A Handbook*, London, pp. 468–77.

IADB/IDB (Inter-American Development Bank) (1966) *European Financing of Latin America's Economic Development*, Washington, DC, mimeographed.

IADB/IDB (1970a) *Socio-Economic Progress in Latin America. Social Progress Trust Fund Annual Reports 1969-1980/81* Washington, DC.

IADB/IDB (1970b) *Proceedings of the Eleventh Meeting of the Board of Governors, Punta del Este, April 1970*, Washington, DC.

IADB/IDB (1980) *Annual Report*, Washington, DC.

ORTIZ MENA, A. (1980) *Development in Latin America: A View from the IDB. Addresses and Documents, 1976-1980*, Washington, DC.

WIONCZEK, M. S. (1967) 'The Central American integration experiment: early success and growing limitations, BOLSA *Review*, 1, 3, March, 127–36.

Index

Abanico, 515,
Abopo-Izozog, 309–10
Acambaro, 57
Acarí, 285
achira (Canna edulis), 196
Aconcagua, 459; province, 465, 466, 475, 489;
  river and valley, 460, 461
Acujutla, 172
Afobaka, 248
agave, 45, 55
agrarian reform, see land reform
Agreste, 347–8
agriculture, 10–13, 14, 15, 38–9, 41–50, 91–2,
  94–5, 109–13, 147, 150–68, 201–5, 207–8,
  218–21, 222–6, 260–2, 264–6, 272–3, 282–5,
  288–90, 305–11, 314–18, 332–5, 344–7,
  347–8, 349–53, 355–6, 358–68, 390–6,
  402–12, 430–47, 449–52, 469, 472–81, 494,
  497–8, 517–18; colonial, 86–9, 91, 201–5,
  207–8, 264–7, 329–35, 339–42, 384–5,
  471–5, 494–7; plantation, 4, 6, 46, 86, 87,
  88–9, 91–2, 103, 109–14, 124, 125, 126, 151,
  157–63, 182, 191, 203–5, 209, 244–5, 250,
  286–7, 315–17, 322–5, 344–7; pre-
  Columbian, 137–9, 193–7, 260–3, 383–4,
  469; shifting, 46, 94, 191, 196, 197, 204, 338,
  355, 359, 383–4, see also conuco, swidden
  cultivation, slash-and-burn cultivation; see
  also livestock production, dairy farming,
  specific crops e.g. bananas
Agrio, lake, 318, 320
Agua Amarga, 504
Aguarí river, 141
Aguas Blancas, 500–1
Aisén, 464, 525–6
Alajuela, 154
Alemán dam, 40
alfalfa, 40, 50, 265, 283, 392, 394, 409, 440,
  495, 497, 498, 507
Algarrobo, 509–10
Algoas, 361, 369
Alicura, 426
Allende, Salvador, 479
Alliance for Progress, 224

alpaca, 260–1, 270
Altar desert, 40
altiplano, Central America, 150, 151, 152,
  154; Central Andes, 2, 253, 256, 258, 261,
  270, 294–6, 298–9, 300–2, 307–8, 309,
  468
Alto Bení, 309
Alvarado, Pedro de, 141
Alvaro Obregón dam, 42
Amapá, 368
Amatitlán, 172
Amazon, 3, 83, 241, 242, 311, 321; basin, 4,
  5, 13, 190, 192, 195, 326, 327, 339, 342,
  356–8, 361, 376
Amazonia, 339, 342, 356–8, 361, 369, 370,
  381
Ambato, 313, 321
Anápolis, 365
Ancud, 520
Andean Pact (Group), 1, 230, 274, 311, 541
Angostura, 198, 210, 215, 216; (Mexico), 42
Angra dos Réis, 370
Anguilla, 97, 122
Anserma, 204
Antigua, 79, 81, 91, 94, 103, 109, 110, 111,
  115, 116, 122
Antilles, 194; Greater, 77, 83, 84, 85, 89, 90,
  113, 124; Lesser, 80, 81, 83, 85, 88, 89, 90,
  95, 103, 106, 107, 115, 122; Netherlands, 79,
  83, 89–90, 92, 115
Antioquia, 206, 217, 220
Antofagasta, 299–300, 458, 460, 462, 466,
  468, 493–4, 497, 499–507, 510–11
Apure, 195, 204, 215, 217
Apurimac, 290
Aracaju, 348,
Aragua valleys, 201, 203; state, 230
Araraquaia river, 325
Aratu, 375
Araucanian: Indians, 468–71, 512, 519–20;
  pine, 327, 369
Arauco, 499, 506, 513–14
Arawak, 83, 84

Arequipa, 264, 270, 278, 280, 285, 286, 290
Argentina, 2, 3, 4, 5, 6, 7, 9, 11, 14, 15, 16,
    17, 19, 20, 46, 49, 66, 259, 267, 301, 302,
    383–453, 465, 495, 518, 523, 525, 526,
    .536
Arias, 394
Arica, 301, 304, 460, 482, 489, 495, 500,
    510
Arima, 100
Aroa, 221
Aroma river, 460
arrowroot, 110, 122
Aruba, 80, 81, 89, 90, 102, 106, 110, 119
Asunción, 383, 384, 411, 451, 452
Atacama, 461, 463, 468, 482, 493–5,
    497–507, 510; Puna de, 495; Salar de, 495
Atitlán, 150, 151
Atoyac river, 36, 44
Atrato river, 189
Aves Swell, 77
Aviles Camacho dam, 44
avocado pears, 196, 260, 265
Ayacucho, 264, 290
ayllus, 262
Aymará Indians, 296
Azapa river, 497
Azcapotzalco, 55, 57
Aztec, 32, 137, 140, 141
Azuero Peninsula, 140, 154

Bagua, 294
Bahama Islands, 80, 90, 92, 97, 116, 123
Bahia, 331, 332, 340, 361, 368, 369
Bahía Blanca, 400, 402, 427
Baja California, 32, 36, 40, 51, 63, 66
Baja Verapaz, 151
Bajío, 38, 43, 50
Balboa, 141
Balmaceda, 526
Balsas river, 36, 44, 52, 57
bananas, 46, 81, 110, 122, 125, 152, 154,
    155, 156, 159–62, 164, 166, 167, 169, 176,
    182, 265, 309, 316–17, 319, 321, 343
Banda Oriental, 340, 392
bandeirantes, 331, 335, 338–9, 342, 381
Barbados, 79, 88, 91, 92, 95, 96, 97, 102,
    103, 106, 107, 108, 110, 111, 113, 114, 116,
    117, 118, 122, 123
Barbuda, 79, 109
Barcelona, 204, 210; Company, 212
Bariloche, 400
Barinas, 203, 204, 212
barley, 258, 265, 498
Barlovento, 201
Barquisimeto, 203, 210, 212
Barranca del Cobre, 36
Barranquilla, 18, 190, 193, 229

barriadas, 269, 276–7; *see also* squatter
    settlements: urban
Batlle y Ordóñez, José, 392
Bay Islands, 140, 142
Bayovar, 286
Beagle Channel, 5, 460
beans, 45, 156, 196, 197, 204, 207, 260, 306,
    355, 384, 469, 480
Belém, 343, 357, 365, 378
Belize, 119, 136, 138, 139, 142, 143, 146,
    150, 168, 170, 173, 175–8, 183
Belmopán, 175
Belo Horizonte, 325, 356, 368, 371, 374, 375,
    376, 378, 380
Bení region and river, 301, 310
Bernigo, 310
Berbice river, 245
Bío-Bío, 458, 460, 461, 463, 468–71, 472,
    474, 511–15, 517, 520
Bluefields, 142, 159
Blumenau, 355, 356, 376
Bocas del Toro, 159
Bogotá, 9, 18, 188, 189, 198, 200, 207, 211,
    214–15, 217, 230, 231, 233, 235, 236–7;
    Viceroyalty of, 143
Bolívar, Simón, 215, 534
Bolivia, 2, 3, 5, 11, 12, 13, 14, 15, 16, 18, 19,
    20, 26, 253–70 *passim*, 295–311, 385, 388,
    486, 493, 499, 511, 541
Bonaire, 80, 89, 90, 119
Bonao, 125
Borborema plateau and escarpment, 326, 334,
    340, 344, 347
Botucatu, 350
boundary disputes, 5, 311, 535
Boyacá, 207
Brasília, 13, 325, 365, 376, 378, 380
Brazil, 2, 3, 4, 6, 7, 9, 13, 14, 15, 17, 18, 19,
    20, 21, 22, 89, 90, 162, 165, 223, 296, 298,
    325–82, 388, 427, 535, 536
breadfruit, 111
Bridgetown (Barbados), 102
British Guiana, 95, 118; *see also* Guyana
British Honduras, 143; *see also* Belize
British Virgin Islands, 92, 93, 97, 106
Brokopondo, 248
Brunswick Peninsula, 522
Bucaramanga, 18
Buenaventura, 189, 192, 206
Bueno river, 461
Buenos Aires, city and province, 6, 9, 32, 267,
    383–96 *passim*, 398–400, 403, 406, 408, 412,
    413–15, 421, 427, 428, 438, 440, 442, 495
Buritica, 206–7
Burton, Sir Richard, 350

coatinga, 328, 340
Cabo, 375

cacao, 4, 81, 87, 110, 125, 141, 142, 143, 166, 203–5, 211, 212, 219, 244, 265, 294, 309, 315–16, 321
Cáceres, 207
Cadereyta, 55
Caguas, 102,
Cajamarca, 260, 264
Cajamarquilla, 264, 275
Cakchiquel Indians, 142
Calabozo, 212
Calama, 501
Caldera, 482, 505
Caledonian Bay, 180
Cali, 9, 18, 198, 230, 235, 236
California, Gulf of, 36, 42, 44, 50
Calima, 195
callampas, 485; *see also* squatter settlements: urban
Callao, 273–6, 285
Callejón de Huaylas, 265, 267
Camaguey, 101
camanchaca, 462–3, 497
Camarones river, 497
Camina river, 497
Camiri, 304
Campana, 402, 430
Campeche, 32, 142; Gulf of, 50
Campina Grande, 340
Campinas, 339, 350, 372, 375
Campo Grande, 364
Campos, 369
cañahua, 258, 260, 265
Canal Zone, 106, 175, 178–83
Cananea, 51
Cangrejera, 57
Cape Horn, 6, 468, 521
Caquetío Indians, 196
Carabaya, 267
Carabobo, 230
Caracas, 9, 12, 188, 192, 201, 203, 210, 211, 212, 214–16, 219, 221, 226, 233, 235, 236–7; *audiencia* of, 213
Caracoles, 504
Caranaví, 301, 308–9
Caratal, 217
Cárdenas (Cuba), 101
Cárdenas, Lázaro, 71; dam, 43
Carib Indians, 84, 90, 176
Caribbean, 2, 4, 5, 77–132, 135, 137, 142, 144, 146, 150, 152, 154, 159, 160, 161, 178, 192, 198, 216, 244; Free Trade Area (CARIFTA), 118, 120; Community and Common Market (CARICOM), 118, 119, 120
Carmópolis, 369
Cariri, 347
Caroní river, 191, 192, 204
Carretera Marginal, 294, 318
Carrizal, 505
Cartagena, 198, 211, 217, 229

Cartago, 205, 217
Casiquiare Canal, 217, 241
Casma, 285
cassava, 84, 111; *see also* manioc
Castilla de Oro, 140
Castries, 98, 99
Castro (Chile), 521
Castrovirreyna, 267
Catamarca, 396, 414, 507
cattle ranching, *see* livestock production
Cauca river, 189, 192, 205, 207, 218, 220, 229
Cautín, 469, 476, 511–12, 519
Caxias, 353
Cayenne, 90, 246, 248
Cayman Islands, 96, 97, 123
Cayo, 176
Cayumbe, 314
Ceará, 327, 340, 347, 368
ceja de la montaña, 258
Central America, 2, 3, 4, 5, 18, 19, 134–85, 187, 260, 265, 316, 317, 539–40, 541
Central American Common Market (CACM), 72, 145, 155, 169–70, 173, 174, 175, 539–40
Cerrejón, 229
Cerro Colorado (Panama), 182
Cerro Grande, 173
Cerro de Pasco, 270, 290, 293
Cerro el Mercado, 52
Cerro Matoso, 229
Cerro Sta Lucía, 483
Cerro Verde, 285
Chacao river, 460
Chacabuco ridge, 461, 463, 501
Chachapoyas, 267
Chaco, 3, 244, 298, 384, 385, 393, 395–6, 406, 408, 410, 411, 416, 420, 433, 434, 441, 445; *see also* Gran Chaco
Chañaral, 460, 505, 506, 510
Chanarcillo, 504
Chancay, 289
Chanchamayo, 293
Chanchan, 263, 264
Chapala lake, 43
Chaparé, 297, 301, 305, 308
Chapultepec, 70
Charcas, *audiencia* of, 296
Chavín culture, 259
Chiapas, 27, 36, 38, 39, 41, 46, 51, 53, 55, 57, 62, 66, 67, 143, 152
Chibcha, 196
Chicama, 261, 266, 270–1
Chiclayo, 278, 280
chicle, 168
Chihuahua, 30, 40, 49, 51
Chile, 3, 5, 6, 11, 12, 14, 15, 17, 19, 20, 255, 259, 267, 296, 299, 301, 384, 385, 400, 457–530, 534, 535, 536, 541
Chile Chico, 526

Chilean Development Corporation (CORFO),
  487, 490, 492, 510–11, 517, 525
chili peppers, 260
Chili river, 286
Chillón river, 276
Chiloé, 464, 520–1, 523, 526
Chimborazo, 256
Chimbote, 278, 285–6
Chimú empire, 263
Chinandega, 153, 162
Chinese immigrants, 30, 43, 94, 104, 284, 286
Chira-Piura irrigation, 286
Chiriqui, 160, 161, 182
Choapa river, 461, 469, 497
Choclococha, 283
Chocó river, 206, 218
Choluteca, 153, 158, 162, 163
Chone, 317
Choritalpa plan, 40
Chorros, los, river, 461
Chubut, 395
Chucamata, 500
Chuquicamata, 495, 505–7, 511
Chuquisaca, 267, 296, 304, 310; *see also* Sucre
Ciboney, 83
Ciego de Avila, 101
Cienfuegos, 101, 104
cinchona, 3, 12, 270, 296
Cisnes river, 464
citronella, 167
citrus fruit, 46, 110, 176, 294, 309, 364, 433,
  444–5
Ciudad Acuña, 43,
Ciudad Bolívar, 191, 218
Ciudad Guayana, 230
Ciudad Juárez, 62
Ciudad Madero, 57
Ciudad Mante, 44
Ciudad Sahagún, 60, 61
climate, 2, 39–41, 81, 157, 191–3, 243,
  254–6, 257–8, 259, 327, 461–4
Coahuila, 40, 51
Coatzacoalcos, 55
Cobán, 152
Cobija, 495
coca, 295, 309, 310
Cochabamba, 265, 267, 297, 299, 301, 304,
  305, 306–7, 308
cocoa, 81, 90; *see also* cacao
coconuts, 110
Cocorote, 200
coffee, 4, 6, 46, 86, 87, 90, 92, 110, 124, 125,
  128, 129, 145, 151, 152, 155, 156, 157–9,
  164–5, 166, 167, 169, 203, 219–21, 223–4,
  229, 244, 271, 294, 309, 310, 315–16, 321,·
  349–53, 357, 361–3, 372, 452
Colchagua province, 466, 485
Colima, 36, 52
Colombia, 2, 3, 5, 9, 11, 13, 14, 15, 17, 19,
  20, 21, 22, 83, 106, 137, 143, 146, 157, 179,
  180, 187–237, 311, 321, 534, 541
Colón (Argentina), 406, 438
Colón (Panama), 172, 181
Colonia do Sacramento, 341
Colonia Továr, 219
colonization and settlement, 4–5, 12–13, 112,
  152, 154, 156, 219–22, 253, 269, 293–5,
  308–11, 317–19, 361–2, 365–8, 392–6,
  511–12, 516–17, 522–4, 525
Colorado river, 42, 43, 44, 48
Columbus, 83, 85, 140
Comayagua, 140, 142
Combarbalá, 503
Commewijne river, 247
Comodoro Rivadavia, 425
communications (including railways, roads and
  river navigation), 14–15, 59, 144–5, 151–2,
  153, 154, 156, 158–9, 160–2, 174, 178–83,
  210–12, 216–18, 221, 229, 257, 260, 270,
  294–5, 300, 301, 317, 318–19, 321–2, 336–8,
  353, 365–7, 376–7, 400–2, 427, 451, 482,
  485, 489, 500–1
comunidades, 262, 270, 306, 519–20
Concepción (Chile), 466, 485, 487, 511,
  512–17, 524
Concepción (Paraguay), 394, 411, 451, 452
Conchos river, 43
Concón, 524
Congonhas do Campo, 336
conquistadores, 32, 85, 86, 113, 140–1, 144,
  198, 263, 383, 469
conuco, 191, 197; *see also* agriculture: shifting
Copán, 139
Copiapó, 458, 460, 461, 482, 497, 503–6
Coquimbo, 461, 463, 474, 476, 482, 493–4,
  497–8, 503–7, 510
cordillera: Central (Colombia), 189,
  (Hispañola), 79, (Puerto Rico), 79; de la Costa
  (Chile), 495–60, 463, 500, 513; Eastern
  (Oriental), 189, 206; de Mérida, 189, 207; de
  Nahuelbuta, 460; Piuche, 460; de Talamanca
  (Costa Rica), 135, 136, 137; Western
  (Occidental), 189, 205
Córdoba (Argentina), 385, 389, 390, 393,
  394, 396, 399, 400, 406, 408, 413, 421, 423,
  427, 428, 430, 438
Córdoba (Mexico), 46
Corinto, 162
Coronel, 507, 514
Corral, 509, 511
Corrientes, 384, 390, 393, 394, 396, 399, 403,
  406, 407, 411, 445
Cortés, 33, 141
Corumbá, 301, 365, 368
Cosoleacaque, 57
Costa de Hermosilla, 44, 48
Costa Rica, 14, 15, 18, 19, 21, 106, 135, 136,
  138, 140, 142, 144, 148, 149, 153–4, 154,

155, 156, 157, 158, 159, 160, 161, 162, 166,
167, 171–5, 182, 539
Cotopaxi, 256
Cottica river, 247
cotton, 4, 6, 47, 84, 88, 89, 90, 92, 109, 151,
152, 156, 162–3, 164, 166, 167, 196, 203,
211, 223, 244, 260, 271, 279, 284, 309, 310,
334, 347, 360, 362, 384, 410, 414, 420, 433,
444–5, 448, 452
Courantyne river, 241, 244
Coya Sur, 502
Coyhalque, 525–6
Cuajone, 285
Cuantitlán, 60
Cuba, 5, 12, 21, 79, 80, 83, 84, 86, 87, 91,
92, 93, 96, 97, 100, 103, 104, 106, 107, 109,
110, 111, 112, 113, 114, 115, 116, 120, 122,
123, 128–30, 135, 166, 181
Cubatão, 370
Cuchillas de Toar, 77,
Cúcuta, 207, 237
Cuenca, 266–7, 313, 318, 319–20
Cueva Indians, 196
Cuiabá, 338, 339
Culiacán river, 38, 43
Cumaná, 190, 198, 203, 204, 208, 210, 212
Cumanagoto Indians, 196
Cundinamarca, 215
Cupica, 195
Curaçao, 80, 89, 90, 102, 106, 111, 119
Curalavá, 471
Curicó province, 446
Curitiba, 355, 356, 362, 375, 378
Cuyo, 384, 385, 389, 390, 392, 394, 396, 399,
420, 431, 468, 481
Cuyuní river, 242, 244
Cuzco, 259, 262, 263, 264, 267, 270, 278,
280, 282, 290, 293

dairy farming, 49–50, 168, 223, 283, 314,
363, 418, 498, 518
Danish West Indies, 92
Darien, 133, 135, 140, 141, 175, 178, 192
Del Monte Corporation, 161
Demerara, 245
Desaguadero, 260
development, levels of 13–19; regional, 19–22;
see also regional development and planning
Devil's Island, 247
Díaz, Porfirio, 26, 71
Dirección Nacional del Banano (Ecuador),
316
Dominica, 80, 81, 85, 91, 92, 97, 98, 103,
111, 118, 122, 123
Dominican Republic, 14, 15, 18, 80, 91, 92,
93, 96, 97, 103, 104, 108, 110, 113, 114, 116,
119, 124, 125–6
donatarias, 331
Durango, 51, 52, 57, 66

Dutch, the, in the Caribbean, 86, 89–90
Dutch Guiana, *see* Surinam
Dutch West India Company, 89, 244, 245
dyewoods, 168

earthquakes, 2, 80–1, 99, 139, 173, 183, 214,
513
Economic Commission for Latin America
(ECLA), 1, 537–8
Ecuador, 2, 3, 5, 9, 11, 13, 14, 15, 18, 19, 20,
140, 160, 227, 253–269 *passim*, 311–22, 534,
540
ejidos, ejidatarios, 47–8, 159, 218
El Arenal, 173
El Dorado, 144, 188, 243, 338, 468
El Laco, 510
El Niño, 284–5
El Romeral, 509–10
El Salvador (Central America), 14, 15, 21,
138, 140, 142, 144, 146, 148, 149, 152–3,
154–5, 157, 158–9, 162–3, 166, 167, 172–5,
183, 534, 540
El Salvador (Chile), 506–7
El Teniente, 498, 506–7
El Toco, 500
El Tocuyo, 200
El Tofo, 509–10
El Toro, 520
Elqui river, 460, 461, 497–8
Encarnación, 411, 450, 451, 452
encomienda, 86, 201, 208, 263, 265, 266,
471–2
enganche, 284
engenhos, 332–5
Ensenada, 388, 430
Entre Ríos, 388, 392–4, 398, 399, 402–3,
406, 407, 408, 416, 423, 438
Erechín, 353
Escuintla, 158, 173
Esmeraldas, 317, 319, 320, 321
Esperanza, 393
Espírito Santo, 332, 336, 363, 368, 374
Essequibo river, 241, 244, 245
estancias, estancieros, 142, 385, 388, 389, 402,
· 403, 411, 472–7, 522
Exótica, 507

Falcón dam, 43, 44
favelas, 379; *see also* squatter settlements:
urban
fazendas, 340, 341
Feira de Santana, 340
fisheries and fishing industries, 50, 113, 176,
197, 200, 205, 208, 249, 254, 271, 278,
284–5, 322, 469, 492–3, 510–11, 518
fishmeal, production and exports, 271, 272–3,
274, 284–5, 286, 510–11, 514
flax, 392, 408, 410, 447
Flores, 152

Florianópolis, 332, 353, 355
Florida Cays, 83
Florida Straits, 90
Fonseca, Gulf of, 140, 141, 152
forestry, 50, 168–9; *see also* timber
Formosa, 392, 395–6, 398, 414, 434, 445
Fort Bulnes, 521
Fortaleza (Brazil), 348, 377
Fortaleza (Peru), 225
Fray Bentos, 403, 427
French Guiana, *see* Guyane
French West India Company, 90
Frontera district (Chile), 511–20
Fuerte river, 38, 42, 43, 48

Galapagos Islands, 311
Garibaldi, 353
garúa, 254
Gatún, 179
Georgetown, 246
ginger, 90
Giradot, 189
Goiânia, 340, 365–80
Goiás, 325, 338–9, 350, 363, 365, 367, 380, 381
Golden Lane (oilfield), 53
Golfo Dulce, 160
Gracias a Dios, cape of, 140, 169
Gran Chaco, 259, 296
Gran Colombia, 188, 215, 216, 218, 311, 534
Gran Sabana, 191
Granada, 141
Gravesande, Laurens Storm van, 245
Grenada, 80, 90, 91, 92, 95, 96, 97, 98, 103, 108, 110, 118, 122, 123
Grenadines, 80, 91, 115
Greytown, 142
Grijalva river, 36, 38, 40
Guadalajara, 33, 58, 59, 60, 61, 62
Guadeloupe, 79, 81, 83, 90, 91, 92, 95, 97, 103, 110, 119
Guaicayarima peninsula, 83
Guaitecas Islands, 460
Guajira peninsula, 83
Guanacasti, 141, 155, 167
Guanajuato, 38, 43, 45, 51, 142
guano, 254, 269–70, 271
Guantánamo, 101, 104, 181
Guaporé, 301
Guaqui, 301
Guaraní, 299, 450
Guarico, 204
Guatemala, 14, 15, 19, 27, 30, 55, 106, 135, 136, 137, 138, 139, 140, 141, 142, 148, 149, 150–2, 153, 155, 157, 158, 160, 161, 162, 163, 166, 167, 169, 171–5, 177–8, 182, 183, 534; City, 138, 143, 144, 146, 148, 151, 152, 158, 167, 172; Captaincy General, 141, 177
Guayacán, 509

Guayana, 188, 191, 192, 193, 201, 221, 233–4, 242
Guayaquil, 9, 261, 264, 265, 268, 313, 314–15, 319–20, 321, 322
Guayas river and basin, 313, 315, 317
Guerrero, 52, 66
Guianas, 2, 3, 5, 83, 85, 88, 90, 241–51
Guipúzcoa Royal Company, 212
Guyana, 118, 241, 246–51
Guyane (French Guiana), 242, 246–51

haciendas, 4, 47, 48, 49, 142, 158, 266, 270, 290–1, 306–7, 472–7
Haiti, 11, 14, 15, 16, 18, 19, 83, 86, 87, 91, 92, 93, 96, 97, 106, 107, 108, 109, 110, 111, 119, 123, 124–5, 126, 130, 345
Havana, 86, 87, 100, 101, 107
henequen, 45, 265
Hidalgo, 61
Hindustani immigrants, 247, 249
Hispañola, 18, 79, 83, 89, 90, 92, 111, 113, 123
Holguín, 101
Honda, 189
Honduras, 11, 14, 15, 16, 18, 77, 106, 135, 136, 139, 141, 142, 143, 144, 146, 148, 149, 150, 153, 154–5, 157, 158, 160, 161, 162, 170, 172–5, 182, 183, 534, 540; Gulf of, 138, 140, 141, 143, 148, 176
Huachipato, 514–15
Huallaga river, 264
Huamanga, 264
Huancavelica, 267, 290
Huancayo, 264, 270, 277–8, 280, 288, 290
Huánuco, 264
Huánuco el Viejo, 264
Huasco river and city, 460, 461, 497, 505, 510
Huastec, 38, 49
Huaylas, 289
Huehuetenango, 151
Humboldt current, 461
hurricanes, 40, 81, 110, 160–1, 175
hydroelectricity, 4, 55, 56, 70, 125, 170–1, 173, 233, 235, 276, 286–7, 304–5, 321, 360, 369–70, 423–4, 426, 449, 452, 515, 520

Ibagué, 207, 237
Ibarra, 313
Ibirima, 355
Ica, 255, 264, 266, 279, 280, 287
Iguaçu, 340
Ijui river, 353
Illapel, 503
Ilo, 285
Imperial river, 461
immigration, 2, 6, 30, 31, 88–91, 95, 96, 97, 103–7, 175–6, 209, 216, 219, 232–3, 287, 309, 335–6, 352, 353–6, 364, 397–9, 414,

451, 511, 518, 523; *see also* population:
  migration, slavery
Inca empire, 259–60, 313, 383
indentured labour, 103
India, East, immigrants, 94, 96, 100, 102
Indian population and cultures, 4, 12, 66–8,
  150–1, 176–7, 193–6, 203–9, 253, 259–63,
  289–91, 308–9, 312–13, 318, 331, 333, 357,
  367–8, 394–5, 468–71; communities, 262
indigenismo, 259
indigo, 4, 87, 90, 142, 143, 203, 219
industrialization, 7, 56–61, 114–15, 120–1,
  126–7, 129, 170–5, 229–31, 268, 273, 274,
  285–7, 321–2, 368–76, 421–5, 428–31,
  486–93
inquilinaje, 472–5
Inter-American Development Bank (IDB), 1,
  538–9
Inter-American Highway, 151, 154, 162, 175
International Coffee Agreement, 165, 315
Ipatinga, 374
Iquique, 460, 500–1, 510
Iquitos, 264, 270, 278
Irapauto, 44
Iron Quadrangle (Brazil), 368
irrigation, 41–5, 124, 125, 129, 196, 255, 276,
  277, 283–4, 360–1, 384, 420–1, 431, 469,
  497–8
Istmo project, 169
Itaipú, 370, 452
Itajai valley, 355
Itapúa, 450
Itata river, 461, 469
Ixtaccíhuatl, 36

Jaboatão, 375
Jacui river, 353, 356, 364, 365
Jalapa, 46
Jalisco, 43
Jamaica, 94, 95, 96, 97, 100, 103, 105, 106,
  107, 108, 110, 112, 114, 115, 116, 117, 118,
  120, 121, 130, 135, 160, 345
Japanese immigration, 309, 364, 451
Jauja, 264
Joinville, 355, 356, 364, 365
Juárez, Benito, 71
Juiz da Fora, 356, 373
Juliaca, 290
Jujuy, 398, 414, 420, 433, 442
Jundiai, 352, 372
Jutiapa, 151

Kaminalijuyu, 138
Kingston, 99, 100, 121
Kingstown, 98
Kyk-over-al, 244

La Amistad dam, 43, 44
La Caridad, 51,

La Convención, 243
La Frontera, 511–20
La Guaira, 211, 221
La Libertad (El Salvador), 162
La Pampa, 396, 399, 406, 408, 421, 425, 430,
  438
La Paz, 258, 264, 268, 296, 297, 299, 300,
  301, 302, 304, 305, 307, 308, 309
La Plata, 423, 427; Viceroyalty of, 5, 296, 385
La Rioja, 414
La Romana, 126
La Serena, 493, 498, 503–4, 509
La Unión (El Salvador), 162
La Venta, 57
La Villa, 69
La Villita, 57
Lago General Carrera, 526
Lagos dos Patos, 365
Lagunera, 43, 44
Laja, 515
Lake District (Chile), 511–20
land reform, 12, 47–8, 112, 128–9, 154–5,
  159, 222, 224–6, 269, 284, 291–3, 306–9,
  314–17, 477–81
land tenure and size of holdings, 4, 11, 12,
  47–9, 86, 87, 94–5, 150–1, 153, 154, 154–5,
  201–5, 218–19, 222–3, 224–6, 264–6, 270,
  290–1, 295, 306, 314–17, 332–3, 359–60,
  364, 408–9, 441–7, 448, 450–1, 472–81,
  517–18, 522–5
landforms, 2–4, 33–9, 77–80, 134–5, 189–91,
  192, 241–2, 255, 256–7, 258–9, 326–7,
  459–61
Las Truchas, 52, 57
Latacunga, 313, 314
latifundia, 4, 47–9, 203–5, 218, 223, 291,
  293, 314–15, 359–60, 472–6, 498, 523
Latin American Free Trade Association
  (LAFTA), 1, 72, 311, 540–1
Lázaro Cárdenas (iron and steel complex),
  52
Leeward Islands, 89, 92, 106, 117
lemon grass, 167
Lempa river, 153, 173
León, 141, 162
Lerma river, 38, 43, 44, 50, 70
Lesseps, Ferdinand de, 178
Ligua river, 461
Lima, 9, 12, 18, 197, 263, 264, 265, 266, 267,
  268, 269, 274–7, 279, 280, 281, 388
Limari river, 461, 497–8
Limay, 395, 426
Linares, 463, 466, 485
Lirquén, 513
Litoral, 388, 389, 390, 426
livestock production, 12, 49–50, 86–7, 113,
  129, 141–2, 167–8, 201, 204, 207–8, 217,
  219, 223–4, 250, 266, 288–9, 291–2, 295,
  310, 314–15, 339–41, 356, 360, 365, 367,

384–5, 388–92, 395, 402–8, 415–18, 434–40, 447–8, 451–2, 472–3, 495, 498, 521, 522–5; *see also* dairy farming
llama, 257, 260–1, 266
llanos, 36, 188, 190, 191, 193, 195, 204–5, 212, 214, 217, 219, 221
Llanquihué, 512, 517, 520
Lluta, 497
Loa river, 460, 497
Loja, 313, 318
Lomé Convention, 119
Londrina, 379
Los Vilos, 505
Lota, 507, 514
Lurín valley, 276

Maceo, 348
Machu Picchu, 261
Magallanes, 458, 464, 499, 515, 521–5, 526
Magdalena river, 189, 192, 205, 207, 211, 217–18, 229
Magellan, Straits of, 460, 521, 522, 525
maguey, 45
mahogany, 142, 168
Maipo river, 461
maize, 45, 46, 47, 138, 150, 152, 166, 167, 193, 194, 195, 197, 204, 207, 258, 260, 265, 306, 309, 314, 319, 338, 355, 363, 364, 384, 408, 415, 420, 432, 440, 445, 447, 448, 452, 469, 472, 498
Malambo, 193
Malleco province, 512
Malpaso dam, 40
Mamoré river, 301
Managua, 148, 153, 158, 162, 172
Manaus, 270, 343, 357–8, 369, 375
manioc, 137, 193, 196, 197, 204, 207, 260, 338, 355, 363, 364, 384, 420, 452
Mantaro valley, 264, 265, 276, 281–2, 288–9
manufacturing industry, 7, 14, 15, 56–61, 114–15, 120–1, 126–7, 129, 169–75, 222, 229–31, 248, 272–3, 274–5, 285–6, 290–1, 305–6, 368–76, 421–5, 428–31, 486–93; aluminium, 114, 235, 248, 250, 372; cement, 58, 126, 172, 174, 286; chemicals, 56–8, 171, 173, 174, 275, 286, 371–2, 428, 430, 490; clothing, shoes, 61–2, 126, 171, 173–4, 230, 275, 370, 371, 421, 428, 449, 487; consumer durables, 273, 274–5, 305, 421, 430, 488–9, 516; electronics, 62, 127, 286, 516; engineering, 58–9, 59, 62, 127, 275, 286, 305, 371–2, 516; fertilizers, 47, 58, 171, 172, 275, 369, 430; food, drink and tobacco, 58, 59, 61, 115, 129, 171, 230, 305, 322, 371, 376, 421, 428, 449, 486–7, 518–19; *frigoríficos*, 403, 405, 406, 409, 416, 434–8, 448, 523; iron and steel, 7, 52, 56–7, 62, 173, 230–5, 285–6, 368, 371, 372–4, 490, 514–15; motor vehicles, 58, 60, 61, 230, 275, 286,

305, 373, 374–6, 430, 489–90; oil-refining, 7, 54, 55, 56–7, 62, 115, 171–4, 230, 304, 318–19, 372, 428, 430, 515; petrochemicals, 57, 62, 230, 305, 322, 369, 428, 430, 489, 515, 524; pharmaceuticals, 58, 127, 275, 322, 371, 372, 421; pulp and paper, 169, 286, 369, 490, 515–16; shipbuilding, 275, 285, 322, 372; textiles, 58, 127, 129, 171, 203, 275, 286, 305, 322, 370, 371, 372, 373, 375, 389, 421, 428, 430, 486–7, 513, 516
Manzanillo, 101
Mapocho river, 461, 483
Mapuche Indians, 469–71, 512, 518–20
Mar del Plata, 427
Maracaibo: lake and basin, 190, 194, 195, 196, 217; Gulf of, 106; city, 210, 212, 216, 219, 237
Maracay, 237
Marajó Island, 357
Maranhao, 343, 361, 369, 370
Marañón, 282, 290
Marcapomacocha, 276, **283**
Margarita Island, 89
marginal sector, 7, 9
Maria Elena, 500–2
Marie Galante, 79
marihuana, 223
market gardening, 151, 166, 283, **288–9**, 307, 497–8
Martinique, 80, 81, 86, 90, 91, 92, 95, 96, 97, 103, 110, 115, 119
Masaya Caldera, 158
mashua, 260
Matagalpa, 158, 168
Matamora, 43,
Matanzas, 101
Mataquito river, 461
Matías de Gálvez, 172, 173
Matilla, 494
Mato Grosso, 327, 335, 338–9, 356, 365, 367, 368, 381, 408, 451; do Sul, 362
Maule: province, 466; river, 460, 461, 472, 486, 513
Maullin river, 461
Maya culture and Indians, 32, 39, 49, 67, 136, 137, 139, 140, 150, 176
Mayo river, 38, 42, 43, 48
Mazaruni river, 244
Mazatec (language), 67
Mazatenango, 167
Medellín, 9, 18, 220, 229, 230, 235, 236
Mejillones, 460
Mendoza, 384, 391, 395, 398, 400, 410, 413, 414, 420, 423, 424, 425, 427, 433, 445
Mennonites, 309, 411
mercedes, 86, 176, 201–2, 471–2
Mérida (Mexico), 41, 49
Mérida (Venezuela), 192, 195, 203, 204, 211
Mesa Central, 38, 39, 41, 43

Mesa del Sur, 67
mescal, 45
Meseta Central (Costa Rica), 141, 149, 153, 154, 156, 158, 159, 161, 173
Mesopotamia, 384, 389, 396, 400, 406, 425, 427, 433
Mexicali, 59, 62
Mexico, 1, 3, 4, 6, 7, 9, 11, 12, 14, 15, 17, 19, 20, 21, 22, 25–75, 87, 135, 136, 137, 138, 139, 140, 143, 146, 151, 166, 170, 171, 176, 177–8, 197, 536, 540; City, 9, 26, 27, 30, 32–3, 38, 44, 46, 49, 55, 56, 57, 58–60, 61, 68–70, 140, 141, 142; Gulf of, 38, 44, 55, 85; Valley of, 36, 38, 40, 45, 48, 70
Mezquital, Valle de, 39, 49, 70
Michoacán, 52
Middle Passage, 90
Miguel Hidalgo dam, 43
Minas Gerais, 325, 327, 336–9, 341, 342, 349, 356, 363, 365, 368, 369, 370, 371, 373, 374, 380
Minatitlán, 55, 57
minifundia, 4, 11, 150, 151, 268, 291, 293, 306, 314–15, 355, 359–60, 364, 475–7, 498, 517, 519–20
mining and mineral resources, 3–4, 50–6, 113–14, 142, 200, 205–7, 208, 226–9, 272–3, 291, 302–5, 335–9, 368–9, 472, 498–510; antimony, 302, 303; barytes, 113; bauxite, 114, 121, 122, 125, 233, 248–9, 250, 369; bismuth, 302; chrome, 113, 114, 129; cinnabar, 267; coal, 51, 56, 229, 369, 373, 498–9, 513–14, 522, 524, 525; cobalt, 113, 114, 129; copper, 3, 6, 51, 52, 113, 125, 129, 182, 200, 221, 270, 271, 285–6, 302, 369, 459, 482, 486, 487, 491, 492, 494, 498, 503–9, 526; diamonds, 6, 368; emeralds, 196; gold, 3, 6, 51, 87, 113, 140, 141, 196, 200, 205–7, 211–12, 217, 221, 247, 266–7, 270, 313, 335–9, 368, 494, 503, 522, 524; iron ore, 3, 51, 113, 114, 129, 201, 226, 229, 271, 285, 286, 302, 305, 368, 373, 494, 509–10; lead, 51, 52, 113, 270, 271, 302, 368, 526; manganese, 113, 114, 129, 368, 374; mercury, 125, 267; natural gas, 53, 56, 61, 63, 129, 229, 302–4, 426; nickel, 113, 114, 125, 229, 369; nitrates, 6, 271, 459, 460, 482, 486, 487, 495, 499–503; oil, 3, 6, 11, 33, 51, 51–6, 100, 114, 121, 123, 129, 176, 222, 226–9, 271, 286, 302, 304–5, 312, 318–19, 320–1, 369, 425–6, 499, 524–5; phosphates, 286, 368; pyrites, 113; salt, 89, 368; silver, 3, 51, 113, 141, 142, 143, 167, 266–8, 270, 271, 302, 303, 388, 494, 503–4, 505; sulphur, 51, 302; tin, 296, 301, 302–4; tungsten, 113, 114, 302, 369; zinc, 52, 113, 125, 270, 271, 302, 369, 526
Miranda state, 230
Misiones, 341, 383, 384, 392, 395–6, 398, 411, 414, 420, 423, 432, 434, 441, 445
mita, 208, 265, 267
Mixtec Indians and language, 30, 67
Moctezuma, 36
Mocuzari dam, 43
Mojos, 261
Mollendo, 270
Momil, 193
Mona Passage, 90
Monclava, 56, 61
montaña, 270, 280, 294, 297
Mont Pelée, 81
Montagua valley, 160
Monterrey, 27, 33, 52, 55, 56, 57, 58, 59, 60, 61, 62
Montevideo, 9, 385, 392, 400, 405, 406, 410, 447, 449
Montserrat, 80, 91, 97, 109, 111, 118, 122
Morelos, 43, 46
Mormon immigration, 30
Morro Velho, 368
Mosquito coast, 140–1, 142, 143, 168
Motilone Indians, 196
Moyabamba, 267
mucambos, 379; *see also* squatter settlements: urban
Muquiyauyo, 289
Mutún, 302, 305

Nacozari, 51
Nahuatl, 67
Natal, 332, 348
Nayarit, 66
Nazas river, 43
Nazca, 285, 286
Neiva, 189
Neuquén, 395, 400, 425, 525
Nevis, 80, 92, 97
New Granada, Viceroyalty of, 141, 197, 215, 311
New Spain, Viceroyalty of, 141, 142
Nicaragua, 2, 11, 14, 15, 77, 106, 135, 136, 140, 141, 142, 144, 146, 148, 153, 155, 157, 158, 159, 161, 162, 166, 167, 168, 169, 170, 172–5, 183; Lake, 144
Nicoya, 141, 154; Gulf of, 138, 140
Nipe Banes, 104
Niteroi, 371
Norte Chico, 482, 493–511
Norte Grande, 462, 482, 493–511
Novo Friburgo, 372
Ñuble province, 465, 466, 485, 520
nuclear power, 4, 370
Nueva Concepción, 152
Nuevitas, 104
Nuevo Laredo, 62
Nuevo León, 46, 66
nutmeg, 110

Oaxaca, 36, 39, 41, 46, 63, 66, 67
Oca, 260
Ocós, 138
Ofqui isthmus, 461
O 'Higgins province, 466, 485, 489, 498
oil palm, 166
okra, 111
Olinda, 368
olives, 418, 420, 445, 497
ollucos, 260
Olmec culture, 39
Olmos, 283
OPEC, 72, 118, 227, 320
Operation Bootstrap, 103, 105, 127
Orange Walk, 176
Orangestad, 102
Organization of American States (OAS), 1, 72, 118, 119, 125, 537
Organization of Central American States (ODECA), 145
Oriente, 253, 296, 298, 299, 309–10, 318
Orinoco, 3, 13, 80, 83, 87, 190, 191, 192, 193, 194, 195, 197, 198, 204, 211, 214, 217, 227, 241, 242
Orizaba, 36, 38
Oroya, 270
Oruro, 267, 297, 299, 300–1, 302, 303, 304–5, 307, 511
Osorno, 464, 512, 517, 518, 520
Otavalo Indians, 312, 321
Otomi (language), 67
Ouro Preto, Vila Rica de, 336, 338, 380
Oyapock river, 244
Oxapampa, 293

Pachuca, 51
Paita, 265
Palena river, 464
Pampas, 2, 3, 384, 389, 390, 392, 394–5, 400–3, 414, 418–20, 431–2, 440–1, 442–4, 445
Panama, 11, 14, 15, 18, 99, 105, 135, 136, 137, 140, 141, 143, 144, 148, 150, 154, 155, 159, 160, 161, 162, 167, 170–3, 175, 178–83, 194, 384; Canal, 6, 124, 133, 160, 178–83, 316, 507; Canal Zone, 144, 167, 178–83; City, 141, 148, 172, 183; disease, 110, 160
Pan-American Highway, 321
Panatal, 327, 335
Panuco, 46
Panuco-Ebano, 53
Papaloapan river, 40, 46
papaya, 260, 294
Pará, 336, 368
paracaidistas, 69; *see also* squatter settlements: urban
Paraguay, 3, 5, 13, 14, 15, 18, 20, 66, 296, 298, 341, 383, 384, 385, 388, 390, 392, 396, 408, 410, 449–52, 541; river, 301, 329, 339, 411, 426

Paraíba valley, 325, 340, 349–50, 353, 359, 369, 372
Paramaribo, 247, 248
Paramonga, 286
páramos, 2, 258, 314
Paraná: river, 329, 339, 340, 370, 392–3, 396, 400, 407, 411, 426, 427, 428, 451–2; state, 327, 349, 353–7, 361–4, 376
Paraná-Paraguay, 3
Paranagua, 363, 451
Paranagua peninsula, 190
Paranam, 248
Paranapanema river, 364
Pardo river, 339
Paría peninsula, 190, 193
Parnaiba river, 370
Parrita, 160
Pasión river, 139
Passo Fundo, 353
Pasto, 189, 211, 266
Patagonia, 2, 3, 384, 385, 394–5, 396, 399, 400, 402–3, 414, 420, 426, 430, 431, 440, 442, 443, 523
Patía river, 211
Pativilca river, 255
Paulo Alfonso falls, 329, 336, 370
Paysandú, 427, 448
Paz del Río, 230
peanuts, 196, 197, 260
pearl fishing, 196, 200, 208, 221
Pecket, 525
Pedro de Valdivia, 500, 502
Peine, 494
Peligre dam (Haiti), 124
Peña Colorado, 52
Peñas, Gulf of, 461
Penco, 513–14
peninsulares, 268, 269
peppers, 45
Pernambuco, 331, 332, 340, 344, 348, 359, 361
Peru, 2, 3, 5, 6, 9, 11, 12, 14, 15, 17, 18, 19, 20, 21, 22, 138, 141, 146, 253–69 *passim*, 269–95, 384, 458, 486, 493, 499–500, 511, 541; Viceroyalty of, 141, 263, 264, 267, 311, 384, 472
Petén, 139, 168, 171, 178
Petorca river, 461
Petrohué river, 461
Petrópolis, 372
physical environment, 2–4, 33–41, 77–83, 134–7, 189–92, 241–3, 254–9, 326–9, 459–64
Piauí, 368, 370
Pica, 494
Piedras Negras, 139
Pilmaiquen, 520
Pinar del Río, 101
pineapples, 110, 196, 260, 294
Pisac, 261

Pisagua, 460, 500–1, 510
Pisayambo, 321
Pisco, 266, 285
Piura, 278, 284
Pizarro, 141, 263, 383
Plate: River, 341; countries, 383–453;
    Viceroyalty of, 385
Poços de Caldas, 372
Point Lisas, 120
Polygono das Secas, 327
Pomeroon, 244
Poncé, 102,
Ponta Pora, 263, 451
Poopo lake, 256
Popayán, 198, 205, 211
Popocatepetl, 36
population, 4, 8, 14, 15, 25–33, 96–109, 122,
    124, 126, 143, 146, 147, 148–53, 154–6, 176,
    177, 208–10, 220, 231–5, 263, 272–3, 290,
    319–20, 342, 344, 352, 358, 377–80, 385,
    396–400, 413–14, 427–8, 446–8, 522; birth-
    rates, 8, 27, 30, 97, 126, 148, 231, 272, 427,
    465; death-rates, 8, 26–7, 97, 126, 148, 208,
    231, 272, 427, 465; density, 14, 146–7,
    149–50, 152–3, 379; distribution, 28–9,
    31–2, 98, 156, 278–82, 296–300, 342–3, 465;
    family planning, 30, 107–8, 126, 231;
    migration, 9, 10, 30, 31, 34, 35, 37, 68–9,
    103–7, 151–3, 156, 232–3, 247, 276, 279–82,
    290–1, 299, 312, 319–20, 335, 352, 353, 355,
    361, 377–9, 397–400, 413, 427, 485, 524,
    525, 536, *see also* immigration
Poraceta, 364
Porce river, 189
Port-au-Prince, 124
Port of Spain, 100, 102, 118,
Port Royal, 87
Porto Alegre, 353, 357, 378
Porto Feliz, 339
Portuguesa river, 195
Porvenir, 524
Posadas, 451
potatoes, 194, 196, 258, 260, 265, 283, 314,
    433, 469, 521
Potosí, 263, 265, 267, 296, 302, 388, 495
Potrerillos, 506–7
Pozo Almonte, 494
Poza Rica, 53
Pozuzo, 270
pre-Columbian cultures, 83–5, 137–40, 192–7,
    259–63, 383–4, 469–71
Proprietary Patents, 103; Systems, 88
Providencia, 142, 143
Prudentópolis, 355
Puangue, 476
Pucallpa, 293
Puebla, 44, 45, 60, 61, 66
pueblos jovenes, 269, 276–7; *see also* squatter
    settlements: urban

Puerto Aisén, 464, 525
Puerto Barrios, 152, 172
Puerto Busch, 301
Puerto Caballos, 141
Puerto Cabello, 194
Puerto Cabeza, 169
Puerto Cortés, 173, 182
Puerto Limón, 159, 161, 172, 182
Puerto Montt, 461, 482, 511–12, 518
Puerto Rico, 5, 77, 79, 80, 83, 84, 86, 87, 91,
    92, 93, 96, 97, 103, 105, 108, 109, 110, 113,
    114, 116, 120, 123, 126–7, 130, 181
Puerto San José, 158
Puerto Suárez, 310
Puerto Varas, 511
Puerto Viejo, 315
Pullinque, 520
pulque, 45
puna, 2, 256–7, 283, 459
Puno, 301
Punta Arenas, 521–5
Punta de Araya, 89
Puntarenas, 172
Putaendo, 475
Putumayo river, 34

quebracho, 393, 396, 411
Quechua Indians, 260, 299, 308
Quetzaltenango, 152
Quiché Indians, 142
Quimbaya, 195
quinoa, 196, 258, 260, 265
Quintana river, 460
Quiriguá, 139
Quisma river, 460
Quito, 9, 198, 205, 215, 260, 264, 266, 268,
    313, 314–15, 317, 318–19, 319–20, 321, 322;
    audiencia of, 34

Rama, 153
Rancagua, 482
Rancho Peludo, 193
ranchos, 235; *see also* squatter settlements
Rapel river, 461, 469
Recife, 334, 340, 341, 343, 346, 348, 368,
    375, 378
Recôncavo, 367, 375
reducciones, 384, 385
regional development and planning, 10,
    19–22, 61–8, 233–5, 274, 278, 280–2, 288,
    290–1, 299, 360–1, 375, 481–3
Reloncaví, Gulf of, 461, 511
Remedios, 198, 207
repartimiento, 86
resguardos, 208, 218
Resistencia, 393, 396, 427
Reynosa, 53, 55, 57
Riberão Preto, 350
rice, 46, 125, 162, 166, 223, 249, 250, 283,

294, 309, 310, 315–17, 319, 356, 357, 363, 365
Riesco Island, 522, 525
Rimac river, 255, 264, 275, 283
Rio Branco, 243
Rio de Janeiro, 6, 332, 338, 343, 346, 347, 349, 369, 370, 371, 372–6, 378
Rio Doce, 336, 368, 374, 376
Río Grande (Mexico), 38, 43, 44, 45, 63
Río Grande, 370
Rio Grande do Norte, 327, 332, 368
Rio Grande do Sul, 353–4, 364–5, 369, 376
Río Lempa, 140
Río Negro, 241, 244, 353, 395, 401, 420–1, 425, 432, 443, 449
Río San Juan, 144
Río Tercero, 430
Riobamba, 266, 313
Roraima, 191, 241
Rosario, 389, 391, 392, 393, 398, 400, 427, 440
Roseau, 98
Ruatán Island, 176
rubber, 3, 12, 166, 219, 270, 294, 296, 357–8
Rubelsanto, 171,
Rupununi, 10–13, 242, 243, 250

Saba, 80, 89, 96, 119
Sabara, 336
Sabinas, basin of, 52, 56
St Christophers, 88; *see also* St Kitts and Nevis
St Croix, 95, 106, 115
St Domingue, 90, 92
St Eustasius, 80, 89, 119
St Georges, 98
St Kitts and Nevis, 80, 88, 90, 91, 95, 96, 103, 106, 109, 110
St Laurent, 247
St Lucia, 80, 85, 91, 92, 93, 95, 97, 98, 99, 106, 108, 110, 118, 122
St Martin, 80, 89, 90, 96, 119
St Thomas, 96, 106, 115
St Vincent, 80, 81, 85, 91, 92, 95, 97, 98, 103, 109, 110, 118, 122, 176
saladeros, 388, 389, 403–5, 406, 408, 411
Salado river, 394
Salamanca, 55, 57, 59
Salar de Atacama, 495
Salina Cruz, 55, 57
Salinas la Blanca, 138
Salta, 388, 410, 414, 415, 420, 425, 433, 434, 442, 495, 511
Saltado river, 43
Saltillo, 61
Salto Grande, 426, 427
Salvador, 334, 336, 338, 346, 348, 369, 375, 378
San Andrés, 142, 143
San Augustín, 195

San Benito, 162
San Cristóbal de las Casas, 151
San Felipe del Morro, 85–6
San Fernando (Trinidad), 100, 102
San Fernando de Apure, 212, 217
San José (Costa Rica), 148, 153, 168, 172
San Juan (Argentina), 409, 413, 433, 444, 445, 507
San Juan (Mexico): river, 43, 44; Llanos de, 38
San Juan (Peru), 285
San Juan (Puerto Rico), 86, 102, 103, 127
San Juan river (Colombia), 189
San Lorenzo, 321, 425, 430
San Luis Potosí, 51
San Miguel, Gulf of, 180
San Nicolás, 424, 430
San Pedro, 515
San Pedro de Atacama, 494, 495, 504
San Pedro de Marcovis, 126
San Pedro Sula, 141, 173
San Rafael, 395, 420–1
San Rosendo, 513
San Salvador, 143, 148, 167, 172, 173
San Sebastián, 204
San Vicente, 513–15
Sancti Spiritus, 101
Santa Ana, 173
Santa Catarina, 341, 353–6, 364–5, 367, 376
Santa Clara, 101
Santa Cruz, 259, 296, 298, 299, 301, 304, 305, 308–10
Santa Cruz (Brazil), 353
Santa Elena (Argentina), 406
Santa Eulalia, 276
Santa Fé, 384, 390, 391, 393, 394, 398, 399, 400, 402, 406, 407, 408, 413, 421, 423, 427, 428, 430, 432, 433, 438, 442
Santa Fé de Bogotá, *see* Bogotá
Santa Fé-La Plata axis, 425, 430
Santa Margarita, 79
Santa Marta, 198, 217
Santa river, 255, 286
Santander, 220
Santiago de Cao, 286–7
Santiago de Chile, 9, 12, 458, 460, 461, 463, 465–6, 468, 482, 483–5, 486–9, 493, 517
Santiago de Cuba, 101, 126
Santiago del Estero, 400, 414
Santo Angelo, 353
Santo Domingo (Ecuador), 317
Santo Domingo (Hispañola), 87, 104, 123, 125, 188
Santo Tomás de Castilla, 182
Santo Tomé, 198
Santos, 301, 352, 361, 374
São Bento, 355
São Bernadino, 374
São Claro, 350
São Francisco river, 329, 336, 340, 360–1, 370

São Joao de Rey, 336, 368
São José dos Campos, 372
São Leopoldo, 353, 355
São Luiz, 343
São Paulo, 6, 304, 331, 336, 338, 339, 340, 341, 343, 347, 349–53, 369–70, 371–6; State, 359, 361–4, 378
São Pedro de Alcantara, 353
São Salvador, 332, 343
São Tomé, 332
São Vincente, 331, 332
Sarstoon river, 168
Sasardi-Marti, 180
Satipo, 293
selva, 258–9, 279, 294
Sergipe, 369
Serra da Mantiqueira, 336
Serra do Espinhaço, 335, 338, 339
Serra do Mar, 326, 327, 328, 349, 370, 374
Serra do Navio, 368
Serra dos Carajás, 368
Serranía de Baudó, 189
sertão, 336, 340–1, 346–8, 357, 381
sesmarias, 332, 340
shanty towns, *see* squatter settlements: urban
Sierra de Cayey (Puerto Rico), 79
Sierra de Córdoba (Argentina), 428
Sierra de Nipe (Cuba), 77
Sierra de los Organos (Cuba), 79
Sierra de Talamanca, 154
Sierra de Trinidad (Cuba), 79
Sierra Grande, 430
Sierra Madre, 33, 36, 38, 41, 42, 43, 46, 50, 67
Sierra Maestra (Cuba), 77
Sierra Nevada de Santa Marta, 190, 192, 195
Sierra Vicuña Mackenna, 460
sigatoka disease, 46, 110, 160, 161, 316
Simpson river, 464
Sinaloa, 46, 48, 66
Sinú river, 189, 192, 193, 195
Siquirres, 154
sisal, 124, 360
slash-and-burn cultivation, 45, 46, 94, 338; *see also* agriculture: shifting
slavery, 4, 83, 86, 87, 88, 89, 90, 91–5, 99, 102, 103, 104, 110, 111, 141, 175, 201, 203, 204, 209–10, 245, 246, 247, 265, 284, 286, 333–5, 345–6, 385, 390, 469
Soconusco, 38, 41
Sogamoso, 189
soils, 149, 156, 157, 160, 330; erosion of, 38–9, 95, 124, 191, 288, 290, 348, 350, 441
Solis dam, 44
Sonora, 38, 40, 42, 44, 48, 49, 51
sorghum, 152, 223, 432, 440, 445
Sorocaba, 371
Soufrière, 81

South American Marginal Highway, *see* *Carretera Marginal*
soya beans, 223, 309, 364, 432, 440, 441, 452
Spanish colonial settlement, 85–7, 140–3, 197–208, 263–8, 383–5, 468–75, 511
squashes, 45, 150, 260, 469
squatter settlements: rural, 295, 308–9; urban, 8–10, 68–70, 99, 102, 235–7, 269, 276–7, 379, 485
Stabroek, 245
Standard Fruit Company, 159–61
Sucre, 265, 267, 296, 304, 308
SUDENE, 360–1, 375
sugar, 4, 6, 46, 81, 86, 87, 88–9, 90, 91–2, 100, 111, 121, 122, 124, 125, 126, 128–9, 130, 151–2, 166, 203, 219, 223, 244, 245, 246–7, 250, 265, 270–1, 279, 283–4, 309–10, 318, 332–5, 344–7, 357, 360–1, 363–4, 384, 389, 410, 420, 433, 444–5, 452
sunflower, 428, 420, 445
Supe, 285
Surinam, 242, 245, 246–51
Suriname river, 244, 247, 248
swidden cultivation 196, 204; *see also* agriculture: shifting

Tabasco, 33, 38, 39, 47, 53, 62, 66, 67
Taitao peninsula, 460
Talara, 286
Talca province, 466, 485
Talcahuano, 513–14, 515
Taltal, 460, 500
Tamarugal, Pampa de, 500
Tamaulipas, 38, 40, 45, 46, 49
Tambo Colorado, 264
Tambo de Mora, 285
Tambopata, 294
Tampico, 53, 55
Tandil, 394
Tanque, 153
Tarapacá, 271, 450, 462, 466, 486, 493–4, 495, 497, 499–503
Tarija, 267, 296, 304, 308, 309
Tarma, 288
Taxco, 51
Tegucigalpa, 142, 173
Tehuacán, 36, 138
Tehuantepec and isthmus of, 33, 36, 38, 41, 45, 52, 56, 57, 67, 135, 141, 152
Tejalpa, 60
Tela, 138
Temuco, 482, 517, 518
Tenochtitlán, 32
Teotihuacán, 32
Tepalcatepec, 36
Tepic, 38, 41,
terra-roxa soils, 350, 361
Texcoco lake, 49, 69
Tiahuanaco, 259

tierra caliente, 40, 46, 154
Tierra del Fuego, 395, 398, 425, 460, 464, 522
tierra fría, 41
tierra templada, 41, 46
Tierradentro, 195
Tiete river, 339, 370
Tijuana, 62
Tikál, 139
timber, 175–7, 249, 265, 309, 369, 393, 396, 492, 515–16
Timoto Indians, 196
Tingo María, 293
Tiquisate, 152, 160, 167
Titicaca lake, 256, 260, 261, 267, 277, 290, 299, 300, 302, 310; basin and region, 260, 261
Tlacotalpan, 40
Tlaxcala, 39, 45
tobacco, 6, 86, 87, 88, 90, 110, 128–9, 141, 166, 203, 211–12, 219, 244, 334, 356, 364, 389, 433, 444
Tobago, 79, 80, 81, 85, 90, 91, 93, 95, 97, 98, 100, 113, 114, 116, 118, 120, 121, 122, 123
Tocantins river and valley, 369, 370
Toconao, 494, 495
Tocopilla, 460, 500, 501
Tocuyo, 190, 203, 204
Todos Santos, 308
Toledo, 176
Toltec, 140
Tolten river, 461, 519
Toluca, 36, 39, 60, 61
Tomé, 486, 513–14
Tonto river, 40
Toquepala, 285
Torreón, 43
Tortuga, 89, 90
tourism, 61, 62, 102, 112, 115–17, 122, 127, 128, 249, 271
Tournavista, 295
Transamazonian Highway, 365–7
transport, *see* communications
treaty of: Ancón, 493; Breda, 245; Montevideo, 540; Nijmegen, 90; Paris, 89, 103; Rijswijk, 90; Utrecht, 245, 385
Tres Marios, 370
Trinidad, 79, 80, 83, 84, 85, 87, 90, 91, 92, 93, 94, 95, 96, 97, 100, 106, 107, 108, 110, 111, 113, 114, 116, 118, 120, 121, 122, 123, 190
Trinidad (Bolivia), 301
Trujillo (Honduras), 141
Trujillo (Peru), 264, 266, 278, 280, 286
Trujillo (Venezuela), 189, 203, 211
Tuburão, 368, 374
Tucumán, 384, 385, 389, 391, 398, 410, 413, 420, 423, 427, 432, 433, 445, 507
Tula, 44, 55

Tumaco, 192, 195
Tumbes, 283
tung oil, 420, 434, 444–5, 452
Tunja, 189, 198
Tuxpan, 53, 55
Tuxtlas, 39, 67
Tuy river, 190, 194, 201
Twelve Year Truce, 89
Tzeltal language, 67
Tzotzil language, 67

Uaxactún, 139
Uberaba, 339
ulluco, 196
Ultima Esperanza, 522–3
Ulúa river, 138, 141, 161
Unare river, 193
Uncia, 302
United Brands, 160–2, 166, 182
United Fruit Company, 152, 155, 159, 160–1
United Provinces of Central America, 143
Urabá, Gulf of, 190, 198, 217
urban planning, 5, 86, 99, 198–200, 262–4, 380, 469–70
urbanization, 7, 8–9, 32–3, 58–60, 68–70, 98, 98–103, 148, 233–6, 274, 278–9, 299, 348–9, 377–80, 421, 465–6, 484
Uruapan (language), 67
Urucum (Mount), 368
Uruguay, 3, 5, 11, 14, 15, 16, 17, 19, 20, 173, 341, 343, 384, 388, 389, 390, 392, 396, 399–400, 403–6, 409–10, 426, 447–9; river, 340, 355, 356, 363, 408, 426, 427, 536, 540, 541
Usumacinta, 38, 139
Uyuni, 301

Vacarias do Mar, 340, 341
Valdivia: city, 464, 487, 517; province, 464, 509, 511–12, 520; river, 461
Valencia, 203, 204, 237; lake, 190, 192, 195
Valle del General, 154
valle region (Bolivia), 253, 265, 296–8, 299, 301
Vallenar, 503–4, 509
Valparaiso, 460, 465–6, 481, 482, 485, 486–9, 493, 498, 499, 517
Valsequillo, 44
vanilla, 219
Vega Real, 87
vegetation, 2–3, 39, 81–2, 136, 168–9, 243, 256–7, 327–9, 462–4
Venado Tuerto, 394
Venezuela, 3, 5, 9, 10, 11, 12, 13, 14, 15, 16, 17, 19, 20, 21, 22, 53, 77, 83, 85, 89, 105, 106, 111, 113, 123, 130, 137, 187, 534, 540, 541
Ventana hills, 394
Vera Cruz, 38, 41, 46, 49, 57, 59, 61, 62, 67
Veragua, 141

Veranópolis, 353
Verapaz, 158
Vicos, 289
Villa Constitución, 430
Villa Hermosa, 55
Villa, Pancho, 71
Villarica, 411
Villazón, 301
Viña del Mar, 466, 481
Virgin Islands, 11, 115, 122; US Virgin
  Islands, 97
viticulture, 265, 283, 356, 364, 389, 410, 420,
  433, 444-5, 448, 480, 493, 497
Vitor river, 497
Vitória, 368, 374
Vitória de Santo Antao, 340
Vizcaino desert, 40
Volta Redonda, 372-4

War of the Chaco, 296, 449
War of the Pacific, 5, 271, 296, 469, 486, 498,
  499, 500, 512
Welsh settlement, 395
wheat, 46, 47, 141, 150, 203, 208, 212, 258,
  265, 306, 310, 363, 364, 388, 392, 408, 415,
  432, 440, 447, 448, 472-3, 480, 486, 498,
  513, 517
Willemstad, 102
Windward Islands, 81, 89, 106, 117

Xingu National Park, 368

Yaciretá, 426, 452
Yacuiba, 301
Yallahs Valley Authority (Jamaica), 112
yams, 111
Yaqui: district, 42, 44; Indians, 36, 41; river,
  38, 43, 48
Yaracuy river, 190, 195
yerba maté, 434, 444-5, 452
yuca, 309, 319; *see also* manioc
Yucatán, 32, 36, 38, 39, 41, 45, 47, 49, 66,
  67, 104, 138, 139, 141, 142, 176; Captaincy
  General of, 176
yungas, 258, 297-8, 301, 305, 308-9
Yuriria lake, 43
Yuruari, 203, 218, 247

Zacatecas, 51
Zaculeu, 139
Zapata, Emiliano, 71
Zapiga, 482
Zapotec, 67, 71
Zaragoza, 207
Zárate, 427, 430, 438
Zaraza, 204
Zaruma, 267
Zócalo, 68
Zulia, 217